Gareth Griffiths

Fine Structure Immunocytochemistry

With Contributions by
Brian Burke and John Lucocq

With 90 Figures

Springer-Verlag
Berlin Heidelberg New York London Paris
Tokyo Hong Kong Barcelona Budapest

Dr. GARETH GRIFFITHS

European Molecular Biology Laboratory
Postfach 102209
W-6900 Heidelberg, Germany

Front Cover: A thawed cryo-section (Tokuyasu technique) of a rabbit kidney cell (RK-13) infected with vaccinia virus. The section was labelled with an antibody against P65, the target protein of the drug rifampicin which blocks virus assembly. This protein is localized to the spherical immature virus but only poorly to the mature, brick-shaped particles.
From a study by B. Sodeik, M. Ericsson, B. Doms, B. Moss, and G. Griffiths.

ISBN 3-540-54805-X Springer-Verlag Berlin Heidelberg New York
ISBN 0-387-54805-X Springer-Verlag New York Berlin Heidelberg

Library of Congress Cataloging-in-Publication Data. Griffiths, Gareth. Fine structure immunocyto-chemistry / Gareth Griffiths with contributions by Brian Burke and John Lucocq. p. cm. Includes bibliographical references and index. ISBN 3-540-54805-X (Berlin)
ISBN 0-387-54805-X (New York) 1. Immunocytochemistry. 2. Electron microscopic immunocyto-chemistry. I. Burke, Brian. II. Lucocq, John. III. Title. QR187.I45G75 1992 574.87'6042--dc20

© Springer-Verlag Berlin Heidelberg 1993
Printed in Germany

Media conversion: Elsner & Behrens GmbH, Oftersheim
31/3145-5 4 3 2 1 0 – Printed on acid-free paper

To Tok and Jan
for their teaching,
inspiration
and friendship

Preface

Electron microscopy in the biological sciences can be divided into two disciplines. The first, concerned with high resolution detail of particles or periodic structures, is mostly based on sound theoretical principles of physics. The second, by far the larger discipline, is interested in the information obtainable from thin sections. The theoretical background to those groups of techniques for preparing and looking at thin sections is often inexact and "loose", for want of a better word. What should be chemistry is often closer to alchemy. This kind of electron microscopy is often enshrined with mystical recipes, handed down from generation to generation. Admittedly, many of the processes involved, such as those required to embed tissue in epoxy resins, involve multiple interconnected steps, which make it difficult to follow the details of any one of these steps. If all these steps are shrouded in some mystery, however, can one really trust the final image that emerges on the EM screen? When we present the data in some semi-quantitative form is there really no better way to do it than to categorize the parameters with ++, +/−, etc? What happens when one labels the sections with antibodies? Does the whole business necessarily need to be more of an "art" than a "science"?

Upon reflecting on these problems in 1981, I had the impression that many of the multi-authored textbooks that existed then (and that have appeared since) tended to exacerbate or at least perpetuate this problem. They are too often "heavy" on recipes and "light" on theoretical background. To be fair, the latter sometimes hardly exists. In many cases, however, I became convinced that there was more theoretical information available than was making its way into the main textbooks.

In 1981, against this background, I decided to write my own book. Having made that decision I found there was no turning back, for the next 10 years. It was during this prolonged effort that I was forced to realize why no such book already existed!

I must admit that I have changed my mind about a number of things since I started this project. The most important concession concerns the use of the new generation plastic resins, especially the Lowicryl resins. For groups such as ours, interested predominantly in the molecular cell biology of membrane proteins, the Tokuyasu cryosection method was the logical first choice for immunolabelling.

However, it is becoming increasingly clear that for many problems these new resins offer a number of advantages over the cryosection method, especially in combination with freeze-substitution. Accordingly, the book aims at giving a balanced view with respect to these two major approaches.

The important theoretical and practical aspects relevant to these techniques are covered in detail; namely freezing, chemical fixation, antibodies, particulate markers and labelling reactions, both with antibodies and with the main, non-immunological approaches, such as the avidin-biotin system.

During the 10 years it took to complete this book, I found the time available for this project decreasing every year, as my responsibilities as part of an active research team in the EMBL Cell Biology Program increased. A few years ago I realized I might never finish the book without some serious help. For this I am extremely grateful to two good friends for stepping in and writing two of the chapters. For the antibody chapter, the choice of Brian Burke was a logical one since Brian, along with Daniel Louvard, had been instrumental in setting up antibody facilities at EMBL. He was also especially aware of the problems involved in immunocytochemistry. For the chapter on particulate markers, John Lucocq was also an obvious candidate since he had published a number of innovative papers on different aspects of the use of colloidal gold.

Three of the chapters I wrote were especially difficult for me. The first, on fixation, is a monster. When one starts to read the vast literature on this subject it is akin to jumping into a bottomless pit. The number of papers is almost infinite. In every citation one follows to a remote journal one finds more interesting references in ever more remote journals. Many of these papers are difficult to ignore. The problem was to see the "forest for the trees", to try to pull out concepts, rules or even guidelines from these hundreds of papers. I am not sure I have succeeded in this and if I did not, it was surely not due to lack of effort. I believe that the problem of chemical fixation is by far the most difficult we face in all of electron microscopy.

The second chapter that gave me difficulties was Chapter 7 on immunolabelling. The idea I had was straightforward and obvious – to give a theoretical background to labelling reactions on sections. Although there is an enormous amount of literature on immunocyto-chemistry, I found it extremely difficult to be able to select out the basic concepts, rules or guidelines. It appears that most of the literature involves practical rather than theoretical descriptions of labelling reactions. I found the existing textbooks far from satisfactory. The appearance of the book by Larrson in 1988 was extremely satisfying and gave me a number of ideas. Before this time, however, I found much of the relevant information in the classical period of immunocytochemistry, especially in the papers by Coons' group in the

1950s. With respect to theoretical principles, I would even go as far as to say that since the 1950s, as the number of papers in the field has increased, the quality of the average paper has deteriorated. Perhaps, **diluted** is a better word. Many of the relevant facts are buried in the Materials and Methods sections of a multitude of papers covering all the biological and medical sciences. In this respect it seems a fair guess that I will have missed a number of important references. For this I apologize and hope that I can rectify this state of affairs in future.

Last, but not least of my problem chapters was the chapter on quantitation which is in two parts – stereology and quantitation of immunolabelling. Both parts had their own problems. For the stereology I must give the background for my rationale. From 1980 I was able to learn and apply basic stereological methods with the help of Hans Hoppeler in Professor Ewald Weibel's laboratory in Bern. The more I realized how powerful the stereological approach could be, the more surprised I became at how little these techniques are used, in particular for many problems which are crying out for their use. This is especially true in molecular cell biology, where this approach is often considered a mysterious one for getting a few esoteric estimates of surface or volume estimates that are not used in an integrated cell biological context. This state of affairs is even more frustrating when one becomes aware of the fact that, during the past decade, the field of stereology has witnessed a true revolution in new concepts. When applied, these concepts can open up totally new directions in many research problems.

I have often wondered why so many biologists in EM are so ignorant of even the basics of stereology. My only suggestion is that it stems, in part, from the fear one feels when one simply glances at an average theoretical stereology paper. Stereology is a very precise, mathematically based discipline, and this is necessarily reflected in the language in which the key theoretical papers are written. There seems little doubt that the strictness of logic, the need to think in three dimensions and the presence of obscure-looking mathematical formulae tend to frighten away many biologists. It is a feeling I have had myself, especially in the early days and it often returns when I am faced with a new concept such as the use of "vertical sections" or the "disector". In my own case, being able to put simple questions to the stereologists usually bridges the barrier between fog and clarity. When one gets the message, that message is almost invariably simple, and very elegant. Then, upon going back and re-reading the original paper the language becomes clear, despite all the fearsome formulae.

In Chapter 11 I have attempted to both focus on the stereological aspects that are most relevant for molecular cell biology and to try to simplify the language. This is a risk, because by simplifying there is always the danger of over-simplifying and muddling up the well-thought-out stereological rules. For this I was careful enough to work closely with top stereologists (see Acknowledgements). My main aim with the stereology part of Chapter 11 was to enable the reader to overcome his/her fear of this group of extremely powerful methods. If I succeed, that reader will then go to the original literature, will consult a stereologist and, ideally, will take one of the many stereology courses that are offered around the world.

Finally, I should emphasize that since I attempted to dig into the theoretical concepts behind the various techniques discussed in this book I could not avoid being very critical of the immunoperoxidase method. I feel that this approach, more than any, exemplifies the problem of alchemy and loose thinking in our field. How often have I heard serious discussions about the idea of punching precise holes in membranes for antibodies to penetrate **without** really affecting the fine structure of cells! Nevertheless, since this approach is still widely used, I have attempted to give some general guidelines to this procedure, as well as the more straightforward "cell ghost" methods, and to delve into the theory behind the use of detergents and other reagents for permeabilization.

At all stages of this book I have tried to consult the best specialists in each field. On this score I believe I have succeeded and to all these people I owe an enormous debt.

GARETH GRIFFITHS

Acknowledgements

My extreme gratitude to all the following people, listed in no particular order:

To all my friends and colleagues in the Cell Biology Program in EMBL for many things, especially ideas. A special thanks to our Cell Biology Program Coordinator, Kai Simons, and our Director General, Lennart Philipson, for providing the atmosphere which has been so important for me. I must also thank Kai for his help with the detergents in Chapter 10.

To Tim Johnson, whom I first met in 1983 and who changed my whole way of thinking about aldehyde fixation. Tim also helped me on a number of drafts of Chapter 3 and contributed enormously to the chapter. Thanks are also due to Werner Baschong for educating me about formaldehyde and for all his help with Chapter 3.

To Brian Burke and John Lucocq for writing Chapters 6 and 8 respectively. Brian also critically read Chapter 7 while John raised important points on Chapter 11 as well as Chapters 4, 5 and 7.

To Kiyoteru Tokuyasu and Jan Slot, to whom this book is dedicated. From both these good friends I have learned an enormous amount relating to many different aspects of this book. In fact, I am still learning from them each time we meet.

To Jaques Dubochet for giving me the benefit of his wisdom on all aspects of freezing and for pointing out how "water is beautiful". Jaques also helped a lot with Chapter 5.

To Rob Parton for critically reading the whole book, and for his many suggestions in the past 3 years.

To Horst Robenek for all his help with the cryo and replica methods for immunolabelling.

To John Gilkey, Eduard Kellenberger, Werner Villiger and Wim Voorhout for all their suggestions and information concerning Chapter 4. To Heinz Schwarz and York-Dieter Stierhof a separate thanks for educating me on the comparison between the different specimen preparation methods with respect to immunolabelling and for critically reading many chapters.

To David Vaux for his help with Chapters 6 and 7 and for writing the section on the use of recombinant DNA technology in Chapter 6.

To Lars-Inge Larrson for writing his book and for all his suggestions in a number of chapters, especially Chapter 7.

To Hubert Reggio and Jan De Mey for their suggestions and criticisms of a number of chapters, especially Chapter 10.

To Stuart Kornfeld and Mathias Uhlen for their expertise and ideas concerning lectins and protein A, respectively.

A special word of thanks to my stereology friends. First, to Hans Hoppeler, Ewald Weibel and Luis Cruz-Orive for teaching me the basics of stereology. Hans was kind enough to read the first four or five drafts of Chapter 11. In retrospect, those early drafts were disasters, and I am grateful to Hans for being so tolerant and understanding. To Torsten Mattfeldt, Terry Mayhew and Luis Cruz-Orive, who, between them, had many suggestions and constructive criticisms over many drafts of this chapter. Finally, to Hans-Jürgen Gundersen, a special thanks for all his time, ideas, his surgical intervention and for **not** worrying about my feelings.

To Kent Christensen, Paul Webster and Norbert Roos for a lot of help and ideas with many chapters. Kent and Paul together helped to write the historical background to the cryosection technique. Paul was also kind enough to allow me to put his unpublished glass-knife method in Chapter 5.

To Edith Elliott and Clive Dennison for proof-reading the whole book. To Edith I am grateful for all her constructive comments and especially for her suggestions for a major reorganization of Chapter 5.

To Francis Barr, Neil Emans and Beate Sodeik for proof-reading the whole book and for making many useful suggestions.

To those specialists who read and corrected selected chapters; Allan Mackenzie (Chapters 2 and 5), John Tooze (Chap. 3), Alasdair McDowall (Chap. 5), Michael Smith (Chap. 5), Herb Hagler (Chap. 5) and Marc Horrisberger (Chap. 8).

To all those who kindly contributed micrographs for the book. They are all individually acknowledged.

To the support staff of EMBL. Foremost, to the many secretaries who typed the multiple drafts of each chapter, especially Julia Pickles, Rachel Wainwright and Anne Walter. This was truly an enormous effort. To our librarians Mary Holmes and Sue Mottram, who found those hundreds of references for me. To our drawing office, especially Petra Riedinger and Sigrid Bednarczyk, for all their help.

A special thanks to Ruth Hollinshead for so many things, especially for her efforts on the computer with the diagrams in Chapters 5, 7 and 11.

To Dieter Czeschlik and Jutta Lindenborn, Springer-Verlag, for their patience and for giving me the pleasure of fooling them into believing that I would never deliver the manuscript for this book.

GARETH GRIFFITHS

Contents

Chapter 1

Introduction to Immunocytochemistry and Historical Background

1.1 Approaches to Immunocytochemistry

The goal of immunocytochemistry, a combination of immunochemistry and morphology, is to define the **cellular location of biochemically defined antigens.** In practice, in addition to antibody-antigen reactions, a number of other high-affinity interactions, such as those between biotin and avidin or between protein A and immunoglobulins, are generally also included under the broad category of "immunocytochemical" reagents.

Immunocytochemistry is necessarily a blend of two different scientific disciplines. On the one hand, biochemistry and immunochemistry provide the specific tools, the antibodies or equivalent high-affinity markers, while on the other, the field of microscopy (including cell biology, pathology and anatomy) determines whether the target molecule is specifically associated with any particular organelle, cell or tissue structure. A number of approaches have long existed for the detection of antigens at the light microscope level as well as many satisfactory EM methods for localizing cell surface molecules. Until the late 1970s, with a few exceptions, there was little success in carrying out ultrastructural localizations of intracellular antigens. This was a consequence of the fact that the various approaches used to make the cells or tissues accessible to the antibody reagents invariably caused extensive destruction of cell structure. Many of these techniques, when carried out carefully and interpreted cautiously, could provide satisfactory answers to some specific questions, but successful examples were generally few and far between. A multitude of papers appeared in histochemical, pathological and cell biological journals where the specificity of the antibody reagents was reasonably well documented but where cellular organization was largely destroyed. There are also, however, many studies showing excellent structural preservations that used poorly characterized antibodies.

In order to preserve fine structural details for immunocytochemical studies, it was natural in the early days to expect the epoxy resins to play a role as important as they had for routine ultrastructural studies. This, however, was not to be, and the use of Epon sections has had only limited success for antibody labelling studies. The major difficulty has been that the specimen preparation procedures for this approach do not allow a significant fraction of antigens in the section to recognize antibodies. There are very few antigens, for instance, that can still bind antibodies on Epon sections after conventional osmium tetroxide treatment. Consequently, this reagent, which greatly facilitates the visualization of membranes, must usually be omitted for immunocytochemistry on Epon sections. There are, however, a few exceptions to this generalization (Singer 1959; Bendayan and Zollinger 1983; see Chap. 3). The denaturing effects of dehydration, infiltration and polymerization are further tissue-processing steps which are detrimental to antigenicity. These steps cannot be avoided in routine Epon embedding. Generally speaking, those relatively few studies where epoxy resin sections have given successful immunocytochemical results tend to be those where the target molecules were present in relatively high concentrations.

Historically, the problem of labelling intracellular structures for electron microscopy has been approached in two different ways. The first, theoretically more attractive way, was to find alternative embedding media to the epoxy resins, especially water-soluble ones which would facilitate labelling of thin sections. The second approach involves **pre-embedding** techniques where antibodies are introduced into detergent-permeabilized cells prior to embedding. Labelling on sections, which has traditionally been referred to as **post-embedding labelling**, was plagued by technical difficulties for many years. Until the mid 1970s, the most favourable procedure, in my opinion, was the cross-linked bovine serum albumin (BSA) technique, developed by Farrant and McLean (1969; see also McLean and Singer, 1970), and subsequently used almost exclusively by Kraehenbuhl, Papermaster and their colleagues (Kraehenbuhl and Jamieson 1972; Kraehenbuhl et al. 1977 Papermaster et al. 1978 a,b). The fine-structural preservation from this technique was generally not satisfactory, a consequence of the excessive shrinkage that occurs when the aldehyde cross-linked BSA is air dried (Griffiths and Jockusch 1980). However, a major contribution of this technique is that, for the first time, it enabled quantitation to be introduced into immunocytochemistry through the use of simple stereological techniques (Kraehenbuhl et al. 1978, 1979).

In the 1970s, two excellent sectioning approaches appeared for localizing both intra- and extracellular molecules. These rely on either the cryo-section technique or on the use of a new generation of hydrophilic embedding resins such as Lowicryl K4M or LR-white. These techniques, which have made the BSA-embedding technique redundant, will be discussed in detail in this book. The application of particulate markers was crucial to the success of these new sectioning approaches for labelling. While the conceptual breakthrough in this area was undoubtedly the introduction of the ferritin as a marker (Singer 1959), it is the colloidal gold conjugates (Chap. 8) that have now become the markers of choice.

The technical difficulties associated with the early post-embedding techniques increased the popularity of the second approach, namely the **pre-embedding methods**. Using solvents or detergents, the idea here is to make "small holes" in cell membranes which are just large enough to let antibody reagents penetrate to all intracellular sites without destroying fine-structural details (see Sternberger 1986). After the antibody reactions the cells or tissues are subsequently treated with osmium, dehydrated and embedded conventionally in epoxy resins. In practice, the most widely used pre-embedding procedure is the immunoperoxidase technique using horseradish peroxidase (HRP), conjugated to a secondary antibody. HRP, like all proteins containing heme groups, will catalyze the oxidation of substrates such as diamino-benzidine (DAB). In its reduced form DAB is water-soluble; after oxidation, however, it forms a brown, insoluble polymer which is highly osmiophilic (Graham and Karnovsky 1966). This method is considered attractive because the HRP effectively amplifies the primary immune reaction by catalytically oxidizing proportionally large amounts of its substrate. A further amplification can be obtained by using antibodies against HRP in the peroxidase-anti-peroxidase (PAP) technique (see Sternberger 1986).

1.2 Criteria for an Electron-Microscopic Immunocytochemical Technique

Having discussed different approaches for immunocytochemistry, it is now relevant to consider the prerequisites of a general EM technique that can be used routinely for localizing any antigen in any cellular location. Below I have listed the six most important theoretical criteria.

- **A specific, high-affinity antibody** (or other high affinity marker) must be available against a cellular antigen that has preferably been well characterized. Ideally, the antibody should be raised against a pure molecule.
- All parts of cells and tissues must be **accessible** to the antibody and to the electron-dense markers. Otherwise false-negative interpretations are possible.
- A significant amount of the **antigenicity** of the component of interest must still be present when the tissue is labelled with the antibody.
- **The fine-structural preservation** should be "adequate". This statement will be discussed in more detail in the next chapter.
- The techniques should facilitate **quantitation** of the labelling. Two criteria must be met to satisfy this requirement. First, **particulate markers** (such as colloidal gold) must be used which can be clearly seen and counted. There is, for example, no satisfactory method for quantitating the reaction product of HRP at the EM level.[1] Second, the degree of labelling should be proportional to the amount of antigen in the structure: this is primarily a consequence of point 2 (above). Only then is it possible to correlate ultrastructural and biochemical data.

---

[1] Dr. Lars-Inge Larrson has brought to my attention the fact that a number of groups, including his own, routinely quantitate the DAB reaction at the light microscopy level using computer aided densitometry (see Reis et al. 1982). He predicts that such approaches will be developed at the EM level also.

- **Double or multiple labelling**. The technique should allow the simultaneous localization of two (or more) different antigens using different electron dense markers which are easily distinguishable from each other.

Closely related to the first three points above is the question of **sensitivity**. One would like the method to be as sensitive as possible so that antigens present in relatively low concentrations can be detected. Obviously, the higher the specificity and affinity of the antibody for its antigen, and of the visualising marker for the antibody, the more sensitive the technique will be. As will become increasingly clear throughout this book, only the approaches for labelling sections can come close to fulfilling all the above conditions.

1.3 Problems with the Pre-Embedding Techniques

The immunoperoxidase techniques are still popular for EM immunocytochemistry. However, with the advent of cryo-sectioning and the new generation of resins for labelling sections, most of the pre-embedding techniques are becoming less attractive for EM studies. The specific reasons for this statement are emphasized below.

- The detergents or solvents required to allow penetration of antibody through membranes inevitably destroy some fine structure.
- Complete accessibility of reagents to all cellular compartments is never completely assured. Hence, the technique does not usually give statistically reliable data.
- Because of the penetration problem, particulate markers can be used only in combination with harsh detergent or solvent treatment ("cell ghost" approach).
- Only antigens which are not extracted can be detected.

Fig. 1. Pre-embedding labelling with peroxidase. A human breast cancer cell line MCF-3 was incubated with a ricin-peroxidase conjugate for 30 min at 37°C. The cells were then fixed in glutaraldehyde and incubated with diaminobenzidine and H_2O_2 before embedding in Epon. The ricin conjugate binds to the plasma membrane (*P*) and is internalized by endocytosis into endosomes (*E*) and elements on the trans side of the Golgi complex (*arrowheads*; *N* nucleus). (Van Deurs et al. 1986). In a subsequent study (van Deurs et al. 1988) using anti-ricin and protein A gold on cryo sections in combination with biochemical data, the ratio of ricin in the plasma membrane:endosomes:*trans*-Golgi at steady state was estimated to be in the ratio of about 100:20:1. This micrograph makes a striking point: while the intensity of the Golgi labelling is variable from one stack to the next, it is difficult to imagine that the ricin concentration is 20 times less in the Golgi than in the endosomes and 100 times less than the plasma membrane. Note that since the peroxidase-conjugate is internalized by these cells, this preparation is actually more favourable for the peroxidase reaction than the more typical immunoperoxidase reaction. In the latter the antibody conjugates have to diffuse into the compartments after permeabilization; in the experiment shown here only the small molecules DAB and H_2O_2 need to diffuse into the cells. ×26000. (Courtesy of Dr. Bo van Deurs, Dept of Anatomy, Panum Institute, Copenhagen, Denmark)

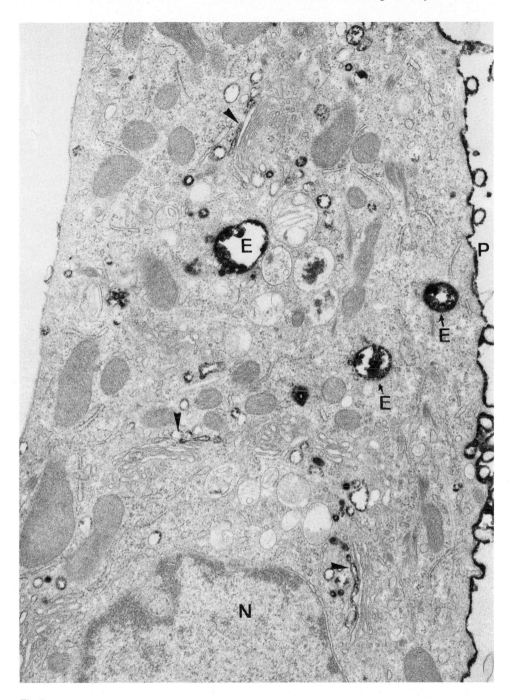

Fig. 1

In practice, except for the "cell ghost" approach, one is restricted with the pre-embedding techniques to using enzymes such as HRP combined with a small soluble substrate such as DAB, whose reaction product cannot be easily quantitated. It is often argued that the HRP method is more sensitive when compared to the use of particulate markers. This argument is blurred by the fact that as the concentration of antigen decreases the positive identification of HRP reaction product becomes more and more subjective. Note that osmium tetroxide as well as other heavy metals such as uranyl acetate are used as contrasting agents in routine EM precisely because they have an affinity for many cellular structures. Even slight variations in the section thickness can alter the amount of electron density due to these reagents. At low antigen concentration it therefore becomes equivocal to decide whether electron density is due to bound antibody or to "general" osmiophilia. In contrast, the detection of a single gold particle is almost always unequivocal because there is no structure in normal cells that even remotely resembles a gold particle. In addition, examples exist of antigens not even being detected (or poorly detected) in compartments that, in fact, have relatively high concentrations of these antigens. Consider, for instance, the many studies that have taken advantage of the heme reaction with DAB in order to localize a peroxidase in those cells that secrete it (e.g. leukocytes). In this case since the peroxidase is already in the organelles of interest, the only technical problem is to get the small DAB and hydrogen peroxide molecules into the intracellular compartments. Nevertheless, many of those studies which could show reaction product in the endoplasmic reticulum (ER) and secretory granules of these cells failed to show it convincingly in the Golgi complex (see Hand and Oliver 1977 for a list of references). Even in many of the studies that did show a Golgi reaction product (e.g. Hand and Oliver 1977) the micrographs clearly suggest that the concentration in this organelle is significantly lower than in the ER. It is now generally accepted that the Golgi complex is an obligatory route from the ER to the cell surface and, moreover, the concentration of secretory proteins is higher in the Golgi complex than in the ER (Bendayan 1984; Posthuma et al. 1988). The variability in the localization in the early studies can surely only be attributed to variations in access of the substrates to the enzyme. An additional well-established artifact of this technique, diffusion of the HRP reaction product (see for example Courtoy et al. 1983), may complicate the interpretation, as well as reduce the effective resolution of the labelling. Further, double-labelling procedures, while they exist in theory (see Norgren and Lehman 1989, and the references cited therein), have not been very successful with this approach. Finally, Fig. 1 shows a striking example of a little appreciated fact, namely, how the immunoperoxidase method can often give results that, though qualitatively correct, may be quantitatively misleading.

Most people who have worked extensively with any of the HRP techniques become aware of these problems and may be frustrated by the presence of many variables that cannot, even empirically, be completely controlled. Even if only qualitative data are required, it is very difficult in practice to find the right compromise between acceptable fine structure preservation and specific labelling. This does not belittle the historical importance of the immunoperoxidase techniques in immunocytochemistry. On the contrary, besides being formerly the most common method for immunolabelling at the EM level, this approach has been extensively used at the light-microscopic level, where it is comparable in sensitivity to immunofluorescence techniques (for a discussion see Larsson 1988).

References

Bendayan M (1984) Concentration of amylase along its secretory pathway in the pancreatic acinar cell as revealed by high resolution immunocytochemistry. Histochem J 16:85–108

Bendayan M, Zollinger M (1983) Ultrastructural localization of antigenic sites on osmium-fixed tissues applying the protein A-gold technique. J Histochem Cytochem 31:101–104

Courtoy PJ, Hunt Picton D, Farquhar MG (1983) Resolution and limitations of the immunoperoxidase procedure in the localization of extracellular matrix antigens. J Histochem Cytochem 31:945–951

Farrant IL, McLean ID (1969) Albumin as embedding media for electron microscopy. Proc 27 Electron Microsc Soc Am: 422–424

Graham RC, Karnovsky MI (1966) The early stage of absorption of injected horseradish peroxidase in the proximal tubules of mouse kidney. Ultrastructural cytochemistry by a new technique. J Histochem Cytochem 14:291–302

Griffiths GW, Jokusch BM (1980) Antibody labelling of thin sections of skeleton muscle with specific antibodies: a comparison of bovine serum albumin (BSA) embedding and ultracryotomy. J Histochem Cytochem 28:969–978

Hand AR, Oliver C (1977) Relationship between the Golgi apparatus, GERL, and secretory granules in acinar cells of the rat exorbital lacrimal gland. J Cell Biol 74:399–413

Kraehenbuhl JP, Jamieson JD (1972) Solid phase conjugation of ferritin to Fab-fragments for use in antigen localization on thin sections. Proc Natl Acad Sci USA 69:1771–1778

Kraehenbuhl JP, Racine L, Jamieson JD (1977) Immunocytochemical localisation of secretory proteins in bovine pancreatic exocrine cells. J Cell Biol 72:406–423

Kraehenbuhl JP, Weibel ER, Papermaster DS (1978) Quantitative immunocytochemistry at the electron microscope. In: Knapp W, Holuber K, Wick G (eds) Immunofluorescence and related staining techniques. North Holland, Amsterdam, pp 245–253

Kraehenbuhl JP, Racine L, Griffiths GW (1979) Attempts to quantitate immunocytochemistry at the electron microscope level. Histochem J 12:317–332

Larsson L-I (1988) Immunocytochemistry: theory and practice. CRC Press Inc, Boca Raton, Florida, 272 pp

McLean JD, Singer ID (1970) A general method for the staining of intracellular antigens with ferritin-antibody conjugates. Proc Natl Acad Sci USA 65:122–127

Norgren RB Jr, Lehman MN (1989) A double-label pre-embedding immunoperoxidase technique for electron microscopy using diaminobenzidine and tetramethylbenzidine as markers. J Histochem Cytochem 8:1283–1289

Papermaster DS, Schneider BG, Zorn MA, Kraehenbuhl JP (1978a) Immunocytochemical localisation of opsin in outer segments and Golgi zones of frog photoreceptor cells. J Cell Biol 78:196–203

Papermaster DS, Schneider BG, Zorn MA, Kraehenbuhl JP (1978b) Immunocytochemical localisation of a large intrinsic membrane protein to the incisures and margins of frog rod outer segment dishes. J Cell Biol 78:415–423

Posthuma G, Slot JW, Veenendaal T, Geuze HJ (1988) Immuno-gold determination of amylase concentrations in pancreatic subcellular compartments. Eur J Cell Biol 46:327–335

Reis DJ, Benno RH, Tucker LW, Joh TH (1982) Quantitative immuno-cytochemistry of tyrosine hydroxylase in brain. In: Chan Palay V, Palay SL (eds) Cytochemical methods in neuroanatomy. Liss, New York, pp 205–228

Singer SJ (1959) Preparation of an electron dense antibody conjugate. Nature 183:1523–1527

Sternberger LA (1986) Immunocytochemistry, 3rd edn Wiley, New York

van Deurs B, Tnnessen TI, Petersen OW, Sandvig K, Olsnes S (1986) Routing of internalized ricin and ricin conjugates to the Golgi complex. J Cell Biol 102:37–47

van Deurs B, Sandvig K, Petersen OW, Olsnes S, Simons K, Griffiths G (1988) Estimation of the amount of internalized ricin that reaches the *trans*-Golgi network. J Cell Biol 106:253–267

Chapter 2

Fine-Structure Preservation

2.1 Introduction

Since the phrase "fine-structure preservation" is central to the theme of this book, it is important to define precisely what it means. It is relatively easy to give a theoretical definition: – cell structure should be preserved exactly as it was in the living state and should be visualized at the resolution limit of the electron microscope.

In routine electron microscopy (EM), this ideal has never been achieved and practical compromises must be made which make it more difficult to define precisely what fine structure preservation entails.

When electron microscopists look at sections of cells or tissues they are seeing a two-dimensional segment of the overall structure. A key question which has always plagued the microscopist is how close is this two-dimensional image to the three-dimensional organization of the in vivo structure, by definition the true 3 D-organization? The necessity of killing the cell in order to assess its living structure is reminiscent of Heisenberg's "uncertainty" principle: there is always the danger that the process of preparing the cell in order to examine it may have destroyed its organization. Indeed, a claim was even made a few years ago that most, if not all, structures seen in EM tissue sections are artefacts of osmium tetroxide staining and other heavy metals used in their processing (I have conveniently lost this reference!). While this is an extreme opinion, I have often heard similar sentiments expressed by sceptical biochemists. Although there are clearly potential artefacts associated with all EM specimen preparation techniques, it is usually possible to recognize them and, therefore, to assess the validity of the results.

Cells have a remarkable degree of structural organization that is undoubtedly the framework for the multitude of integrated functions that they undertake. That structure and function are interdependent is a basic axiom in biology. Hence, the function of a particular cell type will be reflected in its structural organization. Similarly, since many basic functions are common to most cell types, the structures responsible for these functions should be the same or at least similar. Thus, every living cell has a plasma membrane with similar structure and function. With few exceptions, all animal and plant cells have a nucleus containing chromatin and one or more nucleoli surrounded by a nuclear lamina and an envelope that is penetrated by nuclear pores of characteristic structure. They have mitochondria, a region containing the Golgi

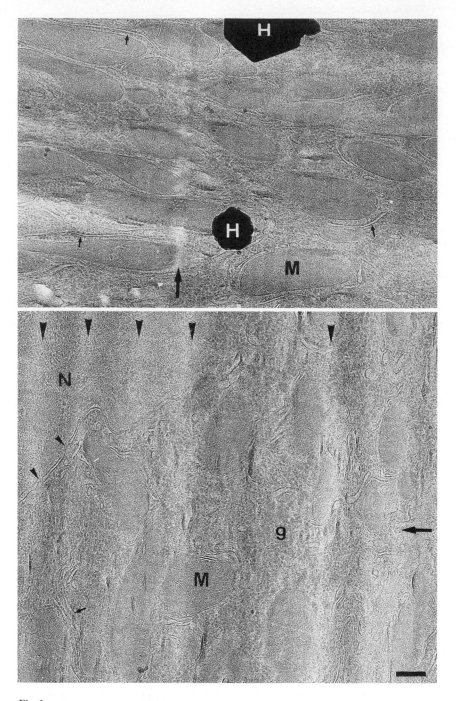

Fig. 1

complex, coated vesicles, endosomes, lysosomes, peroxisomes (microbodies), rough and smooth endoplasmic reticulum, centrioles, microtubules and microfilaments. In addition, many animal cells have junctional complexes as well as large numbers of intermediate filaments, that are tissue specific. Most plant cells in addition possess cell walls, vacuoles and chloroplasts. Even in cases where a certain organelle has not yet been seen in a certain cell, it is reasonable to conclude, on a functional basis, that it must exist. In all cells the structure of these organelles is remarkably similar in vivo and, in most cases, in vitro following a wide variety of methods of preparation.

But, the sceptic may ask, cannot all of these preparation procedures, which necessarily involve chemical modifications, give rise to identical artifacts? Fortunately, a simple and direct argument against this possibility can be put forward based on the recently developed hydrated cryo-section approach. The advantage of hydrated cryosections is that the specimen is rapidly frozen (in fact vitrified, see Chap. 5), sectioned and examined at temperatures below $-150°C$ without any exposure to chemical fixation or stains. Dubochet and colleagues (see McDowall et al. 1983, 1989; Chang et al. 1983) have shown that the major organelles of well-studied cells, such as the rat liver hepatocyte, look essentially the same in a frozen hydrated section of unfixed tissue (Fig. 1) as they do in a conventionally prepared Epon section (Fig. 2), in a cryo-section of chemically fixed cells (Fig. 3) or in a Lowicryl section (Fig. 4); (Figs. 1–4 are a comparison of all these methods for a similar area of rat liver). Although the hydrated cryosection technique has its own artefacts (predominantly related to sectioning; see Fig. 1) the point is nevertheless clear: the conventional methods of preparation give a remarkable degree of fine-structural preservation. This is certainly true at the relatively low resolution that is usually required in EM sectioning studies. What is equally true, however, is that the conventional Epon embedding can rarely preserve structures to the extent that **high resolution**

Fig. 1. Hydrated cryosection (about 120 nm thick) of freshly isolated rat liver. A tiny piece of the tissue (less than 0.5 mm in any dimension) was "slam frozen" on a liquid nitrogen cooled copper block. Cryosections were prepared that were transferred onto 1000 mesh EM grids without supporting films. The sections were flattened with a metal rod and transferred under liquid nitrogen into the "cold stage" of the electron microscope where they were observed using low electron dose conditions. Note that they were never exposed to a temperature above $-140°C$. Electron diffraction indicated that the solid water in the section was vitreous (devoid of crystalline structure). The two figures (which are taken from different areas of the same published micrograph) show remarkable detail, especially when one considers that they were never exposed to chemical fixatives or heavy metal stains. In both figures well preserved mitochondria (M), (in parts with two membranes and crystae visible) are evident. Note also the rough endoplasmic reticulum (*small arrows*), regions containing glycogen (g) and the nucleus (N) with nuclear pores (*small arrowheads*) visible. Obvious artefacts are also evident, namely contaminating ice crystals (H) (which probably resulted from water droplets condensing on the sections in the cryo-chamber of the microtome), knife marks (*large arrow in upper figure*); this also indicated the direction of cutting and "chatter" [periodic differences in thickness (perpendicular to the direction of cutting) – *large arrowheads in lower figure*]. The direction of cutting is indicated by the *large arrow*. ×18000. *Bar* 0.5 μm. (McDowall et al. 1983)

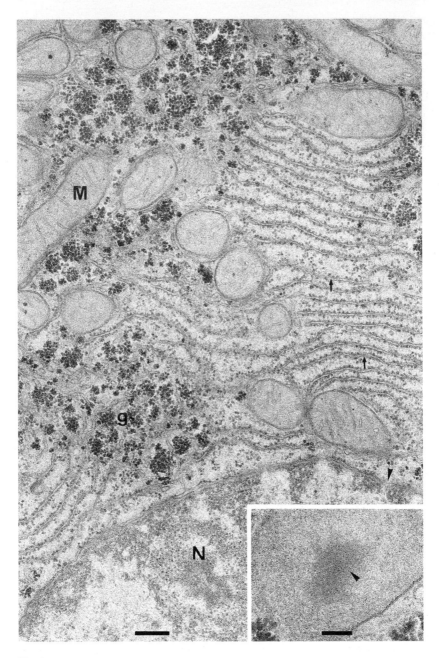

Fig. 2

information can be obtained. This is a consequence of osmication, dehydration, infiltration and heat-induced polymerization.

2.2 Fine-Structure Evaluation in Practice

It is useful at this point to look in more detail at the **practical steps** involved in preparing and analyzing an electron microscopic image of a section; in other words, the steps leading to the fine-structure evaluation. This discussion is based essentially on conventional preparation and evaluation methods.

2.2.1 Fine-Structure Stabilization by Fixation

The first part of fine-structure preservation, the fixation step, is described in detail in the next chapter. This step is the most critical in determining how closely the image seen in the electron microscope resembles the in vivo structure. Later stages of preparation can never regain information lost during fixation.

2.2.2 Contrasting

Aside from the ability to preserve fine structure, it is equally important to be able to **visualize** it. A two-dimensional section of a cell or tissue can only be seen in the EM via **contrast** information provided by either exogenous heavy metal atoms that have precipitated onto certain structures (with poorly defined specificity but in a reproducible way) or via the electron scattering due to the atoms in the structure itself. When sections of tissue fixed in glutaraldehyde and embedded in Epon (without osmium) are examined in a normal transmission EM, they are remarkably devoid of contrast although the actual preservation provided by the fixation may have been adequate. Contrasting these sections

◀───

Fig. 2. Conventional Epon embedded section (about 50 nm thick) of rat liver. This section came from a specimen that is over 20 years old. The original protocol was no longer available but the tissue was almost certainly perfusion-fixed in glutaraldehyde, post-fixed in osmium tetroxide, en bloc stained in uranyl acetate, dehydrated and embedded and the sections stained with a lead salt. This kind of (classical) image is one to which all cell biologists can relate; it may also be the most aesthetic of the four figures presented. The contrast is excellent and details are evident of the rough ER (*arrows*), glycogen granules (*g*), mitochondria (*M*) and nucleus (*N*) *arrowhead* indicates a nuclear pore. Note, however, that in this particular preparation the periodicity in the crystal of the peroxisome (*arrowhead-inset*) is not evident (and presumably destroyed during the preparation (cf. Fig. 3). ×19000, *bar* 0.5 μm (Inset ×90 000, *bar* 100 nm). (Preparation courtesy of Dr. Norbert Roos, EM Laboratory, University of Oslo, Norway)

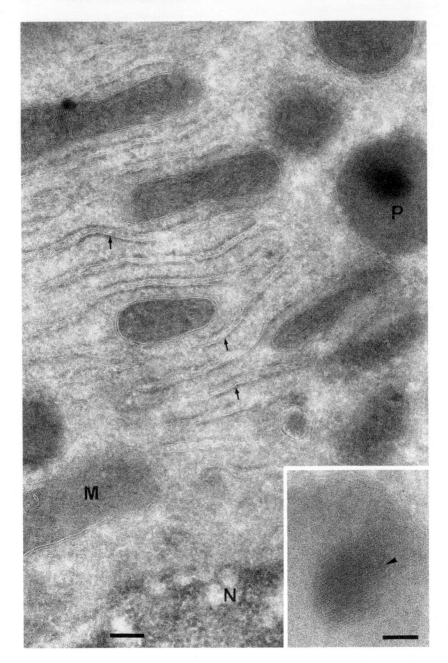

Fig. 3

with, for example, uranium and lead salts allows structures to be seen. Even in the absence of such heavy metal "stains" a similar result can be obtained using a scanning transmission EM (STEM; not to be confused with a normal scanning EM) which, because of its more efficient electron detection, gives far higher contrast (Kellenberger et al. 1986); this is also possible using the energy loss spectrometer that is available on the new Zeiss 902 electron microscope. These procedures enable one to visualize the structures and assess the quality of preservation. Being able to distinguish between the two processes of preservation and visualization can have large practical consequences. If a structure of interest has been destroyed by the initial fixation it is obvious that it cannot be visualized by any contrasting procedure. On the other hand, if a structure has been preserved but has not been stained sufficiently to allow visualization, it is then worthwhile making an effort to visualize it.

2.2.3 Analysis

The third stage, the analysis, begins when the section is first looked at in the EM. The first step in this is to evaluate the quality of the preceding processes, the fixation and contrast and whether the later stages of preparation, dehydration and embedding may have affected the quality of the image. A deductive process allows a low resolution model of the structure to be made. From conventional Epon section electron micrographs, five parameters would appear to be most important for this model, namely **shape, size, position** and **number** of the component structures, as well as overall **contrast** information.

◄──────────────────────────────────

Fig. 3. Cryosection (about 100 nm thick) of rat liver prepared by the Tokuyasu procedure. The liver was perfusion fixed with a mixture of 0.2% glutaraldehyde and 4% paraformaldehyde in 0.1 M phosphate buffer, pII 7.4 (fixed preparation kindly provided by Dr. Jan Slot, University of Utrecht). The tissue was infused with 2.1 M sucrose, frozen in liquid nitrogen and cryosectioned. The sections were embedded in methyl cellulose-uranyl acetate (see Chap. 5). Images, such as this, which is intended to show the distinct appearance of cryosections prepared in this way can be **routinely** obtained without difficulty, even by beginners. Note the "negative/positive" contrast of the membranes. In rat liver, the ribosomes on the rough ER (*arrows*) are usually positively contrasted with this staining protocol. In most other cell types this is true only when significantly thinner sections are prepared (the reason for this difference is not known). Note that the glycogen particles are not clearly visible. This is presumably due to the negative-staining effect of the uranyl acetate which "fills" the cytoplasmic spaces. For visualizing those organelles which are not seen in this section many alternative protocols exist (see Chap. 5). Note that even in this relatively thick section the periodicity (*arrowhead inset*) of the crystal of the peroxisome (*P*) is evident in almost all profiles through this organelles (cf. Fig. 2). Thus, even though images such as this may be less aesthetic than that shown in Fig. 2, they are capable of providing more high resolution information. ×21000, *bar* −0.5 μm. Inset ×105000, *bar* 100 nm

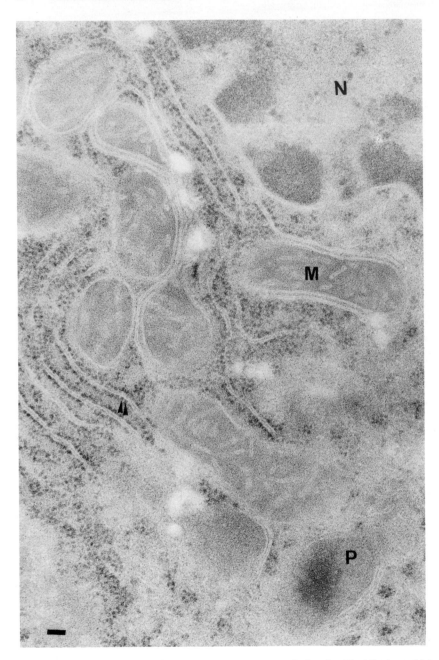

Fig. 4. Lowicryl K4M sections of rat liver. Tissue was perfusion fixed in 1% glutaraldehyde in 0.1M phosphate buffer, dehydrated in ethanol (without osmium treatment) and embedded by progressive lowering of temperature in Lowicryl K4M (see Chap. 4). *M* mitochondrion; *N* nucleus; *P* peroxisome (which shows some periodicity in the crystal; *arrowheads* indicate ribosomes on the rough ER. These are always very prominent in Lowicryl K4M sections ×23000, *bar* 100 nm. Characteristically also, the membranes all appear negatively contrasted in these sections. (Courtesy of Dr. John Lucocq, Anatomy Department, University of Bern, Switzerland)

2.2.4 Interpretation

2.2.4.1 Dimension

More or less simultaneous with analysis begins the process of interpretation. Clearly this is dependent upon the three preceding steps, but is still distinct. Two electron microscopists can interpret the same image completely differently. This last stage is by far the most complex and is hence the most difficult to apply any rules to. At the simplest level, interpretation requires the evaluation of the structure at the **two-dimensional level** without necessarily extrapolating to **three dimensions**. Hence the criteria already mentioned, namely contrast, shape, size, position and number of structures can usually be evaluated on a two-dimensional basis, as can any obvious artefacts such as "breaks" in membranes, "holes" or areas where severe loss of structure has occurred. In this 2-D view of the cell the rough endoplasmic reticulum, for example, is two parallel dense lines with black dots on their periphery. Perhaps the most difficult problem in interpretation is to build up a true three-dimensional model of the structure from two-dimensional images. When we need to know the three-dimensional organization of a structure from two-dimensional sections, the first thing which often comes to mind is to undertake a reconstruction of the structure from serial sections. Before this is considered, however, I would recommend reading an excellent, concise review by Elias (1972---as well as many other references from this author) entitled, *Identification of Structure by the Common Sense Approach* in which the author points out how a little thought about three-dimensional shape can simplify the interpretation of what images of sections through a structure should look like. I can do no better than quote him. Thus:

"The shape of a section through an object depends primarily on the three-dimensional shape of the object cut, and secondarily on the angle of cutting".

He stresses the importance of the principle of dimensional reduction,

"because awareness of it often permits instantaneous identification of shape from single sections".

By dimensional reduction here he is referring to the fact that a section through an n-dimensional object is an (n-1) dimensional figure. In other words, a section through a three-dimensional object is two-dimensional, a section through a two-dimensional object (area) gives a one-dimensional line while a section through a one-dimensional object has 0-dimensions (point). For more details see Weibel (1979) and Chapter 11, Table 1. Elias summarizes this concept accordingly:

"In brief, in most cases a dot in a true section is a section through a fibre; a line is a trace of membrane and a stripe is a profile of a plate or lamina of easily measurable thickness. An area is a profile of a solid".

The importance of the above statements cannot be overemphasized. In this review Elias gives a number of examples of misinterpretation made by failing to take these phenomena into consideration. Above, I referred to the example of the image of sections through the rough ER as two lines with black dots, an example I had arbitrarily chosen before reading the review by Elias. In it he points out how a serious error of interpretation was actually made in the first paper that recognized the rough ER on EM sections.

"The cisterns of the endoplasmic reticulum were originally described as double filaments (Bernhard et al. 1952) although their sections appear in electron micrographs as long, narrow lines. The authors who first described them thought that they had been incredibly lucky to have obtained only longitudinal sections of these supposed double filaments. The principle of dimensional reduction, however, teaches us that sections through such "filaments" should be little pairs of dots. Fortunately, the next author who described the endoplasmic reticulum did not know the work just mentioned. Only a year later he found and described the endoplasmic reticulum correctly (Sjöstrand 1953). He stated that it consists of double membranes. He came to the correct conclusion by intuitive, three-dimensional thinking".

Often, a structure or set of structures in a biological section is more complex than a simple sphere, tube or rod. In theory, two different approaches could be used to facilitate an understanding of the structure. First, a serial section reconstruction and second, an analysis of stereo pairs (from thick, preferably selectively contrasted, tilted sections) prepared using the goniometer of the EM. An important consideration with such techniques is that the thickness of the section may become a limiting factor. This is especially a problem, for small structures, such as vesicles or small tubules, which are as small as or smaller than the thickness of the section. These elements may not be recognized and continuities with other structures missed. Obviously, for a correct interpretation of a structure, whether simple or complex, all the components must be recognizable in the section.

2.2.4.2 Sampling

A potential problem in interpretation is that one mostly takes pictures in a non-random fashion, that is, one selects the micrograph to be taken (the **sampling problem**). Often implicit in this selection is a bias towards or against a certain idealized model. This bias is often continued to the publication stage where the micrographs chosen are really not representative (and perhaps not repeatable!). For a number of studies the obvious solution to this problem is to take a statistical approach (see Chap. 11).

In this context it should not be forgotten that the "sampling problem" has not only negative implications. Most people are aware that the images selected represent only a minute fraction of all possible images through the structure. Less often appreciated, however, is the other side of the coin: if a certain "structural event" is extremely rare, the chances of seeing it in a random section must be infinitesimally small. In other words if one sees this "structural event" (which may be, for example, an organelle or a labelling pattern) of interest even twice(!) it is likely that this "event" will be significant with respect to the whole sample. Note that this statement is a general one that does not take into account the interpretation of the "event". As will become clearer in Chapter 11 the question of the probability of seeing a structure in a section can be dealt with rigorously by stereological approaches.

The process of interpretation is a very subtle one which may be exemplified by a personal example. As an experienced microscopist, I can spend many hours looking at a few immunolabelled sections of familiar cells. An overall impression (perhaps a "theory" in its loosest sense) is usually obtained. Nevertheless, after discussion with a colleague or discovering a key piece of information in the literature I can go back to the microscope and take a fresh look at the same sections. On many occasions, within minutes, I have a different interpretation of the same results! Perhaps, for example, I have decided to take a close look at a part of the cell which I neglected during the first examination. This shows that the interpretation of the image is by far not just related to "technical" parameters but is intricately woven with the background of the scientific problem at hand. This point is often not realized by many scientists at the periphery of electron microscopy, who think that, even in the case of complex biological problems, all that is required is for the operator to put sections into the microscope and simply push the photography button!

2.2.4.3 Artefacts

At this point, a word of caution about artefacts would be appropriate. As for making correct interpretation of three-dimensional structure, there are no hard and fast rules for deciding what structure is an artefact. Artefact here simply means any structure which is not present as such in the living cell. Some artefacts, such as knife marks, "chatter" or other sectioning irregularities are usually easy to recognize. Many others, such as "breaks" in the continuity of membranes (as opposed to oblique sections through the bilayer) can be ignored purely on functional considerations. It is not always that easy, however. Many mistakes have been made, and are probably still being made, both in interpreting an artefactual image as a real structure and, conversely, in calling what is in fact a true structure an artefact. This problem can best be illustrated by reference to some historical examples.

The Golgi complex is recognized as a central organelle in eukaryotic cells with a pivotal role in secretion and membrane biogenesis (Palade 1975). Described by Camillo Golgi in 1898, it was viewed with some scepticism for several decades, culminating in the light microscopical study by Palade and Claude (1949), claiming it to be an artefact of osmium fixation. It was only 5 years later that the authenticity of the Golgi complex was established by electron microscopy (Dalton and Felix 1952). Strictly speaking, it was not really "established" but more "accepted" by the magority in the field since the EM data in the 1950s in no way ruled out the possibility of fixation artefact. After all, these data were based on the use of the same fixative, osmium tetroxide, that had caused such a dispute at the light microscopy level. Thirty-five years later, however, despite tremendous progress in recent years, there is still no comprehensive model of the Golgi complex which can completely explain why it needs a set of flattened cisternae.

A number of new structures have "appeared" along with improvement in specimen preparation techniques, especially during the era following the introduction of glutaraldehyde as a fixative (Sabatini et al. 1963). The previous "absence" of such organelles can now, in a sense, be interpreted as artefactual. Hence, cytoplasmic microtubules (Slauterbach 1963; Ledbetter and Porter 1963), coated vesicles (Roth and Porter 1964) and mitochondrial DNA (Nass and Buck 1963; Nass and Nass 1963) are some classic examples. Each of these new organelles or structures had to survive the test of time and trial, including in vitro studies. In many cases, such as microtubules, the regular structural appearance made it highly unlikely from the outset that the structure could be artefactual. On the other hand, the mesosome, a structure which is a regular infolding of the inner membrane of many bacteria, and to which a number of specialized functions had been attributed (see Fooke-Achterrath et al. 1974; Hayat 1981) is an artefact. In this case, it was perhaps difficult to imagine that the regular, myelin-like folding of the membrane could be an artefact. Although careful EM analysis had made it seem highly likely that these structures were artefacts of chemical fixation (e.g. Ebersold et al. 1981), conclusive evidence for this has again come from the hydrated cryosection technique, with and without fixation (Dubochet et al. 1983; see Fig. 1 Chap. 3). Nevertheless, it still cannot be ruled out that this "artefact" reflects some true physiological difference in those parts of the bacterial membrane that are preferentially susceptible to the effects of the fixative.

The concept of "microtrabeculae" is an example of a structure whose real meaning was much in dispute. Keith Porter is recognized as one of the great pioneers in ultrastructural studies who, along with Sjöstrand, (see above) first described the structure of the endoplasmic reticulum (see Porter 1953; Palade and Porter 1954). In the late 1970s and early 1980s, Porter's group had been studying the structure of "ground cytoplasm" to which no recognizable structure has hitherto been attributed. Using whole cultured cell mounts which have been critical by-point dried and viewed in the high voltage EM (where the electrons can effect greater penetration than at normal voltages), he and his colleagues described a fine network of interconnecting filamentous structures referred to as microtrabeculae (Wolosewick and Porter 1979; Porter and Tucker 1981). Many people strongly dispute this claim and believe this network, which has never been unambiguously visualized by any other preparation technique, to be an artefact of the critical-point drying process which they use in the final step of their preparation technique. In recent years, this concept that all or most organelles are interconnected by filamentous structures is not often discussed. Nevertheless, these observations have at least helped to draw attention to the fact that the cytoplasm is not simply an amorphous "soup".

It should be stressed here that artefacts may sometimes be informative. For example, it is clear that in normal plastic embedded cells, variable degree of extraction of many components always occurs. This extraction improves the contrast of many structures such as membranes and thus helps to resolve them, even at relatively low magnifications. The 8–10-nm railroad-track appearance of membranes themselves as seen after osmium tetroxide treatment, the so-called unit membrane, is itself a form of artefact. It is now established, from a number of independent techniques, that membranes are ≈ 4-nm lipid bilayers into which are intercalated large numbers of spanning and peripheral membrane proteins. Many of these are globular structures, whose dimensions are large enough to make them theoretically resolvable in sections. However, up until now, there has been little convincing demonstration of any structures interrupting the uniform railroad-track appearance (but see, for example, Carlemalm and Kellenberger 1982; Rash et al. 1982). It is indeed evident from independent physical techniques, such as X-ray diffraction studies, that the structure of the membrane is largely destroyed after osmium tetroxide treatment (Parsons 1972). Under these conditions it is obvious that molecular interpretation of membrane structure is not possible. Nevertheless, at the usual low resolution level, the uniform railroad-track structure serves as an excellent morphological indication of a membrane. It is also a useful guide to section thickness: as sections become thicker than about 80 nm it becomes less and less likely that one can can clearly resolve the two dense lines.

2.2.5 Selective Destruction

For structural or immunocytochemical studies there may be a distinct advantage in either destroying certain structures in order to visualize others, or removing components which mask the visualization of the structures of interest. Such selective destruction, or extraction, is perfectly acceptable as long as one realizes what is destroyed and what is preserved. This is the classical approach, for example, for studying the cytoskeleton: membranous structures and interfering cytoplasmic material are selectively extracted with detergents or solvents allowing fine details of filaments, which are not affected by such treatment, to be visualized. It may be dangerous, however, to extrapolate directly from such studies to follow events such as membrane-cytoskeletal interactions. The recent introduction of mechanical means to disrupt tissue culture cells without homogenizing them will probably become a powerful method for looking at membrane organelles (Simons and Virta 1987; Balch et al. 1987). In this approach a wet filter is placed briefly over the monolayer and then removed, taking with it pieces of the plasma membrane. Although the cytoplasm leaks out, some functions such as intracellular transport from ER to Golgi (see the above references) will continue under appropriate conditions. This is also an excellent method to visualize clathrin-like coats on budding vesicles. Other methods of permeabilization which may facilitate selective extraction will be discussed in Chapter 10. Even in normal tissue culture cells

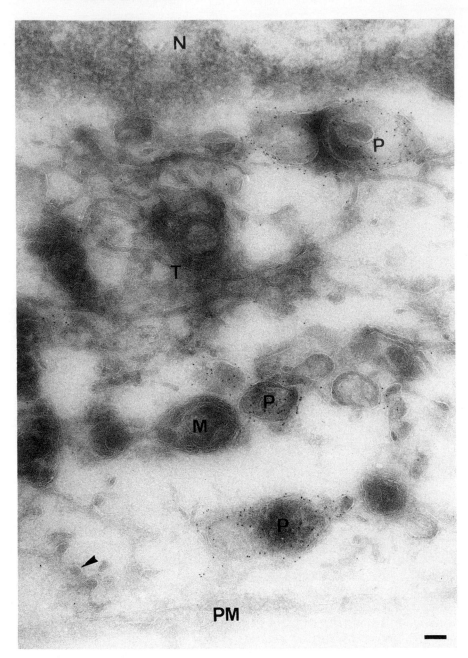

Fig. 5

the occasional appearance of a lysed cell can often reveal details that are not evident after normal fixation (see Fig. 5).

2.3 Summary

In summary, in most cases where one studies thin sections by electron microscopy one aims to preserve the structure of cells as closely as possible to the native state, as well as to visualize the profiles of organelles. In practice, there are techniques that allow this to be carried out, at least for low resolution studies, and there are also guidelines which allow the quality of the preservation to be subsequently assessed. At all stages in the process caution must be exercised, and if doubt exists it makes sense to use as many independent methods of specimen preparation as possible.

This conceptual framework for fine-structural evaluation, the essence of which most microscopists would probably accept as being important for conventional fine-structural studies, must not be abandoned when the additional complication of antibody labelling is thrown into the equation. Unfortunately, this has often been the case in the past. From the point of view of the microscopist, it is important to state that immunocytochemistry is merely a **selective contrasting** EM technique that enables the target molecule to be visualized, usually indirectly with a secondary marker such as colloidal gold or ferritin. Moreover, it could be argued that fine-structural preservation becomes even more important for immunocytochemistry than for conventional structural studies since an antigen hitherto unrecognizable in the section must be preserved and indirectly visualized. While the loss of such a molecule during specimen preparation may not be noticed in routine EM studies, it would clearly be disastrous for an immunocytochemical study of that molecule.

A final important point to emphasize again is that although subtle differences exist in the appearance of organelles from one specimen preparation technique to the next the overall appearance of most cell structures is strikingly

◄──

Fig. 5. Example of selective extraction to improve visualization of membrane organelles. A cryosection of an 8% paraformaldehyde-fixed NRK cell labelled with an antibody against the cation-independent mannose 6-phosphate receptor and protein A gold. This is a fortuitous section through a cell that was probably lysed during fixation. In contrast to the majority of cells that appeared normal, the cytoplasm of this cell is clearly extracted. The smaller tubular projections of many organelles become more visible in such preparations. Note the complex tubular-cisternal organization of what is probably the trans Golgi network (*TGN – T*) with relatively little gold label. The receptor rich prelysosomal compartment (*P*) is physically close to the Golgi/TGN but functionally distinct. The *arrowhead* indicates what is most likely a part of an early endosome, next to the remains of the plasma membrane (*PM*). *N* nucleus; *M* mitochondrion. ×60000, *bar* 100 nm. (For more details see Griffiths et al. 1988, 1990)

similar between the different sectioning methods. The four quite different specimen preparation protocols used for the preparation of rat liver epithelial cells seen in Figs. 1–4 will be discussed in detail in later chapters.

References

Balch WE, Wagner KR, Keller DS (1987) Reconstitution of transport of vesicular stomatitis virus G protein from the endoplasmic reticulum to the Golgi complex using a cell-free system. J Cell Biol 104:749–760

Bernhard W, Haguenau F, Gautier A, Oberling Ch (1952) La structure submicroscopique des elements basophiles cytoplasmiques dans le foie le pancréas et les glandes salivaires. Z Zellforsch Mikrosk Anat 37:281–289

Carlemalm E, Kellenberger E (1982) The reproducible observation of unstained embedded cellular material in thin sections: visualisation of an integral membrane protein by a new mode of imaging for STEM. EMBO J 1:63–67

Chang JJ, McDowall AW, Freeman R, Walter CA, Dubochet J (1983) Freezing sectioning and observing artefacts of frozen hydrated sections for electron microscopy. J Microsc 132:109–123

Dalton AI, Felix MD (1952) Studies of the Golgi substance of the epithelial cells of the epidymis and duodenum of the mouse. Am J Anat 92:277–305

Dubochet J, McDowell AW, Menge B, Schmid EN, Lickfeld KG (1983b) Electron miscroscopy of frozen-hydrated bacteria. J Bact 155:381–390

Ebersold HR, Cordier IL, Lüthy P (1981) Bacterial mesosomes: method dependent artefacts. Arch Microbiol 130:19–22

Elias H (1972) Identification of structure by the common-sense approach. J Microsc 95:59–68

Fooke-Achterrath M, Lickfeld KG, Reutsch VM Jr, Aebi U, Tschöpe U, Menge B (1974) Close-to-life preservation of *Staphylococcus aureus* mesosomes for transmission electron microscopy. J Ultrastruct Res 49:270–285

Griffiths G, Hoflack B, Simons K, Mellman I, Kornfeld S (1988) The mannose 6-phosphate receptor and the biogenesis of lysosomes. Cell 52:329–341

Griffiths G, Matteoni R, Back R, Hoflack B (1990) Characterization of the cation-independent mannose 6-phosphate receptor-enriched prelysosomal compartment in NRK cells. J Cell Sci 95:441–461

Hayat MA (1981) Fixation for electron microscopy. Academic Press, New York

Kellenberger E, Carlemalm E, Villiger W (1986) Physics of specimen preparation and observation of specimens that involve cryoprocedures. In: Science of biological specimen preparation. SEM Inc, AMF O'Hare (Chicago) USA, pp 1–20

Ledbetter M, Porter K (1963) A "microtubule" in plant cell fine structures. J Cell Biol 19:239–250

McDowall AW, Chang JJ, Freeman R, Lepault L, Walter CA, Dubochet J (1983) Electron microscopy of frozen hydrated sections of vitreous ice and vitrified biological samples. J Microsc 131:1–9

McDowall AW, Gruenberg J, Römisch K, Griffiths G (1989) The structure of organelles of the endocytic pathway in hydrated cryosections of cultured cells. Eur J Cell Biol 49:281–294

Nass MMK, Buck CA (1963a) Intramitochondrial fibres with DNA chartacteristics. I. Fixation and electron staining reactions. J Cell Bio 19:593–612

Nass S, Nass MMK (1963b) Intramitochondrial fibres with DNA characteristics. II. Enzymatic and other hydrolytic treatments. J Cell Bio 19:613–621

Palade GE (1975) Intracellular aspects of the process of protein synthesis. Science 189:347–357

Palade GE, Claude A (1949) The nature of the Golgi apparatus. Parallelism between Golgi apparatus and intracellular myelin figures. J Morphol 85:35–112

Palade GE, Porter KR (1954) Studies on the endoplasmic reticulum. I. Its identification in cells in situ. J Exp Med 100:641–650

Parsons DF (1972) Status of X-ray diffraction of membranes. Ann NY Acad Sci 145:321–328

Porter KA (1953) Observations on a submicroscopic basophilic component of the cytoplasm. J Exp Med 97:727–735

Porter KR, Tucker JB (1981) The ground substance of the living cell. Sci Am 244:40–46

Rash JE, Johnson TJA, Hudson CS, Giddings FD, Graham WF, Eldefrawi ME (1982) Labelled-replica techniques: post-shadow labelling of intramembrane particles in freeze-fracture replicas. J Microsc 128:121–138

Roth TF, Porter KR (1964) Yolk protein uptake in the oocyte of the mosquito *Aedes aegypti*. J Cell Biol 27:313–320

Sabatini DD, Bensch K, Barrnett RJ (1963) Aldehyde fixation for morphological and enzyme histochemical studies with the electron microscope. J Cell Biol 17:19–58

Simons K, Virta H (1987) Perforated MDK cells support intracellular transport. EMBO J 6:2241–2247

Sjöstrand F (1953) Systems of double membranes in the cytoplasm of certain tissue cells. Nature 171:31–35

Slauterbach D (1963) Cytoplasmic microtubules 1. Hydra. J Cell Biol 18:367–388

Weibel ER (1979) Stereological methods. Practical methods for biological morphometry, vol. 1. Academic Press, New York

Wolosewick JJ, Porter KR (1979) Microtrabecular lattice of the cytoplasmic ground substance-artefact or reality. J Cell Biol 82:114–139

Chapter 3

Fixation for Fine Structure Preservation and Immunocytochemistry

Before any cell or tissue section can be viewed in the electron microscope it is always necessary to stabilize or "fix" the sample. The ultimate aim of "fixation" is to "freeze" cell and tissue organization in a particular time frame so that every molecule and, ideally, every ion present in that cell or tissue remains in their original location during visualization. The use of "freeze" is appropriate because the only fixation procedure which could come close to satisfying the above conditions would be rapid cryo-fixation (= physical fixation of native tissue) followed by cryosectioning and observation of the frozen hydrated sections by cryoelectron microscopy (Chang et al. 1981; McDowall et al. 1983, 1989; see Fig. 1). Unfortunately, for cytochemical or immunocytochemical methods, which require enzyme activity or an antibody-antigen reaction, this cryo-fixation approach cannot be used because these procedures must be carried out at temperatures where water is a liquid. Hence for immunocyto-chemistry, as well as enzyme cytochemistry, chemical fixation is currently an unavoidable prerequisite. As discussed in Chapter 4, an exception to this statement is the freeze-substitution approach, without using conventional fixatives. In this technique, however, the tissue must still be exposed to chemical solvents. Chemical fixation should ideally serve two functions: first, to **preserve** cell and tissue organization as near as possible to the native organization, and second, to **protect** the tissue against all later stages of preparation with minimal deterioration of fine-structure.

3.1 Fine Structure Preservation

Chemical fixation for EM has been extensively studied and there are many books, reviews and research papers on the subject. However, there is no

Fig. 1A,B. Frozen hydrated sections of *Staphylococcus aureus* bacteria. In **A** the bacteria were fixed with 1% OsO4 before freezing. Typical mesosome (*M*) structures are evident in electron-transparent areas devoid of ribosomes. For comparison, a bacteria conventionally fixed and embedded is shown in the *inset*. In **B**, the cells were vitrified without fixation. Note the uniform appearance of ribosomal areas in the cytosol and complete absence of mesosomes. ×47000, *bar* 1 μm. Courtesy of Dr. Alasdair McDowall (Howard Hughes Institute, Dallas, Texas) and Dr. Jaques Dubochet (EM Institute, University of Lausanne, Switzerland) from their paper in J. Bact. 155, 381–390 (1983)

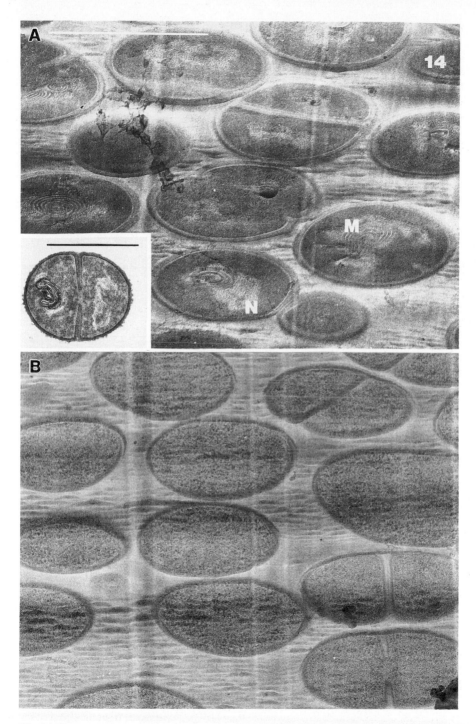

Fig. 1 A, B

definitive or even comprehensive treatise on this subject. Studies in vitro and in vivo have indicated parameters affecting the quality of fixation. The complexity of the interactions between these various parameters, however, makes it impossible with our present level of understanding to predict what the ideal fixation conditions should be for any particular cell or tissue. Most of the literature on fixation has dealt with the role of this process in **fine structure preservation**. For this reason, the greater part of this chapter will deal with this aspect. The problem of fixation for immunocytochemical labelling will be dealt with separately at the end.

Theoretical considerations are vague when one considers fixation in practical terms. It may be easier to say what one does **not** want in fixation, and concentrate on keeping five phenomena to a minimum, namely:

- Physical loss of epitopes such as by extraction of molecules and structures – this category would also cover cleavage and loss of epitopes, such as occurs with O_sO_4.
- Volume and shape changes, which lead to structural distortions and may have a significant effect on any quantitative studies.
- Steric hindrance of epitopes by fixative-induced cross-links.
- Significant chemical alteration of epitopes by direct reaction with fixatives: in this category fixative-induced denaturation effects, especially for proteins, could also be included.

A biological specimen is routinely fixed in its living state. While this is obvious and may appear trivial, the potential to ruin the fixation process **before** tissues come in contact with the fixative cannot be overestimated. It is important that the time between the death of the animal (or removal of a biopsy) and the initial contact of the fixative with the cells of the tissue should be as brief as possible. Even a few minutes can cause irreversible structural and biochemical alterations. Preparation of the tissue is thus a critical part of the fixation process. Optimal methods of initial preparation vary greatly from tissue to tissue.

Before the parameters that affect the quality of fixation are discussed, it is useful to summarize briefly the whole preparation schedule for classical epoxide plastic embedding, which can then serve as a reference for other techniques. The most commonly used primary fixation employs a buffered aldehyde. In this first and most critical step in the whole process an inter- and intra-molecular network of cross-links is formed, mostly between amino groups, that stabilize cells and tissue. Aldehyde fixation is followed by a buffer rinse which may also be an important process, especially with respect to extraction of molecules and to volume changes. In routine electron microscopy this step is followed by a secondary or post-fixation with osmium tetroxide which effectively cross-links unsaturated lipids and nucleophilic ligands. Osmium tetroxide, being a heavy metal, also increases the contrast of structures to which it binds, especially membranes. Whereas with regard to proteins, the effects of aldehyde fixation is relatively mild, osmium tetroxide, a strong oxidant, cleaves proteins adjacent to tryptophan residues. Regions between tryptophan may be extracted when osmium tetroxide is used as a primary fixative resulting in a real loss of epitopes (see p. 54). This step is followed by two additional procedures that may also have rather drastic effects on the chemical composition of the tissue: dehydration and infiltration/embedding. The latter will be discussed in Chapter 5. The next part of this chapter will focus on the **aldehyde fixatives** in order to assess their known effects on fine-structure preservation and discuss, where possible, influences on antigenicity. Other fixatives will be mentioned briefly.

3.1.1 Glutaraldehyde

3.1.1.1 Glutaraldehyde Chemistry

Glutaraldehyde (MW 100) is an aliphatic dialdehyde which forms colourless crystals that are highly soluble in water, ethanol and most organic solvents. Aqueous solutions, which are relatively stable, have a mildly acidic pH, are clear and have a distinct, pungent odour. While not extremely toxic it can lead to skin problems upon contact. Since its introduction by Sabatini et al. (1963), this dialdehyde has become the fixative of choice for routine electron microscopy of almost all cells and tissues. The success of glutaraldehyde as a fixative lies in its ability to cross-link proteins rapidly, effectively and irreversibly.

At room temperature in water, glutaraldehyde is present in four distinguishable states of hydration which are in a temperature-dependent equilibrium with each other (Korn et al. 1972; Whipple and Ruta 1974) (Fig. 2). These are the non-hydrated, the mono-hydrate, the dihydrate and the cyclic forms. The latter represent the predominant species (65–80% below 37°C; Whipple and Ruta 1974). Since all of these forms are uncharged, they should freely pass through biological membranes. A recent reference indicates that, in addition to the monomeric species, small polymers of glutaraldehyde (up to seven units) may also exist in solution (Tashima et al. 1987). The significance of these forms for the cross-linking reactions described below is not clear.

Reactions with Amino Groups. Glutaraldehyde reacts with many nucleophiles in the cell generating numerous cross-linked structures (Bowes and Cater 1968; Habeeb and Hiramoto 1968). Of these nucleophiles, amines are probably present in highest concentration. Amines also react very rapidly with glutaraldehyde to form numerous products (see Johnson 1985 for a list of references). These reactions with primary amines are reviewed here, but the reader should be aware that sulfhydral groups from cysteine and imidazole side chains of histidine are also participants in cross-linking reactions (Habeeb and Hiramoto 1968).

The pKa of the amine-aldehyde adducts are lower than those of the parent amines by 2–4 pH units. While the parent amines are predominantly in the protonated form at pH 7 to 7.5, the amine-aldehyde products are protonated to a much lesser extent. Therefore, according to Johnson (1985), protons are released during the formation of imines (Schiff bases), dihydroxymethylamines

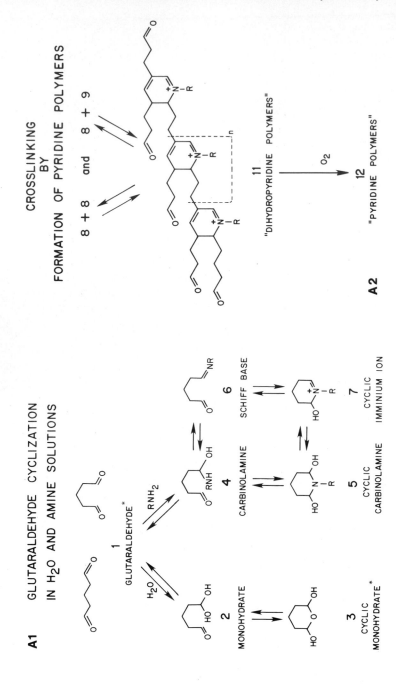

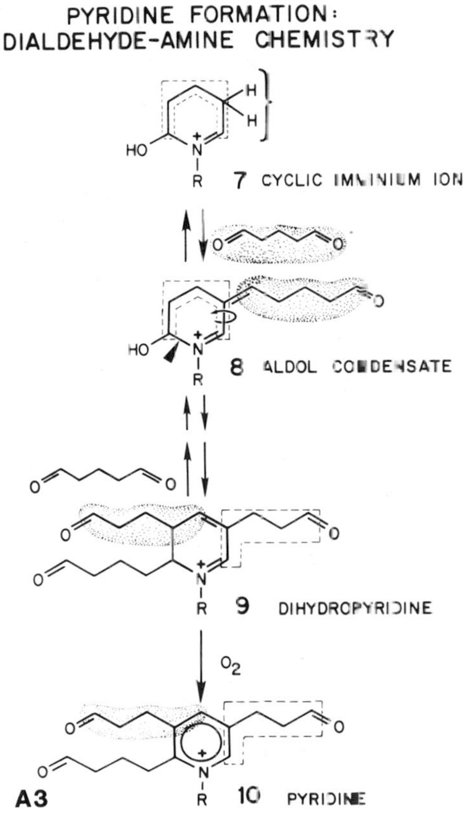

PYRIDINE FORMATION:
DIALDEHYDE-AMINE CHEMISTRY

7 CYCLIC IMINIUM ION

8 ALDOL CONDENSATE

9 DIHYDROPYRIDINE

O_2

10 PYRIDINE

A3

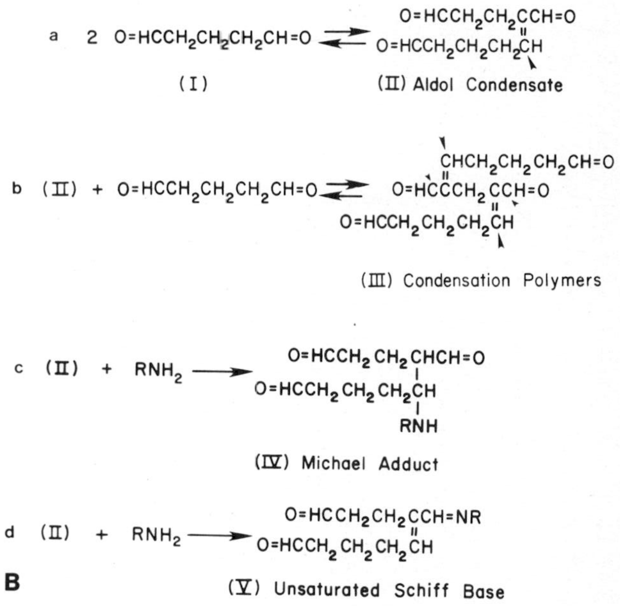

a 2 O=HCCH$_2$CH$_2$CH$_2$CH=O ⇌

O=HCCH$_2$CH$_2$CCH=O
O=HCCH$_2$CH$_2$CH$_2$CH

(I)   (II) Aldol Condensate

b (II) + O=HCCH$_2$CH$_2$CH=O ⇌

CHCH$_2$CH$_2$CH$_2$CH=O
O=HCCH$_2$CH$_2$CCH=O
O=HCCH$_2$CH$_2$CH$_2$CH

(III) Condensation Polymers

c (II) + RNH$_2$ ⟶

O=HCCH$_2$CH$_2$CHCH=O
O=HCCH$_2$CH$_2$CH
RNH

(IV) Michael Adduct

d (II) + RNH$_2$ ⟶

O=HCCH$_2$CH$_2$CCH=NR
O=HCCH$_2$CH$_2$CH

(V) Unsaturated Schiff Base

B

Fig. 2A,B. Cross-linking reactions of glutaraldehyde. (Johnson 1986). A (1–3) shows the fast reactions (which occur in seconds) while B shows the slow reactions (which may take many minutes, if not hours)

and cyclic enamines (Fig. 2A). The consequences of the resulting drop in pH are potentially of great importance in the fixation reaction (Johnson 1985), as will be discussed below.

The reactions then proceed rapidly (Fig. 2A), from the cyclic iminium ion, by consuming extra glutaraldehyde molecules to form first an aldol condensation product (8), and later dihydropyridine (9). The latter incorporates three glutaraldehyde molecules. Up to this point, all reactions are, in principle, **reversible**. In the presence of excess glutaraldehyde, dihydropyridine (9) consumes oxygen and forms a substituted pyridine around the amino group (10), an irreversible step. This reaction is **unproductive** for fixation since the amino group is irreversibly blocked and not cross-linked.

A scheme for **rapid** cross-linking makes use of a "limiting" concentration of glutaraldehyde. Aldol condensates (8) and dihydropyridine (9) condense to form branched dihydropyridine polymers with a wide range of sizes (11). Again, consumption of oxygen makes the reactions irreversible during the formation of "pyridine polymers" (12). This repeating unit of pyridine polymer contains two glutaraldehyde molecules per amine. It is important to note that the effect of oxygen at this final step appears to be that of limiting polymer growth which could otherwise be larger (Fig. 2). Hence, in an anaerobic condition larger polymers should be formed. Theoretically, even cations such as silver (Ag^+), lead (Pb^{2+}) or gold (Au^{3+}) ions could also catalyze reaction 11 to 12 (Johnson 1987).

It is important to recognize that a **slower** cross-linking process occurs simultaneously. That is, aldol condensation (Fig. 2B) between any free aldehyde moities (Rasmussen and Albrechtson 1974). The latter reaction is pH- and temperature-dependent. If the rapid cross-linking cannot occur because the primary amine concentration is low then aldol condensation becomes the default mode of cross-linking. The stoichiometry and kinetics of O_2 uptake in glutaraldehyde-amine reactions suggests that pyridine formation predominates as long as precursor materials are available (Johnson 1987). Whereas the rapid cross-linking reactions occur on the scale of seconds to a few minutes these slow reactions occur over many hours,if not days.

Johnson (1986) has suggested that two different kinds of compounds carrying amino groups should be considered in the reactions with glutaraldehyde, namely, "free" soluble amines such as amino acids and "solid state" amines such as those in proteins that constitute membranes and filaments as well as amino lipids of membranes. In vitro studies on model compounds have indicated that, of the two classes, the soluble amines are far more reactive with glutaraldehyde and that they form the precursors for the pyridine polymer bridges between "solid state" amines.

The reaction of proteins with glutaraldehyde leads to a shift in absorbance in the UV region from a weak absorbance maximum at 280 nm to a strong absorption maximum in the range of 265 nm (Korn et al. 1972).

The rate of glutaraldehyde cross-linking is affected by pH. At higher pHs, the rate of condensation increases significantly (Johnson, personal communication). This is supported by observations showing more significant cross-linking at pH 8 than at lower pHs (Bowes et al. 1965, 1966; Hopwood et al. 1970; Salema and Brandao, 1973; Rasmussen and Albrechtsen, 1974).

The results of Bowes and Cater (1966) indicated that, at pH 8, for every 10000 molecular weight units of collagen, 10 amino groups were cross-linked out of a maximal 15 or 16. Total modification was not obtained even after 18 h at room temperature. Likewise Cheung and Nimni (1982), found that one third of the lysine residues in collagen were refractory to reaction with glutaraldehyde over a range of glutaraldehyde/lysine ratios. Thus it is likely that not all primary amino groups on a protein are accessible for reaction with glutaraldehyde.

The final result of glutaraldehyde cross-linking is that the cytoplasm becomes gel-like and insoluble. This gelation process has been extensively studied in vitro and the importance of many of the fixation parameters was systematically studied by using this process as an assay (e.g. Millonig and Marinozzi 1968; Kellenberger et al. 1982). The gel is exceedingly resistant to protein denaturing agents such as urea, wide extremes of temperature, pH or ionic strength (Richards and Knowles 1968), to freezing and thawing, and to sonication (Pentilla et al. 1974).

Summarizing Glutaraldehyde Chemistry. Many biologists understandably "switch-off" when faced with a series of complex chemical reactions. For fear the message may be lost, I shall summarize here the critical chemical points in the fixation of cells and tissues.

- Glutaraldehyde in solution is uncharged and can thus rapidly cross all biological membranes. It can produce significant intracellular cross-linking in a matter of seconds and forms a large, three-dimensional network of cross-links throughout the cytoplasm in tenths of seconds to minutes. These cross-links are irreversible.
- The ratio of the concentration of glutaraldehyde to the concentration of amines (especially "soluble" amines) in the tissue of interest is critical in determining the kinetics, size and the nature of the products formed. According to in vitro studies (Johnson 1985 and pers. commun.), the highest degree of cross-linking as determined by the size of the cross-linked products occurs with a ratio of glutaraldehyde to free amines of 2:1. Too high a concentration of glutaraldehyde can actually **inhibit** the formation of the rapid cross-links. The cross-linking process can then default to the slower aldol condensation reactions. The concentration of amines clearly varies from one tissue to the next and, ideally, should be determined biochemically (Johnson 1986). It follows that the concentration of glutaraldehyde required for rapid cross-linking of each tissue will, likewise, vary. One should also note that below a critical concentration of proteins, no significant cross-linking can occur (Hopwood 1970, 1972).
- The initial reaction of amines with glutaraldehyde results in a significant release of protons and an ensuing drop in pH. This phenomenon should be countered by adequate buffering. However, the buffers routinely used for

EM are made up of charged molecules that do not cross membranes in significant amounts (see below).
- The fixation process consumes oxygen in the proportion of 0.5 mol oxygen per mol amine (Johnson 1986, 1987). This is a respiration-independent process which may, in fact, compete with respiratory processes. Note that in tissue pieces the peripheral layers may be reacting with the aldehyde while the deeper layers are consuming oxygen by respiration (see below).

3.1.1.2 Effect of Glutaraldehyde on Protein Structure

From a number of studies it can be concluded that glutaraldehyde has a measurable effect on the secondary and tertiary structure of many proteins in solution. However, these effects are generally relatively small (e.g. Lenard and Singer 1968). The influence of glutaraldehyde on the crystallographic structure of a few proteins has also been studied and, again, the effects are only seen at high resolution in most cases (Quiocho and Richards 1966). Some enzymes are biologically active in the crystal even after glutaraldehyde treatment (Richards and Knowles 1968).

During the reaction of glutaraldehyde with proteins, there is a transient loss of charge. In forming the carbinolamine molecule (step 4 in Fig. 2A), the NH_3^+ groups on proteins will lose their positive charge. This is compensated for at step 7 (Fig. 2A) by the positive charge on the nitrogen of the cyclic iminium ion. It is conceivable that this phenomenon could alter protein conformation (Johnson, pers. commun.).

3.1.1.3 Physiology of Glutaraldehyde Fixation

Compared to rapid freezing (which can arrest living processes on the milli-second scale) chemical fixation with aldehyde is at least an order of magnitude slower. Nevertheless, the action of glutaraldehyde is still reasonably rapid once it comes in contact with the cell plasma membrane. It kills cells within seconds. The rapid cross-linking chemistry (polypyridine formation) is largely complet-ed within a few minutes after the fixative contacts the plasma membrane of the cell. In a study of isolated mitochondria, Wakabayshi (1972) even concluded that the structure preservation with glutaraldehyde was "complete" within 1–2 s at 0°C! For a tissue slice, of course, the rate of diffusion through the tissue becomes a limiting factor. The first observable physiological event is a drop in membrane potential as intracellular ions tend to move out of the cell and equilibrate with the extracellular milieu. This has traditionally been interpreted to mean that extracellular and intracellular ions completely equilibrate (Maunsbach 1966; Fozzard and Dominguez 1969; Pentilla et al. 1974). Simultaneously, in Ehrlich tumour cells, there was a simultaneous loss of measurable intracellular magnesium and ATP (Pentilla et al. 1974). According to the latter authors, between 85–90% of intracellular potassium was released within 2 min of fixation with 5% glutaraldehyde. In other systems studied, this equilibration of ions does not occur so rapidly following cell death, which we can assume occurs in seconds: in some systems it may take more than 1 hour for potassium ions to equilibrate with the extracellular milieu (Elbers 1966; Carstensen et al. 1971). The electrophysiological studies of Elbers (1966) indicated that although the resistance of fixed cells was considerably lower than that of unfixed cells, the fixed membrane can still behave as an insulating barrier to the flow of current. Also glutaraldehyde fixation clearly does not fix small ions such as Na^+ or Mg^{2+}.

Pentilla et al. (1974) used the resistance of cultured cells to physical stress as a functional assay for cross-linking. The idea here was to see if even brief exposure to glutaraldehyde would protect cells against high-speed centrifugation (100 000 g), freeze-thawing or sonication. A surprising finding was that when cells were fixed for 10 s in glutaraldehyde, rinsed, then centrifuged (or sonicated), followed by osmication, dehydration and embedding in Epon, the morphology was reasonably well preserved (although not "ideal"). In contrast, unfixed cells subjected to the same treatment were found to be completely destroyed. It seems that few studies in the literature have looked in detail at very brief fixation times (1–2 min, or less).

The above conclusion, that glutaraldehyde passes freely through the plasma membrane, is supported by many observations that show that the concentration and hence osmolarity of the glutaraldehyde per se (as opposed to that of the buffer vehicle) has little, if any, osmotic effect on cell volume (Bone and Denton 1971; Morel et al. 1971; Rasmussen and Albrechtsen 1974; Schultz and Karlsson 1972; Tisdale and Nakajima 1976). On the other hand, in a careful morphometric study on lung tissue, Mathieu et al. (1978) concluded that as the glutaraldehyde concentration increased (while maintaining the same concentration of the buffer vehicle) a significant shrinkage of this tissue was observed. In this tissue, the extracellular matrix may, perhaps, have provided a significant permeability barrier to the free passage of the fixative. Alternatively, one should also consider the balance between the rate of cross-linking versus shrinkage (dehydration) caused by a higher concentration of aldehyde outside versus inside the cells. In vitro studies on tissue homogenates have suggested that, when compared to muscle or liver of the rat, the cross-linking reactions of glutaraldehyde with lung tissue are significantly slower (Johnson, pers. commun.). This would allow more time for osmotic changes to occur before mechanical stability has occured.

The tumour cells studied by Pentilla et al. (1974) became permeable to the vital dyes methylene blue, nigrosin and trypan blue within 2 min of fixation with 3% glutaraldehyde. With longer fixation times, however, the dyes were excluded, presumably because the cross-linking of cytoplasm hinders penetration. Similarly, 1% glutaraldehyde-fixed Chinese Hamster Ovary cells are permeable to sucrose (Griffiths et al. 1984). Other studies show, however, that the plasma membrane maintains its semi-permeable properties after fixation, that is, that cells are still osmotically sensitive (Millonig and Marinozzi 1968; Bone and Denton 1971; Tisdale and Nakajima 1976; Lee et al. 1981). In such studies, the cells, after fixation, will reversibly shrink or swell when placed in hypertonic or hypotonic medium, respectively, (Bone and Ryan 1972; Bone and Denton 1971). Contradicting this are the studies of Pentilla et al. (1974) and Carstensen et al. (1971), which indicate that some cells are osmotically inactive within minutes of fixation. A possible explanation for these differences is that the plasma membrane retains its semipermeable properties in all fixed tissues, but the latter vary in their intracellular concentration of protein, and specifically of available free amine groups. The more cross-linked the cell, the more it would be able to mechanically resist osmotic changes.

There are many studies showing that buffer molecules, unlike fixative molecules, have a significant osmotic effect during and after fixation. Collectively, these argue strongly that charged buffer molecules cannot pass freely through the plasma membrane or that they pass through it relatively slowly, whereas uncharged molecules are known to enter cells freely. Harris and Peters (1953), for example, showed that methylene blue, which has one charged group, cannot pass through the plasma membrane, whereas the uncharged leucobase of the dye passes freely, even in the absence of fixation. Taylor (1987) showed in a plant system that fixed cell membranes are, in fact, impermeable to the routine buffers used in EM.

The idea that charged buffer ions do not freely cross either living or dead cell membranes is in disagreement with the commonly accepted idea that the buffer enters the cell before the fixative (Palade 1952). While originally proposed for OsO4 it seems most unlikely to hold for aldehyde fixatives. The importance of the buffer in conventional electron microscope fixation appears to be to maintain **extracellular** pH and isoosmotic conditions and that the buffering of intracellular pH may simply be a function of the natural buffering capacity of the cytoplasm (which is not very large – see Johnson 1985) and passage of protons out through the fixed membrane. This explanation is hardly new. In John Baker's classic textbook, **Principles of Biological Techniques** (1968), and in his many papers dealing with the mechanism of fixation studied by light microscopy, it was always maintained that a buffer vehicle was unimportant for the fixation of intracellular milieu but recommended that the vehicle have a composition similar to extracellular fluid of tissue (see below).

3.1.2 Formaldehyde

The first use of formaldehyde as a fixative was by Blum (1893), who noticed its hardening properties on the skin of his fingers; (for a summary of the history of the use of formaldehyde as a fixative see Puchtler and Meloan 1985; Fox et al. 1985). Formaldehyde was generally considered to be the poor relative of glutaraldehyde, especially following the introduction of glutaraldehyde in 1963 (Sabatini et al. 1963), which supplanted the use of osmium tetroxide and, to a lesser extent, formaldehyde as the routine primary fixative for electron microscopy. It is paradoxical that, although the basic chemistry of formaldehyde has been known for decades, including its action in cross-linking amino groups in proteins, its role as a fixative is poorly appreciated. According to dogma, formaldehyde either does not cross-link proteins at all or it cross-links them ineffectively compared with glutaraldehyde. The explanation given is that formaldehyde, unlike glutaraldehyde, has only one functional aldehyde group. Yet many examples where the formaldehyde has given better immunocytochemical labelling than glutaraldehyde have, in recent years, led to a rebirth of interest in its use both for light- and electron microscope immunocytochemistry.

3.1.2.1 Formaldehyde Chemistry

At room temperature pure, dry formaldehyde is a colorless gas. It boils at $-19°$ C and freezes at $-118°C$ (Walker 1964). Both liquid and gas polymerize spontaneously at low temperature. They can be kept in the pure monomeric state for a very limited time, and only at temperatures approaching $100°C$ (Walker 1964). As a consequence, formaldehyde is commercially available only as a concentrated aqueous solution or in a polymerized state, as a dehydrated powder.

Liquid Formaldehyde. In aqueous solutions formaldehyde reacts with water to form methylene glycol

$$CH_2O + H_2O \leftrightarrow HO \, CH_2 \, OH$$

Above concentrations of 1–2% by weight, hydrated formaldehyde polymerizes to form polyoxymethylene glycols having the formula $HO(CH_2O)_nH$. The concentration of polymers as well as the average degree of polymerization (n) increases progressively with concentration. Polymerization proceeds according to the following scheme:

$$CH_2O + H_2O \leftrightarrow HO \, CH_2 \, OH$$

$$2 \, HO \, CH_2 \, OH \leftrightarrow HO(CH_2O)_2H + H_2O$$

$$HO(CH_2O)_2 + HO \, CH_2 \, OH \leftrightarrow HO(CH_2O)_3H + H_2O \text{ etc.}$$

The polyoxymethylene glycols decrease in solubility with increasing molecular weight until they start to precipitate out of solution. At room temperature, this point is reached above 30%, w/v. Increase in temperature shifts the equilibrium in favour of smaller polymers as does dilution of a concentrated HCHO solution. If diluted to less than 1–2% it becomes completely monomeric, a process that occurs relatively slowly however; e.g. more than 24 h is required when a 37% w/v solution is diluted to 1%. Similarly freshly diluted solutions of highly concentrated (10–30%) solutions will initially have a higher proportion of the larger oligomers than will the same

Table 1. Average molecular weight of solution. (Data from Walker 1964)

% Formaldehyde	Average MW at 20°C
1	30
5	32
10	35
15	39 (42.5 at 0°C)
20	42

Table 2. Degree of polymerization of formaldehyde at 35°C.
Relative proportion of different polymers as a function of concentration[a]

Concentration	Number of CH_2O units					
	1	2	3	4	5	6
5%	82	14	3	0.5	0.09	0.01
10%	66	21	8	2.7	0.98	0.2
15%	55	24	12	5	2.3	0.9
35%	31	21	17	11	8	5

Data from Walker (1964). For a more complete list see this reference.
[a] Although not stipulated in this reference, it is presumed that these values represent the analysis after the solutions have reached equilibrium. If so, freshly diluted solutions will tend to have relatively higher amounts of the higher molecular weight forms (W. Baschong, pers. commun.).

solution when it reaches equilibrium, taking up to 24 h to reach this equilibrium.

The degree of polymerization at any particular concentration and temperature is constant. This can be estimated by measuring the average molecular weight of the solution (Table 1). The relationship between the size of the polymers and the concentration of formaldehyde is given in Table 2.

Upon storage, formaldehyde forms traces of formic acid, which accounts for the acidity of formaldehyde solutions (pH 2.5). These traces of formic acid are easily neutralized by a weak buffer solution, however.

Formaldehyde is commercially available in three forms, two liquid and one solid.

- 37% formaldehyde solution (wt/wt) plus ≈10% methanol, added as a stabilizer. This solution (termed "formalin") is relatively stable even at temperatures approaching 0°C.
- 35% solution without methanol (or less than 1%). These will tend to form larger polymers that precipitate out of solution. This is especially true when the solutions are kept at 4°C.

● The solid polymer, termed paraformaldehyde. This is a mixture of poly-oxymethylene glycols containing from 8 to 100 formaldehyde units per molecule. The powder dissolves slowly in cold water and more rapidly at elevated temperatures hydrolyzing and depolymerizing as it dissolves. This process is markedly increased by the addition of dilute alkalis (to pH ~10) or acids (Walker 1964). Following the recommendation of Karnovsky (1965), paraformaldehyde is routinely dissolved by adding to water at 60°C, mixing, then adding a few drops of 1 M NaOH until the solution clears.

3.1.2.2 Mechanism of Formaldehyde Fixation

There appear to be two different modes by which formaldehyde cross-links amino groups of proteins, depending on the concentration of fixative.

Low Concentration Cross-Linking of Proteins. At concentrations where the monomer methylene glycol predominates (below 2%) the major type of cross-links are methylene bridges. These are formed by a two-step reaction where formaldehyde reacts first with uncharged but not with charged, protonated NH_3^+ groups (Gustavson 1956; Puchtler and Meloan 1985)

1. $R-NH_2 + CH_2O \rightarrow R-NH-CH_2OH$

These methylol compounds condense with amide or other groups in a second step to yield methylene bridges ($\boxed{}$).

2. $R-NH-CH_2OH + NH_2-CO-R \rightarrow R-NH-\boxed{CH_2}-NH-CO-R' + H_2O$

Another way to visualize these reactions is as follows where an NH_2 group on protein molecules or free soluble amines reacts with methylene glycol and with the release of a water molecule (Johnson pers. commun.):

$$H-\underset{R}{N} \quad + \quad CH_2-OH$$

(with H and OH connected by dotted line above)

Removal of H_2O indicated by dotted line

↓

$$H-\underset{R}{N}-CH_2\!-\!OH + H\!-\!\underset{R}{N}-CH_2-OH$$

Continued condensation leads to *linear* cross links

↓   ↓

$$H-N-\overline{CH_2}\left[N-CH_2\right]OH$$
$$\quad\;|\qquad\;|$$
$$\quad R\qquad \left[R\right]$$

The concentration of amine and formaldehyde must be nearly the same. Note that R can represent a protein molecule or the remainder of a low molecular weight amine. As the formaldehyde concentration increases the size of the amine-methylene cross-link would be expected to decrease such that, at $2:1$ formaldehyde:amine the dominant structure ought to be-
$$HOCH_2-N-CH_2OH.$$
$$\qquad\quad\;|$$
$$\qquad\quad R$$

The repeating unit shown in brackets ($- CH_2 -$) is the methylene cross-bridge. According to French and Edsall (1945), three of these CH_2 NR units can condense to form cyclic compounds:

$$
\begin{array}{c}
CH_2 \\
/ \quad \backslash \\
R\,N \qquad N-R \\
| \qquad\quad | \\
H_2C \qquad CH_2 \\
\backslash \quad / \\
N \\
R
\end{array}
$$

There is evidence that other groups besides the amino group of lysine are cross-linked in proteins, especially amide and sulfhydryl groups and the imidazole nitrogen of histidine (Fraenkel-Conrat et al. 1945; Fraenkel-Conrat and Olcott 1948). There are also indications that peptide bonds may be involved but this is more controversial (French and Edsall 1945; Bowes and Cater 1966; Martin et al. 1975). The corn protein, zein, for example, can be cross-linked by formaldehyde solution despite the fact that it contains no lysine residues (French and Edsall 1945). Nevertheless, in normal proteins, amino groups of lysine residues appear to be one major target of formaldehyde. As for glutaraldehyde, many of these reactions of formaldehyde appear to be pH-dependent with much more reaction in the alkaline range (French and Edsall 1945), a condition in which more amino groups are in the non-protonated state. Bowes and Cater (1966) calculated that there were an estimated six cross-bridges per 10000 Da of collagen at pH 8.5 and at pH 6.5, and only four at pH 5. There is still some disagreement on this point, however, since Gustavson (1956) indicated that maximal tissue cross-linking occurred at pH 4–5.5 and that the increasing amount of formaldehyde bound at higher pH was simply due to blocking of further reactive groups without effective cross-linking (see Puchtler and Meloan 1985). In support of this, Larsson (1988) noted that when gut tissue was fixed with formaldehyde at neutral or basic pH, the antigen of interest, gastrin 17, was extracted, yet at acidic pH the antigen was retained (this does not formally rule out that the former fixation does not block accessibility for labelling but see the author's discussion on p. 47 of this reference).

There appears to be a general agreement that methylene cross-bridges formed by formaldehyde are less stable than the cross-links formed by glutaraldehyde to heat, extremes of pH and chemical denaturation (French and Edsall 1945; Fraenkel Conrat et al. 1945; Bowes et al. 1965; Bowes and Cater 1966). The claim by Barka and Anderson (1963) and Pearse (1968) that the majority of these cross-bridges could be reversed by simple washing is not supported by all the evidence in the literature (see French and Edsall 1945; Puchtler and Meloan 1985 for references). In support of this idea, however, Larsson (1988, p. 45) mentioned that prolonged washing of formalin-fixed, deparaffinized tissues in running tap water overnight does, in some cases, "restore" antigenicity "lost" during the initial fixation (see also Puchtler and Meloan 1985; Gustavson 1956). Nevertheless, significant stability to acid (Fraenkel-Conrat et al. 1947; Fraenkel-Conrat and Meecham 1949), heat and denaturing agents clearly occurs even after treatment with low concentrations of formaldehyde (Bowes and Cater 1966), as well as to milder conditions such as buffer rinses (Flitney 1966). On this point one should note that the general impression that formaldehyde gives inadequate cross-linking for EM is mostly based on studies where relatively low concentrations of formaldehyde were used. In this respect, the classical studies of Baker and McCrae (1966) deserve mention. Using formaldehyde fixation, without osmium post-fixation, they embedded pancreatic tissue in araldite. The quality of their micrographs, while not optimal, was better than one might have expected. In this study a washing step was omitted and the preparations were first rinsed with 50% ethanol. Even 1% formaldehyde for 1 min appeared to give some degree of fixation, whereas 0.25% gave very little. Only subtle differences were noted, in general, when they compared concentrations ranging from 1 to 10%, pHs ranging from 4–9, times from 1 min to 8 days, and temperatures between 1 and 45°C. Similarly, Harris and Farrell (1972) showed that concentrations between 0.1 to 1.4% formaldehyde could render collagen resistant to acid and to the action of collagenase in vitro; in this study extensive washing appeared to have little effect on the cross-linking. It should also be noted that early studies by Stanley (1944) and others (see French and Edsall 1945 for references) showed that concentrations of formaldehyde as low as 0.01–0.1% could inactivate the activity of viruses without affecting antigenicity. These studies, as well as others by Carson et al. (1972), suggest that significant cross-linking by even low concentrations of formaldehyde may occur in a relatively short time.

The bulk of the evidence suggests, however, that methylene bridges form relatively slowly and inefficiently. The study by Hopwood (1970) showed that 4% formaldehyde cross-linked bovine serum albumin slowly and poorly, as determined by an in vitro viscometric assay. Similarly, in their study of collagen, Bowes and Cater (1966) showed significantly less cross-linking (about half) by formaldehyde compared to glutaraldehyde. Habeeb (1969) demonstrated that significant cross-linking of BSA by 3% formaldehyde required several days, while Fox et al. (1985) showed in a quantitative assay that formaldehyde required 24 h for maximal binding to tissue (note here that binding does not necessarily mean cross-linking). Further, Grillo et al. (1971) showed that significantly less insulin was preserved in pancreatic tissue after formaldehyde fixation and routine embedding for EM when compared to glutaraldehyde fixation. Fox et al. (1985) quantitated the binding of 4% ^{14}C-formaldehyde to 16-μm sections of tissue. The results showed clearly that it took 18 h at 37°C and over 24 h at 25°C for the binding to reach equilibrium. In other words, it may take 24 h at room temperature for 4% formaldehyde to fix completely, although it should again be noted that fixing, cross-linking and saturating all sites with formaldehyde **may not** mean the same thing. A study by Edidin et al. (1976) should also be cited in this context. These authors used the photobleaching recovery technique to look at the lateral diffusion of fluoresceinated membrane proteins of cultured cells. A 90-min fixation with 0.5% formaldehyde in phosphate buffer did not affect the lateral mobility of these proteins, whereas when a 5% solution was used for the same period, the diffusion was arrested, presumably due to cross-linking.

It should be noted that complete cross-linking may not be essential for many purposes; while the cross-linking may often be inadequate for conventional plastic or paraffin embedding (and most often formaldehyde fixation is judged only after such protocols) the degree of cross-linking required for cryo-sectioning or freeze substitution may be far lower (see Chap. 4 and 5 and the example in Fig. 5). Similarly, although it is often considered that formaldehyde causes tissue shrinkage, these observations have mostly been made after plastic or paraffin embedding. In one study where careful measurements were made on the fixation process itself, it was concluded that concentrations from 1 to 20% formaldehyde had little effect on tissue volume (Fox et al. 1985). These considerations make it seem plausible that, in contrast to glutaraldehyde, some of the formaldehyde-protein cross-links may be depolymerized by the heating necessary for the processing for plastic or paraffin embedding. It seems just as possible, however, that these phenomena simply reflect the lesser degree of crosslinking with formaldehyde compared to glutaraldehyde.

As with glutaraldehyde the reaction of formaldehyde (even a 1% solution) with either amino acids or tissue homogenates produces a significant drop in pH (Johnson 1985; Kallen and Jencks 1966). Two molecules of methylene glycol can react with primary amines to produce dihydroxymethylamines and hydrogen ions according to the following scheme:

$$RNH_2 + 2CH_2(OH)_2 \rightleftharpoons RN(CH_2OH)_2 + 2H_2O$$

$$RN(CH_2OH)_2 + RNH^+_3 \rightleftharpoons RN(CH_2OH)_2 + H^+ + 2H_2O$$

High Concentration Formaldehyde Cross-Linking of Proteins. Polymers that form at higher concentrations of formaldehyde are seen to be more efficient at cross-linking proteins than methylene glycol (Baschong et al. 1983, 1984, and pers. commun.). The polyoxymethylene cross-bridges, in contrast to methylene bridges, are formed rapidly. Presumably the same target groups are attacked that are cross-linked by methylene groups at low formaldehyde concentrations. Baschong et al. (1983) used the ability of fixatives to inhibit depolymerization of a bacteriophage coat protein as an assay of cross-linking. These authors found that compared to 1% glutaraldehyde (which cross-linked the most), a freshly diluted 5% solution of formaldehyde (from 35% formaldehyde stock solution) was almost as effective after a 5 min period (giving cross-linking values $\approx$ 80% of the glutaraldehyde ones). In a histochemical study of the effect of 10% formaldehyde solution on enzyme activity, Seligman et al. (1950) found, for example, no effect of a 24 h buffer rinsing step on the retention of enzyme activity by tissues. Baschong et al. (1983) showed, however, that these cross-bridges can be reversed by a low pH treatment (below pH 6) or by borohydride treatment.

The rapidity with which high concentrations (>10%) of formaldehyde can cross-link proteins in vitro as well as the stability of these cross-links to washing steps is further demonstrated by the data of Flitney (1966), documented in more detail below (pp. 55–56).

Fixation with Formaldehyde at Different pH. An interesting approach for fixation was introduced by Berod et al. (1981), who argued that formaldehyde should cross-link more effectively at high pH, but under this condition the more effective cross-linking would hinder the penetration of the fixative, especially through membranes. These authors therefore allowed the initial fixation of their tissue slices with 4% formaldehyde in phosphate buffer to occur at pH 6.5 (5 min) followed by a subsequent fixation (15 min) with 1% formaldehyde in sodium borate buffer pH 11. Using immunofluorescence, these authors showed that tyrosine hydroxylase could be much better visualized following the pH shift protocol when compared to routine neutral pH fixation. The latter protocol gave an insignificant immunofluorescence signal which the authors claimed was due to incomplete cross-linking and consequent loss of antigen during rinsing. Although the latter claim was not convincingly supported by the data, they presented the potential of this pH shift protocol to improve cross-linking of formaldehyde has been clearly demonstrated in a later study by Eldred et al. (1983) and in a recent study by Bacallao et al. (1989; and pers. commun.). Using the confocal microscope with the polarized epithelial cell line MDCK, it was shown in the latter reference that the in vivo height of the monolayer was maintained using the fixation protocol of Berod et al. (1981). At neutral pH, on the other hand, formaldehyde fixation led to a significant collapse of the monolayer. Recent data suggests that the pH shift protocol also provides good preservation of MDCK cells at the electron microscope level, using epon-embedding or cryosections (R. Parton, unpublished data).

Aldehyde-Amine mixtures. Luther and Bloch (1989) have recently shown that mixtures of formaldehyde with primary amines improved fixation of cultured myocytes. These authors have described the reaction by the following scheme;

Fig. 3A–H. Use of light microscopy to assess the quality of fixation/use of amine-formaldehyde mixture. Fixation of cultured *Xenopus* muscle cells in cyclohexylamine-formaldehyde (**A-D**) compared with formaldehyde alone (**E–H**). Each set of micrographs depicts the cell before (**A, E**) and after (**B, F**) fixation, and after permeabilization and immunolabelling (**C, G**). Fixation in cyclohexylamine-formaldehyde resulted in alterations in mitochondria (*arrows* in **A** and **B**), but no other structural alterations were obvious. Fixation in formaldehyde alone caused contraction and loss of striations (compare **E** and **F**), and changes in mitochondria (**M**), organelle position (**O**), and cell shape (**S**). Differences in phase contrast are evident after mounting the permeabilized cells in glycerol (**C, G**), but no further structural changes are evident in either cell. **D** and **H** show immunofluorescence labelling for vinculin. Both cells show localized labelling near the substrate (*arrows*) and a high background in the nucleus (**N**). The formaldehyde-fixed cell also has a high cytoplasmic background fluorescence. Labelling of focal contacts (**FC**) in a non-muscle cell is also seen in **H**. The 10 μm *bar* in **H** applies to all figures. (Courtesy of Dr. Paul W. Luther, Department of Physiology, University of Maryland. For more details see Luther and Bloch 1989)

$$\begin{array}{ccc} R & & R \\ | & & | \\ H_2C = O + NH_2 + O = CH_2 & \rightleftharpoons & HOCH_2 \cdot N \cdot CH_2OH. \end{array}$$

These groups can subsequently react with an amino acid side chain, thereby forming a cross-link derived from formaldehyde and the primary amine (Fraenkel-Conrat and Mecham 1949):

$$\begin{array}{ccccc} \text{protein} & R & & \text{protein} & R \\ | & | & & | & | \\ XH + HOCH_2 \cdot N \cdot CH_2OH + HX & \rightarrow & X \cdot CH_2 \cdot N \cdot CH_2 \cdot X + 2H_2O. \\ & | & & | \\ & \text{protein} & & \text{protein} \end{array}$$

The mixture which was recommended for fixation was 0.45% (150 mM) formaldehyde (w/v) and 75 mM cyclohexylamine (again, see p. 39 for discussion of methylene-amine cross-links).[1] Excellent light micrographs were provided in this study to show the improved preservation of organelles when compared to formaldehyde by itself (see Fig. 3). This improvement in fine structure was obtained without any loss of immunocytochemical signal.

Boyles et al. (1985) have used glutaraldehyde-diamine mixtures to preserve actin filaments. The glutaraldehyde-amine mixture was especially useful in this in vitro system to prevent the cleavage of actin by OsO_4.

Mixtures of Glutaraldehyde and Formaldehyde. The EM literature is laden with studies where glutaraldehyde and formaldehyde are mixed together, often in

---

[1] Paul Luther (pers. commun.) has described their recipe in more detail than was present in the original reference. Importantly, the amine should be kept in glass, not plastic, containers. Then put 172 μl cyclohexylamine in 10 ml H_2O. Add 0.8 ml 0.25 M Pipes buffer, pH 7.4. Bring to pH 7.4 with HCl. Adjust the volume to 18 ml. Immediately before use add 2 ml paraformaldehyde solution (4.5 % in H_2O). (It would appear likely that at this pH Hepes would be a better buffer – GG).

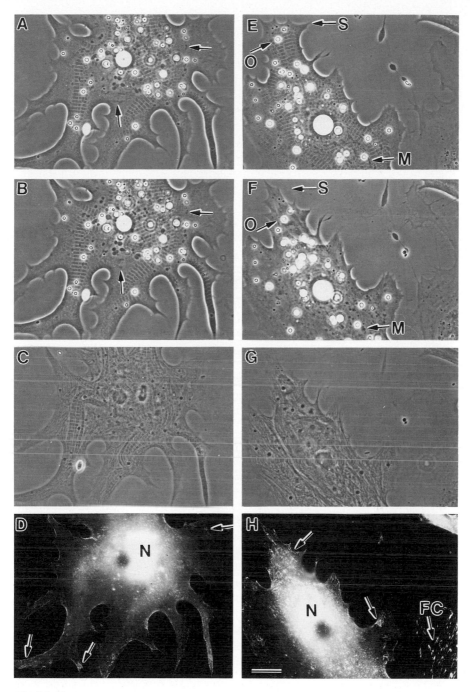

Fig. 3A–H

recipes that almost become laboratory patents. The idea stems from the original paper by Karnovsky (1965), where a mixture of 4% glutaraldehyde and 6% formaldehyde was recommended for fixation. The rationale usually given to justify using such mixtures is the old dogma that formaldehyde penetrates quickly but fixes slowly, whereas glutaraldehyde does the reverse. While this idea may be essentially correct, it does not take into account the possibility that the two fixatives may compete with each other or alternatively interact with each other in the formation of cross-bridges. For the latter, Johnson (1985, 1986 and pers. commun.), has shown that when mixtures of these two fixatives are used, their chemistry does, in fact, appear to interact by mechanisms not yet understood. A number of striking conclusions can already be drawn. First, the rate of consumption of oxygen observed with glutaraldehyde and amines above is rapidly suppressed when formaldehyde is added to the mixture. Second, formaldehyde changes the molecular size range of the products formed in a complex, concentration-dependent manner. Preliminary studies with monoamines (of which most amino acids are examples) indicate that the average size of the glutaraldehyde cross-linked products is shifted to the lower molecular weight range when formaldehyde is also present. In contrast, however, with diamines (lysine is an example) formaldehyde increases the size of the glutaraldehyde cross-linked product (suggesting that more extensive cross-linking occurs). One possibility, suggested by Johnson (1986) to explain many empirical findings of "improved fixation" with mixtures of the two fixatives, is that the formaldehyde reduces the effect of anoxia. That is, it may allow more oxygen to become available for respiration in parts of the tissue that have not yet contacted the fixatives. If this theory is correct, the same effect should be observed when oxygen is bubbled through a fixative solution containing only glutaraldehyde.

Despite these warnings based purely on theoretical considerations, it must be admitted that, in practice, empirically determined mixtures are widely used with success for both structural and immunocytochemical studies.

3.1.3 Effects of Aldehydes on Cell Components Other than Proteins

Although immunocytochemistry deals primarily with localizing protein antigens, in principle, antibodies can be made against any cell component. Therefore a brief discussion of the effects of aldehydes on other cell components is warranted. Again, most of the available data pertains to glutaraldehyde. A summary is given in Table 3 of the qualitative effects of fixatives on different cellular macromolecules (from Johnson, pers. commun.).

3.1.3.1 Lipids

Both glutaraldehyde and formaldehyde can react with, and in principle cross-link, the primary amino groups on amino lipids, such as phosphatidyl

Table 3. Fixation of cellular components. (Compiled by T. Johnson, Sept. 1989)

	Osmium tetroxide	Glutaral-	Formal-	Uranyl salts UO_2^{++}
Proteins	$+^a$	$++$	$+$	
Amino acids	$+$	$+$	$+$	
Other low MW primary amines	$+$	$+$	$+$	
Lipids[b] containing:				
Saturated fatty acids	$-$	$-$	$-$	$-$
Unsaturated fatty acids	$++$	$-$	$-$	$-$
Phospholipids with primary amino head groups	$+$	$+$	$\pm$	$++^c$
Non-primary amino headgroups	$-$	$-$	$-$	$++^c$
Carbohydrates				
Simple (low MW sugars)	$-$	$-$	$-$	$-$
complex (glycogen)	$\pm$	$?^d$	$-$	$-$
Nucleid acids				
RNA	$?^d$	$?^d$	$?^d$	$+$
DNA	$?^d$	$?^d$	$?^d$	$+$
Ions (Na$^+$, K+, Ca^{2+}, etc.)	$-$	$-$	$-$	$-$

[a] O_sO_4 cross links proteins but also cleaves proteins at tryptophan. Protein fragments can be washed away so antibody binding sites can be lost.
[b] The functional group of the lipid responsible for immobilization is indicated. Note that an unsaturated lipid can be cross linked by glutaraldehyde if it has a primary amino group to react with glutaraldehyde.
[c] For UO_2^{++} ions, the target group is the phosphate group ($-$ charge).
[d] Glycogen and nucleic acids are not well cross linked. The association of these molecules with proteins such as histones (DNA) and the synthetic and degenerative enzymes (glycogen), which are easily cross linked, probably traps the molecule in question. These molecules are thus poorly preserved unless significant amounts of protein are complexed with them.

ethanolamine and phosphatidyl serine, to each other or to amino groups in proteins (Gigg and Payne 1969; Wood 1973). Accordingly, other phospholipids should not be cross-linked. There is convincing evidence that many phospholipids are still freely mobile in the plasma of membranes, even though mobility may be reduced after glutaraldehyde fixation even at low temperature (Jost et al. 1973; van Meer and Simons 1986). This is clearly a serious restraint for doing immunocytochemical studies on lipids. A method for the isolation of plasma membrane vesicles actually took advantage of the fact that exposure to aldehydes induces blebbing of this membrane (Scott 1976). The vesicles were enriched in cholesterol and sphingomyelin as well as some glycoproteins. Thus, not only lipids but also some proteins may be lost in this vesiculation process.

A serious problem with respect to cross-linking the amino lipids in cell membranes is first, that the total amount of lipid varies greatly from one membrane to the next (Table 4) and second, the fact that the majority of these lipids are on the inner, cytoplasmic, leaflet of the membrane (Tanford 1980; van

Table 4. A) Analytical protein and lipid content of several membranes. (Tanford 1980)

	Percent of dry weight	
	Protein	Lipid
Myelin	18	79
Human erythrocyte	49	43
Bovine retinal rod	51	49
Mitochondria (outer membrane)	52	48
Acholeplasma laidlawii	58	37
Sarcoplasmic reticulum	67	33
Gram-positive bacteria	75	25
Mitochondria (inner membrane)	76	24

B) Lipid compositions of some biological membranes (expressed as percent by weight of total lipid). (Tanford 1980)

	Human erythrocyte	Human myelin	Beef heart mitochondria	*E. coli*
Phosphatidic acid	1.5	0.5	0	Trace
Phosphatidylethanolamine	18	20	27	65
Phosphatidylglycerol	0	0	0	18
Phosphatidylinositol	1	1	7	0
Phosphatidylserine	8.5	8.5	0.5	Trace
Cardiolipin	0	0	22.5	12
Sphingomyelin	17.5	8.5	0	0
Glycolipids	10	26	0	0
Cholesterol	25	26	3	0

C) Typical lipid composition of some intracellular organelles of rat liver. (Zambrano et al. 1975; see also van Meer 1989)

	Percentage of total phospholipid phosphorus			
	Mitochondrial membrane	ER	Plasma membrane	Lysosomal membrane
Phospholipids:				
SPH	0.5	2.5	16.0	20.3
PC	4.6	10.1	7.7	4.5
PS	0.7	2.9	9.0	1.7
PE	34.6	21.8	23.3	14.1
CL	17.8	1.1	1.0	1.0
LBPA	0.2			7.0
Cholesterol/phospholipid (mol/mol):	0.03	0.08	0.40/0.76	0.49

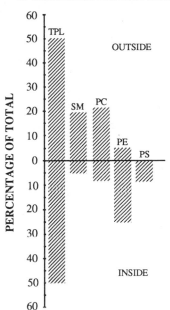

A HUMAN RED BLOOD CELL MEMBRANE

Fig. 4. Majority of fixed lipids are on the inner, cytoplasmic leaflet of plasma membrane. TPL total phospholipids; *SM* sphingomyelin; *PC* phosphatidyl choline; *PE* phosphatidyl ethanolamine; *PS* phosphalidyl serine. PE and PS are the only two primary amino lipids that can be cross-linked by aldehydes

Meer 1989; see Fig. 4) and are only accessible to fixative molecules that can penetrate the bilayer. This makes it difficult to use protocols where mixtures of aldehydes with amino acids are used that are capable of forming large polymers and more extensive cross-linking of amino groups. It is unlikely that such large polymers would be able to pass through the membrane.

There are also other problems. According to Baker (1968), there is a considerable loss of phospholipids when tissues are left in formaldehyde for long periods. He points out, for example, that in the test tube, phosphatidyl serine actually dissolves in a 4% aqueous formaldehyde solution. Also, during or after fixation, phospholipids in aqueous solutions have a strong tendency to form artifactual myelin bodies (Baker 1965), or bubbles (Crawford and Barer 1951) which are familiar to most electron microscopists. Neutralization of the positive charge of primary amines may be a large contributory factor in this phenomenon (Johnson, pers. commun.).

Glutaraldehyde, however, does fix phospholipids to some extent in that it protects them against some effects of dehydration by lipid solvents. Thus, after 1% glutaraldehyde fixation for 24 h, 38% of retinal rod membrane phospholipids were not extractable by lipid solvents (as compared to unfixed membranes) whereas only 7% were resistant to extraction following 4% formaldehyde fixation for the same period (Nir and Hall 1974). Higher concentrations of formaldehyde were not tested in the latter study. Ward and Gloster (1976) determined that only 2–3% loss of labelled lipid occurred after glutaraldehyde fixation of myocardium with a similar amount occurring after the buffer wash and after osmium tetroxide. Similar results were obtained by Roozemond

(1969) for the rat hypothalamus and Wood (1973) for myelin and mitochondrial fractions. According to Wolman and Greco (1952), formaldehyde also reacts with double bonds in unsaturated lipids (see also Jones 1969a).

3.1.3.2 Nucleic Acid

There is no convincing evidence that, under normal fixation conditions, glutaraldehyde reacts appreciably with either DNA or RNA molecules and any stabilization of these components is probably due to the cross-linking of proteins that are firmly bound to them (Hopwood 1970, 1975, 1985; Langenberg 1980). There is more data arguing that formaldehyde reacts with, and presumably may cross-link, amino groups in purines and pyrimidines of nucleic acids (Haselkorn and Doty 1961; von Hippel and Wong 1971; Vologodski and Frank-Kamenetskii 1975; Stevens et al. 1977; Li 1972). The ability of aldehydes to react with and stabilize DNA and RNA, either directly or indirectly, is obviously of great potential importance for in situ hybridization methods.

3.1.3.3 Carbohydrates

In addition to carbohydrate moieties on proteins that will be fixed with aldehydes, there is some evidence that glutaraldehyde can immobilize carbohydrate, such as glycogen (Millonig and Marinozzi 1968). The mechanism of immobilization is far from clear, but it may be that the fixative simply cross-links closely associated proteins thus, indirectly, protecting the glycogen. For thawed cryosections it appears to be especially difficult to maintain glycogen in sections, in most cases glycogen washed out of the sections leaving electron-transparent spaces. This can be especially problematic for tissues high in glycogen such as liver (rats are often fasted overnight to overcome this problem – J. Slot, pers. commun.) or MDCK cells (R. Parton, pers. commun.).

Hydroxyl groups of carbohydrate may be oxidized by periodate to form aldehyde groups. If lysine molecules are present they should be able to cross-link these aldehyde groups. This was the theoretical principle of the method developed by McLean and Nakane (1974) in order to cross-link carbohydrate moieties in tissues. An additional complication in the method, however, was that paraformaldehyde was also added to the periodate lysine mixture (hence referred to as PLP mixture). The complication stems from the fact that formaldehyde and lysine will react with each other in vitro (and the lysine will polymerize). The authors noted a decrease in the pH of the fixative from pH 7.4 to values approaching 6 when the formaldehyde was added. This probably reflects the formation of protons by the formaldehyde amine reaction in a fashion similar to that seen when glutaraldehyde reacts with amino groups (Johnson 1985). As pointed out by Hixson et al. (1981), the mechanism of reaction of the PLP mixture appears to be more complex than that envisaged by McLean and Nakane. Hixson et al. (1981), in fact, suggested that lysine-formaldehyde mixtures alone would be beneficial for immunocytochemistry, in that high molecular weight polymers would be formed. However, lysine and the polymers of formaldehyde-lysine are not likely to penetrate the cell, so cytoplasmic fixation should be accomplished only by whatever concentration of formaldehyde that is able to penetrate the cell. The presence of lysine would also effectively reduce the concentration of free formaldehyde.

In our laboratory, the PLP mixture has not given satisfactory results for cryo-section immunocytochemistry. In contrast, Brown and Farquhar (1984) found it useful for preembedding labelling of a glutaraldehyde-sensitive antigen and excellent fine structure preservation was obtained. It should be noted, however, that after the labelling step the authors post-fixed their tissues in 1% glutaraldehyde before osmication, dehydration and embedding and the initial PLP fixation only played a role of temporary stabilization.

3.1.4 Cross-Linking Agents Other than Glutaraldehyde and Formaldehyde

3.1.4.1 Other Aldehydes

A list of other aldehydes that can, and have been, used to cross-link proteins is presented by Hayat (1981). Of these, acrolein is the most widely used for structural studies (Jones 1969b; Izard and Libermann 1978). It is a very toxic, highly volatile, and reactive aldehyde that has mostly been used in mixtures with glutaraldehyde and/or formaldehyde (see comment above). It cross-links proteins at least as rapidly and irreversibly as glutaraldehyde (Flitney 1966); see Table 5.

3.1.4.2 Non Aldehyde Cross-Linking Reagents for Proteins and Nucleic Acids

In addition to aldehydes, a battery of other bifunctional reagents are known which will cross-link proteins (see Wold 1972 for review). Unlike glutaraldehyde and formaldehyde, these compounds form cross-bridges of fixed lengths. Examples are the imidoesters (Hunter and Ludwig 1972) and peroxy-disulfate (Needles 1967). Many of these reagents have been introduced for "nearest neighbour" protein-protein studies rather than for tissue stabilization. I shall restrict the discussions here to those reagents whose utility for either fine structural studies or for immunocytochemistry has already been shown.

The most promising group of reagents are the bifunctional imidoesters which react with amino group of lysine. The imidoesters do not alter the net charge of proteins (Wold 1972; Hunter and Ludwig 1972; Hassel and Hand 1974). Wolfsy and Singer (1963) showed that extensive amidation of lysine groups in bovine serum albumin and rabbit IgG with ethyl acetimidate resulted in little alteration of their chemical and physical properties. Later, the same workers used dimethyl malonimidate as a cross-linking reagent for the same proteins and showed that 85% of the total free lysines could be imidated without destroying the antigenic determinants. Hartman and Wold (1967) used dimethyl apidimidate to cross-link ribonuclease without losing its enzymatic activity. McLean and Singer (1970) then introduced the use of diethylmalonimidate as a fixative for EM and immunocytochemistry using erythrocytes embedded in cross-linked bovine serum albumin. A more extensive study was subsequently made by Hassel and Hand (1974) and Hand and Hassel (1976). They initially showed that diimidoesters, in particular dimethylsuberimidate (DMS), were comparable to aldehydes in their ability to cross-link proteins in vitro. In the latter publication they showed that DMS gave acceptable fine structure preservation for plastic embedding and that biochemical activity and cytochemical localization of a number of enzymes as well as glycogen were improved over glutaraldehyde fixation.

The diimidoesters, such as DMS, cross-link optimally at high pH (e.g. 9.5 in the study by Hassel and Hand 1974). Further, these compounds are usually only slightly soluble in water and are dissolved in DMSO. In order to be able to fix under more physiological conditions, Yamamoto and Yasuda (1977) introduced ethyl-3 ('3 dimethyl-aminopropyl)-carbodiimide, a water soluble carbodiimide (WSC) as a fixative. A concentration of 4% in phosphate or cacodylate buffers gave in vitro cross-linking equivalent to 2% glutaraldehyde. Fine structural preservation was clearly acceptable with this fixative and, again, the biochemical activities of a number of enzymes were significantly higher than when glutaraldehyde or formaldehyde were used. Two of the enzymes tested, alcohol dehydrogenase and glutamate dehydrogenase, were especially interesting. The activity of the former was totally lost after exposure to glutaraldehyde, formaldehyde and DMS, but 15% of its activity remained when using WSC. Conversely, 55% of glutamate dehydrogenase activity was preserved with DMS, whereas the activity of this enzyme was abolished by all of the other three fixatives. This fixative is clearly a good candidate for further evaluation and should be carefully tested for immunocytochemical studies.

In addition to protein cross-linking, reagents are available that cross-link nucleic acids. Noteworthy are the flurocoumarin derivatives of the psoralens. These will penetrate cells rapidly and intercalate into double-standard DNA. In the presence of ultraviolet light they will cross-link pyrimidine residues (Pathak and Kramer 1969; Cole 1971; Wiesenhahn et al. 1977; Wiesenhahn and Hearst 1978; Hanson et al. 1976). Psoralen derivatives are also available that cross-link RNA (Wollenzien et al. 1978). With the growing interest in the in situ hybridization methods, it seems likely that nucleic acid cross-linkers may have to be considered in future labelling studies.

Finally, within this discussion of non-aldehyde cross-linking mechanisms it is worth citing McBeath and Fujiwara (1984), who also used a photosensitive cross-linking agent 1,3,5,triazido-2,4,6 trinitrobenzene (TTB) as a fixative. Although this study was exclusively at the light microscopy level, the authors showed impressive immunofluorescence labelling in tissue culture cells for a range of cytoskeletal proteins.

Osmium Tetroxide. The characteristic appearance of many structures, in particular membranes, in conventional plastic sections depends to a large extent on the effect of osmium tetroxide (OsO_4), used as a secondary fixative after glutaraldehyde. Osmium tetroxide reacts most rapidly with unsaturated acyl chains of membrane lipids with one mole osmium being bound per mole double-bond (see Behrman 1983, for a review). Osmium tetroxide is also highly reactive to nucleophiles such as amino and sulphhydryl groups.

In addition to its role in contrasting structures, especially membranes, it has been well documented that this fixative can facilitate the retention of tissue lipids (see Riemersma 1963; Stoeckenius and Mahr 1965; Korn 1966; Pilfors and Weibull 1985). Of those studies that have looked at the fixation of lipids by fixatives, that by Cope and Williams (1969), is especially worthy of mention. These workers showed that whereas glutaraldehyde alone led to considerable losses of phosphatidyl choline and phosphatidyl ethanolamine, glutaraldehyde fixation followed by O_sO_4 gave a significant retention of these phospholipids. Similar protection of phospholipids by osmium was also reported by Silva et al. (1968) and Weibull et al. (1983). In the paper by Silva (1968), evidence was also presented for a role of uranyl acetate in protecting lipids against solvent extraction. It is possible that future immunocytochemical studies on lipids may need to consider the use of osmium more seriously.

Although it has traditionally been thought that osmium tetroxide is a lipid fixative, more recent data show clearly that it can also cross-link proteins in vitro (Nielson and Griffiths 1979). Clearly, however, while the chemistry is complex and poorly understood, its effects are often destructive. It was shown by Parsons (1972), for example, that the low angle X-ray diffraction patterns of membranes, slightly affected by glutaraldehyde, were drastically altered by OsO_4, while Lenard and Singer (1968), showed that this fixative drastically affected the conformation of a number of proteins. It follows from this that the railroad track appearance of "unit" membranes in plastic sections, though a useful guide to their presence, has little bearing on the true molecular structure of those membranes. Note, for example, that completely delipidated membranes will still show the "unit" membrane appearance (Korn 1966, 1968; Morowitz and Terry 1969), so that the two electron dense lines cannot generally represent the two lipid leaflets.

Maupin-Szaimer and Pollard (1978) showed using in vitro studies that OsO_4 treatment completely destroys actin filaments within minutes. Further, OsO_4, in contrast to glutaraldehyde, caused rapid loss of secretory proteins from zymogen granules in the pancreas (Amsterdam and Schramm 1966). The most likely explanation for these phenomena comes from studies by Behrman's group (Deetz and Behrman 1981; Emerman and Behrman 1982), who have shown that OsO_4 cleaves some peptide bonds in proteins and that this cleavage occurs predominantly at tryptophan residues. Significantly, these authors have found that the cleavage can be blocked in vitro by tertiary amines which suggests a possible strategy for preventing some of the degradative effects of the fixative in vivo. While a similar protection by tertiary amines against the effects of osmium was also observed by Baschong et al. (1984), these authors additionally showed that these amines by themselves can lead to structural distortions in a model protein. In spite of the potentially harmful effects of osmium on antigens, especially proteins, it is remarkable, and fortunate for the electron microscopist, that the appearance of osmicated tissues in micrographs is, in fact, similar to its appearance in unfixed hydrated cryosections (Chang et al. 1981; see Chap. 1).

3.2 Factors Affecting the Quality of Fixation for Fine Structure Preservation

I have already described many of the important factors that influence the quality of fixation with respect to fine structure preservation, but it is the interactions between these various factors that makes the overall fixation process an extremely complex one. Some of the contradictions that exist in the literature have already been pointed out. Here I shall list these factors, along with some of the inconsistencies. My aim is to make clear that, for fine structural preservation, and even more for immunocytochemistry, there is tremendous potential for working out new and improved fixation procedures for different tissues and different antigens. Even though finding an acceptable fixation has always been an empirical procedure, there are certain rules or at least relevant observations that can guide us.

Again, it is necessary to repeat the statement that with respect to fine structural preservation, the practical aim should be to reduce **extraction**, **denaturation, steric hindrance, chemical alteration of epitopes** and **changes in volume and shape**. In this section I have tried to select key references and important concepts rather than providing a complete literature study (which can be found until 1981 in Hayat 1981). It is also necessary to state that most of the information on aldehyde fixation pertains to glutaraldehyde, because it has been most extensively studied. Some of this information will also be directly applicable to formaldehyde, but some will not. It is indeed ironic that so much less is understood about the process of formaldehyde fixation than that of glutaraldehyde, even though formaldehyde has been used for many more years than glutaraldehyde. Finally, in the list given below, it should be noted that no attempt has been made to arrange the various factors in order of their importance.

3.2.1 Concentration and Length of Fixation

For fine structural studies, one could summarize by saying that as long as a threshold concentration of glutaraldehyde and a certain minimal time of fixation are used, these two parameters have surprisingly little influence on preservation of ultrastructure in general. A chemical reason for not using too high a concentration of glutaraldehyde has already been given (see p. 33). As far back as 1966, Maunsbach showed, for example, that 0.25% glutaraldehyde gave excellent preservation of the proximal tubules of the kidney. Even with formaldehyde, Baker and McCrae (1966) obtained a surprising degree of cross-linking with a 1% concentration applied for only a few minutes, even though ultrastructural preservation could by no means be described as ideal. Whereas the time required to fix cultured cells with glutaraldehyde is relatively brief, a limiting factor for tissues is, of course, the time required for the fixative to diffuse throughout the block (see Sect. 3.2.7).

Assuming complete accessibility of the fixative to all parts of tissues, the elegant study by Flitney (1966) helps to give us an appreciation of the effect of time, concentration, and temperature of aldehyde fixatives on their cross-linking ability. Flitney used thin (20 μm) gels of a mixture of gelatin and albumin in which a fluorescein label had been attached to the albumin. In the absence of aldehyde cross-linking, label was completely washed away very rapidly by buffer solutions. This could be quantitated fluorimetrically. Using this model system, various aldehydes were examined as a function of times and temperatures for their cross-linking effects. Some of Flitney's data (his Table 1) have been reproduced in Table 5. Calculations in this system (which assumed that the fixatives diffused at a rate comparable to glycerol) suggested that the aldehydes would completely diffuse through the albumin-gelatin gel in a matter of seconds. Note also that, with respect to the concept of Johnson (1986), in this system only "solid-state" amines (precursors for inter-protein cross-links – see p. 32) are available to the fixative. There are no 'free' soluble amines present, and the degree of cross-linking in Flitney's study would therefore be much less than that expected in a tissue. On the other hand, it may be representative of tissues which have low concentrations of soluble primary amines.

The first striking conclusion drawn from these results is the rapidity with which the three aldehydes (as well as others tested) gave sufficient cross-linking to retain the albumin in the gel. A note of caution should be added that phosphate buffer was used here (its effect on extraction of proteins will be described later in this chapter) and is one possible reason why the label appeared to be retained less with increasing time. There is no difference between 0.6 and 6% glutaraldehyde, nor is there any difference between 0 and 20°C, except for formaldehyde, where less retention was observed at the higher temperature. In all the experiments, the highest retention of albumin was observed after acrolein, although it should be pointed out that the concentration used (10%) was extremely high. Note that the formaldehyde concentration used (4%) gave a high degree of retention of albumin after 1 min but surprisingly, with longer periods, significantly more albumin leached out of the gel.

In a second series of experiments, Flitney studied the effect of rinsing the fixed gels in phosphate buffer. The relevant data for formaldehyde and glutaraldehyde are reproduced here in Table 6.

Accepting the assumption that more extraction might occur with increasing times in phosphate (as opposed to, say, PIPES buffer, see p. 60), it is clear that 0.6% glutaraldehyde gave effective cross-linking in 1 min at 20°C (but not at

Table 5. Percent retention of albumin in 20 μm sections of albumin gels following fixation. (Data of Flitney 1966). For more detail see p. 55

	Fixation time and temperatures							
	1 min		30 min		60 min		120 min	
	0°C	20°C	0°C	20°C	0°C	20°C	0°C	20°C
Acrolein (10%=	98	97	98	96	97	97	96	97
Glutaraldehyde (6%)	98	92	83	90	73	82	75	89
Glutaraldehyde (0.6%)	96	90	83	83	81	82	82	83
Formaldehyde (4%)	93	88	–	62	55	46	37	24

Table 6. Elution of the conjugates from sections following exposure to fixative. (Data of Flitney 1966)

Values represent percentage of albumin eluted from model gels

	Fixation time	Fixation temp.	Time exposed to buffer solution after fixation (h)				
			0.25	1	2	4	24
Glutaraldehyde	1	0	16	24	28	35	52
0.6%	1	20	5	8	11	11	26
	30	0	3	3	6	9	12
	30	20	2	3	5	6	17
Formaldehyde	1	0	42	50	55	58	70
4%	1	20	40	54	62	64	78
	30	0	31	44	54	60	79
	30	20	6	7	9	10	20
	60	0	18	35	38	41	55
	60	20	6	7	8	8	18

0°C) and (by the criterion assayed here) was almost complete after 30 min (at either temperatures). Neither 10% acrolein nor 6% glutaraldehyde gave any improvement on this condition (results not shown). For 4% formaldehyde, fixation was essentially complete after 30 min at 20°C. For this fixative there was a significant effect of temperature on the degree of cross-linking: even at 2 h (the longest time tested; not shown here), cross-linking was only about 70–80% completed at 0°C.

Even though most EM fixations are probably carried out with excessive concentrations and incubation times, it is fair to point out that the minimal concentration needed to give adequate cross-linking is not infinitely low. This is illustrated by the work of Heller et al. (1971), who studied the effect of glutaraldehyde on mitochondria in vitro. These workers estimated that the minimal concentration of glutaraldehyde required to "fix" mitochondria, as measured by its ability to prevent volume changes induced by the detergent Triton X-100, was 0.2%. In other words, 2 mg/ml glutaraldehyde was required to stabilize an estimated protein concentration of 2 mg/ml. At very low concentrations, glutaraldehyde may even be metabolized (Heller et al. 1971).

3.2.2 Temperature of Fixation

Few studies that are easily interpretable have focussed directly on the effects of temperature on the fixation process. Theoretical arguments for low versus physiological temperatures have been given by Hayat (1970,1981). Hence, one can argue that a low temperature is preferable because autolysis and other unwanted physiological processes are less active at lower temperatures. On the other hand, at more physiological temperatures (20–37°C) the rate of diffusion (penetration) as well as the rate of fixative reaction is significantly increased (Flitney 1966; Hopwood 1970; see Table 3). Biological considerations may be important in considering which temperature to use. For example, if microtubules are of interest, then lower temperatures should be avoided since in vitro studies show that microtubules tend to depolymerize under this condition (Olmstead and Borisy 1975). Conversely, in cell culture work, low temperature is often preferred to ensure that many physiological processes, such as endocytosis, are blocked prior to the onset of fixation.

3.2.3 pH and the Buffer Vehicle

Over the years it has been customary in EM studies to use a buffered fixative at a pH in the range of 6.8–7.4. For most cells the intracellular pH is about 7.0–7.4 (Roos and Boron 1983), so at first glance, it would appear logical to buffer the fixative to a similar pH. The situation is far more complex than it seems, however. In his classic 1952 paper, Palade showed that a wave of acidification, visualized by using phenol red as a pH indicator, occurred in tissues preceding their blackening by osmium tetroxide. He further showed empirically that when the fixative was buffered at a pH around neutrality, a significant improvement in fine structure was seen. Veronal acetate (Michaelis) buffer at pH 7.4 gave the most acceptable results. The veronal acetate mixture became the standard buffer in primary fixative for over a decade and is still extensively used as a buffer for secondary fixative, after glutaraldehyde. However, Baker (1965) pointed out that veronal acetate buffer at pH 7.4 has little or no buffering capacity! The reason for this is that under the conditions in which this buffer is usually used the buffering species veronal (pKa = 7.43) is present only at 0.028 M. At pH >7 acetate is so far away from its pKa that it is little more effective than NaCl as a buffer. Furthermore, Claude (1962) had previously claimed that equivalent primary fixation with osmium tetroxide was attainable whether a buffer or just distilled water was used (although the latter is surely not to be recommended for aldehydes because of osmotic effects; see below). Similarly, there are many published reports for aldehydes used with different tissues that pHs ranging from 4 to 9 are required for optimal fixation (see Hayat 1981). Hyde and Peters (1970) even claimed that optimal fixation of fowlpox virus capsid was best with unbuffered glutaraldehyde at pH 3!

As already stated, most of the available data indicate that at least in the first seconds and minutes of fixation the aldehyde fixative is free to enter cells, whereas most charged buffer ions are not. Two consequences of this statement are the following:

- The protons released during fixation in the cytoplasm will have to be neutralized by the inherent buffering capacity of cytoplasmic components. According to Johnson (1985 and pers. commun.) this capacity is inherently low and will allow the intracellular pH to drop to about 6.
- Buffer ions, and not fixative molecules, will be primarily responsible for the effective osmotic pressure of the buffered fixative solutions.

Accordingly, the effects of the charged buffer ions in current use must be primarily extracellular. Nevertheless, a large number of studies suggest that the appearance of **intracellular** fine structure is improved when the fixative is buffered in the range of pH 6–8 (see Hayat 1981). Clearly, more studies are needed to gain a better understanding of these complex processes.

Many traditional buffers are composed of a mixture of charged species and a non-charged one. It appears likely that only the uncharged species may penetrate biological membranes in significant amounts. Consider some examples (compiled by T. Johnson, pers. commun.). In all cases the acidic species is listed first.

- Acetic acid (HAc)-acetate (Ac⁻) buffer
  HAc – uncharged; Ac⁻ – charged.
- Tris-Tris HCl buffer
  Tris H⁺ – charged
  Tris – uncharged
- Imidazole-amidazole HCl buffer
  Imidazole H⁺ – charged; amidazole, base – uncharged.
- Cacodylic acid-cacodylate buffer
  Cacodylic acid – uncharged; cacodylate – charged
- Carbonic acid-bicarbonate buffer
  Carbonic acid – uncharged; bicarbonate – charged.

In theory, uncharged, basic species of buffers could minimize the wave of acidification in the cytoplasm induced by aldehydes reacting with amines.

Some buffers have both forms charged, e.g. phosphate buffer $H_2PO_4^-$, HPO_4^{-2} as well as most of the buffers introduced by Good et al. (1966) (e.g. PIPES, HEPES). These aliphatic amines are zwitterionic like the amino acids (which can also act as buffers) in that they contain both positive and negative charges in the same molecule.

3.2.3.1 Characteristics of an Ideal Buffer for Fixation

- The pKa of an ideal buffer should be 0.2–0.3 pH units less than the required pH in order to maximize buffering capacity (Johnson 1985).
- The buffer should be as soluble in water as possible and be chemically stable. To minimize the pH decrease at equilibrium, a buffer concentration of at least 0.1 M and preferably 0.2 M be used (Johnson 1985).
- The buffer should not react with the fixative.
- The buffer should not form complexes to any significant degree with metal ions that are required to stabilize many intracellular structures.
- In order to neutralize the protons released during fixation, it may be beneficial to use a buffer whose basic form is unchanged and passes easily through membranes while the acidic form is charged. This possibility is at present only a theoretical one (Johnson, pers. commun.).

With respect to this list, it is interesting to look at the characteristics of the buffers commonly used in EM. Both cacodylate (pKa 6.2) and carbonate (6.4) have pKas which are too low for most fixation purposes (Johnson 1985). Phosphate buffers, while having the necessary pKa, do not have the required solubility properties in the presence of divalent cations and conversely Mg^{2+} and Ca^{2+} have very low solubilities in the presence of phosphate. $CaHPO_4$ has a solubility limit of only 2–3 mM and $MgHPO_4$ only 12 mM (Weast and Astle 1979). Further, as pointed out by Good et al. (1966), phosphate buffer is not an ideal buffer for biological purposes in general since it tends to precipitate most polyvalent cations and often acts as a metabolite or an inhibitor in many biochemical processes.

With respect to the balance of charged to uncharged species in the buffer, cacodylate is clearly undesirable. Aside from its low pKa, the problem is that the acidic form is uncharged while the basic form is charged. This suggests that the fixation-induced pH drop in the cytoplasm may be decreased even further with this fixative. We cannot, however, rule out that cacodylate being an arsenate-based compound, may have other beneficial effects, such as rapidly shutting off respiration (enabling more oxygen to become available for cross-linking, a suggestion by T. Johnson, pers. commun.). The latter is at present investigating the use of buffers consisting of imidazole derivatives that do not react with glutaraldehyde, in the hope of finding better buffers that have an unchanged, basic species.

The available evidence suggests that the tertiary amine heterocyclic buffers introduced by Good et al.(1966) are better candidates for use in fixation (Salema and Brandao 1973; Baur and Stacey 1977; Schiff and Gennaro 1979; Johnson 1985). These include PIPES, HEPES and MOPS. For buffering in the pH range 7.2–7.4 MOPS (pK 7.1), HEPES pK 7.3 are theoretically the best candidates (Johnson 1985). For a useful discussion of common pitfalls in the preparation of buffers for electron microscopy, see Kalimo and Pelleniemi (1977). For a general references on recipes for making buffers see Gomori (1957) and Hayat (1986).

3.2.3.2 Extraction of Cellular Components by Buffers

Many relevant EM papers can be cited with respect to the effects of different buffers on the extent of extraction of cellular components both during and after fixation. In general, these tend to support the above theoretical arguments. Hence, Kuran and Olszewska (1974) found that phosphate buffer led to significant extraction of total protein in both fixed and unfixed nuclei. Cacodylate resulted in less extraction. Interestingly, different classes of proteins appeared to be extracted by the different buffers. Salema and Brandao (1973) measured the total extraction of chlorophyll, protein and phospholipid during aldehyde fixation, washing and dehydration of plant cells and compared phosphate, bicarbonate and PIPES buffers. Again, phosphate gave the most extraction of all components under all conditions tested. At pH 6.8, for example using phosphate as the buffer, there was a net loss, after all the preparation steps, of 38% chlorophyll, 43% protein and 58% phospholipids. In contrast, with PIPES buffer at pH 6.8 the equivalent values for the three same parameters were 4.5, 5.2 and 0% respectively. Similarly, Schiff and Gennaro (1979) showed significantly less extraction of phospholipid when using PIPES buffer as compared to cacodylate. In the latter study, PIPES buffered glutaraldehyde, in fact, retained 7% more lipid **after** fixation rinsing and dehydration than cacodylate buffer did **before** dehydration.

In contrast to the above observations, Coetzee and van der Merwe (1984) found that out of ten buffers tested, phosphate gave the **least** extraction of ions, amino acids and protein during fixation of bean leaves. This observation points out the danger of making too many generalizations when considering complex biological systems.

With respect to extraction, one should also consider the role of metal ions in protecting proteins and nucleic acids. If, for example, a protein requires metal ions in order to acquire its stable functional state, any buffer which tends to form complexes with such ions may have a drastic effect on that protein. In such cases it may be important to add the required ions in the presence of a buffer that does not form complexes with, or precipitate, those ions.

If, as we have argued above (see p. 36), the charged buffer molecules do not enter the fixed plasma membrane, it is hard to imagine how different buffers lead to differences in extraction of intracellular components. The only speculation one can make is that small amounts of buffer molecules may cross the membrane where they facilitate the extraction phenomena.

3.2.4 Osmolarity and Ionic Strength of the Buffer Vehicle

Of all the factors that affect the fixation process, the effects of buffer osmolarity and ionic strength are perhaps the most difficult to understand and to summarize in terms of a simple model. The vast literature on this subject is the most confusing of all the literature on fixation. The main reason is that this topic covers a complex interplay between many interrelated processes, namely, the rate of fixative penetration, the rate and extent of cross-linking (stabilization), the rate of solvent movement (leading to shrinkage or swelling) and the rate of equilibration of fixative between the outside and inside of cells (Johnson, pers. commun.).

In order to start as simply as possible, consider the cross-linking of protein in vitro by aldehydes. As pointed out by Schiff and Gennaro (1979), at a defined pH, the solubility, isoelectric point and dispersion of all organic molecules are affected by the ionic strength and composition of the vehicle. Similarly, Korn and Weissman (1966) showed that the retention of phospholipid and neutral lipids depends very much on the ionic composition of the fixative vehicle. When we try to make the conceptual leap from organic molecules in vitro to cells and tissues, immense complications enter into the formula. To begin with, consider the terms **osmolarity** and **tonicity**, which are expressions often used loosely in describing the fixative. Implicit in the use of these terms is the idea that the cells can be considered as simple bags of known osmotic pressure surrounded by a semi-permeable membrane. Osmolarity is defined as a physico-chemical property of a solution that depends in the total number of particles (ions, molecules) in that solution, whereas tonicity refers to the biological effects on cell volume by solutions of known osmotic pressure. An alternative concept is the difference between the total osmotic pressure (as defined above) and the effective osmotic pressure (Hayat 1970) that depends on the concentration of those molecular species which are unable to pass the membrane. In the critical early seconds of the fixation process it appears likely that the molecules of the fixative can themselves contribute to the effective osmotic pressure. If there is a physical barrier for diffusion, such as a dense extracellular matrix, then these effects can presumably be prolonged. Once significant intracellular cross-linking (or equilibration of the fixative on both sides of the membrane) has occurred (which in most cells will happen in seconds) the concentration of the charged buffer ions should determine the effective osmotic pressure and the concentration of the fixative plays a smaller role.

Baker (1968) was firmly convinced that no correlation existed between the osmotic pressure of a fixing solution and its swelling or shrinkage action upon a cell. It therefore made little sense to him to make a fixative isotonic to cytoplasm. He pointed out, for example, that a 5% acetic acid solution of relatively high osmotic pressure induces swelling (rather than shrinkage as expected) of cells and tissues whereas 0.5% chromium trioxide with relatively low osmotic pressure, induced a significant shrinkage. These two compounds were common histological fixatives of the light microscope era. By the same token, 4% formaldehyde, with an osmotic pressure five times that of mammalian blood, often causes cells and tissues to swell rather than shrink (Baker 1968). This author, as already mentioned, suggested the use of fixative solutions with the vehicle (buffer) having the same osmotic pressure as the **extracellular** medium in the hope of reducing volume changes. In the simplest instance, for example, this would mean dissolving the fixative in sea water for marine animals. Many later workers have adopted this principle (e.g. Arborgh et al. 1976). As far back as 1942, however, Cannan et al. (1942) rightly pointed out that a living cell may maintain osmotic equilibrium with an external environment of low or high osmotic pressure by means of energy-dependent processes. Killing the cell is bound to affect this balance. The situation cannot, therefore, be as simple as Baker proposed. Even considering the erythrocyte, which is the simplest cell model for fixation, Ponder (1942) showed that the unfixed cell did not, in fact, behave as a simple osmometer. There are, nevertheless, many observations in the past 20 years that show that cells and tissues, in general, tend to swell when fixed (or rinsed after fixation) in a solution of low osmotic pressure and have a lesser tendency to shrink in a high osmotic pressure medium (see Hayat 1981). As mentioned, the most critical factor to consider here is the concentration of the buffer. One cannot ignore, however, those contrasting studies that indicate a significant osmotic effect due to the fixative itself (e.g. Iqbal and Weakley 1974; Barnard 1976 and Mathieu et al. 1978). Again the simplest explanation for this discrepancy would be that, in the cells and tissues used by the latter authors, the membrane presents a significant permeability barrier to the fixative, perhaps due to the nature of the extracellular matrix. An additional explanation could be that other processes, besides the osmotic changes, are operating at the same time. We have already referred to some of these processes, namely:

- Permeability changes induced by the fixative. This phenomenon is complicated by the fact that post-mortem changes (indirect effects of the fixative) are also occurring. Amongst these effects will be dissipation of intracellular pH gradients (normally maintained by energy-dependent processes) in various organelles (e.g. mitochondria, endosomes and lysosomes). Hence the osmotic pressure of the cytoplasm may be changing during the fixation period.
- Cross-linking of the cytoplasm by the fixative reduces the effect of osmotic changes. This effect would be expected to be more significant for glutaraldehyde than formaldehyde. Also, as intracellular proteins are cross-linked their effective molecular weights will increase, the number of osmotically active particles decreases, while resistance to shrinkage and swelling is increased.
- When positive charges in the cytoplasm are removed by aldehydes this will increase the net negative charge and water molecules will move to maintain the so-called Gibbs-Donnan equilibrium, causing either swelling or shrinking. In this respect, Millonig and Marinozzi (1968) observed that even drops of fixed plasma albumin (i.e. not surrounded by a plasma membrane) will swell or shrink in appropriate sodium chloride solutions.
- The presence of additional semi-permeable membrane compartments in the cytoplasm, such as mitochondria and endoplasmic reticulum, and the possibility that free water and solutes may be heterogeneously distributed are additional complications to be considered (Clegg 1982; Franks 1982).

The overall effect of these phenomena occurring simultaneously makes it clear that the model of a cell behaving as a simple osmometer is very limited.

The best practical advice would be to choose a buffer at a concentration of at least 0.1 M (Johnson 1985) and to adjust fixative concentrations to maximize the rate of cross-linking. Note again, that maximum cross-linking in vitro occurs in the range glutaraldehyde-amine ratio of 2:1 to 4:1). We now routinely use 0.2 M PIPES buffer pH 7.0, or 0.2 M HEPES pH 7.4 for most of our immunocytochemical studies. Alternatively, one should use relatively large volumes of buffers if lower concentrations are used. Buffering equivalents are greatest near the pKa, but increasing the volume (relative to the size of the tissue piece) and/or the perfusion flow rate will also effectively increase the buffering capacity. The above arguments are all aimed at getting maximal cross-linking and minimal extraction. It should again be noted that there may be an advantage in getting significant extraction, such as for the visualization of certain organelles.

3.2.5 Purity of the Aldehyde Fixative; Storage and Disposal

The standard procedure for preparing formaldehyde solutions is to make them from the dry, polymerized powder, paraformaldehyde. The main arguments put forward in the past against using commercial 37% solutions are: (1) the presence of traces of formic acid and (2) the unwanted presence of methanol added as a stabilizer. The formic acid, as already indicated, can be easily buffered. The second problem can be avoided by using methanol-free solutions. The 37% solutions contain quite significant concentrations of the higher molecular weight polymers which most likely are the desired species for effective cross-linking. The stock solution should preferably be diluted just before use in order to have as high a concentration of the oligomers as possible. Conversely, if for any reason the monomer is preferred, the diluted solution should be left to stabilize for at least 24 h before use (Walker 1964). An argument in favour of the use of paraformaldehyde rather than the 37% solutions is that the solutions should be more reproducible from one experiment to the next. Some EM supply companies provide 20% formaldehyde in sealed ampoules.

For glutaraldehydes, purity may be critical. Some commercial solutions of glutaraldehyde contain pure monomer, that has a weak absorbance maximum at 280 nm, and other species, believed to be the α, β unsaturated aldol condensation products which have a much stronger absorbance maximum at 235 nm (Richards and Knowles 1968; Gillett and Gull 1972; Schulz and Karlsson 1972). It became clear from Anderson's (1967) work that these 235 nm products are, in general, more detrimental than the pure monomer as far as enzyme activity is concerned. The 280 nm species is generally recommended for enzyme cytochemical studies and is now commercially available from many sources. It should be noted, however, that a report by Tashima et al. (1987) indicates that, contrary to dogma, oligomers of glutaraldehyde can also show the same 280 nm absorbance maximum.

There are observations in the literature suggesting that the 235 nm species may be preferred for effective cross-linking. Robertson and Schulz (1970) deliberately treated commercial glutaraldehyde for over 2 h at 96°C in order to convert the monomer to the 235 nm species, a process evident by a yellowing of the solution. The fine structure of brain tissue was clearly improved with the 235 nm products compared to the pure 280 nm species. A similar claim was made by Richards and Knowles (1968). A possible explanation of this phenomenon has been put forward by Johnson (pers. commun.), who suggests that the aldol condensates (235 nm), while not themselves likely to pass through biological membranes freely, are in equilibrium with a small amount of monomer which will be freely permeable. It may be that the concentration of monomer originally used in the above studies was too high for effective cross-linking (see p. 33). This theory predicts that lowering the concentration of pure monomer would give the same "improvement" in preservation. Clearly, however, these interesting observations should be studied in more detail with respect to both fine structure and antigen preservation.

Both glutaraldehyde and formaldehyde stock solutions, as well as the diluted solutions are best stored at −20°C (Gillett and Gull 1972; Rasmussen and Albrechtsen 1974; Tokuyasu pers. commun.; our own unpubl. results).

As pointed out by Johnson (pers. commun.) the simplest and cleanest method to inactivate used aldehyde fixatives is to pour them into a container containing a concentrated (1 M) solution of commercial grade glycine.

3.2.6 Tissue Type

Electron microscopists empirically and qualitatively search for the "best" fixation procedures which give images of cells and tissues that they believe most closely represents the "native" condition. Having obtained an acceptable procedure for a particular tissue, they are often forced to start the process again when faced with a new tissue, which presumably reflects the chemical and physical differences between tissues. One important parameter which has been investigated in preliminary studies is the amount of free amines in different tissues that are available to react with glutaraldehyde (Johnson 1985). This parameter, which can be assayed in homogenates of tissues, appears to vary considerably from one tissue to the next (Johnson, pers. commun.). As discussed above, a molar ratio of glutaraldehyde:amine in the range of 2:1 to 4:1 appears to give the most effective cross-linking in vitro.

3.2.7 Methods of Fixation

After an animal is killed, changes occur almost immediately; probably the most significant is anoxia, or lack of oxygen. A significant example in this respect is the work of Landis and Reese (1981), who studied astrocytes, specialized glial cells in central nervous tissue. These workers were interested in specialized assemblies of intramembraneous particles seen in freeze-fracture images of the plasma membranes of these cells. Whereas chemical fixation per se, before freezing, had no effect on the presence of these assemblies, as little as 30 s anoxia led to their progressive disappearance. In the presence of oxygen the fixed preparations could be maintained for as long as 30 min without loss of assemblies.

In studies where a delay is expected before the fixative makes contact with cells of the tissue, measures should be taken to ensure oxygenation. In this respect, it is again relevant to note that, in addition to cellular respiration, oxygen is also required for the cross-linking reactions with glutaraldehyde (Johnson and Rash 1986; Johnson 1986 1987).

Clearly, the "optimal" method of fixation for tissues is one that ensures cell contact with the fixative as rapidly as possible which is probably as important as are the constituents of the fixation mixture. The most commonly used methods include immersion, perfusion and, less frequently, injection of the fixative. The beginner is well advised to consult the literature for appropriate references for each tissue of interest (see below).

For many tissues in experimental animals **perfusion fixation** is often the method of choice. This is a surgical procedure which requires some technical expertise and practice. For a detailed and comprehensive discussion of perfusion methods for the brain, which is probably the most difficult tissue to work with see Friedrich and Mugnaini (1981). These authors emphasize that there are often significant variations in results due to slight differences in procedure. Accordingly, perfusions done under the same conditions should routinely give about the same results. Only when the techniques have been perfected so as to be reproducible, can one confidently judge a new protocol. For examples of technical details for perfusing testis, kidney and liver, see Forssman et al. (1979), Elling et al. (1977) and Wisse et al. (1984), respectively.

Johnson (pers. commun.) has recently suggested three useful tips for perfusion fixations in general. The first is to warm the perfusing solution to about 42°C before starting the perfusion. This can help to remove small bubbles that can block capillaries: by the time the solution has passed through the peristaltic tubing its temperature should be close to 37°C. Second, by Millipore filtering the solution before use, small particulate matter (deriving from all salts used in solution preparation) is removed which can also block passage of fixatives through capillaries. The third point is to bubble oxygen through the fixative before use (for reasons described above).

A recent innovation has been the use of **microwave energy** as a means to speed up the rate at which relatively large tissue blocks can be fixed. First introduced by Mayers (1970) as a physical method for light microscopy avoiding the use of chemical fixatives, recent work for EM has concentrated on its use in combination with aldehyde fixatives (see Merritt and Frazer 1977 and Hopwood 1985 for reviews).

The emerging consensus is that this method combines local heating in the specimen (to about 40–55°C) and lesser understood effects of the microwave energy that enable the aldehyde fixatives to provide adequate cross-linking in a few seconds only (Login et al. 1986). An additional advantage appears to be the ability to adequately fix relatively large pieces of tissues (up to 1 cm3) (Login and Dvorak 1988; Login et al. 1990). Recent papers by Hopwood et al. (1988) and Kok et al. (1987) suggest that the critical feature of good microwave fixation in practice is the accurate control of heat delivery by the microwaves. It should also be noted that recent publications by Hjerpe et al. (1988) and Zondervan et al. (1988) suggest that the rate of reaction of antibodies with tissue sections may also be increased by using microwaves.

A final approach which should be mentioned here is **phase-partition fixation**. This approach, in which an aqueous solution of a chemical fixative is mixed with a water insoluble solvent, was originally used for insect eggs (Zalokar and Erk 1977). The impermeable outer shell of such specimens makes it difficult for normal aqueous fixatives to penetrate. The principle of this method is to shake the mixture of aqueous solution, containing fixative, with a suitable solvent such as heptane. When the two solutions separate, the fixative then partitions between the two phases. When hydrophobic specimens, such as *Drosophila* eggs were suspended in this mixture the fixation was much improved over simple aqueous fixation. This approach has subsequently been used for more "normal" tissues and a study by Hattel et al. (1983) suggests that the approach could be used to reduce tissue shrinkage. Later studies indicated that both protein and, surprisingly, lipid loss could be reduced with this approach, compared to aqueous fixation (Mays et al. 1981; Leist et al. 1986).

3.2.8 Rate of Fixative Penetration

As has already been pointed out, the action of aldehyde fixatives as well as OsO_4, once they come into contact with living cells, is extremely rapid, and extensive cross-linking will occur in a matter of seconds. Whereas for isolated cells in tissue culture this contact is instantaneous, in the case of tissues sufficient time must be allowed for the fixative to diffuse through all parts of the tissue. A number of studies have actually described formulae for estimating the diffusion times in tissues for different fixatives (e.g. Dempster 1960; Johannessen 1978; Hayat 1981). Such predictions are complicated by the fact that as the cross-linking network increases in peripheral zones of the block, the rate of penetration of fixative into deeper parts becomes significantly reduced (see, for example, Cheung et al. 1985).

For most EM studies, however, it appears sufficient to state that the pieces of tissue should be prepared as small as possible, and preferably less than 1 mm in any one dimension. Below this size most fixatives should usually equilibrate freely with all the cells (in most tissues) within 30–60 min at 20–37°C. Note, however, that the time for completion of cross-links can be considerably longer than this, especially for formaldehyde (see above). Some additional points relative to the rate of fixative diffusion will be mentioned below.

When one cannot avoid using pieces of tissues significantly larger than 0.5–1 mm^3 it may be worthwhile to consider the use of microwave energy to speed up the fixation process (see above).

3.3 Fixation Artefacts

Any structural difference between a section of a dead cell and a hypothetical perfect section through a living cell could be defined as an artefact. Although the following comments apply mostly to fixation-induced changes, they also apply to any artefact induced by the preparative procedure.

The overriding practical consideration with respect to artefacts must be "how does it affect the ability of the investigator to answer the biological question of interest"? As already mentioned in Chapter 1, artefacts may be useful as well as harmful. When an artefact is deleterious, it is important that the investigator recognizes that fact. However, this is not always easy because we have no way of observing the ultrastructure of a living cell.

When possible, the simplest and most direct way to recognize an artefact is to study the effect of adding fixative to living cells observed with a light microscope. This was the approach first demonstrated effectively by Strangeways and Canti in 1928 and subsequently championed by Baker (see the 1968 monograph). Many of these observations laid the groundwork for the techniques in use today. In an era where the use of the electron microscope often dominates structural studies, it is easy to forget the power of the light microscope. A few pertinent examples from the more recent literature will be mentioned below in order to demonstrate how useful this approach can be.

Foremost amongst the recent studies were those of Mersey and McCully (1978), who looked at the effects of fixation of an elongated hair cell found on tomato plant stems. These cells are unusually large, 0.17 to 1.8 mm long and 0.05 to 0.15 mm in diameter. The hair cells are covered by a fixative-impermeable layer of cuticle. When the hairs are cut off the plant, the only access for the fixative is through the cut base of the cells. Hence, the pathway of a diffusing fixative could be followed in a living cell under a light microscope. The first noticeable effect of the fixative was a local arrest of cytoplasmic streaming. With experience, the investigators were able to measure the rate of diffusion of the various fixatives through the cytoplasm. Some of their major conclusions are recapitulated below:

- The penetration rate of commonly used fixatives in many different buffers such as phosphate, cacodylate or HEPES was the same, about 140 µm/min at 20°C.
- This rate was halved at 4°C.
- When calcium was added to the fixative it appeared to "arrest cytoplasmic streaming 50 µm ahead of the fixation front".
- Sucrose added to the fixative reduced the rate of penetration to 100 µm/ min.
- The most striking and worrisome result was that a complex system of cytoplasmic membranes which they referred to as the "pleomorphic canalicular system" became vesiculated by all the fixative mixtures tested. This is a striking result at the light microscope level: clearly, images of this structure at the EM level would bear no relationship to the real structure.

A similar observation was made in an analogous study by Buckley (1973a,b) on cultured chicken cells. These cells contain a highly pleomorphic "lysosomal system" which vesiculated in glutaraldehyde fixatives. Only by adding high concentrations of calcium, preferably in association with magnesium, to the fixative could this structure be maintained. The same artefact was shown recently by Swanson et al. (1987) for an extensive tubular lysosomal structure in macrophages. Light microscopical observations were used to monitor the effects of different fixatives on the organization of this structure. In this case, fixation with glutaraldehyde at 37°C prevented the vesiculation.

Other light microscopic studies have shown that many organelles, such as mitochondria (Bereiter-Hahn and Vöth 1979) chromosomes (Bajer and Mole-Bajer 1971; Skaer and Whytock 1976) or cytoskeletal elements (Luther and Bloch 1989) undergo significant changes in shape or position in the early seconds of fixation, before gelation of cytoplasm. It is to be expected that the recent introduction of the confocal microscope will be a powerful tool to study in more detail how fixatives affect living cells (see Bacallao et al. 1989).

It is obvious, however, that the use of a light microscope to study fixation events is limited to special cells and to special problems. A positive effect, such as a distinct visible alteration, is significant, whereas a negative result is meaningless. Baker (1968) appreciated this point when he wrote: "so far as submicroscopic objects are concerned, we have no direct means of knowing whether a fixative is reliable or not" (Baker 1968 p. 28). Similar arguments were put forward by Palade (1954). (Note that these statements were made before the advent of the hydrated cryo-section method (see Fig. 1).) As an example, consider the routine use of Triton X-100 as a permeabilizing agent for immunofluorescence studies with cultured cells. When formaldehyde-fixed cells are treated with 0.1–0.2% (v/v) Triton X-100 the effects of this treatment are hardly noticeable in routine light microscopy studies at the usual magnification used. When these cells are further prepared for electron microscopy, in either plastic or cryo-sections, it becomes obvious, however, that their fine structure has been largely destroyed (Griffiths, unpubl. data).

Baker's (1968) suggestion to control for the ultrastructural effects of fixatives was to compare many different fixation protocols. This is clearly sound advice and indeed modern concepts of cell structure have relied heavily on this comparative approach. This strategy is nevertheless highly equivocal and subjective. Palade (1954) emphasized the power of X-ray crystallography in comparing the fixed and unfixed states. However, this approach is limited to periodic objects.

During the past 20 years the freeze fracture (freeze etch) method has been commonly used to assess the effects of fixatives on cell structures, especially membranes. The literature contains many references, for example, to differences noted in the numbers or distribution of intra-membranous particles (IMP) when fixed and unfixed membranes were compared by this technique. Because the first step in the freeze-fracture method usually involves a sophisticated rapid-freezing step it has often been accepted **at face value** that this method must give the real in vivo image. This is not necessarily true, however, for many reasons, some of which are listed below:

- There is no unequivocal method of assessing the state of solid water in the tissue, and especially, of distinguishing between the vitreous and cubic states of ice (see Chap. 5). Often, the fractured tissue is routinely warmed from liquid nitrogen temperature ($-196°C$) to about -90 to $-110°C$ in order to "etch" or sublime water. According to theory, even if "freezing" were perfect, vitrified tissue must undergo a transition to cubic ice when the tissue is warmed to these temperatures. In other words, devitrification occurs before sublimation.
- In the study by Roof and Heuser (1982), the number of IMPs in the membrane bore no clear quantitative relationship to, and was significantly less than the number of spanning membrane proteins present in that membrane. Their data argued strongly that aggregation of the groups of membrane proteins occured to form a single IMP.
- Interpretation of freeze-fracture images are complicated by the fact that what is observed in the EM is a metal replica of the fracture plane. As shown by Peters (1984), the most commonly used metal, platinum, is prone to "decoration artefacts", a tendency to accumulate more on some structures than on others. Further, the commonly used rotary shadowing methods tend to have rather poor resolution because of the thick layer of metal (usually platinum) that is usually deposited. Such a thick layer of platinum may also be a cause of aggregation of membrane proteins (J. Costello Duke University, pers. commun.).

It is neither possible nor useful here to list all the different kinds of fixation artefacts which have been described. There are, however, three kinds of artefacts that are especially common in electron microscopy which deserve special mention.

Anoxia. When a fixative contacts the peripheral part of a tissue, lack of oxygen in the deeper parts of this tissue is often a major cause of artefacts (Webster and Ames 1965; Landis and Reese 1981). This is also true for post-mortem or biopsy material in medical research. It makes sense, where possible to oxygenate the pre-fixative medium as well as the fixative itself. Alternative fixation protocols are available that might indirectly increase the supply of oxygen; the first goal in these tissues should be to minimize structural changes associated with anoxia which precede crosslinking. These include the addition of hydrogen peroxide which may result in an increase in the level of oxygen (Peracchia and Mittler 1972; see Johnson 1986, 1987) or the addition of respiratory inhibitors such as azide (Minassian and Huang 1979) or cyanide (Johnson 1986, 1987). Note again that oxygen is consumed by the intermediates of glutaraldehyde-amine reaction (Johnson and Rash 1980; Johnson 1986, 1987).

"Physiological" Artefacts. A prior knowledge of the biology and biochemistry of tissues, cells and organelles of interest can often facilitate the design of an

optimal fixative protocol. Again, it is often important to maintain the structure of interest under optimal conditions **before** effective cross-linking by the fixative. Hence, for microtubules, the use of calcium and low temperatures should be avoided (Olmstead and Borisy 1975).

If quantitative studies are to be made of cell or organelle volumes or surface areas, it is obviously imperative that the effective osmotic pressure of the buffer be such that significant volume changes are avoided before, during, and after the fixation process.

Membrane Artefacts. These are probably the most common and insidious artefacts in electron microscopy. A major reason for them is, as mentioned, the inability of aldehyde fixatives to effectively cross-link membrane lipids (see pp. 46–50). According to Baker (1968), lipids have a tendency to "take up" water and increase their surface area during fixation. The result is artefactual myelin bodies (Baker 1965) "blobs" (Bereiter-Hahn and Vöth 1979) or "blisters" (Hasty and Hay 1978; Shelton and Mowczkow 1978) and perhaps also mesosomes (see p. 27). There are also many records of artificial vesiculation occurring as well as fixation-induced fusions of pre-existing vesicles (Doggenweiler and Heuser 1967; Hausmann 1977; Bretscher and Whytcock 1977). Indeed, as mentioned above, vesiculation induced by aldehydes has been used to purify plasma membranes (Scott 1976).

If artefacts of this nature are to be avoided, or at least reduced, it makes sense to carry out the fixation and subsequent steps on ice. At this temperature, the rate of lateral diffusion of membrane lipids will be reduced and more importantly, all vesicular transport, including endocytosis and exocytosis, will be stopped.

3.4 Effect of Fixatives on Enzyme Activity

The effects of aldehydes on the biological activities of enzymes has been extensively documented. It is useful to summarize the main points from these observations since they serve as a sensitive indication of the effects of these fixatives on protein function. As for their effects on antigen-antibody reactions, two different effects can be envisaged when enzyme activity is inhibited by a fixative. The first is that the cross-linking has chemically modified the catalytic (or binding) site for the substrate. The second is that the cross-links sterically hinder free access of the substrate to the enzyme.

3.4.1 Effect of Glutaraldehyde on Enzyme Activity

The early work of Sabatini et al. (1963) showed that the activities of many enzymes were only slightly affected by glutaraldehyde treatment. A scan of the extensive literature on this subject leads one to conclude that, in the range of glutaraldehyde concentrations used in routine fixations (up to 4%) the activity of most enzymes is reduced (up to 80%) but, with only a few exceptions, some activity remains.

The most convincing data showing that enzyme activity may be compatible with glutaraldehyde fixation are two observations on more complex systems in vitro. The first (Utsumi and Packer 1967) showed that glutaraldehyde-fixed mitochondria can still maintain proton gradients, as well as the capacity to oxidize substrates such as pyruvate and succinate. The second comes from the studies of Park et al. (1966), who observed that glutaraldehyde had no noticeable effect on isolated chloroplasts: there was no difference in freeze fracture images of the membranes, in their optical rotary dispersion properties or in the absorption spectra between unfixed and fixed chloroplasts. Significantly, the "Hill reaction", which culminates in the release of oxygen, was also unaffected.

3.4.2 Effect of Formaldehyde on Enzyme Activity

Formaldehyde fixation is, in general, more favourable for maintaining enzyme activity than glutaraldehyde fixation. In a comparison of the two aldehydes, Janigan (1965) found approximately 15% activity remaining for different phosphatases assayed after glutaraldehyde fixation for 6 h, but about 50% with formaldehyde after the same time. As with glutaraldehyde, the formaldehyde inhibition tends to be concentration- and time-dependent (Janigan 1965). It is notable, however, that Seligman et al. (1950) were able to preserve between 26 and 87% of the activity of a wide variety of enzymes after 10% formaldehyde fixation for 24 h. There are, nevertheless, examples of enzymes such as alcohol dehydrogenase and glutamate dehydrogenase whose activities are completely abolished by formaldehyde (see Yamamato and Yasuda 1977). Figure 5 shows that good preservation after cryo-sectioning and labelling is even obtainable after cross-linking with as little as 0.5% formaldehyde for 30 min.

3.5 Fixation for Immunocytochemistry

Most of the foregoing discussion on fixation is relevant for the **preservation** of antigens. When it comes to the question of the **detection** of antigens, however, we must change the "rules". The goal of preserving as many as possible of the molecules, especially proteins, in cells does not necessarily facilitate the conditions required for antibodies to react with antigens. In fact, it is a well known observation that, in general, the higher the cross-binding the lower the tendency for the immunocytochemical reaction to occur. It is thus exceedingly difficult to ascribe any "rules" with respect to fixation and immunocytochemis-

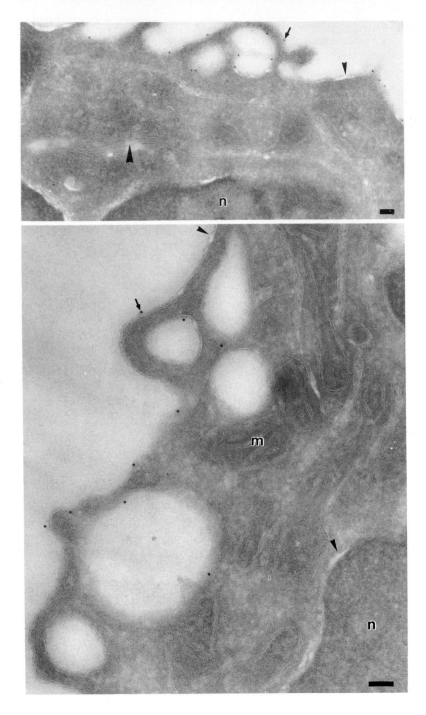

Fig. 5

try: it is necessarily a compromise between preserving the antigen as well as the structure of the organelles of interest on the one hand, and maintaining the ability of the antigen to recognize the antibody, on the other. This compromise obviously depends a great deal on the biological question being addressed.

With respect to loss of immunoreactivity following fixation, two possibilities must be considered.

- The first is that this process chemically alters the structure of epitopes.
- The second is that the extensive cross-linking sterically hinders the access of the antibody to the antigen, even when the latter is in a thin section.

While it is difficult to clearly distinguish between these effects, the fact that the aldehydes have relatively minor effects on the activities of many enzymes and on the three dimensional structure of those proteins that have been studied makes it more likely that the steric hindrance factor is the more serious problem for immunoreactivity. This is also consistent with the fact that immunoreactivity of formaldehyde fixed tissues is generally higher than that of glutaraldehyde fixed tissues. Aside from the fact that formaldehyde cross-links tend to be less stable than those of glutaraldehyde, it also appears likely that the overall pattern of cross-links will be different for the two fixatives. Johnson (pers. commun.) has pointed out that whereas glutaraldehyde forms a dense, three-dimensional network of **branched** cross-links, the available evidence suggests that formaldehyde will mostly give rise to **linear** cross-links. The latter would presumably be less likely to hinder the access of antibodies to antigens. While the steric hindrance effect is potentially more serious for preembedding labelling protocols, it should be noted that the degree of cross-linking can also significantly affect the access of antibodies to antigens that are within a thin section (see Chap. 11).

3.5.1 Effect of Glutaraldehyde on Antigenicity

At present it is difficult to come to a generalization on the effects of glutaraldehyde on antigenicity. There are, however, a small number of studies,

◄ ——

Fig. 5. Thawed cryo-section of R 1.1 thymoma cell line (which express thymus leukemia (TL) antigen at two different magnifications. The cells were fixed for 30 min in 0.5% paraformaldehyde in 0.2 M PIPES pH 7.0 at room temperature. The cells were infused with 0.3% PVP in 2 M sucrose for 45 min. The sections were labelled first with anti TL monoclonal antibody (culture supernatant undiluted), second with a rabbit anti mouse IgG (from Organon Teknika-Cappel) and finally with protein A-gold. There was some dispute in the literature as to the localization of this antigen: earlier reports had claimed it was present over mitochondria. These data show clearly, however, that the antigen is exclusively localized over the plasma membrane (*arrows*). Mitochondria (*m*), nucleus (*n*) and Golgi complex (*large arrowhead*) are well preserved but unlabelled. The small arrowheads indicate areas of membrane which have split during the final drying step in methyl cellulose (0.3% uranyl acetate in 2% methyl cellulose). This is a commonly seen artefact in cryo-sections which does not correlate with any particular fixation conditions. (Courtesy of Drs. Cheryl Hatfield and Alasdair McDowall, Howard Hughes Medical Institute, UT Southwestern Medical Center, Dallas, Texas)

where qualitative EM immunolocalization studies have been complemented with quantitative biochemical characterization of the antigenicity before and after glutaraldehyde fixation. Kraehenbuhl et al. (1972, 1977) measured in vitro the effect of glutaraldehyde on the ability of five pancreatic enzymes to bind their respective antibodies. One of these, a ribonuclease, was quantitatively unaffected by 2% glutaraldehyde, even after 16 h treatment. The four other enzymes behaved roughly the same after treatment with 0.25% glutaraldehyde (for 10 h), with about 75% of antibody-binding activity remaining. When the glutaraldehyde concentration was increased to 2%, the level of activity dropped to around 50%. Considering the relatively long fixation times used, the overall conclusion is that, for these five enzymes, routine glutaraldehyde fixation is compatible with a good qualitative immunocytochemical reaction. Similarly, Kyte (1976) showed that the large subunit of Na^+, K^+ ATPase of the distal tubules of the kidney, when fixed in 0.5% glutaraldehyde for 1 h ($0°C$) bound about 30% less specific antibody in vitro than control unfixed tissue. When the glutaraldehyde concentration was increased to 2%, then 70% of the antigenicity was lost. Using a sensitive radioimmunoassay to measure the effect of different fixation conditions on the antigenicity of the three cytoskeletal proteins vinculin, a-actinin and tropomyosin, Geiger et al. (1981) observed a striking difference between the three antigens in their behaviour towards glutaraldehyde. When a-actinin and tropomyosin were treated for 1 h in vitro with 0.1% glutaraldehyde (in addition to 3% formaldehyde), only 16% of their antigenicity was lost compared with the unfixed antigens. Vinculin, on the other hand, was drastically affected: 0.1% glutaraldehyde (either alone or in combination with 3% formaldehyde) led to a 95% loss of antigenicity after 1 h fixation. The striking observation made by these authors was that a 2–10 min pre-fixation with 20 mM ethylacetimidate plus 3% formaldehyde protected the antigen, both in vitro and in vivo against a subsequent fixation by glutaraldehyde. This procedure resulted in only 20% loss of antigenicity. The authors concluded that the acetimidate, with only one imidate group, bound some key amino groups in the antigen and protected them from cross-linking by glutaraldehyde which, by itself, either chemically altered the antigen or blocked accees to it. This is clearly an important observation and it will be interesting to see whether other antigens that are very sensitive to the effects of glutaraldehyde can also be protected.

Another more recent observation by Dinchuk et al. (1987; and pers. commun.) is worth mentioning here. These authors were interested in the labelling of plasma membrane immunoglobulins of lymphocytes. Normal fixation with 1% glutaraldehyde (< 30 min) essentially abolished immunoreactivity. However, when the 1% glutaraldehyde was quenched after 1–5 min with 1 M glycine (giving a final concentration of 50–100 mM glycine) before the (pre-embedding) labelling reactions a significant cell surface labelling pattern was obtained. For a detailed quantitative comparison of the effects of glutaraldehyde on surface antigens see Leenen et al. (1985).

A very rough generalization would be to say that there are three types of antigens with respect to the effects of glutaraldehyde; first, antigens that are minimally affected by glutaraldehyde (from 0.5–2% concentration), such as amylase or ribonuclease (Kraehenbuhl et al. 1977) or the spike proteins of Semliki Forest virus (Griffiths et al. 1983); second, antigens that are quantitatively affected by glutaraldehyde in a concentration-dependent manner; and third, antigens which do not react with their antibodies even after fixation with very low glutaraldehyde concentrations (less than 0.2%), e.g. vinculin (Geiger et al. 1981).

The effect of glutaraldehyde clearly differs greatly from one antigen to the next. When immunoreactivity is lost after glutaraldehyde fixation, it is commonly assumed that this is due to a direct effect of the cross-linking on the three-dimensional structure of the epitope. While this may be true in some cases, a more likely explanation would be that the dense, branched meshwork of glutaraldehyde cross-links sterically hinder accessibility of antibodies (which are, after all, relatively large molecules) to the antigen. This phenomenon may also apply even when that antigen is close to the surface of an unembedded cryo-section.

3.5.2 Effect of Formaldehyde on Antigenicity

The ability of most antigens to bind to their antibodies is far less affected by formaldehyde when compared to glutaraldehyde. This statement is supported by studies on the few antigens that have been tested in vitro (e.g. Kyte 1976; Geiger et al. 1981; van Ewijk et al. 1984; Leenen et al. 1985) as well as the many hundreds that have been tested in vivo, especially for light microscopical immunocytochemical studies. In fact routine immunofluorescence studies are almost always done using low concentrations (3–4%) of formaldehyde for brief periods (~ 20 min). Hence, the use of high ($\geq 8\%$) concentrations of formaldehyde for EM, which we routinely use for our cryo-section studies, has great potential and will also be discussed in other parts of this book.

With respect to preservation of antigen activity, formaldehyde is usually less "deleterious" than glutaraldehyde. In our experience, many antigens whose immunoreactivity is lost with glutaraldehyde concentration as low as 0.01–0.05% will often withstand at least a 5–30 minute fixation in 8–10% formaldehyde. Consider, for example, the "G" protein, the membrane spanning glycoprotein of vesicular stomatitis virus (VSV). The antigen of the normal wild type virus is insensitive to formaldehyde, and can even be left for weeks in an 8% solution at 4°C without loss of antigenicity. It is, however, extremely sensitive to glutaraldehyde (0.1% for 15 min masks its antigenicity). In one experiment, when a preparation that was left for 4 days in 8% formaldehyde was post-fixed for 15 min with 0.5% glutaraldehyde, the ability of this protein to bind our polyclonal antibody was completely lost. Similar observations were made by Smit et al. (1974) for antigens on the surface of lymphocytes. The temperature sensitive mutant of VSV, 045, has a single amino acid difference from the wild type G protein and its transport is blocked in the endoplasmic reticulum at the restrictive temperature. In contrast to the situation with the wild type virus, the 045-G protein appears to be insensitive to the effect of 2% glutaraldehyde (Bergman and Singer 1983; Griffiths et. al. 1985; and unpubl. observ.). Such a result is not surprising when one considers that in sickle cell anaemia, for example, a single amino acid substitution in haemoglobin leads to major structural and consequently functional differences from normal haemoglobin. Formaldehyde (4–8% for 1–24 h) is now the fixative of choice in our laboratory for routine immunocytochemistry using cryosections.

For an interesting comparison of the effects of formaldehyde and glutaraldehyde on different cell surface antigens using a quantitiative micro ELISA assay see Leenen et al. (1985).

With respect to immunocytochemistry, it is difficult at present to make any clear conclusions about the effects of mixtures of glutaraldehyde and formaldehyde on antigenicity. I am not aware of any convincing in vitro data on this point.

3.5.3 Effect of Acrolein on Antigenicity

This powerful cross-linker has not been extensively used for immunocytochemistry. One reason may be that this highly volatile compound is very unpleasant to work with as well as to store. A report by Boonstra et al. (1985) showed that 1% acrolein, in combination with 4% formaldehyde, preserved more antigenicity of epidermal growth factor than did glutaraldehyde, as assessed by an in vitro assay, and gave satisfactory immunolabelling combined with very good fine structure preservation on thawed cryo-sections. This fixative was also successfully used in immunocytochemical studies of pituitary hormones and neuropeptides by Smith and Keefer (1982) and King et al. (1983) respectively. Similar results have been obtained by Geuze and Slot (pers. commun.) and in our laboratory. In contrast, Farr and Nakane (1981) and van Ewijk et al. (1984) found that even low concentrations ($\approx$0.02%) of this fixative significantly reduced immunoreactivity.

3.5.4 Effect of Osmium Tetroxide on Antigenicity

It is surprising that some antigens will react with their antibodies in plastic sections even after primary fixation in glutaraldehyde, secondary fixation in OsO_4, dehydration and embedding in a hydrophobic resin. Bendayan and Zollinger (1983) showed that sodium metaperiodate treatment could reverse some deleterious effects of OsO_4 and allow many different antibodies to recognize their antigens in osmicated Epon sections. There are other examples in the literature, usually of antigens that are present in high concentrations, such as those in secretory granules. In most of these studies the tissues were fixed in glutaraldehyde initially and one can envisage that this helped to protect the antigen against cleavage by osmium (see, for example, Baschong et al. 1984).

There are examples, however, of antibodies being able to recognize osmium treated proteins in vitro (Spendlove and Singer 1961).

The above examples of spared antigenicity after OsO_4 treatment are the exception rather than the rule, however. This is supported by a quantitative ELISA assay by van Ewiyk et al. (1984), who found that for five cell surface antigens tested, the immunoreactivity of four of them was abolished below 0.1% OsO_4 with one still showing some reactivity at 2% of this fixative. For quantitative EM immunocytochemistry there appears in general to be little justification at present to use OsO_4 as a routine fixative. Other heavy metals can be used to give adequate contrast: in cryo-sections, for example, uranyl acetate can be effectively used to contrast membranes. Admittedly, not everyone will agree with these statements.

3.5.5 Effect of Factors Affecting Cross-Linking on Antigenicity

In contrast to purely structural studies, for immunocytochemistry (as well as enzyme cytochemistry) the concentration and time of fixation may be critical parameters. A wealth of studies, both in vitro and in vivo, have shown that for many antigens the degree of antibody-binding capacity is inversely proportional to the degree of cross-linking. This is especially true for glutaraldehyde which inactivates many antigens at very low concentrations. This point is well recognized. Less appreciated is the problem of antigen loss from sections due to incomplete cross-linking (and possibly from loss of membrane via vesiculation; Scott 1976). This has been most strikingly demonstrated for amylase in the pancreas. Amylase is a very resistant antigen that can withstand long glutaraldehyde treatments in vitro and still bind antibodies to it (Kraehenbuhl et al. 1977). Nevertheless, Posthuma et al. (1984, and pers. commun.) failed to convincingly demonstrate an immunocytochemical reaction for amylase in the endoplasmic reticulum (ER) of pancreatic exocrine cells when brief fixation times (1 h or less) with glutaraldehyde were used. When the tissues were fixed with 2% glutaraldehyde for longer periods (4–24 h), the ER amylase was quantitatively retained (Posthuma et al. 1984, 1985). This phenomenon is likely to be even more important when formaldehyde fixation is used. It must also be seriously considered when dealing with small molecular weight antigens.

I have found no convincing data on the effect of temperature of fixation on antigenicity.

The effects of crude versus purified forms of formaldehyde and glutaraldehyde were studied by Leenen et al. (1985) in their extensive quantitative immunocytochemical analysis of a range of cell surface antigens. Whereas, in general, little quantitative differences between the two fixative forms were seen for most antigens examined in a few cases the crude fixative caused a more drastic loss of immunoreactivity.

Finally, it should be noted again that the amount of free amines available for cross-linking may differ considerably from one tissue to the next. This fact may help to guide the experimenter in choosing the best protocol which is almost always empirically decided. A better understanding of the chemistry of the cross-linking reactions may, in future, help us to better control the cross-linking reactions and thereby design more optimal fixation protocols.

3.5.6 Assessing the Effects of Fixation on Antigenicity

In any immunocytochemical experiment it is likely that a standard fixation protocol will be used first. When one fails to obtain the desired result, it is reasonable to assume that the fixation protocol may have been inadequate. As mentioned in other parts of this chapter, aside from the possibility that the amount of antigen in the section was too small to be detected (see Chap. 11), four different effects of fixation should be considered with respect to loss of antigenicity.

- A direct chemical modification of the antigen.
- Access of the antibody to the epitope is blocked by excessive cross-linking of the cell or tissue.
- Insufficient cross-linking resulting in loss of antigen during exposure to aldehydes (Scott 1976) or in the washing steps.
- Denaturation of antigen.

Clearly, by comparing the effects of different fixation protocols the immunocytochemical study itself can facilitate an understanding of the effects of fixation on the immunocytochemical reaction. The use of light microscopy is highly recommended for this purpose. Nevertheless, there are limits to this approach and for some antigens it may be necessary to use an independent immunochemical approach to assess directly the effects of the fixative. These different methods can be considered for this purpose.

3.5.6.1 Immunoblotting

The antigen can be transferred from an SDS gel onto nitrocellulose. Strips of the latter can then be incubated with the antibody (following a suitable blocking step such as newborn calf serum) and the antibody can be visualized with a suitable secondary reagent such as radioactive protein A, protein A gold or a second antibody coupled to horseradish peroxidase or alkaline phosphatase (which can be visualized directly or after silver enhancement). (For more details, the textbook by Mayer and Walker 1987 is recommended; see also Chap. 6). By treating the nitrocellulose strip with different fixative conditions before the antibody reaction, the effects of the fixative can be qualitatively or quantitatively assessed. A potential problem with this approach is that the SDS treatment may significantly modify the antigen. It should be noted, however, that many proteins will renature on nitrocellulose after the SDS has been removed (Mayer and Walker 1987). For a discussion of the use of this approach in immunocytochemistry see Larsson (1988).

3.5.6.2 Dot Blotting

A recent article by Riederer (1989) has described the use of a simple and elegant dot-blot assay to assess the influence of fixatives on antigens available in pure form. In this assay, which is based on a method first described by Hawkes et al. (1982), small volumes of the diluted protein are dropped onto marked areas on the blotting paper. After drying, the antigen is treated with a blocking step and then with antibody. The latter can be visualized by the same methods used for immunoblotting (above). In the modification by Riederer (1989), 20 µg/ml of the protein of interest is mixed with an equal volume of either PBS (control) or of double-concentrated fixative for 20 min at 4°C. Aliquots of 0.5 µl are then spotted onto nitrocellulose filters: a distinct theoretical advantage of this approach is that the antigen is fixed **before** adsorption on the nitrocellulose, which reduces the possibility that the adsorption or drying steps modify the antigenic determinant. Using the amido black protein stain the author showed that, in the case of their model protein, brain spectrin, the fixative did not appear to have any effect on the binding of the antigen to the filter. After a 30 min blocking step in normal horse serum containing 0.1% Triton X100, the nitrocellulose strip was incubated overnight with antibody, washed and treated with either HRP-conjugated second antibody or with 125I protein A. For spectrin, 4% formaldehyde blocked about half of the activity, whereas 0.5% glutaraldehyde abolished 90%.

3.5.6.3 Cryostat Sections

In addition to the use of dot blots, Riederer (1989) also described the use of 10 µm cryostat sections for quantitative assessment of antibody bound to the antigen in situ following different fixation protocols. The results were in reasonable agreement with the dot-blot data. Another approach used by Riederer (1989) was to use SDS gels in order to assess the degree of cross-linking of brain proteins by different fixation protocols. The idea here was to fix mouse brains and then homogenize the tissue and separate the proteins on SDS gels. The dot blot assay was used in parallel to compare the total immunoreactivity for each condition. Using Coomassie blue staining of the gels, the author could show that, by increasing the degree of cross-linking, the ability of the protein to enter the gel was retarded, as expected. Proteins of higher molecular weight were more readily cross-linked than the smaller proteins. A similar approach was also used by Hopwood (1969), Maupin-Szaimer and Pollard (1978) and Baschong et al. (1983). By introducing the use of several different assays to study the effects of fixatives both in situ and in vitro, the paper by Riederer (1989) can be considered a model study.

3.5.6.4 ELISA Test

A powerful approach to test the use of fixatives on cell surface antigens was introduced by van Ewijk et al. (1984) and Leenen et al. (1985). In this method cells are grown on microtitre wells and a fixative is applied. Subsequently the antibody is applied, and a number of methods can be used to detect the bound antibody, such as immunofluorescence or immunoenzymatic methods (as in screening for monoclonal antibodies). The method chosen by these authors was fluorimetry that was detected using a scanning microfluorimeter. The strength of the approach is that quantitative data are provided and the signal from unfixed cells (this should be done on ice to prevent internalization of antibodies and antigens!) can be compared with that obtained with different fixatives. The authors provide a wide range of data on different antigens.

3.6 A Concluding Remark

There is an enormous literature on fixation per se and a still larger one where fixation methods have been used in the course of routine ultrastructural studies. Anyone starting studies with a new cell or tissue type would be well advised to make a thorough literature search to see what protocols have been found most effective by previous workers in studies on the cells or tissues of interest.

References

Amsterdam S, Schramm M (1966) Rapid release of the zymogen granule protein by osmium tetroxide and its retention during fixation by glutaraldehyde. J Cell Biol 29:199–207

Anderson PJ (1967) Purification and quantitation of glutaraldehyde and its effects on several enzyme activities in skeletal muscle. J Histochem Cytochem 15:652–661

Arborgh B, Bell P, Brunk U, Collins, V P (1976) The osmotic effect of glutaraldehyde during fixation. A transmission electron microscopy, scanning electron microscopy and cytochemical study. J Ultrastruct Res 56:339–350

Bacallao R, Antony C, Dotti C, Karsenti E, Stelzer EHK, Simons K (1989) The subcellular organization of Madin-Darby Canine Kidney cells during the formation of a polarized epithelium. J Cell Biol 109:2817–2832

Bajer A, Mole-Bajer J (1971) Architecture and function of the mitotic spindle. Adv Cell Mol Biol 1:213–265

Baker JR (1965) The fine structure produced in cells by fixatives. J R Microsc Soc 84:115–131

Baker JR (1968) Principles of biological microtechnique: a study of fixation and dyeing. Wiley, New York

Baker JR, McCrae JM (1966) The fine structure resulting from fixation by formaldehyde: the effects of concentration, duration and temperature. J R Microsc Soc 58:391–399

Barka T, Anderson PJ (1963) Histochemistry: theory, practice and bibliography. Hoeber, New York

Barnard T (1976) An empirical relationship for the formulation of glutaraldehyde-based fixatives. J Ultrastruct Res 54:478–786

Baschong W, Baschong-Prescianotto C, Kellenberger E (1983) Reversible fixation for the study of morphology and macromolecular composition of fragile biological structures. Eur J Cell Biol 32:1–6

Baschong W, Baschong-Perscianotto C, Wurtz M, Carlemalm E, Kellenberger C, Kellenberger, E (1984) Preservation of protein structures for electron microscopy by fixation with aldehydes and/or OsO4. Eur J Cell Biol 35:21–26

Baur PS, Stacey TR (1977) The use of PIPES buffer in the fixation of mammalian and marine tissues for electron microscopy. J Miscrosc 109:315–327

Behrman EJ (1983) The chemistry of osmium tetroxide fixation. Sci Biol Spec Prep SEM Inc, AMF O'Hare, pp 1–5

Bendayan M, Zollinger M (1983) Ultrastructural localization of antigenic sites on osmium fixed tissues applying the protein A-gold technique. J Histochem Cytochem 31:101–109

Bereiter-Hahn J, Vöth M (1979) Metabolic state dependent preservation of cells by fixatives for electron microscopy. Microsc Acta 82:239–250

Bergman JE, Singer SJ (1983) Immunoelectron microscopic studies of the intracellular transport of the membrane glycoprotein (G) of Vesicular Stomatitis virus in infected chinese hamster ovary cells. J Cell Biol 97:1727–1735

Berod A, Hartman BK, Pujol JF (1981) Importance of fixation in immunohistochemistry: use of formaldehyde solutions at variable pH for the localization of tyrosine hydroxylase. J Histochem Cytochem 29:844–850

Blum F (1893) Der Formaldehyd als Härtungsmittel. Z Wiss Mikrosk 10:314–315

Bone Q, Denton EJ (1971)The osmotic effects of electron microscope fixatives. J Cell Biol 49:571–581

Bone Q, Ryan KP (1972) Osmolarity of osmium tetroxide and glutaraldehyde fixatives. Histochem J 4:331–347

Boonstra J, Maurik P, Detize LHK, de Laat SW, Leunissen JLM, Verkley A (1985) Visualization of epidermal growth factor receptor in cryosections of cultured A431 cells by immuno-gold labelling. Eur J Cell Biol 36:209–219

Bowes JH, Cater CW (1966) The reaction of glutaraldehyde with proteins and other biological materials. J R Microsc Soc 85:193–200

Bowes JH, Cater CW (1968) The interaction of aldehydes with collagen. Biochim Biophys Acta 168:341–352

Bowes JH, Cater CW, Ellis MJ (1965) Determination of formaldehyde and glutaraldehyde bound to collagen by carbon assay. J Am Leather Chem Assoc 60:275–285

Doyles J, Anderson L, Hutcherson P (1985) A new fixative for the preservation of actin filaments: fixation of pure actin filament pellets. J Histochem Cytochem 33:1116–1128

Bretscher MS, Whytock S (1977) Membrane associated vesicles in fibroblasts. J Ultrastruc Res 61:215–217

Brown WJ, Farquhar MG (1984) The mannose 6-phosphate receptor for lysosomal enzymes is concentrated in cis Golgi cisternae. Cell 36:295–307

Buckley IK (1973a) Studies in fixation for electron microscopy using cultured cells. Lab Invest 29:398–410

Buckley IK (1973b) The lysosomes of cultured chick embryo cells. A correlated light and electron microscopic study. Lab Invest 29:411–421

Cannan RK, Palmer AH, Kibrick AC (1942) The hydrogen ion dissociation curve of beta-lactoglobulin. J Biol Chem 142:803–822

Carson F, Lynn JA, Martin JA (1972) Ultrastructural effects of various buffers, osmolarity and temperature on paraformaldehyde fixation of the formed elements of blood and bone marrow. Tex Rep Biol Med 30:125–142

Carstensen EL, Aldridge WG, Child SZ, Sullivan P, Brown HH (1971) Stability of cells fixed with glutaraldehyde and acrolein. J Cell Biol 50:529–537

Chang JJ, McDowell AW, Lepault J, Freeman R, Walter CA, Dubochet J (1981) Freezing, sectioning and observation artefacts of frozen hydrated sections for electron microscopy. J Microsc 132:109–123

Cheung DT, Nimni ME (1982) Mechanism of crosslinking of proteins by glutaraldehyde II. Reaction with monomeric and polymeric collagen. Connect Tissue Res 10:201–216

Cheung DT, Perelman N, Ko EC, Nimni ME (1985) Mechanism of crosslinking of proteins by glutaraldehyde III. Reaction with collagen in tissues Connect Tissue Res 13:109–115

Claude A (1962) Fixation of nuclear structures by unbuffered solutions of osmium tetroxide in slightly acid distilled water. Proc 5th Int Congr Electron Microsc 2:L-14

Clegg JS (1982) Alternative views on the role of water in cell function. In: Franks F (ed) Biophysics of water. Wiley, New York

Coetzee J, van der Merwe CF (1984) Extraction of substances during glutaraldehyde fixation of plant cells. J Microsc 135:147–158

Cole RS (1971) Psoralen monoadducts and interstrand cross-links in DNA. Biochim Biophys Acta 254:30–39

Cope GH, Williams MA (1969) Quantitative studies on the preservation of choline and ethanolamine phosphatide during tissue preparation for electron microscopy. J Microsc 90:31–46

Crawford GNC, Barer R (1951) The action of formaldehyde on living cells as studied by phase-contrast microscopy. J Microsc Sci 92:403–423

Deetz JS, Behrman EJ (1981) Reaction of osmium reagents with amino acids and proteins. Int J Pept Protein Res 17:495–502

Dempster WT (1960) Rates of penetration of fixing fluids. Am J Anat 107:59–69

Dinchuk JE, Johnson TJA, Rash JE (1987) Postreplication labeling of E-leaflet molecules: membrane immunoglobulins localized in sectioned, labeled replicas examined by TEM and HVEM. J Electron Microsc Tech 7:1–16

Doggenweiler CF, Heuser JE (1967) Ultrastructure of the prawn nerve sheaths. J Cell Biol 34:407–420

Edidin M, Zagyansky Y, Lardner TJ (1976) Measurement of membrane protein lateral diffusion in single cells. Science 191:466–468

Elbers PF (1966) Ion permeability of the egg of *Lymnala stagnalis* L. on fixation for electron microscopy. Biochem Biophys Acta 112:318–329

Eldred WD, Zucker C, Karten HJ, Yazulla S (1983) Comparision of fixation and penetration enhancement techniques for use in ultrastructural immunocytochemistry. J Histochem Cytochem 31:(2):285–292

Elling F, Hasselager E, Friis C (1977) Perfusion fixation of kidneys in adult pigs for electron microscopy. Acta Anat 98:340–342

Emerman M, Behrman EJ (1982) Cleavage and cross-linking of proteins with osmium (VIII) reagents. J Histochem Cytochem 30:395–397

Farr AG, Nakane PK (1981) Immunohistochemistry with enzyme labeled antibodies: a brief review. J Immunolog Meth 47:129–144

Flitney FW (1966) The time course of the fixation of albumin by formaldehyde, glutaraldehyde, acrolein and other higher aldehydes. J R Microsc Soc 85:353–364

Forssman WA, Ito S, Werbe E, Aoki A, Dym M, Fawcett DW (1979) An improved perfusion method for the testis. Anat Rec 198:307–314

Fox CH, Johnson FB, Whiting J, Roller PP (1985) Formaldehyde fixation. J Histochem Cytochem 33:845–853

Fozzard HA, Dominguez G (1969) Effect of formaldehyde and glutaraldehyde on electrical properties of cardiac purkinje fibers. J Gen Physiol 53:530–540

Fraenkel-Conrat H, Mecham DK (1949) The reaction of formaldehyde with proteins. VII Demonstration of intermolecular cross-linking by means of osmotic pressure measurements. J Biol Chem 177:477–486

Fraenkel-Conrat H, Olcott HS (1948) Reaction of formaldehyde with proteins. VI Cross-linking of amino groups with phenol, imidazole or indole groups J Biol Chem 174:827–843

Fraenkel-Conrat H, Cooper M, Olcott HS (1945) The reaction of formaldehyde with proteins. J Am Chem Soc 67:950–956

Franks F (1982) Physiological water stress In: Franks F (ed) Biophysics of water. Wiley, New York

French D, Edsall JT (1945) The reactions of formaldehyde with amino acids and proteins. Adv Protein Chem 2:277–335

Friedrich VL, Mugnaini C (1981) Electron microscopy. Preparation of neural tissues for electron microscopy. In: Heimer L, Robards MJ (ed) Neuroanatomical bract-tracing methods. Plenum Press, New York, pp 345–375

Geiger B, Dutton AH, Tokuyasu KT, Singer SJ (1981) Immunoelectron microscopy studies of membrane-microfilament interactions: distribution of a-actinin, tropomyosin, and vinactin in intestinal epithelial brush border and chicken gizzard smooth muscle cells. J Cell Biol 91:614–628

Gigg R, Payne S (1969) The reaction of glutaraldehyde with tissue lipids. Chem Phys Lipids 3:292–295

Gillett R, Gull K (1972) Glutaraldehyde, its purity and stability. Histochemie 30:162–167

Gomori G (1952) Microscopic histochestry: principles and practice. Univ Chicago Press

Gomori G (1955) In: Colowick P, Kaplain NO (eds) Methods in enzymology. Academic Press, New York, vol 1. pp 138–146

Good NE, Winget GD, Winter W, Connolly TN, Izawa S, Singh RMM (1966) Hydrogen ion buffers for biological research. Biochem J 5:467–477

Griffiths G, Quinn P, Warren G (1983) Dissection of the Golgi Complex. 1. Monensin inhibits the transport of viral membrane proteins from medial to trans Colgi cisternae in baby hamster kidney cells infected with semliki forest virus. J Cell Biol 96:835–850

Griffiths G, McDowall A, Back R, Dubochet J (1984) On the preparation of cryosections for immunocytochemistry. J Ultrastruc Res 89:65–78

Griffiths G, Pfeiffer S, Simons K, Matlin K (1985) Exit of newly synthesized membrane proteins from the trans cisterna of the Golgi complex to the plasma membrane. J Cell Biol 101:949–963

Grillo TAI, Ogunnaike PO, Faoye S (1971) Effects of histological and electron microscopical fixatives on the insulin content of the rat pancreas. J Endocrinol 51:645–649

Gustavson KH (1956) Aldehyde tanning. In : The chemistry of tanning processes. Academic Press, New York, pp 244–282

Habeeb AFSA (1969) A study of the antigenicity of formaldehyde and glutaraldehyde-treated bovine serum albumin and ovalbumin-bovine serum albumin conjugate. J Immunol 102:457–465

Habeeb AFSA, Hiramoto R (1968) Reaction of proteins with glutaraldehyde. Arch Biochem Biophys 126:16–26

Hand AR, Hassel JR (1976) Tissue fixation with diimidoesters as an alternative to aldehydes. II. Cytochemical and biochemical studies of rat liver fixed with dimethylsuberimidate. J Histochem Cytochem 24:1000–1009

Hanson CV, Shen CKJ, Hearst J (1976) Crosslinking of DNA in situ as a probe for chromatin structure. Science 193:62–64

Happich WF, Taylor MM, Feairheller SH (1970) Amino acid composition of glutaraldehyde-stabilized wool. Text Res J 40:768–775

Harris ED, Farrell ME (1972) Resistance to collogenase: a characteristic of collagen fibrils cross-linked by formaldehyde. Biochem Biophys Acta 278:133–141

Harris JE, Peters A (1953) Experiments on vital staining with methylene blue. Q J Microsc Sci 94:113–124

Hartman FC, Wold F (1967) Crosslinking of bovine pancreatic ribonuclease A with dimethyl adipimate. Biochemistry 8:2439–2448

Haselkorn R, Doty P (1961) The reaction of formaldehyde with polynucleotides. J Biol Chem 236:2738–2745

Hassel J, Hand AR (1974) Tissue fixation with diimodoesters as an alternative to aldehydes. I. Comparison of crosslinking structure obtained with dimethylsuberimidate and glutaraldehyde. J Histochem Cytochem 22:223–239

Hasty DL, Hay ED (1978) Freeze-fracture studies of the developing cell surface. II. Particle-free membrane blisters on glutaraldehyde-fixed corneal fibroblasts are artefacts. J Cell Biol 78:756–768

Hattel L, Nettleton GS, Longley JB (1983) A comparison of volume changes caused by aqueous and phase partition fixation. Mikroskopie 40:41–49

Hausmann K (1977) Artefactual fusion of membranes during preparation of material for electron microscopy. Naturwissenschaften 64:95–96

Hawkes R, Niday E, Gordon J (1982) A dot-immunobinding assay for monoclonal and other antibodies. Anal Biochem 119:142–147

Hayat MA (1970) Principles and techniques of electron microscopy. vol I. Biological Applications. Van Nostrand Reinhold Co, New York

Hayat MA (1981) Principles and techniques of electron microscopy: biological applications, 2nd edn, vol 1. Univ Park Press, Baltimore, Maryland

Hayat MA (1986) Basic techniques for transmission electron microscopy. Academic Press, Harcourt Brace Jovanovich, New York, pp 411

Heller J, Ostwald TJ, Bok D (1971) The osmotic behaviour of rod photoreceptor outer segment discs. J Cell Biol 48:633–649

Hixson DC, Yep JM, Glenney JR, Hayes T, Wallborg EF (1981) Evaluation of periodate-lysine-paraformaldehyde fixation as a method for crosslinking plasma membrane glycoproteins. J Histochem Cytochem 29:561–566

Hjerpe A, Boon ME, Kok LP (1988) Microwave stimulation of an immunological reaction (CEA/anti-CEA) and its use in immunohistochemistry. Histochem J 20:388–396

Hopwood D (1969) A comparison of the crosslinking abilities of glutaraldehyde, formaldehyde and a-hydroxyadipaldehyde with bovine serum albumin and casein. Histochemistry 17:151–161

Hopwood D (1970) The reactions between formaldehyde, glutaraldehyde and osmium tetroxide, and their fixation effects on bovine serum albumin and on tissue blocks. Histochemie 24:50–64

Hopwood D (1972) A review: Theoretical and practical aspects of glutaraldehyde fixation. Histochem J 4:267–303

Hopwood D (1975) The reactions of glutaraldehyde with nucleic acids. Histochem J 7:267–276

Hopwood D (1985) Cell and tissue fixation, 1972–1982. Histochem J 17:389–442

Hopwood D, Allen CR, McCabe M (1970) The reactions between glutaraldehyde and various proteins. An investigation of their kinetics. Histochem J 2:137–150

Hopwood D, Yeaman G, Milne G (1988) Differentiating the effects of microwave and heat on tissue proteins and their cross-linking by formaldehyde. Histochem J 20:341–346

Hunter MJ, Ludwig ML (1972) Amidination. In: Hirs CHW, Timasheff SN (eds) Methods in enzymology, vol 25. Academic Press, New York, pp 585–615

Hyde JM, Peters D (1970) The influence of pH and osmolality of fixation on the fowlpox virus core. J Cell Biol 46:179–183

Iqbal A, Weakley HF (1974) The effects of different preparative procedures on the ultrastructure of the hamster ovary. 1. Effects of various fixative solutions on ovarian oocytes and the granulosa cells. Histochemie 28:95–102

Izard C, Libermann, L (1978) Acrolein. Mutat Res 47:115–138

Janigan DT (1965) The effects of aldehyde fixation on acid phosphatase activity in tissue blocks. J Histochem Cytochem 13:473–483

Johannessen JV (1978) Electron microscopy in human medicine, vol 1. McGraw-Hill, New York

Johnson TJA (1985) Aldehyde fixatives: quantification of acid-producing reactions. J Electron Microsc Tech 2:129–138

Johnson TJA (1986) Glutaraldehyde fixation chemistry. A scheme for rapid crosslinking and evidence for rapid oxygen consumption. In: O Johari (ed) Science of biological specimen preparation. SEM Inc, AMF O'Hare, Chicago, pp 51–62

Johnson TJA (1987) Glutaraldehyde fixation chemistry: oxygen-consuming reactions. Eur J Cell Biol 45:160–169

Johnson TJA, Rash JE (1986) Glutaraldehyde chemistry: fixation reactions consume O_2 and are inhibited by tissue anoxia. J Cell Biol 87:234a

Jones D (1969a) The reaction of formaldehyde with unsaturated fatty acids during histological fixation. Histochem J 1:459–491

Jones D (1969b) Acrolein as a histological fixative. J Microsc 90:75–77

Jost PC, Brooks UJ, Griffith OH (1973) Fluidity of phospholipid bilayers and membranes after exposure to osmium tetroxide and glutaraldehyde. J Mol Biol 76:313–318

Kalimo H, Pelleniemi LJ (1977) Pitfalls in the preparation of buffers for electron microscopy. Histochem J 9:241–246

Kallen RG, Jencks WP (1966) Equilibria for the reaction of amines with formaldehyde and protons in aqueous solution. J Biol Chem 241:5846–5878

Karnovsky MJ (1965) A formaldehyde-glutaraldehyde fixative of high osmolarity for use in electron microscopy. J Cell Biol 27:137A

King JC, Lechan RM, Kugel G, Anthony ELP (1983) Acrolein: a fixative for immunocytochemical localization of peptides in the central nervous system. J Histochem Cytochem 31(1):62–68

Kok LP, Visser PE, Boon ME (1987) Histoprocessing with the microwave oven: an update. Histochem J 20:323–328

Korn AH, Feairheller SH, Filachione EM (1972) Glutaraldehyde: nature of the reagent. J Mol Biol 65:525–529

Korn ED (1966) Structure of biological membranes: the unit membrane theory is reevaluated in light of the data now available. Science 153:1491–1498

Korn ED (1968) Structure and function of the plasma membrane. J Gen Physiol 52:257–277

Korn ED, Weissman RA (1966) I. Loss of lipids during preparation of amoebae for electron microscopy. Biochim Biophys Acta 116:309–316

Kraehenbuhl JP, Jamieson JD (1972) Solid phase conjugation of ferritin to Fab-fragments of immunoglobulin G for use in antigen localization on thin sections. Proc Natl Acad Sci USA 69:1771–1776

Kraehenbuhl JP, Racine L, Jamieson JD (1977) Immuno-cytochemical location of secretory proteins in bovine pancreatic exocrine cells. J Cell Biol 72:406–414

Kuran H, Olszewska MJ (1974) Effect of some buffers on the ultrastructure dry mass content and radioactivity of nuclei of Haemanthus katharinae. Folia Histochem Cytochem 12:173–182

Kyte J (1976) Immunoferritin determination of the distribution of (Na⁺ + K⁺) ATPase over the plasma membranes of renal convoluted tubules. J Cell Biol 68:287–297

Landis DMD, Reese TS (1981) Astrocyte membrane structure: changes after circulatory arrest. J Cell Biol 88:660–663

Langenberg WG (1980) Glutaraldehyde nonfixation of isolated viral and yeast RNAs. J Histochem Cytochem 28:311–315

Larsson L-I (1988) Immunocytochemistry: theory and practice. CRC Press, Boca Raton, Florida, pp 272

Lee RMKV, McKenzie R, Kobayashi K, Garfield RE, Forrest JB, Daniel EE (1981) Effects of glutaraldehyde fixative osmolarities on smooth muscle cell volume and osmotic reactivity of the cells after fixation. J Microsc 125:77–88

Leenen PJM, Jansen AMAC, van Ewijk W (1985) Fixation parameters for immunocytochemistry: The effect of glutaraldehyde or paraformaldehyde fixation on the preservation of mononuclear phagocyte differentiation antigens In: Bullock GR, Petrussz P (eds) Techniques in imunocytochemistry. Academic Press, London, pp 1–24

Leist DP, Nettleton GS, Feldhoff RC (1986) Determination of lipid loss during aqueous and phase partition fixation using formalin and glutaraldehyde. J Histochem Cytochem 34:437–441

Lenard J, Singer SJ (1968) Alterations of the conformation of proteins in red blood cell membranes and in solutions by fixatives used in electron microscopy. J Cell Biol 37:117–121

Li IIJ (1972) Thermal denaturation of nucleohistones-effects of formaldehyde reaction. Biopolymers 11:835–839

Login GR Dwyer BK, Dvorak AM (1990) Rapid primary microwave-osmium fixation. I Preservation of structure for electron microscopy in seconds. J Histochem Cytochem 38:755–762

Login GR, Dvorak AM (1988) Microwave fixation provides excellent preservation of tissue, cells and antigens for light and electron microscopy. J Histochem 20:373–387

Login GR, Stavinsha WB, Dvorak AM (1986) Ultrafast microwave energy fixation for electron microscopy. J Histochem Cytochem 34:381–387

Luther PW, Bloch RJ (1989) Formaldehyde-amine fixatives for immunocytochemistry of cultured xenopus myocytes. J Histochem Cytochem 37:75–82

Martin CJ, Lam DP, Marini M (1975) Reaction of formaldehyde with the histidine residues of protein. Bioorg Chem 4:22–26

Mason P, Griffiths JC (1964) Cross-linking fibrous proteins by formaldehyde. Nature 203:484–486

Mathieu O, Claassen H, Weibel ER (1978) Differential effect of glutaraldehyde and buffer osmolarity on cell dimensions. A study on lung tissue. J Ultrastruct Res 63:20–34

Maunsbach AB (1966) The influence of different fixatives and fixation methods on the ultrastructure of rat kidney proximal tubule cells. II. Effects of varying osmolarity, ionic strength, buffer system and fixative concentration of glutaraldehyde solutions. J Ultrastruct Res 15:283–309

Maupin-Szaimer P, Pollard TD (1978) Actin filament destruction by osmium tetroxide. J Cell Biol 77:837–847

Mayer RJ, Walker JH (1987) Immunochemical methods in cell and molecular biology. Academic Press, London

Mayers CP (1970) Histological fixation by microwave heating. J Clin Path 23:273–5

Mays ET, Feldhoff RC, Nettleton GS (1981) Determination of protein loss during aqueous and phase partition fixation using formalin and glutaraldehyde. J Histochem and Cytochem 32:1107–1112

McBeath E, Fujiwara K (1984) Improved fixation for immunofluorescence microscopy using light-activated 1,3,5-triazido-2,4,6-trinitrobenzen. J Cell Biol 99:2061–2073

McDowall AW, Chang JJ, Freeman R, Lepault J, Walter CA, Dubochet J (1983) Electron microscopy of frozen hydrated sections of vitreous ice and vitrified biological specimens. J Microsc 131:1–9

McDowall AW, Gruenberg J, Römisch K, Griffiths G (1989) The structure of organelles of the endocytic pathway in hydrated cryosections of cultured cells. Eur J Cell Biol 49:281–294

McLean JW, Nakane PK (1974) Periodate-lysine-paraformaldehyde fixative. A new fixative for electron microscopy. J Histochem Cytochem 22:1077–1083

McLean JD, Singer SJ (1970) A general method for the specific staining of intracellular antigens with ferritin-antibody conjugates. Proc Natl Acad Sci USA 65:1228–1234

Medawar PB (1941) The rate of penetration of fixatives. J R Microsc Soc 61:46–52

Merritt JH, Frazer JW (1977) Microwave fixation of brain tissue as a neuroanatomical technique - a review. J Microwave Power 12:133–145

Mersey B, McCully ME (1978) Monitoring of the course of fixation of plant cells. J Microsc 114:49–76

Millonig G, Marinozzi V (1968) Fixation and embedding in electron microscopy. Adv Opt Electron Microsc 2:251

Minassian H, Huang S-N (1979) Effect of sodium azide on the ultrastructural preservation of tissues. J Microsc 117:243–253

Morel FMM, Baker RF, Wayland H (1971) J Cell Biol 48:91–100

Morowitz HJ, Terry TM (1969) Characterisation of the plasma membrane of *Mycoplasma laidlawii*. Effects of selective removal of protein and lipid. Biochem Biophys Acta 183:270–294

Needles HL (1967) Crosslinking of gelatin by aqueous peroxydisulfate. J Polym Sci A5:1–13

Nielson AJ, Griffiths WP (1979) Tissue fixation by osmium tetroxide. J Histochem Cytochem 27:997–999

Nir I, Hall MO (1974) The ultrastructure of lipid-depleted rod photoreceptor membranes. J Cell Biol 63:587–598

Olmstead JB, Borisy GG (1975) Microtubules. Annu Rev Biochem 42:507–523

Palade GE (1952) A study of fixation for electron microscopy. J Exp Med 95:285–297

Palade GE (1954) The fixation of tissues for electron microscopy. Proc 3rd Int Congr EM London:p 129–143

Park RB, Kelley J, Drury S, Sauer K (1966) The Hill reaction of chloroplasts isolated from glutaraldehyde-fixed spinach leaves. Proc Natl Acad Sci USA 55:1056–1064

Parsons DF (1972) Status of X-ray diffraction of membranes. Ann NY Acad Sci 145:321–326

Pathak MA, Kramer DM (1969) Photosensitization of skin in vivo by furocoumarins (Psoralens). Biochim Biophys Acta 195:197–199

Pearse AGE (1963) Some aspects of the localization of enzyme activity with the electron microscope. J R Microsc Soc 81:107–118

Pearse AGE (1968) Histochemistry: theoretical and applied, 3rd edn, vol. Churchill, London

Pentilla A, Kalimo H, Trump BF (1974) Influence of glutaraldehyde and/or osmium tetroxide on cell volume, ion content, mechanical stability, and membrane permeability of Ehrlich ascites tumor cells. J Cell Biol 63:197–214

Peracchia C, Mittler BS (1972) Fixation by means of glutaraldehyde-hydrogen peroxide reaction products. J Cell Biol 53:234–238

Peters KR (1984) Continuous ultra thin metal films. In: Johari O (ed) Science of biological specimen preparation. SEM Inc, O'Hare, Chicago, pp 221–233

Pilfors L, Werbull C (1985) The consumption of osmium tetroxide by components of the cytoplasmic membrane of *Acheoteplasma laidlawii* and its morphological implications. Micron Micros Acta 16:77–83

Ponder E (1942) The red cell as an Osmometer. Cold Spring Harbor Symp. Quant Biol 8:133–143

Posthuma G, Slot JW, Geuze HJ (1984) Immunocytochemical assays of amylase and chymotrypsinogen in rat pancreas secretory granules. J Histochem Cytochem 32:1028–1034

Posthuma G, Slot JW, Geuze HJ (1985) The validity of quantitative immunoelectron microscopy on ultrathin sections as judged by a model study. Proc R Microsc Soc 20:IMS

Puchtler H, Meloan SN (1985) On the chemistry of formaldehyde fixation and its effects on immunohistochemical reactions. Histochemistry 82:201–204

Quiocho FA, Richards FM (1966) The enzyme behavior of carboxypeptidase-A in the solid state. Biochemistry 5:4062–4067

Rasmussen KE, Albrechtsen J (1974) Glutaraldehyde. The influence of pH, temperature, and buffering on the polymerization rate. Histochemistry 38:19–26

Richards FM, Knowles JR (1968) Glutaraldehyde as a protein crosslinking reagent. J Mol Biol 37:231 233

Riederer BM (1989) Antigen preservation tests for immunocytochemical detection of cytoskeletal proteins: influence of aldehyde fixatives. J Histochem Cytochem 37:675–681

Riemersma JC (1963) Osmium tetroxide fixation of lipids Nature of the reaction products. J Histochem Cytochem 11:436

Robertson EA, Schultz RL (1970) The impurities in commercial glutaraldehyde and theri effect on the fixation of brain. J Ultrastruct Res 30:275 287

Roof DJ, Heuser JE (1982) Surfaces of rod photoreceptor disk membranes: integral membrane components. J Cell Biol 95:487–500

Roos A, Boron WF (1983) Intracellular pH. Physiol Rev 61:296–434

Roozemomd RC (1969) The effect of fixation with formaldehyde and glutaraldehyde on the composition of phospholipids extractable from rat hypothalamus. J Histochem Cytochem 17:482–486

Rothman JE, Lenard J (1977) Membrane asymmetry: the nature of membrane asymmetry provides clues to the puzzle of how membranes are assembled. Science 195:743–753

Sabatini DD, Bensch K, Barrnett RJ (1963) Aldehyde fixation for morphological and enzyme histochemical studies with the electron microscope. J Cell Biol 17:19–58

Salema R, Brandao I (1973) The use of PIPES buffer in the fixation of plant cells for electron microscopy. J Submicrosc Cytol 5:79–96

Schiff RI, Gennaro JF Jr (1979) The role of the buffer in the fixation of biological specimens for transmission and scanning electron microscopy. Scanning 2:135–148

Schultz RL, Karlsson UL (1972) Brain extracellular space and membrane morphology variations with preparative procedures. J Cell Sci 10:181–195

Scott RE (1976) Plasma membrane vesiculation: a new technique for isolation of plasma membranes. Science 194:743–745

Seligman AM, Chauncey HH, Nachlas MM (1950) Effect of formalin fixation on the activity of five enzymes of rat liver. Stain Technol 26:19–23

Shelton E, Mowczko WE (1978) Membrane blisters: a fixation artefact. A study in fixation for scanning electron microscopy Scanning 1:166–173

Silva MT, Carvalho Guerra F, Magalhnes MM (1968) The fixative action of uranyl acetate in electron microscopy. Experientia 24:1070–1076

Skaer RS, Whytock S (1976) The fixations of nuclei and chromosomes. J Cell Sci 20:221–231

Smit JW, Meijer CJLM, Decary F, Feltkamp-Vroom TM (1974) Paraformaldehyde fixation in immunofluorescence and immunoelectron microscopy: preservation of tissue and cell surface membrane antigens. J Immunol Methods 6:93–98

Smith PF, Keefer DA (1982) Acrolein/glutaraldehyde as a fixative for combined light and electron microscopic immunocytochemical detection of pituitary hormones in immersion-fixed tissue. J Histochem Cytochem 30 (12):1307–1310

Spendlove RS, Singer SJ (1961) On the preservation of antigenic determinants during fixation and embedding for electron microscopy. Proc Natl Acad Sci USA 47:14–18

Stanley WM (1944) The preparation and properties of influenza virus vaccines concentrated and purified by differential centrifugation J. Exp Med 8:193–199

Stevens CL, Chay TR, Loga S (1977) Rupture of base pairing in double-stranded poly (riboadenylic acid). Poly (ribonuclylic acid) by formaldehyde: medium chain lengths. Biochemistry 16:3727–3739

Stoeckenius W, Mahr SC (1965) Studies on the reaction of osmium tetroxide with lipids and related compounds. Lab Invest 14:1196–1207

Strangeways TSP, Canti RG (1928) The living cell in vitro as shown by darkground illumination and the changes induced in such cells by fixing reagents. Q J Microsc Sci 71:1–12

Swanson J, Bushnell A, Silverstein SC (1987) Tubular lysosome morphology and distribution within macrophages depend on the integrity of cytoplasmic microtubules. Proc Natl Acad Sci USA 84:1921–1925

Tanford C (1980) The hydrophobic effect: formation of micelles and biological membranes. 2nd edn Wiley, New York

Tashima T, Kawakami U, Harada M, Sakata T, Satoh N, Nakagawa T, Tanaka H (1987) Isolation and identification of new oligomers in aqueous solution of glutaraldegyde. Chem Pharm Bull 35:4169–4180

Taylor DP (1987) Direct measurement of the osmotic effects of buffers and fixatives in *Niletta flexilis*. J Microsc 159:71–80

Tisdale AD, Nakajima Y (1976) Fine structure of synaptic vesicles in two types of nerve terminals in crayfish stretch receptor organs: influence of fixation methods. J Comp Neurol 165:369–386

Utsumi K, Packer L (1967) Glutaraldehyde-fixed mitochondria. I. Enzyme activity, ion translocation, and conformational changes. Arch Biochem Biophys 121:633–640

van Ewijk W, van Soest PL, Verkerk A, Jongkind JF (1984) Loss of antibody binding to prefixed cells: fixation parameters for immunocytochemisty. Histochem J 16:179–193

van Meer G, Simons K (1986) The function of tight junctions in maintaining differences in lipid composition between the apical and the basolateral cell surface domains of MDCK cells. EMBO J, 5:1455–64

van Meer G (1989) Biosynthetic lipid traffic in animal eukaryotes. Annu Rev Cell Biol 5:247–275

Vologodskii AV, Frank-Kamenetskii MD (1975) Theoretical study of DNA unwinding under the action of formaldehyde. J Theor Biol 55:153–166

von Hippel PH, Wong K-Y (1971) Dynamic aspects of native DNA structure: Kinetics of the formaldehyde reaction with calf thymus DNA. J Mol Biol 61:587–613

Wakabayshi, T (1972) Ultrastructure and functional states of mitochondria – effect of fixatives on the stabilization of unstable configurations. Nagoya J Med Sci 35:1–17

Walker JF (1964) Formaldehyde, 3rd edn. van Nostrand-Reinhold, New York

Ward BJ, Gloster JA (1976) Lipid losses during processing of cardiac muscle for electron microscopy. J Microsc 108:41–50

Weast RC, Astle MJ (1979) CRC Handbook for chemistry and physcis. CRC Press, Boca Raton, Florida

Webster H, Ames A (1965) Reversible and irreversible changes in the fine structure of nervous tissue during oxygen and glucose deprivation. J Cell Biol 26:885–909

Weibull C, Christiansson A, Carlemalm E (1983) Extraction of membrane lipids during fixation, dehydration and embedding of *Acholeplasma laidawii* cells for electron microscopy. J Microsc 129:201–207

Whipple EB, Ruta M (1974) Structure of aqueous glutaraldehyde. J Org Chem 39:1666–1668

Wiesenhahn G, Hearst JE (1978) DNA unwinding induced by photoaddition of psoralen derivatives and determination of dark-binding equilibrium constants by gel electrophoresis. Proc Natl Acad Sci USA 75:2703–2707

Wiesenhahn G, Hyde JE, Hearst JE (1977) Photoaddition of trimethylpsoralen to *Drosophila melanogaster* nuclei: a probe for chromatin structure. Biochemistry 16:925–932

Wisse E, De Wilde A, De Zanger R (1984) Perfusion fixation of human and rat liver tissue for light and electron microscopy: a review and assessment of existing methods with special emphasis on sinusoidal cells and microcirculation. In: Johari O (ed) The science of biological specimen preparation for microscopy and microanalysis. SEM Inc, AMF O'Hare, IL 60666, pp 31–38

Wolfsy L, Singer SJ (1963) Effects of the amidination reaction on antibody activity and on the physical properties of some proteins. Biochemistry 2:104–109

Wold F (1972) Bifunctional reagents. In: Hirs CHW, Timasheff N (eds) Methods in enzymology. vol 25, part B, Academic Press, New York, pp 623–649

Wollenzien PL, Youvan DC, HearstJE (1978) Structure of psoralen-crosslinked ribosomal RNA from *Drosophila melanogaster*. Proc Natl Acad Sci USA 75:1642–1646

Wolman M, Greco J (1952) The effect of formaldehyde on tissue lipids and on histochemical reactions for carbonyl groups. Stain Technol 29:317–324

Wood JG (1973) The effects of glutaraldehyde and osmium on the proteins and lipids of myelin and mitochondria. Biochim Biophys Acta 329:118–122

Yamamoto N, Yasuda K (1977) Use of a water-soluble carbodiimide as a fixing agent. Acta Histochem Cytochem 10:14–37

Zalokar M, Erk I (1977) Phase-partition fixation and staining of *Drosophila* eggs. Stain Technol 52:89–95

Zambrano F, Fleischer S, Fleischer B (1975) Lipid composition of the Golgi apparatus of rat kidney and liver in comparison with other subcellular organelles. Biochim Biophys Acta 380:357–369

Zondervan PE, De Jong A, Sorber CWJ, Kok LP, De Bruijn WC, Van der Kwast ThH (1988) Microwave-stimulated incubation in immunoelectron microscopy: a quantitative study. Histochem J 20:359–36

Chapter 4

Embedding Media for Section Immunocytochemistry

In those techniques grouped under the label "post-embedding", it is obviously necessary to section cells or tissues in some manner before the labelling reactions are made. The ideal situation would be, perhaps, to use a device such as a tissue chopper or vibratome, which could section unembedded, fixed tissues as thin as, say, 100 nm. This is currently not possible and it is doubtful, to say the least, whether undamaged sections of such unsupported tissue could ever be made. Since there is no method capable of doing this at present, an embedding medium becomes a prerequisite. Clearly, there are stringent conditions attached to any embedding medium that is to be of general use in fine-structure immunocytochemistry.

These are:

- It must easily infiltrate tissues and then be uniformly hardened without significant swelling or shrinkage.
- The blocks should combine hardness with plasticity so that they can be sectioned smoothly, at least as thin as 100 nm. For stereological studies this statement could be extended to "as thin as possible". For most purposes an additional requirement is to be able to section relatively large sections (up to 1 mm^2, or even more). The sections should also be flat on the support and free from compression.
- The infiltration (including the steps leading up to it), polymerization or sectioning processes should not adversely affect the ability of antigens, at least on the surface of sections, to recognize antibodies nor modify fine structure.
- After labelling, the sections must be dried before entering the electron microscope column. The fine structure should not be affected by this process.
- The sections and the structures contained in them must be resistant to irradiation by the electron beam.
- It must be possible to adequately contrast and recognize the structures present in the sections.

Many different strategies have been used for post-embedding labelling, not all specifically designed for the purpose. In fact, none meet all of the above requirements completely. In the past decade, however, significant advances have been made in two different kinds of approaches. For convenience in Fig. 1, I have classified the different methods known into five groups depending on three interrelated factors:

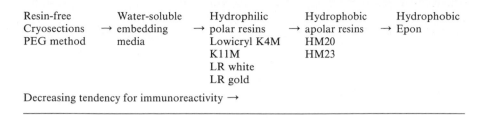

Resin-free Cryosections PEG method	Water-soluble $\rightarrow$ embedding media	Hydrophilic $\rightarrow$ polar resins Lowicryl K4M K11M LR white LR gold	Hydrophobic $\rightarrow$ apolar resins HM20 HM23	Hydrophobic $\rightarrow$ Epon

Decreasing tendency for immunoreactivity $\rightarrow$

Fig. 1. Classification of embedding media for immunocytochemistry

- The degree of dehydration that is necessary before the labelling reactions.
- The hydrophobicity of the embedding medium (if present) during the labelling reaction.
- The degree to which the antigens are covalently bound to the embedding medium.

These three criteria, whose magnitude increases from left to right, collectively describe the degree to which one expects antigens in the tissue to recognize antibody during the labelling reaction. It is emphasized that this is based on theoretical arguments only and, in practice, can only be taken as a rough guide (as will be pointed out below, recent data indicate, for example, that the more hydrophobic lowicryl HM20 can in many cases give very similar labelling signals to the more hydrophilic K4M and to cryo-sections). For all protocols, the water, which must be present for the antibody reactions, has to be removed before the section is introduced into the electron microscope. The four different types of approaches will now be discussed.

4.1 Resin-Free Sections – Temporary Embedment

This is the basis of the Tokuyasu (1973, 1978) cryo-section technique, as well as the less well known polyethylene glycol (PEG) embedding technique of Wolosewick (1980). In both these methods, tissues are temporarily embedded in a medium which is solid either at cold temperatures or at room temperature (PEG) but easily removed when the sections are floated on aqueous solutions. After the labelling, they are dehydrated and dried in a simple embedment step. Both methods involve a partial and temporary dehydration before the first embedding in that the water is partly replaced by sucrose in the cryo-method and PEG in the other.

Since the cryo-section technique will be discussed in detail in Chapter 5, only the points relevant to a comparison with the other methods will be mentioned here. In this "temporary embedment" approach, it is the initial aldehyde cross-linking which is responsible for stabilizing the biological material during the labelling reactions. Once the sections have been put onto an aqueous medium they are, in fact, identical (ignoring possible differences in structure brought about by the different pre-embedding and sectioning procedures). In both cases we are faced with the same problem at the end – how to dry and contrast the labile sections without destroying them by surface-tension forces. A different resin-free approach has also been introduced by Gorbsky and Borisy (1985, 1986) which involves the use of polymethylmethacrylate. This will be described briefly below.

4.1.1 The PEG Method

PEG embedding for electron microscopy was first introduced by Richards et al. (1942). As developed by Wolosewick (1980, 1984), the method involves infiltrating and embedding tissues in PEG 3350. The polyethylene glycols, also known by a variety of commercial names such as Carbowax, are a class of liquid and solid polymers of general formula $H(OCH_2CH_2)_n OH$ where n is 4 or larger. As n increases, the solidification temperature also increases. For example, PEG 400 has an average n value between 8.2 and 9.1 and solidifies between 4 and 8°C: it is therefore liquid at room temperature. On the other hand, PEG 4000 with an average n value in the range 68–84, is solid below 54–58°C. As the molecular weight increases, so does the viscosity and therefore the time required to infiltrate a given tissue. PEG 3350, which Wolosewick recommends for the method, is freely miscible with both water and ethanol.

The method as used in practice is as follows: small pieces of tissue are dehydrated sequentially in increasing concentrations of PEG 3350 in water, at room temperature, namely 25 and 40% (v/v). They are then put into increasing concentrations of PEG at 55°C, e.g. 50, 70, 95% and three changes of 100%. A warm piece of filter paper is recommended for changing the specimens from one vial to the next. About 1 h at each concentration is recommended for complete infiltration, although this may be considerably reduced for some tissues. The tissues are embedded in pure molten PEG (at 55°C) in gelatin capsules which are then cooled by plunging into liquid nitrogen. Such, relatively rapid cooling is preferred over slow cooling at room temperature in that the blocks become more homogeneous (Wolosewick, pers. commun.).

The blocks are carefully trimmed with a razor blade and can be mounted onto an ultramicrotome specimen holder using dental wax or rapid-drying glue. Sectioning is done using either glass or diamond knives in the absence of a water trough at room temperature. The biggest difficulty, in practice, is to be able to cut sections thin enough for electron microscopy. The blocks are extremely brittle; in other words, they lack the plasticity required for thin sectioning. It is doubtful whether sections thinner than 100 nm can be sectioned by this procedure.

The sections can be transferred from the knife onto grids using a loop of either 2,3 M sucrose or 40% PEG in phosphate buffered saline, in exactly the same way as for the cryo-section method (see Chap. 5). Labelling and post-embedding is then identical to the latter method.

That this method is capable of providing good fine-structural preservation is evident in many, but not all, of Wolosewick's published pictures. The less than adequate structural preservation evident in some of his early published micrographs can, in my opinion, be attributed to air drying artefacts, especially when sections have been critical point-dried. The latter procedure, still used routinely for scanning electron microscopy, is documented to be a potentially harsh procedure which can cause severe volume changes in tissues (see Boyde and Maconnachie 1983). The ensuing air-drying artefacts closely resemble those seen in thawed cryo-sections which have been air-dried in the absence of methyl cellulose protection (Griffiths et al. 1984). When PEG sections were dried by embedding them in Epon, the structural preservation was, in fact, indistinguishable from conventional epon sections (Wolosewick 1983) (or from cryo-sections embedded in Epon see p. 179). Further, when protected against air drying using methyl cellulose embedment, the images can appear similar to thawed cryo-sections (Wolosewick 1983).

The single advantage of the PEG method over the cryo-section method is that it can be done using technology present in any EM laboratory. It is a very simple technique which may, in fact, have even more applications for making semi-thin sections for light microscopy labelling. The disadvantages of the method are first, as mentioned, the difficulty of thin sectioning and second the fact that the physico-chemical effects of PEG on structural preservation and antigenicity have not been extensively studied. For recent modifications in the embedding protocol see Bard and Ross (1986) and Mowary et al. (1989). An example of the use of this approach for plant tissues is given in Hawes and Horne (1985).

4.1.2 The Poly-Methylmethacrylate Method

A method developed by Gorbsky and Borisy (1985; 1986) is to embed cells or tissues in poly-methylmethacrylate dissolved in dichlormethane. The solvent is allowed to evaporate at room temperature and thereby the blocks become hard enough to cut very thin sections. According to the authors, sectioning as thin as 20 nm is possible using glass or diamond knives fitted with a water trough. There is no chemical polymerization, the linear polymers are apparently held together by hydrogen bonding. After sectioning, the plastic can be removed using a suitable solvent such as acetone, and the sections can be rehydrated before antibody labelling. The possibilities for drying the section should be similar to sections prepared by the PEG method or by cryoultramicrotomy. The authors have described a variety of different methods for drying the sections, including negative staining, and the use of critical-point drying followed by rotary shadowing.

The published micrographs using this technique, which the authors refer to as Reversible Embedment Cytochemistry are impressive, especially with respect to cytoskeletal elements.

4.2 Permanent Embedding Media

4.2.1 Water-Miscible Media

Thin-section electron microscopy as a tool for studying ultrastructure came of age with the introduction of the epoxy resins in the 1950s. Prior to this time, methacrylate resins had been used with limited success. The significant advantage of the epoxy resins was their ability to become cross-linked extensively without losing the plasticity required to enable them to be sectioned as thin as 5–10 nm. They are also relatively resistant to the damaging effects of the electron beam. These properties were generally lacking in the early methacrylate resins.

The major disadvantage of the epoxy resins, both for high-resolution structural preservation and especially for immunocytochemistry, is their extremely hydrophobic nature. Tissues must therefore, be completely dehydrated in protein denaturing solvents before infiltration. Further, the resins can only be polymerized by heating above about 50°C. A further problem of their relatively high viscosity was overcome by the development of Spurr's medium (Spurr 1960). As suggested by Kellenberger (1987), the high degree of covalent interactions between these resins and the biological material (see Causton 1986) is also a problem for immunocytochemistry.

Based on these facts it is not surprising that efforts have been made to develop completely water-miscible embedding media for electron microscopy. Although acceptable preservation has been published using a variety of methods (see, for example, Pease 1966; Peterson and Pease 1970; Heckman and Barrett 1973; Spaur and Moriarty 1977), the practical difficulties of sectioning have restricted these techniques to a few laboratories, and usually for a few years only.

For immunocytochemistry, the most extensively used approach in this class of media has been the Bovine Serum Albumin (BSA) method introduced by Farrant and Mc lean (1969) and McLean and Singer (1970). In this method, the fixed tissue is infiltrated with a 20% solution of BSA which is then extensively cross-linked with glutaraldehyde before being dried and sectioned (see Kraehenbuhl and Jamieson 1972; Kraehenbuhl et al. 1977; Papermaster et al. 1978; Griffiths and Jockusch 1980). The technique is worth mentioning because it was the first in which a systematic attempt was made to quantitate antibody labelling (Kraehenbuhl et al. 1978; 1980). The major problem with this approach is the severe shrinkage which occurs when the cross-linked BSA blocks are dried, prior to sectioning (Griffiths and Jockusch 1980).

The water-miscible media become less hydrophilic after polymerization and sectioning. Hence, in most of the techniques referred to, the blocks are sectioned, like epoxy blocks, on a water trough. If surfaces of the section were truly hydrophilic, they would be expected to sink in the water. This was, in fact, often but not always a problem in practice with the BSA method.

A class of water-soluble embedding mixture which involves a mixture of melamine and formaldehye was introduced by Bachhuber and Frösch (1983).

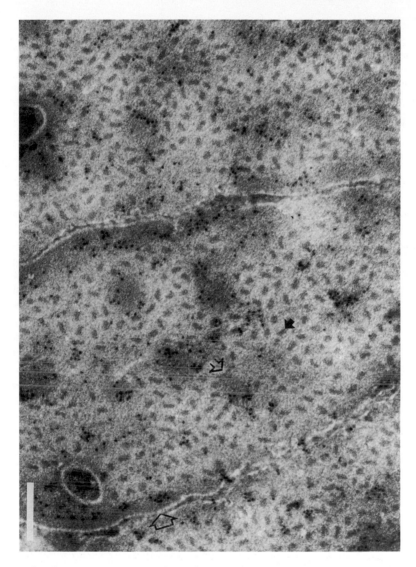

Fig. 2. Embedding in a water soluble resin. Chicken gizzard muscle was fixed with 3% formaldehyde and embedded in polyvinylalcohol (PVA) as described by Malek and Small (Protoplasma 139, 160–169, 1987). The sections were labelled with an antibody against the actin-binding protein, filamin and an anti rabbit (10 nm) gold conjugate. The sections were positively stained with uranyl acetate in 95% methanol at 60 °C for 5 min followed by lead citrate (3 min). In this section the positively stained myosin filaments (*dense arrows*) are quite distinct and unlabelled. Most of the label is associated with the periphery of dense bodies (*open-ended arrow*) and attachment plaques (*open arrow*). *Bar* 0.2 μm. (Courtesy of Dr. Victor Small, Institute of Molecular Biology, Salzburg, Austria.)

These can be polymerized by either heat (50°C) or, at room temperature, by acid. Although some shrinkage accompanies the polymerization, this apparently varies with the tissue. Hence for rat liver, this shrinkage was severe (25%) whereas for muscle it was only 5% (Bachhuber and Frösch 1983). The polymerized blocks are apparently easy to section and the published micrographs using this method show remarkably good fine structural detail. This approach has not yet been extensively exploited for immunocytochemistry. Preliminary results were apparently not promising, however (Bachhuber and Frösch 1983; Schwarz and Humbel, pers. commun.).

Another approach, introduced by Small (1984), was to embed tissues in polyvinyl alcohol (PVA- 10000 MW). The fixed tissue is infused with a 20% solution of this compound for 24–48 h at 40°C in closed tubes. Subsequently, the process was continued in an open container at the same temperature, a condition that allows the material to dry. The complete process, which includes a drying step at 60°C, takes about 1 week. The blocks must be sectioned using glycerol as a flotation medium since the sections sink in water. A subsequent publication by this group showed good preservation and labelling of smooth muscle tissue at both the light and electron microscope level (Small et al. 1986; see Fig. 2).

Finally, a recent publication by Escolar et al. (1988) has introduced a new water-soluble acrylic compound (Immunobed A) which, when combined with methyl methacrylate, was compatible with thin sectioning and immunolabelling for tubulin.

With the possible exception of the PVA method and the Immunobed A approaches, however, water-miscible or truly hydrophilic permanent embedding media, while superficially attractive for immunocytochemistry, have not lived up to their expectations in practice.

4.2.2 New Acrylic Resins

In the past 5–10 years there have been dedicated attempts by a few chemists, interested in biological problems, to develop hydrophilic acrylic-type resins for EM. Although not water–miscible, these resin mixtures are designed so that they combine at least a degree of hydrophilicity with good cutting properties and electron-beam resistance. These "tailor-made" mixtures are rapidly proving to be powerful tools for immunocytochemistry. One of these is polymerized with heat, in the conventional way, while the others can be hardened at low temperatures.

4.2.2.1 Low Temperature Embedding: Lowicryl Resins

The major breakthrough in the use of acrylic resins has come by designing at least partly hydrophilic resins that can infiltrate tissues and can be polymerized at low temperature. The theoretical background for this idea came from studies by physical chemists on the effects of solvents at low temperature on protein structure and function (Petsko 1975; Douzou 1977). The aim of these kinds of studies was to slow down the rates of enzymatic reactions in order to facilitate kinetic studies and to identify reaction-intermediates.. Hence, solvents were used as a form of "anti-freeze". The "take-home" message from these studies was that low temperatures tended to stabilize proteins during the removal of the solvent (i.e. water) by dehydration (Petsko 1975). At temperatures above 0° C, removal of this outer water shell characteristically leads to "**denaturation**" of proteins (a term usually associated with unfolding and loss of activity; see Kauzmann 1957). At temperatures significantly below 0°C, however, these effects can often be reduced and sometimes avoided completely. The magnitude of the effects appears to depend critically on the dielectric constant, polarity and pH of solvent, as well as on the physic-chemical properties of the protein itself. Finding the optimal combination of these factors is extremely complex and has to be empirically determined for each protein in vitro (Douzou 1977).

Based on this knowledge, Kellenberger's group at Basel have made a serious effort to develop an embedding medium which could infiltrate tissue and be polymerized at sub-zero temperatures below 0°C. The concept had been tried, albeit with limited success, by other workers (see Sjöstrand 1976). The effort by the Basel group has been very successful and, in a relatively short time span, these new resins, known commercially as the Lowicryl resins, have justifiably obtained a wide popularity for immunocytochemical studies.

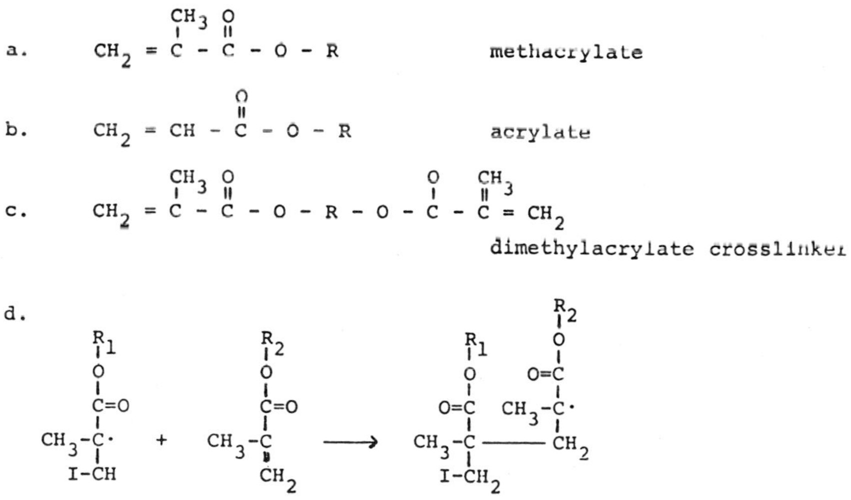

Fig. 3a–d. A schematic formula showing the basic monomers and the reaction pathways for polymerization of lowicryl resins. In **a** and **b**, R represents the position of the side group in the monomer. For **c**, R represents a dialcohol linker between the two methacrylate groups. In **d**, a methacrylate monomer has been activated with an initiator (I) to form an unpaired electron at C2. This radical now attacks another monomer to begin the growth of the polymer chain. (Carlemalm et al. 1982). The composition of K4M and HM20 is shown in Table 1

Table 1. Resin mixtures K4M and HM20. (Carlemalm et al. 1982)

wt.%	Monomer	Side chain (R)
K4M (polar)		
48.4	Hydroxypropyl methacrylate	$- CH_2 - CH_2 - CH_2 - OH$
23.7	Hydroxyethyl acrylate	$- CH_2 - CH_2 - OH$
9.0	n-Hexyl methacrylate	$- CH_2 - (CH_2)_4 - CH_3$
13.9	Cross linker[a]	–
HM20 (apolar)		
68.5	Ethyl methacrylate	$- CH_2 - CH_3$
16.6	n-Hexyl methacrylate	See above
14.9	Cross-linker[a]	–

[a] The cross-linker used is triethylene glycol dimethacrylate (TEGDMA)

$$H_2C{=}\overset{\overset{\textstyle H_3C}{|}}{C}{-}\overset{\overset{\textstyle O}{||}}{C}{-}O{-}CH_2CH_2O{-}CH_2CH_2O{-}CH_2CH_2O{-}\overset{\overset{\textstyle O}{||}}{C}{-}\overset{\overset{\textstyle CH_3}{|}}{C}{=}CH_2$$

The amount of cross-linker used here will yield blocks of medium hardness. The relative amount of cross-linker can be varied within a certain range.

This class of resin is made up of mixtures of aliphatic acrylate and methacrylate esters (Fig. 3). The chemical rationale behind their design was that the resin "backbone" and polymerization reaction, once optimized, was left unchanged, but the physical and chemical nature of the resin, both before and after polymerization could be altered by simple modification of side groups off the backbone (Carlemalm et al. 1982; Table 1). The resins have been designed to give the following useful characteristics:

- They have low freezing points.
- They have low viscosities, even at low temperatures (although viscosity increases in inverse proportion to temperature). Of the two best known resins which first became commercially available HM20 shows water-like viscosity even at −35°C, whereas K4M is more viscous.
- They can be polymerized at low temperatures by ultraviolet light. They can also be polymerized using chemical accelerators (like epoxy resins) at temperatures above 0°C.
- The polymerized blocks have very good sectioning properties.
- The sections are stable to the electron beam. They also have relatively low electron scattering properties which are important for some high resolution studies.
- The side chains can be modified, for example, to vary the hydrophobicity of the resins. Lowicryl K4M is relatively more hydrophilic and blocks can be polymerized in the presence of up to 5% water at temperatures down to −40° C. HM20, while more hydrophobic than K4M, can still accept up to 0.5% water in the blocks and can be polymerized down to −50°C. Two new resins,

the polar K11M and the less polar HM23, can be polymerized down to about −60 and −80°C respectively (Acetarin et al. 1987) but have not yet been extensively used. K4M and HM20 have been shown to be capable of preserving structural information at very high resolution (Armbruster et al. 1982; Garavito et al. 1982). K4M, because of its hydrophilicity, has now become the most widely used resin for immunocytochemistry (see p. 107).

A note of caution with the use of the Lowicryl resins is that a number of workers who have dealt with these compounds on a routine basis have developed serious skin allergies and related problems. In extreme cases some of these people have had to stop working with these resins altogether. It is essential to wear gloves (vinyl type or similar) and to take as much precaution as possible to avoid contact with either the liquids or the vapours. The use of a fume hood is recommended.

For both resins, the polymerization reaction is initiated by benzoyl methylether and long wave (360 nm) ultra-violet irradiation. Within the temperature range 0 to −40°C, the activator starts a free radical reaction uniformly when added (0.5–0.6%, w/w) to the liquid resin. The polymerization reaction proceeds without measurable temperature increase if the volume of the resin is small enough (0.5–1 ml). For higher temperature UV-polymerization, up to +30°C, benzoin ethylether (0.5% w/w) can be used as an initiator (Carlemalm et al. 1982). However, significant shrinkage can occur during this process. Volume decreases between 15 and 20% have been measured; in the case of crystals embedded in K4M or HM20, this resulted in a 5% linear shrinkage in the specimen (Carlemalm et al. 1982). For uniform polymerization with UV light it is important that the tissue be free of interfering pigments, especially yellow ones; osmium tetroxide in the concentrations normally used for electron microscopy (1%) will also interfere with the reaction (Carlemalm et al. 1982). Lower concentrations can be used, however (Villiger 1990). The important point appears to be that the tissue should only have a light brown colour after the osmium treatment. For freeze-substitution studies normal levels of osmium can be used (Humbel et al. 1983; Verkleij et al. 1985). It is apparently essential that all the monomers used in the Lowicryl resins are relatively pure (Carlemalm et al. 1982). Finally, the polymerization reactions are inhibited by oxygen and it has been recommended to de-gas the resin using mild vacuum or by passing nitrogen through the resin before starting the infiltration process (Carlemalm et al. 1982; Fryer and Wells 1983). As seen below, however, the use of tightly fitting tubes may obviate the need to take these special precautions.

Dehydration Prior to lowicryl Embedding. Aside from the universal problems of fixation, the most critical step with respect to preserving molecular structure, and therefore antigenicity, in Lowicryl embedding, is the dehydration step. For the theoretical reasons given above, this is best done at lowered temperatures. The potential of this step to denature proteins appears to be far greater than the infiltration and polymerization step in the resins themselves. This is evident from the studies of Carlemalm et al. (1982) in which two model protein crystals were dehydrated with different solvents and then infiltrated and embedded identically in K4M or HM20. The structural preservation of the embedded crystals at the different stages was directly assayed by X-ray crystallography (i.e. without sectioning). Under the best conditions, remarkably high resolution data were obtained, down to 6 Å. This is an important point to note. A surprising and perhaps worrying result, however, was that the two crystals chosen, aspartate aminotransferase (AAT) and catalase, were completely different in their response to different solvents. Ironically, the solvent that affected the structure of AAT the least, ethylene glycol (giving 6 Å resolution), completely destroyed the periodicity of the catalase crystals. Therefore, for immunocytochemistry, the choice of dehydration solvent could, at least in theory, be of critical importance for labelling. In practice, so far, this does not appear to be a major problem and in the laboratories where the Lowicryl resins are used routinely for immunocytochemistry, ethanol appears to have become the solvent of choice. In the crystal study mentioned, this solvent gave the best preservation of catalase (8–10 Å) and a very acceptable (12–14 Å) resolution of AAT. Recently, however, Bendayan et al. (1987) have recommended the use of methanol (see below). In cases where an antigen present in relatively high concentrations cannot be detected, it may be worthwhile to try different dehydration solvents.

Kellenberger (1987 and pers. commun.) has argued that the lower the temperature of dehydration, the better is the preservation of the "water shells" around antigens during the subsequent infiltration process (see also Humbel and Müller 1986).

Practical Procedures for Embedding in Lowicryl K4M. Following routine fixation in aldehyde, tissues are dehydrated, as for Epon embedding, in an ascending series of ethanol while the temperature is progressively lowered. This protocol is referred to as the PLT (progressive lowering of temperature). A typical dehydration schedule using ethanol and Lowicryl K4M is shown in Table 2. Following the two changes of ethanol the tissue is infiltrated with Lowicryl:ethanol (1:1) mixture (for 1 h) then by Lowicryl:ethanol (2:1) (for 1 h)

Table 2. A typical dehydration scheme for the 'PLT' technique, with ethanol as dehydration agent. (Carlemalm, pers. commun.)

	Temperature (°C)			Time (min)
	K4M	HM20	K11M + HM23	
30% ethanol	0°	0°	0°	30
50% ethanol	−20°	−20°	−20°	60
70% ethanol	−35°	−35°	−35°	60
95% ethanol	−35°	−50°	−50°	60
100% ethanol	−35°	−50°	−60°	60
100% ethanol	−35°	−50°	−60°	Overnight

PLT = progressive lowering of temperature. For more details see Roth 1989.

and finally into pure Lowicryl (two changes). The tissues are left to infiltrate for at least 12 h at −35°C. Normally, samples are kept at −35°C in the pure resin mixtures overnight and then transferred to a suitable container. Following the advice of the Basel groups (Roth et al. 1981; Carlemalm et al. 1982) the use of Beem gelatin (e.g. Lilly no. 1) capsules has been recommended. According to Humbel et al. (1983) and Humbel and Müller (1986), however, since the Lowicryl polymerization is inhibited by oxygen it makes sense to use a tightly sealed container. This avoids any unpolymerized resin at the end of the polymerization reaction. When open tubes are used there is often a zone of unpolymerized resin on the top of the tube; presumably this is due to oxygen in the air. Degassing the resin, or bubbling nitrogen through it before the infiltration has no apparent effect on this phenomenon according to Schwarz (pers. commun.) who recommends the use of the 0.7-ml plastic tubes from Sarstedt (no. 72.699) which are air-tight. Eppendorf tubes are not recommended since they cannot be sealed in an air-tight fashion. With these Sarstedt tubes Schwarz is able to completely polymerize volumes as small as 100 μl. Note that the absence of unpolymerized resin also means less health hazard. Sealed containers are also available for flat embedding with the Cambridge-Reichert CS Auto processor. Another reason for sealing the tubes is that, at least for IIM23, some evaporation of the unpolymerized resin can occur, even at −45°C; this can be clearly seen as a loss of volume of resin if open flat embedding moulds are used (Voorhout, pers. commun.).

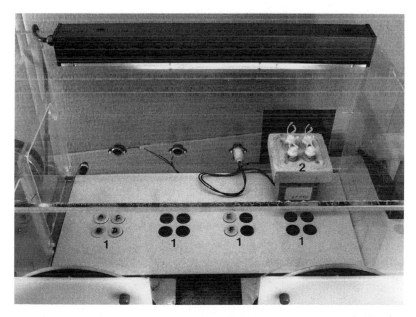

Fig. 4. Photograph of the sample cooling chamber designed by Dr. Werner Villiger for infiltration of Lowicryl resins. *1* Sample cooling chambers; *2* stirring apparatus (not mounted over cooling chambers). (Courtesy of Dr. Werner Villiger, Biocenter, Basel, Switzerland)

A home-made cooling plate, designed by Werner Villiger in Basel, is shown in Fig. 4 in order to give a clearer visual impression of how the actual transfer is made in practice.

When dimethyl formamide (DMF) is used with K4M the recommended procedure is to infiltrate the specimen first with two parts DMF and one part resin for 60 min at $-35\,^{\circ}$C followed by one part DMF to two parts resin for 60 min at the same temperature. The tissue is then placed in pure resin for 1 h and finally into a new change of resin overnight at $35\,^{\circ}$C (Carlemalm et al. 1982).

A few modifications to the standard Lowicryl K4M embedding protocol were recently recommended by Bendayan et al. (1987). First, methanol was used for dehydration as follows: after fixation the tissue pieces were placed sequentially in 30% methanol (5 min at $4\,^{\circ}$C), then 50% (5 min at $4\,^{\circ}$C), 70% (5 min at $-10\,^{\circ}$C) and then 90% (30 min at $-20\,^{\circ}$C). Infiltration is then performed at $-20\,^{\circ}$C in 90% methanol: Lowicryl (1:2) for 60 min followed by pure K4M (two changes) overnight. The polymerization is also made at $-20\,^{\circ}$C in a deep-freeze machine modified by addition of four ultraviolet lamps, positioned above and below the specimens (see below). When many samples need to be embedded

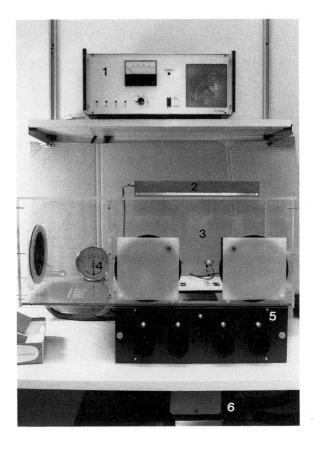

Fig. 5. Photograph of the prototype of the PLT-apparatus for embedding in Lowicryl resins. *1* Temperature control unit; *2* standard fluorescence tube; *3* dehydration- and embedding chamber; *4* hygrometer; *5* Peltier cooling device; *6* pre-cooling device for Peltier elements. (Courtesy of Dr. Werner Villiger, Biocenter, Basel, Switzerland)

Fig. 6. Photograph of the UV-polymerization chamber for Lowicryl resins designed by Dr. Werner Villiger. *1* UV-lamps, 360 nm (behind UV-shield); *2* UV-shield; *3* UV-chamber (door removed); *4* gelatine or BEEM capsule holder; *5* temperature sensor; *6* embedding medium storage container (dark brown); *7* sample rotator mounted in freezer with minimal temperature of −50°C. (Courtesy of Dr. Werner Villiger, Biocenter, Basel, Switzerland)

simultaneously in Lowicryl a modified specimen carrier has recently been recommended by Templeman and Wira (1986).

For more technical details of embedding in Lowicryl resins see Carlemalm and Villiger (1989), Hobot (1989) and Villiger (1990).

Polymerization and Sectioning. The capsules containing tissue and 100% resin must be polymerized with indirect UV irradiation. For this a deep-freeze machine capable of cooling down to at least −35°C fitted with a UV lamp emitting 360 nm light must either be constructed or bought commercially (Balzers Union or Cambridge Instruments). Again, in order to help the reader to get a better feel for how the polymerization is carried out, the prototype machine designed by Villiger is shown in Fig. 5. When the UV source hits the resin directly, the polymerization reaction proceeds too rapidly and non-uniformly. Hence, aluminium foil is conveniently put around the container which holds the tubes or capsules in such a way that it reflects the light uniformly onto the specimen (Fig. 6). Since the polymerization reaction with the Lowicryl resins is strongly exothermic the use of a liquid cooling bath as "heat sink" is highly recommended (Humbel et al. 1983; Humbel and Müller 1986). As pointed out by Ashford et al. (1986), simply air-cooling the polymerization tubes in the freezer does not appear to be sufficient to prevent the significant temperature increases during polymerization. These authors, in fact, proposed that this phenomenon was responsible for the difficulties encountered in some laboratories with respect to reproducibility of polymerization . Note that the use of open tubes may also have been a factor, as discussed above. Ashford et al. (1986) recommended the use of an insulated, stainless steel bath containing methanol or 70% ethylene glycol (see Fig. 1 of this reference). According to Schwarz (pers. commun.), a simple stainless steel tray containing ethanol will also suffice; enough solvent is put into the bath so that it covers the level of resin in the specimen tubes. It should be noted, however, that two commercial machines are now available for automatic tissue processing which appear to work well in practice (from Cambridge-Reichert and Balzers). In practice, following 24 h of polymerization at low temperatures, the blocks are further hardened for 2–3 days at room temperature in order to facilitate sectioning. A procedure recommended by Schwarz (pers. commun.) is to put the open tubes outside in daylight for the final hardening of the blocks. On bright sunny days, a few hours will suffice; on cloudy days 1–2 days are required. After this process K4M blocks are yellowish in colour whereas HM20 blocks are reddish. Upon storage away from sunlight the latter become more transparent while the K4M blocks remain yellowish.

Well-embedded blocks are easily sectioned on a normal ultra-microtome using either glass or diamond knives fitted with a water trough. Since the K4M resin is hydrophilic, it may be easier to section when the average level of water in the knife boat is kept slightly below the knife level. Care should be taken to avoid getting water on the block face and the blocks should be stored under dry conditions, preferably in a desiccator since they may absorb water during storage (Roth 1989). Following the "daylight" polymerization procedure of Schwarz, however, even K4M blocks can apparently be stored without precaution; perhaps the natural UV irradiation causes a more complete polymerization.

There appears to be a consensus emerging that HM20 resins are easier to section than K4M. Further, the quality of the HM20 sections are often better than those of K4M. Sections of the latter tend to expand greatly on the surface of water, an effect that is especially pronounced when the sections are very thin (Schwarz and Humbel 1989). Thus, HM20 is also suitable for the use of large, histological-type sections. The sectioning of these resins offers three clear advantages when compared to the use of cryo-sections. First, serial sectioning is possible (see Fig. 10). Second, the blocks can be stored and transported at room temperature. Third, the sections can be stored dry (at least for 1 year) on grids before the immunolabelling reactions; they are simply allowed to wet on the surface of PBS for 10–15 min before the labelling reactions (Schwarz, Voorhout, pers. commun.). The grids may also be briefly examined in the EM before the labelling reactions or after labelling before staining.

4.2.2.2 Very Low Temperature Lowicryl Resins

As mentioned, two new Lowicryl resins which can be polymerized at significantly lower temperatures than K4M and HM20 have been developed by the Basel group (Acetarin et al. 1986). While the hydrophilic K11M can be polymerized down to $-60\,°C$, the hydrophobic HM23 will do so down to $-80\,°C$. The crucial factor for these resins is their very low viscosity, enabling infiltration at these very low temperatures.

There are two major reasons why these modifications are potentially very important. First, the adverse effects of dehydration on structural preservation should be significantly reduced at these low temperatures (Hobot et al. 1984; Kellenberger et al. 1986b; Kellenberger 1987). From a theoretical point of view, it has been argued that despite the presence of the chemical solvent the water of hydration around macromolecules may be completely intact at these low temperatures (Kellenberger 1987). Further lipid extraction should become negligible at these temperatures (Weibull et al. 1983a, b). Clearly, these phenomena are desirable for both high resolution structural preservation and for immunolabelling. Although the published micrographs look excellent (and unpubl. data support the theory; M. Müller, pers. commun.), it is too early to say whether all the theoretical predictions will also become evident in practice.

The second, and perhaps more important aspects of these new resins is that they are especially suited for the approach of rapid cooling and freeze-substitutions (see below).

Sectioning – The Surface of Lowicryl Sections. Since sectioning of the Lowicryl resins is no different to sectioning conventional plastic resins practical details need not be discussed here. Lowicryl sections have often given surprisingly high levels of immunolabelling when one considers that, unlike unembedded cryo-sections, the presence of the dense resin would be expected to prevent **any** penetration of label into the depth of the section. Indeed, by reembedding and sectioning labelled Lowicryl resins, a number of studies have shown that immunogold particles are restricted to the upper surface of the sections (Bendayan et al. 1987; Stierhof et al. 1987; Stierhof and Schwarz 1991). In a comparative quantitative study of Lowicryl and cryo-sections for a defined antigen, we previously concluded that the labelling of Lowicryl resin was too high to be explained by a surface reaction only (Griffiths and Hoppeler 1986; see Griffiths et al. 1989 for a correction of the latter paper). Recently, Kellenberger et al. (1986a,b) and Kellenberger (1987) (and pers. commun.s) has proposed that the "water shells" around antigens are removed when the temperature is raised during the polymerization. In the methacrylates there is less tendency for the resin to form covalent linkages with antigens (Causton 1986). The result, he argues, is that after complete hardening of the block, a gap, free of resin, is left where the hydration shells had been. Subsequently, during sectioning the knife would follow the surface of least resistance and would tend to cleave (as in freeze-fracture) over the surface of antigens such as proteins. The latter would thus be much more exposed on the surface of the section than would be expected if they were embedded (i.e. cross-linked to the resin). By metal shadowing the surface of sections of Lowicryl resins, this group has, in fact, shown that these sections show a surface relief of 2–6 nm. In contrast, the surface of epon sections show a relief that is two or three times smaller than that of Lowicryl sections

(Fig. 14A,B, Kellenberger et al. 1987; Stierhof and Schwarz 1991). Kellenberger (pers. commun.) has also suggested that the hydrophilicity of the embedding medium may actually be less important for the immunolabelling; rather, the critical parameter may be the degree of covalent interaction (cross-linking) between the antigen and the embedding medium. The highly reactive epoxy resins would form tight interactions with antigens. Thus, during cutting, the plane of section would be just as likely to go through a molecule as over its surface. This theory is attractive because it can also explain the recent observations from many laboratories that the more hydrophobic HM20 gives very similar amounts of labelling as does K4M. The reactions of both classes of resin to, say, proteins would be expected to be similar (and quite different to epon resins). A second possibility to explain the higher labelling efficiencies in Lowicryl (especially K4M) sections in some cases is that a cleavage process (akin to that in freeze fracture) may occur perpendicularly to the section surface (Kellenberger et al. 1987). This would also effectively increase accessibility of antigens in the depth of the section.

Assuming this cleavage hypothesis is the correct interpretation of the data, Kellenberger et al. (1987) further speculated that the degree of antigen cross-linking by fixatives may modify the way in which the antigen is cleaved. If true, this would be yet another complex aspect of the effect of fixatives on antigens that we would need to consider in the future.

4.2.2.3 Rapid Embedding Method for Lowicryl

In cases where the low temperature dehydration and polymerization may not be essential for preserving significant antigenicity, advantage may be taken of a recent modification introduced by Altman et al. (1984). Here, the whole procedure from fixation to sectioning can be reduced from 4–6 days to 4 h by doing all the steps except the polymerization at room temperature and the polymerization itself at 4°. The latter is apparently completed in 45 min or less at this temperature. A 15 W UV lamp is brought relatively close (10 cm) to the sample. These authors empirically found that dimethylformamide was the optimal dehydrating solvent. In this study, quantitative analysis of labelling density of kidney and retina with anti-ATPase and anti-opsin, antibodies respectively, indicated that there was no loss of labelling when compared to the same tissues embedded in Lowicryl in the conventional way at low temperatures. The sensitivity of these antigens obviously differs from that of tubulin in the study by Armbruster et al. (1983b). A modification of this method for brain tissue has been published by Valentino et al. (1985). Another rapid embedding method for Lowicryl has recently been described by Simon et al. (1987).

4.2.2.4 Immunolabelling Using Lowicryl sections

Immunolabelling of Lowicryl sections does not differ essentially from labelling for any other sectioning technique (see p. 173). Until recently, the K4M resin was the most widely used resin for routine labelling studies. For examples of the use of this technique in practice, the reader is referred to the papers of Roth et al. (1981); Roth and Berger (1982); Armbruster et al. (1982; 1983a); Thorens et al. (1982); Bendayan 1983; Orci et al. (1984); Ernst et al. (1986); Freudl et al. (1986); Lethias et al. (1987); van Tuinen and Riezman (1987); Bendayan and Ørstavik (1987); Desjardins and Bendayan (1989); Hobot (1989); and Roth (1989). Recently, however, the use of the more hydrophobic HM20 has become more popular; for many antigens the labelling with this resin has been similar (if not higher – see, for example, Bendayan et al. 1987) to that seen on K4M sections (Engfeldt et al.1986; Schwarz and Humbel 1989; Voorhout et al. 1989a, b; see below). Schwarz and Humbel (1989, and pers. commun.) have suggested that an important reason for the lack of popularity of HM20 for immunolabelling studies was that it tends to give higher background labelling than K4M. With suitable blocking reagents, however, this problem can usually be avoided (see Chap. 7) and the superior sectioning qualities of this resin can be taken advantage of (see also Dürrenberger 1989).

The study of Armbruster et al. (1983b), showed that low temperature dehydration and embedding gave significantly higher labelling of tubulin in microtubules when compared to room temperature embedding. This effect cannot be generalized, however since Bendayan et al. (1987) failed to observe any difference between temperatures around $-20\,^\circ$C and those above $0\,^\circ$C.

The only published data available concerning the effect of different dehydration solvents with the PLT (progressive lowering of temperature) approach on labelling are the study of Roth et al. (1981), where a slight reduction in labelling density over some intracellular compartments was seen using either dimethylformamide or ethylene glycol instead of methanol. Bendayan et al. (1987) recently found that, for some antigens, the intensity of immunolabelling was increased by adding up to 10% (vol/vol) of water to the polymerization mixture. In many studies in the literature there has been a tendency to incubate Lowicryl sections with antibodies and/or gold conjugates for relatively long periods often overnight at $4\,^\circ$C. This may be more a matter of convenience, however, and laboratories such as the Villiger-Kellenberger group in Basel routinely use incubation times of 1 h (pers. commun.). Similarly, Schwarz and Stierhof use times between 30 and 60 min (pers. commun.). It is not widely appreciated that the Lowicryl resins can also be used for routine light microscopic immunolabelling studies such as immunofluoresence. As with cryo-sections (see Chap. 5) even the same section can be used for both light and electron microscopic localizations. Both K4M and HM20 can be used but the latter is preferred because of its sectioning qualities (Humbel and Schwarz 1989; Schwarz and Humbel 1989; Schwarz pers. commun.; Voorhout et al. 1989a, b).

The groups of Schwarz, Humbel and Stierhof have made extensive (often unpublished) comparisons between K4M and HM20 (see Humbel and Schwarz

1989; Schwarz and Humbel 1989). They have also compared these resins with cryo-sections. Further, for the Lowicryl resins they have looked in detail at the effects of temperature during dehydration in that they have compared the PLT approach with freeze-substitution (see below). The latter offers the only possibility for avoiding aldehyde fixation, and the effects of this parameter can also be looked at (Humbel et al. 1983; Humbel and Müller 1986; Engfeldt et al. 1986). A striking message from the work of this group is that, if one approach fails with respect to immunolabelling it makes sense to try other approaches. I can best demonstrate this by quoting a few unpublished examples from this group. Chicken erythroblasts, for example, were either fixed in a mixture of glutaraldehyde (0.05%) and formaldehyde (2%) or left unfixed. They were then either prepared for cryo-sectioning (in the case of the fixed preparations) or prepared for Lowicryl K4M or HM20, either following rapid freezing and freeze-substitution or using the PLT procedure. With respect to immunolabelling of haemoglobin, a major antigen in these cells, both freeze-substitution of unfixed material in K4M and cryo-sections after fixation gave an equivalent strong signal. The signal with HM20 (unfixed, freeze-substitution), however, was about ten times lower. For histone H5, however, all three protocols gave a qualitatively similar signal. With respect to the Lowicryls, these results were only true when freeze-substitutions were done without fixation. When the cells were fixed in aldehyde and embedded, following either freeze-substitution or the PLT procedure, the immunolabelling for both antigens was very poor. The significant increase in labelling in the fixed cryo-sections as compared to the fixed, Lowicryl-embedded sections could not be explained by an increase in penetration of labelling into the section interior in the former, since the gold label was essentially restricted to the cryosection surface in these examples (Stierhof and Schwarz 1991). Thus, the effect of the fixation here must have been different with respect to the two protocols.

Of the many antigens that this group has tested, the immunocytochemical signal is usually, but not always, slightly higher with the cryo-sections than with either of the Lowicryls. They have a number of examples (as above), with identical preparation procedures where K4M has been significantly better than HM20 as well as a few where the reverse was true. It is to be expected that, for smaller antigens which may be difficult to effectively cross-link, the resin approach may give better immunolabelling since the resin itself can also stabilize the antigen and prevent it from leaching out during the labelling and rinsing step. In cryo-sections **only** the aldehyde cross-links can serve this purpose.

Examples of immunolabelling studies using Lowicryl K4M sections using one antibody are shown in Figs. 7–9 while Fig. 10 shows the labelling of two consecutive serial sections with two different antibodies. In Figs. 11 and 12 examples of HM20 labelling are shown.

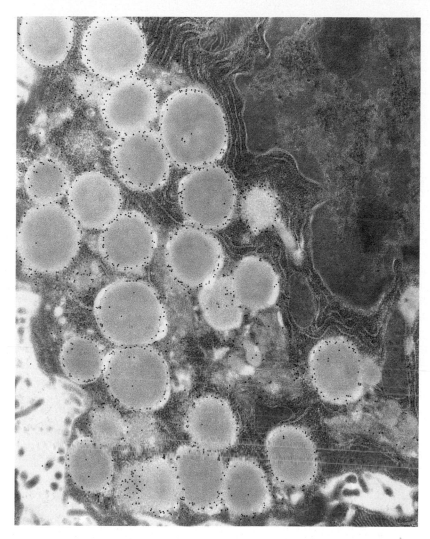

Fig. 7. Example of labelling of Lowicryl K4M section. Normal adult Balb-C mouse parotid gland fixed by perfusion with Karnovsky's full-strength fixative at room temperature, dehydrated with ethanol at 0 and at -20°C, and embedded in Lowicryl K4M at -20°C. Specimens were polymerized with UV light for 16 h at -20°C and for 36 h at room temperature. Ultra-thin sections were mounted on Collodion-coated nickel grids and incubated with rabbit anti-mouse proline-rich protein (IgG serum fraction containing polyclonal PRP antibodies) preabsorbed with human salivary α amylase for 4 h at room temperature followed by protein A gold (15 nm) for 1 1/2 h at room temperature. The IgG fraction and protein A were diluted with phosphate buffered saline containing 0.1% ovalbumin, 0.05 M glycine and 0.15% Tween 20. Grids were jet washed with this same solution and with distilled water and contrasted with aqueous uranyl acetate and lead citrate before examination. The micrograph clearly shows the localization of proline-rich proteins at the periphery of the secretion granules. Sections incubated with preabsorbed IgG fraction from the same rabbit taken prior to immunisation were unlabelled. $\times$21000 (larger). (Courtesy of Dr. Geoffrey Cope, Department of Biomedical Science, University of Sheffield, England)

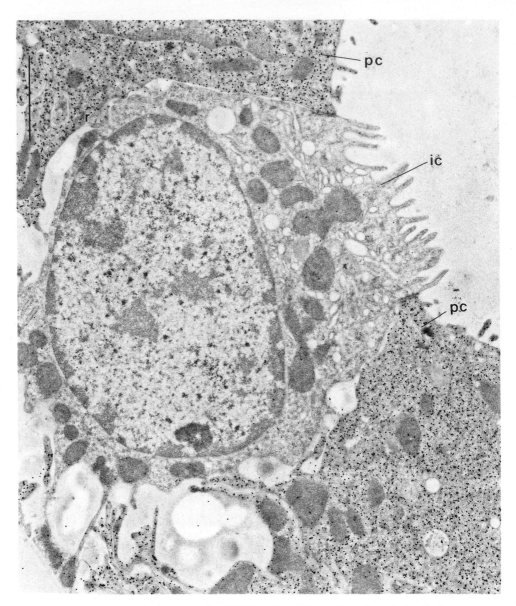

Fig. 8. Example of labelling of Lowicryl K4M section. Thin section of cortical collecting duct from rat kidney. For immunolabelling, the kidney tissue was fixed by perfusion with 1% glutaraldehyde in phosphate buffered saline for 10 min, followed by immersion fixation at room temperature for a further 2 h. Tissue was then dehydrated in ethanol at progressively lower temperatures and finally embedded overnight in 100% Lowicryl K4M at −35°C. Polymerization was carried out by irradiation of tissue in gelatine capsules by indirect UV light, in a chest freezer for 24 h at −35°C, followed by final hardening under UV light at room temperature for 1–2 days. Thin sections were cut and were first incubated for 2 h with a specific rabbit polyclonal antibody against vitamin D-dependent protein (28 kDa − calbindin), followed by a 1-h incubation with protein A-gold complexed diluted in PBS containing 1% bovine serum albumin. Very few gold

4.2.2.5 Contrasting of Lowicryl Resins

In general, the same approaches have been used for contrasting Lowicryl resins as are used for routine epoxy resin sections. Thus, the most important contrasting agents are uranyl and lead salts. It is important to use the uranyl salt before the lead since the contrast is far better than when the staining order is reversed (Schwarz, pers. commun.). Lead acetate is often preferred to lead citrate, since the latter tends to give undesirable granularity on the section. The procedure recommended by Roth (1989) is to stain the sections for 4–5 min on 3% aqueous uranyl acetate followed by a water jet rinse. The grids are dried on filter paper and then floated for 45 s on Millonig's lead acetate solution (Millonig 1961). HM20 sections are generally easier to contrast than K4M. Thus, Schwarz and Humbel (1989 and pers. commun.) recommend 2–3 min uranyl acetate for HM20 and 5–10 min for K4M.

An alternative approach is to use a similar methyl cellulose-uranyl acetate mixture to that used for cryo sections (see Chap. 5), following the ideas of Tokuyasu. Roth (1989) recommends a mixture of 1.8% uranyl acetate, 0.2% (w/v) methyl cellulose (25 centipoise) made from a 2% stock solution of the latter. As for cryo sections there is certainly room in future for novel contrasting procedures to be tried with the Lowicryl resins. The recently published micrographs with Lowicryl K4M sections embedded in methyl cellulose, uranyl acetate (Roth et al. 1990) are first-rate and reminiscent of the PVA-embedded cryo sections (Tokuyasu 1989). This approach was not, however, successful with HM20 sections (Voorhout, pers. commun.).

A recent paper by Bénichou et al. (1990) reports that the use of uranyl acetate treatment (0.5% for 30 min) of bacteria, after fixation and before dehydration and embedding in Lowicryl improved the sectioning qualities of the blocks. In the absence of this treatment, the bacteria separated from the resin during sectioning. This addition of uranyl acetate had no apparent effect on the degree of immunolabelling for a number of antigens. Significantly, these authors also found that the use of Lowicryl HM20 was much better than K4M for the sectioning properties/preservation of the bacteria

4.2.2.6 LR White

This polyhydroxy-aromatic acrylic resin, originating from Causton (1984) is commercially available from the London Resin (LR) company. It is a low-viscosity medium, whose viscosity is slightly higher than Lowicryl K4M, which

---

particles are present in the cytosol of the intercalated cell (ic), whereas the cytosol of two adjacent principal cells (pc) is heavily labelled with gold particles. The high resolution of this technique is shown by the paucity of labelling over mitochondria, and by the labelling of slender cellular processes on the basolateral side of the positive cells. *Bar* 2 μm. (Courtesy of Dr. Dennis Brown, Renal Unit, Massachusetts General Hospital, Boston)

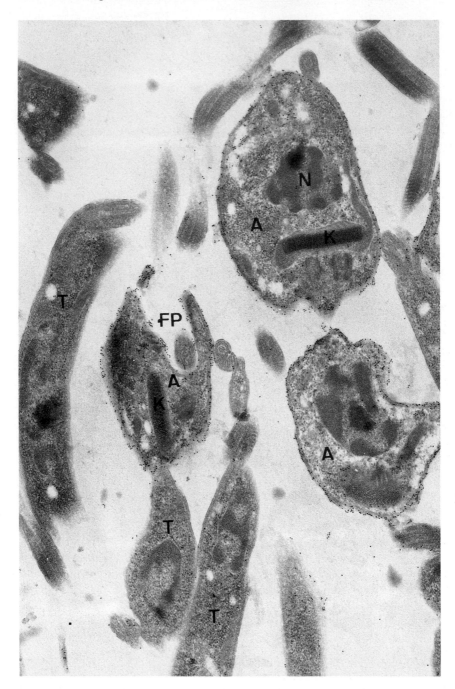

Fig. 9

requires tissue dehydration with ethanol prior to polymerization. This medium can be polymerized in one of three ways. First, by heat (50°C), second by ultraviolet light and third, chemically, by the use of an aromatic tertiary amine accelerator. This polymerization must be done under anaerobic conditions. The resin appears to section as well as epoxy resins. It has also been used as an alternative to methyl cellulose as a protectant for cryo-sections (Keller et al. 1984). Although LR White has been less extensively used than the Lowicryl resins it has been used in a number of successful immunocytochemical studies (e.g. Newman and Jasani 1984; see refs. below):

The monomer is not completely insoluble in water since up to 12% water is miscible with it (Newman 1987). However, when polymerized, it is hydrophilic and swells on the water trough after sectioning. The LR white blocks appear to be more hydrophobic when compared to Lowicryl K4M; the blocks appear less able to take up water from the atmosphere and the sections are less likely to sink in the water trough (W. Villiger, pers. commun.). Although most embedding with LR white has been done at room temperature or by heat, it is possible to infiltrate with it at temperatures down to −20°C (Newman 1987).

As with the Lowicryl resins, it is generally recommended to avoid the use of high concentrations of osmium tetroxide prior to ultraviolet light polymerization. The simplest recommendation would be to avoid osmium altogether. Similarly, pigmented tissues which absorb UV may also affect the polymerization. A recent publication from Moore and Staehelin (1988), however, shows that for some plant tissues the use of 1% OsO4 for 1 h prior to dehydration and embedding in LR white was still compatible with both embedding fine structural appearance and the immunoreactivity of cell wall matrix polysaccharides. Newman et al. (1983) have recommended going directly from the 70% ethanol stage to a mixture of 70% ethanol and the pure monomer, that is, to avoid higher concentrations of ethanol. According to a recent report by Wright

---

Fig. 9. Thin section of amastigote and trypomastigote forms of *Trypanosoma cruzi* incubated in the presence of anti-cysteine proteinase and subsequently in the presence of gold-labelled antibodies. The surface of amastigotes (*A*) is heavily labelled, whereas very few particles are seen on the surface of the trypomastigotes (*T*). Labelling of vesicles located close to the flagellar pocket (*FP*) is evident. ×21000. The parasites were collected by centrifugation (2000 g for 10 min at 4°C), washed twice with phosphate-buffered saline (PBS) and fixed for 60 min in a solution containing 0.1% glutaraldehyde, 2% formaldehyde (freshly prepared from paraformaldehyde) in 0.1 M phosphate buffer, pH 7.2. After fixation the parasites were washed twice with PBS, dehydrated in methanol and embedded in Lowicryl K4M at −20°C (Bendayan et al. 1987). Thin sections were collected on 300-mesh nickel grids, incubated subsequently for 60 min at room temperature in a PBS solution, pH 8.0, containing 5% non-fat milk and 0.01% Tween 20 (PBS-MT), and then in the same solution containing the polyclonal antibody. After incubation the grids were washed three times in PBS containing 1% bovine serum albumin and 0.01% Tween 20, pH 8.0, and incubated for 60 min at room temperature in PBS-MT solution containing gold-labelled goat anti-rabbit IgG (E-Y Laboratories, USA) diluted 1:20. Some grids were incubated only in the presence of the gold-labelled antibody. After incubation the grids were washed with PBS and distilled water, stained with uranyl acetate and lead citrate and observed in a JEOL 100 CX or Zeiss 902 transmission microscope. (Courtesy of Drs. Thais Souto-Padrón and Wanderley de Sousa, Instituo de Biofisica Carlos Chagas Filho, Centro de Cicias da Saude, University of Rio de Janeiro)

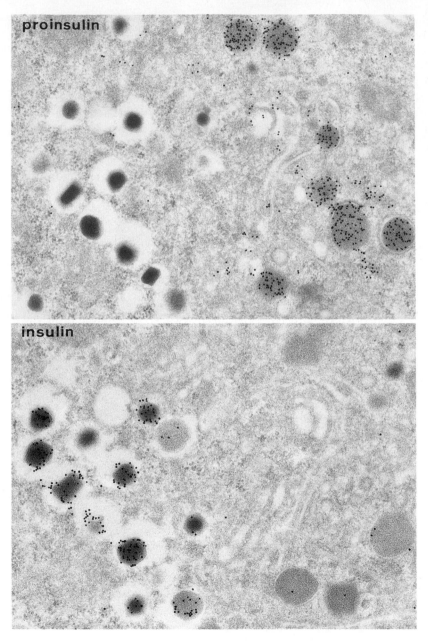

Fig. 10. Serial-section immunolabelling of Lowicryl K4M sections. Islet of Langerhans isolated from the rat pancreas by collagenase digestion, were fixed with 1% glutaraldehyde in 0.1 M sodium phosphate buffer, pH 7.4 and embedded at low temperature in Lowicryl K4M resin. Thin sectioning was carried out on a LKB ultramicrotome fitted with a glass knife (which produced sections of better quality than a diamond knife). The floating sections were picked on 50 mesh nickel grids. Two consecutive thin serial sections were first floated for 10 min on a drop of 0.5% egg albumin in PBS; one section was transferred for 2 h on a mouse monoclonal antibody to

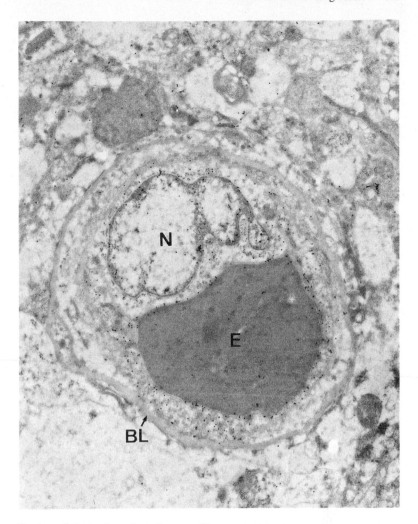

Fig. 11. Localization of the brain endothelium-specific protein HT7 in a cerebral capillary in chick telencephalon. Tissue was fixed with 4% formaldehyde, 0.05% glutaraldehyde for 16 h at 4°C, dehydrated at progressively lower temperatures in ethanol and embedded in Lowicryl HM20 at −35°C. Rabbit anti-HT7 antibodies were detected with protein A-15 nm gold. The HT7 expression correlates strongly with blood-brain barrier function. *N* nucleus of endothelial cell; *E* enthrocyte filling the entire lumen of the capillary; *BL* basal lamina. ×20000. For more detail see Albrecht et al. 1990. (Courtesy of Dr. Heinz Schwarz, Max-Planck Institut für Entwicklungsbiologie, Tübingen, F.R. Germany)

---

insulin (Mab 3 diluted 1:100 in PBS), the other to a proinsulin antibody (GS-4G9 diluted 1:5000). The sections were washed with PBS and sequentially exposed to rabbit antimouse IgG (1:200) for 1 h, then to protein A-Gold (15 nm gold particles) for another 1 h (with PBS rinses after each antibody step. The grids were sequentially stained with uranyl acetate (7 min) and lead citrate (1 min) with distilled water rinses in between. ×34000. (Courtesy of Dr. Lelio Orci, Department of Histology, University of Geneva, Switzerland). For more details see Cell, 49, 865, 1987

Fig. 12. Section of *E. coli* bacteria over-expressing the membrane protein OmpA that was fixed in 2% formaldehyde, 0.05% glutaraldehyde block stained in 1% uranyl acetate in water, dehydrated and embedded in Lowicryl HM20 by the PLT method. The sections were labelled with antibodies against OmpA followed by 15 nm gold-protein A. The section was unidirectionally shadowed (at an angle of 15° *in the direction of the arrow*) with carbon-platinum. While most of the gold particles have been shadowed, many (indicated by *arrowheads*) have not, and must have been on the under-surface of the section. Presumably they must have passed through breaks in the section. The *asterisks* indicate ruptured areas in the section (sectioning artefacts) which enable the surface of the support film to be shadowed (this can give an indication of the section thickness; see Fig. 14B for comparison). ×30000. (Courtesy of Dr. Heinz Schwarz, Max-Planck Institut für Entwicklungsbiologie, Tübingen, F.R. Germany). For more details see Schwarz and Humbel (1989)

and Rine (1989), it is apparently important that the temperature is regulated very precisely (2°C) during the polymerization. Although the manufacturers recommend a temperature of 60°C, these authors find that under this condition considerable shrinkage ensues. They recommend, instead, that the initial polymerization should be done at 45°C overnight followed by 50°C for a further 24 h.

As for Lowicryl sections immunogold labelling is restricted to the surface of LR white sections (Newman and Hobot 1987). When horseradish peroxidase was used as a marker, however, the DAB reaction product was able to penetrate significantly into the section (Ellinger and Pavelka 1985; Newman and Hobot 1987). The authors argued that this phenomenon may be the result of some swelling of the hydrophilic plastic resin on aqueous solutions. According to Posthuma and Slot (pers. commun.), this resin was not suited for their quantitative approach since labelling efficiencies were not the same from one block to the next. Specifically, there was significantly more labelling of their model antigen amylase when the latter was mixed in a matrix of 5% gelatin, as opposed to 10% gelatin, before embedding in LR white (for more details of this approach see Chap. 11). In this respect, LR white was similar to K4M (Posthuma et al. 1987) and to LR gold (see below, Posthuma and Slot, unpubl. data).

According to Villiger (pers. commun.), the LR resins are far less toxic than are the Lowicryl resins, presumably because the monomers are longer chains than are the Lowicryls.

For additional examples of EM immunolabelling studies using LR white see Graber and Kreuzberg (1985) Ellinger and Pavelka (1985); Abrahamson (1986); Wynford-Thomas et al.; 1986 Ring and Johanson (1987); White et al. (1988), and Mutasa and Pearson (1988). In the latter reference lactoferrin was successfully localized even after osmium tetroxide-treatment before infiltration and heat-induced polymerization. Also significant was the study by Bendayan et al. (1987), who compared in a quantitative study most of the commonly used resins with respect to the immunolabelling of a model antigen amylase. The LR white resin, along with the epoxy resins, gave the lowest levels of labelling, with K4M and LR gold (see below) giving slightly higher values. Significantly, in this example, HM20 gave about twice the levels of labelling seen for all the other resins and this labelling was even slightly higher than that estimated in cryo sections. An example of LR white labelling of plant tissue is shown in Fig. 13.

4.2.2.7 LR GOLD

The London Resin Company has also introduced a resin analogous to Lowicryl K4M called LR gold (Causton 1984). This is a hydrophilic aromatic acrylic resin which can be cross-linked at low temperatures by blue light. For a recent reference on the use of this resin for EM immunocytochemistry see Trahair et al. (1989).

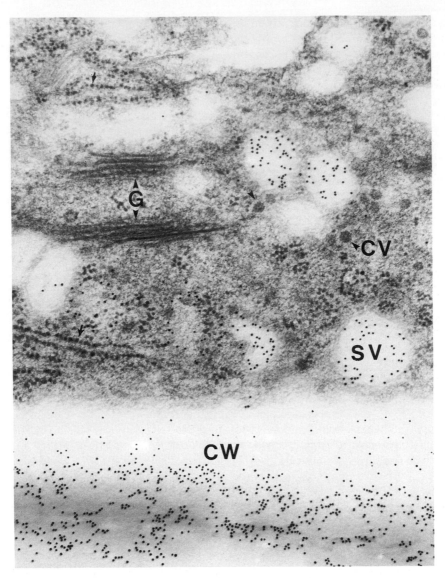

Fig. 13. Immunolabelling of plant tissue using LR white resin. Cloned root tip epidermal cells were labelled with anti-polygalacturonic acid/rhamnogalacturonan I antibodies. The terminal 3–5 mm of the root tip was fixed in 2.5% glutaraldehyde in 10 mM sodium phosphate, pH 7.2 for 2 h. After washing in buffer, tissue was postfixed for 1 h in 2% osmium tetroxide in buffer, washed again in buffer, and dehydrated through an ethanol series. All the above steps were done at room temperature. Tissue was infiltrated at 4°C with gentle agitation in 1:2 (v/v) LR white resin:ethanol for 2 h, 2:1 (v/v) LR white:ethanol for 2 h and 100% LR white for 24 h. Samples were polymerized in a 60°C vacuum oven for 20 h. Ultra-thin sections were blocked in 5% low-fat milk in PBST (500 mM NaCl, 0.1% Tween-20, 0.02% sodium azide, pH 7.2) for 30 min, immunolabelled in a 1:8 dilution of primary antiserum for 2 h, washed in PBST + 0.5% Tween-20, incubated in protein A-colloidal gold (10 nm) for 30 min, washed in PBST + 0.5% Tween-20, washed in distilled H_2O, counterstained in 2% aqueous uranyl acetate for 5 min and Reynolds

Also noteworthy is the immunocytochemical study by Berryman and Rodewald (1990) using this resin. The morphological preservation, and especially the contrast of the tissue studied, the jejunal epithelium of the neonatal rat, was far superior to previously published micrographs using LR resins and it seems pertinent to mention some details. The initial fixation, a mixture of glutaraldehyde, formaldehyde and picric acid, was followed by a post-fixation in uranyl acetate. The latter, plus the subsequent dehydration in acetone should, the authors argued, help to preserve phospholipids (see p. 47 Chap. 3). As for the Lowicryl resins, the temperature was decreased during dehydration and the infiltration and UV polymerization was made at $-20\,^{\circ}$C. Significantly, these authors found that, following the uranyl acetate step, it was necessary to use a Tris rather than a phosphate buffer in order to avoid precipitates.

Following the immunogold labelling the sections were fixed with 2% glutaraldehyde and contrasted for 45 min with 2% osmium tetroxide, followed by 3 min with Reynold's lead citrate tetroxide, a procedure which gave excellent contrast. Finally, although the authors found the blocks easy to section and handle, they noted that the sections were often unstable under the electron beam. To overcome this problem they coated their sections with Formvar as a final step in the preparation. For a quantitative comparison of labelling of MDCK cells with Lowicryl HM20, LR gold and cryosections see Table 3 in Chapter 9.

4.2.3 Epoxy Resins

I have already referred to the difficulties associated with the use of epoxy sections for immunolabelling. Since many workers are still routinely using these sections for this purpose, I must expand on this criticism.

Even in the absence of osmium, dehydration at temperatures above $0\,^{\circ}$C must be expected to have severe effects on protein conformation (although aldehyde cross-linking will reduce the effect). It is well documented that many lipids, as well as proteins, will be extracted. The infiltration and heat-induced polymerization steps provide a further possibilities for the antigenicity to be affected. Many proteins do survive these treatments and many papers have been presented showing significant labelling, especially for proteins which are in very high concentrations, such as in secretory granules (see, for example, Roth et al.

---

lead stain for 15 s. All incubations were done at room temperature. Sections were viewed with a Philips CM 10 electron microscope.

The micrograph illustrates abundant labelling of the Golgi complex (*G*) as well as Golgi associated vesicles and the cell wall, especially the outer layer, which is in direct contact with the external environment. For more biological background see Swords and Staehelin (1989). *CV* coated vesicles; *SV* secretory vesicles. (Courtesy of Drs. Margaret Lynch and Andrew Staehelin, Dept. Molecular, Cellular and Developmental Biology, University of Colorado at Boulder, USA)

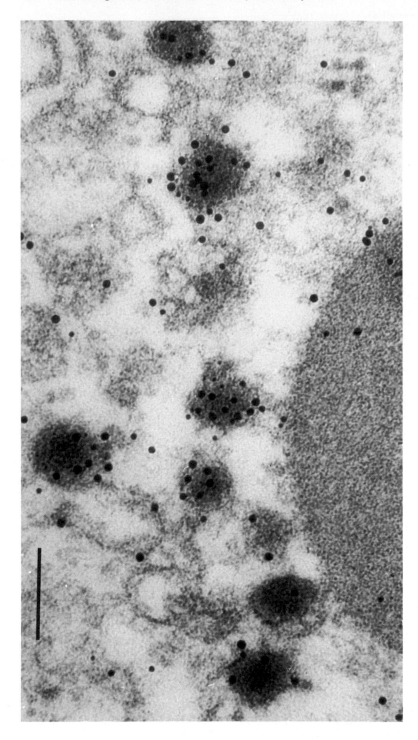

Fig. 14

1978; Bendayan et al. 1980). Often it is necessary to use "etching" methods which physically dissolve parts of the plastic and have the effect of making the plastic more hydrophilic (see Baskin et al. 1979; Erlandsen et al. 1979; Causton 1984 for a review). Less often appreciated is the ability of epoxide groups to react directly with peptide groups in proteins (Causton 1984, 1986). This may both reduce antigenicity and be the cause of the higher background labelling which is inevitably seen with Epon sections (Roth et al. 1981).

With respect to both fine-structure preservation especially high resolution information, and to the signal to noise ratio of immunolabelling, the classical Epon-embedding technique using osmium tetroxide post-fixation cannot be the **general** method of choice for immunocytochemistry. Nevertheless, in the past few years some notable exceptions must be mentioned, especially the study by Bendayan and Zollinger (1983), who introduced the use of sodium metaperiodate which somehow "unmasks" the effects of osmium tetroxide. In this paper a number of antigens were localized on sections that showed "classical" morphology. Other groups have also obtained impressive results with this approach, usually for antigens in high concentrations (see Hearn et al. 1985, Viale et al. 1985, Yokota et al. 1985, 1986 for examples with animal tissues, and Northcote et al. 1989 for an example using plants). The study by Viale et al. (1985) is also an exception to the general statement I made earlier (pp. 4, 5) against the use of horseradish peroxidase as a post-embedding label. In this case, acceptable qualitative results were obtained with an avidin-biotin-peroxidase method. Examples of immunolabelling of epon sections are shown in Figs. 14 and 15A, B. In the examples shown the labelling was comparable to the levels obtained with Lowicryl K4M sections.

The most significant results I have seen using epoxy resin sections comes from the immunolabelling of small molecular weight neurotransmitters in nervous tissue. At present, some of the best published localization data on these compounds comes from classical EM preparation methods involving glutaraldehyde, osmium tetroxide and conventional embedding in epoxy resins (see Ottersen 1987, 1989 for reviews). It seems likely that such harsh procedures of specimen preparation are beneficial for keeping these small molecular weight compounds in the tissue block, up to and including the labelling step, on the section. In the case of Ottersen's work this statement is supported by elegant quantitative data (see Chap. 11). Significantly, in the papers of Ottersen, the

◀ ───

Fig. 14. Immunolabelling of Epon section of cultured Leech neurons. The cells were fixed with a mixture of 0.6% glutaraldehyde and 0.4% paraformaldehyde in 80 mM cacodylate buffer, pH 7.4 for 10 min followed by post-fixation for 10 min in 0.01% OsO4. After conventional embedding in Epon, the sections were etched with a mixture of sodium ethylate and hydrogen peroxide prior to labelling with anti-serotonin antibody (for 20 h at 4°C) followed by 8 nm protein A gold (for 4 h at room temperature). The dense core synaptic vesicles are significantly labelled. It is remarkable that this protocol was successful whereas Lowicryl K4M resins gave negative results. *Bar* 100 nm. (Courtesy of Dr. Damien Kuffler, Biocenter, University of Basel, Switzerland). For more details see Kuffler et al. J. Comp. Neurol., 256, 516–526 (1987). Present address: Institute of Neurobiology, University of Puerto Rico, San Juan, Puerto Rico

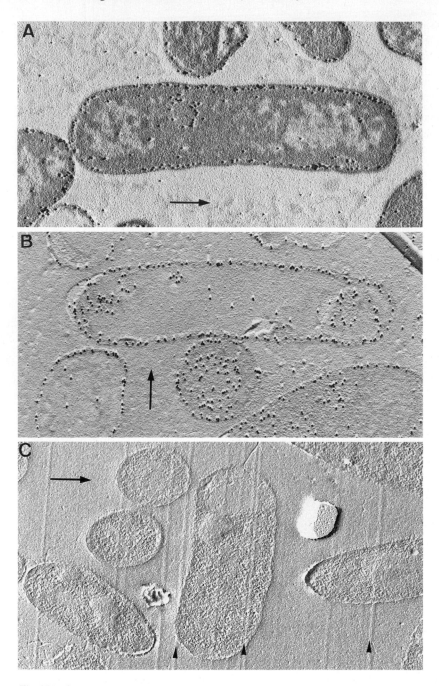

Fig. 15A–C

osmicated sections are treated with the metaperiodate method of Bendayan and Zollinger (1983). Importantly, Ottersen (1989) showed that in the case of glutamate, the use of osmium had no quantitative effect on the labelling density.

Two other interesting approaches using Epon sections were also recently published. The first by Mar and White (1988) used a solution of alcoholic sodium hydroxide to remove Epon from the section. The sections are then immunolabelled and re-embedded in a thin layer of Epon plastic by the method of Keller et al. (1984; see p. 179) which was developed for cryo-sections. The second method by Baigent (1990) introduced the idea of evaporating a film of carbon onto the epon sections (lying on a Formvar film) before extracting the resin, again with alcoholic sodium hydroxide. This film protects the biological material during the removal of the plastic, and during the subsequent immunolabelling and final drying of the sections. Both these approaches could also be classified under "temporary embedment".

4.3 Freeze Substitution

In the past few years approaches combining rapid freezing followed by **freeze substitution** and embedding in Lowicryl or other plastic media have become more prominent for both structural and immunocytochemical studies. The essence of these methods is to vitrify specimens by rapid cooling, then to chemically fix them at low temperature by using mixtures of fixative and of solvents. Subsequently the embedding medium is introduced with the solvent (with or without fixative). The initial polymerization is usually carried out at sub-zero temperatures and the final hardening at room temperature. Although the freeze-substitution approach in itself has been around for a very long time, it has only recently been applied for immunocytochemistry; the first group to see the potential of this approach was that of Humbel and Müller (see Humbel et al. 1983; Humbel and Müller 1986). The results obtained have, in many cases, been very impressive and it is clear that for a number of studies this will become the

---

Fig. 15A–C. *E. coli* bacteria that overproduce OmpA were fixed in 2% formaldehyde, 0.5% glutaraldehyde, block stained with uranyl acetate in water, dehydrated at room temperature in water and embedded in Epon (**A** and **B**). One thicker (**A**) and thinner (**B**) section from the same block (and on the same grid) were labelled for OmpA (as in Fig. 12), contrasted with uranyl acetate and lead citrate and unidirectionally shadowed with platinum/carbon (15°; *arrows* indicate direction of shadowing). The section in **A** was estimated to be 50–70 nm whereas that in **B** was only 15 nm thick. In the absence of shadowing such a thin section would be very difficult to visualize. The amount of label in these Epon sections was comparable to that obtained with Lowicryl. Both **A** and **B** show surface relief. In **C**, the bacteria were embedded in Lowicryl HM20 (as in Fig. 12) and unlabelled uncontrasted sections were shadowed as above. The shadowed surface shows knife marks (*arrowheads*); note that such knife marks would be very hard to observe in the absence of shadow. Note also the surface relief, especially over the bacteria which is more pronounced than in **A** or **B**. **A–C** all ×25000. (Courtesy of Dr. Heinz Schwarz, Max-Planck Institut für Entwicklungsbiologie, Tübingen, FRG)

method of choice. In some cases there are advantages in chemically fixing at conventional temperatures (4–37°C) before freeze substitution, thereby avoiding room temperature dehydration and infiltration. Finally, as shown by recent studies the freeze-substitution approach offers the only possibility for completely avoiding routine chemical fixation.

For recent reviews of the freeze-substitution approach for both structural and immunocytochemical studies see Humbel and Schwarz (1989), Schwarz and Humbel (1989) and Nicolas (1989). Additional excellent examples of the use of freeze substitution for structural studies can be found in Hunziker et al. (1984); Gilkey and Staehelin (1986); Hobot (1987); Dahl and Staehelin (1989) and Studer et al. (1989). For a recent excellent study on yeast see Takeshige et al. (1992) J Cell Biol 119:301.

4.3.1 Rapid Freezing

It is widely accepted that the use of liquid nitrogen is not compatible with rapid cooling of fresh, unfixed, tissues in order to avoid ice crystals (see Roos and Morgan 1990). As for the hydrated cryo-section method, vitrification must be the goal of this rapid cooling process. In contrast to the latter method, however, with the freeze-substitution approach, as for the freeze-fracture method, there is no way to assess the state of freezing unequivocally in the final image, except to carry out hydrated cryo-sectioning and electron diffraction in parallel. This remains a serious theoretical difficulty at high resolution. Nevertheless, for the resolution required in most immunocytochemical-structural studies, this problem can usually be ignored if care is taken at the cooling step (see Humbel and Schwarz 1989; Schwarz and Humbel 1989 and Voorhout et al. 1989a, b for reviews). Since details of freezing will be discussed in more detail in Chapter 5, only the information relevant to the freeze-substitution approach will be mentioned here.

The simplest method which is compatible with "good" rapid freezing is immersion of small pieces of tissue into liquid ethane (McDowall et al. 1983), Freon (Nagele et al. 1985) or propane (Inoue et al. 1982). More sophisticated, and technically more demanding methods involve (1) slamming the specimen against a cold metal-block (van Harrevald et al. 1965; Heuser et al. 1979; Murata et al. 1985), (2) using a liquid jet of propane (Müller et al. 1980) or (3) high pressure-freezing (Hunziker and Schenk 1984; Hunziker et al. 1984; Müller and Moor 1984; Moor 1987). With the exception of the latter approach, it is unlikely that any of these methods will enable vitrification of more than about 20 µm of the peripheral zone around the specimen (Venetie et al. 1981; Elder et al. 1982; Plattner and Bachman 1982; McDowall et al. 1983). From recent studies using the high-pressure freezing approach very impressive structural detail has been shown for specimens as large as 600 µm (Dahl and Staehelin 1989; Studer et al. 1989). Although not directly shown in these studies, recent unpublished data using hydrated cryo-sections indicate that these large blocks can, in fact, be vitreous (Müller, pers. commun.). Since the water is eventually replaced by embedding medium in all freeze-substitution studies, the only distinction which is possible is between (peripheral) areas where no effects of ice crystals are seen and other, deeper areas where noticeable effects are seen. This distinction is always somewhat subjective. A close inspection of many recently published micrographs in fact reveals typical net-like effects of ice crystal damage.

4.3.2 Freeze Substitution and Immunolabelling

The next stage in the technique is to put the rapidly cooled specimen into a mixture of solvent and fixative. An important early study in freeze substitution was that of MacKenzie (1972), who studied the effects of different dehydration protocols on isolated macromolecules after freezing. The significant finding was that structures collapsed during the substitution process and that this phenomenon could be avoided if the temperature was kept below $-58\,°C$ (215 K).

Since this period, most studies have aimed at keeping the tissues colder than this temperature in the first hours of the substitution (see Humbel and Schwarz 1989 for a detailed discussion). Until recently, in most approaches the temperature was allowed to increase gradually until the final infiltration in plastic was done at room temperature and the polymerization at higher temperatures. With the advent of the low-temperature resins, however, it is now possible to proceed with all the steps at temperatures below $0\,°C$, an approach first recommended on theoretical grounds by Fernandez-Moran (1960). Again, from a theoretical point of view, it has been argued that the key advantage of the low temperature is that the water in the immediate vicinity of the structure is either not removed at all or is removed in a way which is less disturbing to the structure (see Kellenberger 1987 and Humbel and Schwarz 1989 for a discussion). The important point is that, in practice, significant improvements have been reported in specific cases, both with respect to fine structural preservation (Humbel et al. 1983; Hunziker et al. 1984; Humbel and Müller 1986) and immunocytochemical labelling (Inoue et al. 1982; Nicolas et al. 1987a, b; 1989).

Classical studies using the freeze-substitution approach have usually used osmium tetroxide. Although high concentrations of osmium must be avoided when used at normal temperatures with the Lowicryl resins, it is possible to use this fixative in freeze-substitution protocols using these resins since there is significantly less blackening of tissue with osmium at these low temperatures (Humbel et al. 1983; Verkleij et al. 1985). The recent data of Nicolas et al. (1989a, b) indicate that, when combined with the freeze substitution approach and Epon embedding, osmium can in some cases even be compatible with good immunolabelling. In an earlier study by Inoue et al. (1982), however, this approach did not allow the localization of luteinizing hormone (LH) in the endoplasmic reticulum and Golgi complex of cells in the anterior pituitary, although it did allow detection of this antigen in the secretion granules, which obviously contain relatively high concentrations of LH.

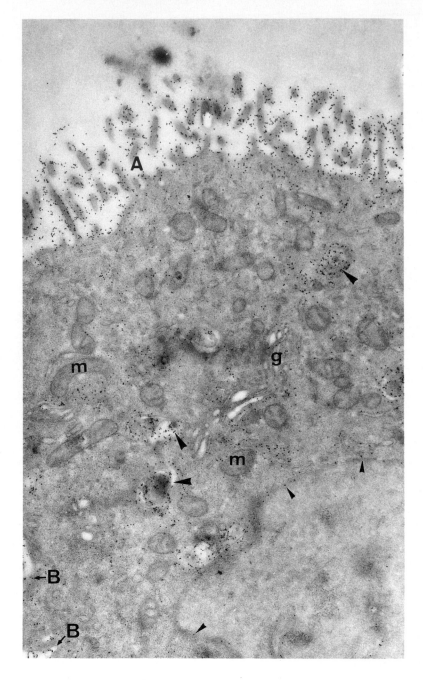

Fig. 16

While it appears that every group has its own preferred recipe for the initial fixation cocktail, most of the recent studies that have used the freeze-substitution approach for immunocytochemistry have either used glutaraldehyde or acrolein. In the study by Hunziker et al. (1984) for example, following initial freezing at high pressure, 3% glutaraldehyde and 0.5% uranyl acetate in pure methanol were used in three stages 17 h at −90°C, 13 h at −60°C and then 12 h at −35°C. Note that such long periods of infiltration are characteristic of the freeze substitution approach which is obviously a practical disadvantage. This was followed by a number of steps where increasing concentrations of embedding medium were added over a period of 4 days in this case either Lowicryl K4M or HM20. The latter was cross-linked with UV light at room temperature for 2 days. Although no immunocytochemical labelling was done, this, and a parallel study (Hunziker and Schenk 1984), showed excellent morphological preservation of cartilage tissue. In another study by Murata et al. (1985), acrolein was the preferred fixative. In this case the tissue, rat stomach, was slam frozen using liquid nitrogen cooled copper blocks, then fixed and substituted using 5% acrolein in acetone at −80°C for 2 days followed by −20°C for 2 h. The tissues were then embedded in either Epon or Lowicryl K4M. The visualization of fine structure, as well as the immunolabelling of pepsinogen, was in this case better with Epon than with Lowicryl K4M. An important practical tip for freeze substitution in general is that the volumes of substitution media used should always be at least 1000 times larger than the sample size: for most purposes the latter should not exceed 0.5 mm in any dimension (Voorhout, pers. commun.).

In a recent study by Nicolas et al. (1989), a detailed comparison was made of different fixation, freeze-substitution and embedding protocols on both the fine structure and immunolabelling of luciferase in a bioluminescent bacterium. Whereas conventional fixation and Epon embedding was poorly compatible with immunolabelling, fast freezing followed by various freeze-substitution

◄─────────────────────────────────────   ──────

Fig. 16. Example of freeze-substitution immunolabelling using Lowicryl HM20. Immunogold labelling of Forssman glycolipid on freeze-substitution embedded MDCK cells. Transwell filters with a confluent monolayer of MDCK cells were fixed for 1 h in 2% paraformaldehyde in 0.1 M phosphate buffer pH 7.4. Cells were cryo-protected by immersing the filters in 30% glycerol in 0.1 M phosphate buffer containing 1% paraformaldehyde for 30 min. Filters were cut in small squares and frozen in liquid propane (−180°C). Frozen samples were freeze substituted at −90°C in methanol supplemented with 0.5% uranyl acetate for 36 h in a Reichert Cs-auto. After raising the temperature to −45°C at a rate of 5°C/h and washing several times with pure methanol, the samples were infiltrated with Lowicryl HM20 at −45°C. Infiltration was done in the following graded series of Lowicryl-methanol mixtures: 1:1 for 2 h, 2:1 for 2 h, pure Lowicryl for 2 h and pure Lowicryl overnight. Samples were polymerized by UV-light at −45°C for 2 days in a closed flat embedding mold in the Reichert Cs-auto. Sections were incubated with primary antibody diluted in 20 mM Tris buffered 130 mM saline pH 8.2 supplemented with 0.1% BSA and 0.1% cold fish gelatin and protein A-gold (12 nm) diluted in the same buffer. Sections were stained for 5 min with 3% uranyl acetate and for 1–2 min with Reynolds lead citrate and observed in a Jeol 1200EX electron microscope at 60 kV. ×21000. *Bar* 1 μm. (Courtesy of Dr. Wim Voorhout, Department of Cell Biology, University of Utrecht). For more details see Van Genderen et al. (1991) J Cell Biol 115:1009–1019

protocols gave excellent structural preservation and labelling. As mentioned above, some of these protocols contained osmium tetroxide. While LR white embedding following freeze-substitution was considered inferior to Epon embedding for ultrastructural features, it gave almost four times higher density of immunolabelling than did Epon.

A quantitative comparison of different protocols was also made by Schwarz and Humbel (1989). Using a bacterial system and the labelling of a membrane protein, OmpA, these authors showed significantly higher labelling after freeze substitution as compared with conventional (progressive lowering of temperature) embedding in Lowicryl. As mentioned above, because of its better sectioning and contrasting properties, Lowicryl HM20 was the preferred resin; for most protocols tested there was little difference in labelling intensity between the HM20 and K4M for most antigens, although the K4M gave slightly higher values for all protocols. Note again, however, that for some antigens striking differences can be seen between the two resins and between different protocols (see comments p. 108).

A significant finding by Humbel et al. 1983; Nicolas (1989); Nicolas et al. (1989) and Schwarz and Humbel (1989) was that the freeze substitution approach could also be used without conventional fixatives. Hence, fast freezing may be followed by freeze-substitution in acetone (Nicolas 1989) or methanol (Humbel et al. 1983; Schwarz and Humbel 1989) and then embedding in either Epon or LR white (Nicolas 1989) or the Lowicryl resins (Schwarz and Humbel 1989). Both groups showed acceptable fine structure and immunolabelling. This is probably the only approach for immunocytochemistry at the EM level where "classical" fixatives may be avoided. In this case, however, it can be argued that the solvent now plays same role in fixation as it does in classical light microscopy, albeit at low temperature. A potential problem of this approach is that antigens may be displaced after the sectioning procedures (Schwarz and Humbel 1989; Schwarz, pers. commun.). Using freeze substitution another possibility to consider is to use uranyl acetate as a fixative, with or without pre-fixation in aldehydes. In the study by Schwarz and Humbel (1989) cryo fixation followed by uranyl acetate in methanol gave immunolabelling for OmpA in bacteria that was as high as all other protocols tested (see also Voorhout et al. 1989a, b).

An alternative, approach to the conventional cold fixation following rapid freezing is to fix tissues initially in aldehydes in the conventional way and then to freeze. In this way cryo-protectants such as $\approx$ 2 M sucrose or dimethylformamide (Meissner and Schwarz 1990) may be used and the specimen can be substituted and embedded at low temperatures. According to Humbel (1984), it is difficult to remove sucrose during the substitution process. Although this approach does not avoid classical fixation problems, it may avoid problems associated with dehydration at room temperature (Humbel and Schwarz 1989; van Genderen et al. 1991; see Fig. 16). It also avoids the necessity for the often technically demanding fast-freezing methods; further, the whole of the block can be easily vitrified, in contrast to most fast-freezing approaches. In the study by Schwarz and Humbel (1989), however, this approach gave a significant lower labelling for the model antigen when compared to the cryo-fixation, freeze-substitution approach.

The recent introduction of a new generation of very low temperature Lowicryl resins (see above) should be very useful, in combination with the freeze-substitution approach, in future studies. Although it is too early to make a full judgement on these resins, they are especially attractive candidates for carrying out immunocytochemistry of lipids since, as mentioned, the data indicate that lipids are quantitatively retained after dehydration and infiltration at these low temperatures (Weibull et al. 1983a,b; Weibull and Christiansson 1986).

References

Abrahamson DR (1986) Post-embedding colloidal gold immunolocalization of laminin to the lamina rara interna, lamina densa, and lamina rara externa of renal glomerular basement membranes. J Histochem Cytochem 34:847–853

Acetarin JD, Carlemalm E, Villiger W (1986) Developments of new lowicryl resins or embedding biological specimens at even lower temperatures. J Microsc 143:81–88

Acetarin JD, Carlemalm E, Kellenberger E, Villiger W (1987) Correlation of some mechanical properties of embedding resins with their behaviour in microtomy. J Electron Microsc Tech 6:63–79

Albrecht U, Seulberger H, Schwarz H, Risau W (1990) Correlation of blood-brain barrier function and HT7-protein expression in chick brain circumventricular organs. Brain Res (in press) 535:49–61

Altman LG, Schneider BG, Papermaster DS (1984) Rapid embedding of tissues in Lowicryl K4M for immunoelectron microscopy. J Histochem Cytochem 32:1217–1223

Armbruster BL, Carlemalm E, Chiovetti R, Gavavito RM, Hobot JA, Kellenberger E, Villiger W (1982) Specimen preparation for electron microscopy using low temperature embedding resins. J Microsc 126:77–85

Armbruster BL, Kellenberger E, Carlemalm E, Villiger W, Garavito RM, Hobot JA, Chiovetti R, Acetarin JD (1983a) Lowicryl resins – present and future applications. In: Johari O (ed) Science of biological specimen preparations. SEM Inc, AMF O'Hare, Chicago, pp 77–81

Armbruster BL, Garavito RM, Kellenberger E (1983b) Dehydration and embedding temperatures affect the antigenic specificity of tubulin and immunolabelling by the protein A-colloidal gold technique. J Histochem Cytochem 31:1380–1384

Ashford AE, Allaway WG, Gubler F, Lennon A, Sleegers J (1986) Temperature control in Lowicryl K+M and glycol methacrylate during polymerization is a low temperature embedding method. J Microsc 144:107–126

Bachhuber K, Frösch D (1983) Melamine resins, a new class of water soluble embedding media for electron microscopy. J Microsc 130:1–9

Baigent C (1990) Carbon based immunocytochemistry (CarBIC): a new approach to the immunostaining of epoxy resin-embedded material. J Microsc 158:73–80

Bard JB, Ross AS (1986) Improved method for making high-affinity sections of soft tissue embedded in polyethylene glycol (PEG): its use in screening monoclonal antibodies. J Histochem Cytochem 34:1237–1246

Baskin DG, Erlandsen SL, Parsons JA (1979) Influence of hydrogen peroxide or alcoholic sodium hydroxide on the immunocytochemical detection of growth hormone and prolactin after osmium fixation. J Histochem Cytochem 27:1290–1292

Bendayan M (1983) Ultrastructural localization of actin in muscle epithelial and secretory cells by applying the protein A-gold immunocytochemical technique. J Histochem 15:39–58

Bendayan M, Ørstavik TB (1982) Immunocytochemical localization of Kalikrein in the rat exocrine pancreas. J Histochem Cytochem 30:58–66

Bendayan M, Zollinger M (1983) Ultrastructural localization of antigenic sties on osmium tetroxide-fixed tissues applying the protein A-gold technique. J Histochem Cytochem 31:101–109

Bendayan M, Roth J, Perrelet A, Orci L (1980) Quantitative immunocytochemical localization of pancreatic secretory proteins in subcellular compartments of the rat acinar cell. J Histochem Cytochem 28:149–60

Bendayan M, Nanci A, Kan FWK (1987) Effect of tissue processing on colloidal gold cytochemistry. J Histochem Cytochem 35:983–996

Bénichou JC, Fréhel C, Ryter A (1990) Improved sectioning and ultrastructure of bacteria and animal cells embedded in Lowicryl. J Electron Microsc Techn 14:289–297

Berryman MA, Rodewald RD (1990) An enhanced method for post-embedding immunocyto-chemical staining which preserves cell membranes. J Histochem Cytochem 38:159–170

Boyde A, Maconnachie E (1983) Not quite critical-point drying. In: Johari O (ed) The science of biological specimen preparation for microscopy and microanalysis. SEM Inc, AMF O'Hare, Chicago, pp 71–77

Carlemalm E, Garavito RM, Villiger W (1982) Resin development for electron microscopy and an analysis of embedding at low temperature. J Microsc 126:123–143

Carlemalm E, Villiger W (1989) In: Bullock GR, Petrutz P (eds) Techniques in immunocyto-chemistry, vol 4. Academic Press, London, pp 29–45

Causton BE (1984) The choice of resins for immunocytochemistry. In: Polak JM, Varndell IM (eds) Immunolabelling for electron microscopy. Elsevier, Amsterdam, pp 29–37

Causton BE (1986) Does the embedding chemistry interact with tissue? In: Müller M, Becker RP, Boyle AB, Wolosewick JJ (eds) Science of biological specimen preparations. SEM Inc, AMF O'Hare, Chicago, pp 209–214

Craig S, Miller C (1984) LR white resin and improved on-grid immunogold detection of vicilin, a pea seed storage protein. Cell Biol Int Rep 8:879–888

Dahl R, Staehelin LA (1989) High-pressure freezing for the preservation of biological structure: Theory and practice. J Electron Microsc Tech 13:165–174

Desjardins M, Bendayan M (1989) Heterogenous distribution of type IV collagen, entactin, heparan sulfate proteoglycan, and laminin among renal basement membranes as demon-strated by quantitative immunocytochemistry. J Histochem Cytochem 37:885–897

Douzou P (1977) Enzymology at sub-zero tempatures. Adv Enzymol 45:157–272

Drenckhahn D, Dermietzel R (1988) Localization of the actin filament cytoskeleton in the renal brush border: a quantitative and qualitative immunoelectron microscope study. J Cell Biol 107:1037–1048

Dürrenberger M (1989) Removal of background label in immunocytochemistry with polar Lowicryls by using washed protein A-gold-precoupled antibodies in a one-step procedure. J Electron Microsc Tech 11:1–8

Elder HY, Gray CC, Jardine AG, Chapman JN, Biddlecombe, WH (1982) Optimum conditions for the cryo-quenching of small tissue blocks in liquid coolants. J Microsc 126:45–61

Ellinger A, Pavelka M (1985) Post-embedding localization of glycoconjugates by means of lectins on thin sections of tissues embedded in LR white. Histochem J 17:1321–1336

Engfeldt B, Hultenby K, Müller M (1986) Ultrastructure of hyaline cartilage. Acta Pathol Microbiol Immunol Scand Sect A 94:313–323

Erlandsen SL, Parsons JA, Rodning CB (1979) Technical parameters of immunostaining of osmicated tissue in epoxy sections. J Histochem Cytochem 27:1286–1289

Ernst SA, Palacios JR, Siegel GJ (1986) Immunocytochemical localisation of Na⁺, K⁺, ATPase catalytic polypeptide in mouse choroid plexces. J Histochem Cytochem 33:189–195

Escolar G, Sauk JJ, Bravo ML, Krumwiede M, White JG (1988) Development of a simple embedding procedure allowing immunocytochemical localization at the ultrastructural level. J Histochem Cytochem 36:1579–1582

Farrant IL, McLean ID (1969) Albumin as embedding media for electron microscopy. Proc 27th Electron Microsc Soc Am: 422–424

Fernandez-Moran H (1960) Low temperature preparation techniques for electron microscopy of biological speciments based on rapid freezing with liquid helium II. Ann NY Acad Sci 85:689–713

Freudl R, Schwarz H, Stierhof Y-D, Gamon K, Hindennach I, Henning U (1986) An outer membrane protein (OmpA) of *Escherichia coli* K-12 undergoes a conformational change during export. J Biol Chem 261:11355–11361

Fryer PR, Wells C (1983) Technical difficulties overcome in the use of lowicryl 4KM EM embedding resin. Histochemistry 77:141–143

Garavito RM, Carlemalm E, Colliex C, Villiger W (1982) Separate junction ultrastructure as visualised in unstained and stained preparations. J Ultrastruct Res 80:344–353

Gilkey JC, Staehelin LA (1986) Advances in ultrarapid freezing for the preservation of cellular ultrastructure. J Electron Microsc Tech 3:177–189

Gorbsky G, Borisy GG (1985) Microtubule distribution in cultured cells and intact tissues: improved immunolabeling resolution through the use of reversible embedment cytochemistry. Proc Natl Acad Sci USA 82:6889–6893

Gorbsky G, Borisy GG (1986) Reversible embedment cytochemistry (REC): A versatile method for the ultrastructural analysis and affinity labeling of tissue sections. J Histochem Cytochem 34:177–188

Gräber MB, Kreutzberg GW (1985) Immuno gold staining (IGS) for electron microscopical demonstration of glial fibrillary acidic (GFA) protein in LR white embedded tissue. Histochemistry 83:497–500

Griffiths G, Hoppeler H (1986) Quantitation in immunocytochemistry: correlation of immuno-gold labeling to absolute number of membrane antigens. J Histochem Cytochem 34:1389–1398

Griffiths G, Jockusch J (1980) Antibody labelling of thin sections of skeletal muscle with specific antibodies. J Histochem Cytochem 28:969–978

Griffiths G, McDowell A, Back R, Dubochet J (1984) On the preparation of cryosections for immunocytochemistry. J Ultrastruct Res 89:65–78

Griffiths G, Fuller SD, Back R, Hollinshead M, Pfeiffer S, Simons K (1989) The dynamic nature of the Golgi complex. J Cell Biol 108:277–297

Hawes CR, Horne JC (1985) Polyethylene glycol embedding of plant tissues for transmission electron microscopy. J Microsc 137:35–45

Hearn SA, Silver MM, Sholdice JA (1985) Immunoelectron microscopic labelling of immunoglobulin in plasma cells after osmium fixation and epoxy embedding. J Histochem Cytochem 33:1212–1218

Heckman GA, Barrett RJ (1973) GACH: a water-miscible, lipid retaining embedding polymer for electron microscopy. J Ultrastruct Res 42:156–179

Heuser JE, Reese TS, Dennis MJ, Jan Y, Jan C, Evans L (1979). Synaptic vesicle exocytosis captured by quick freezing and correlated with quantal transmitter release. J Cell Biol 81:275–300

Hobot JA (1987) Use of on-section immunolabelling and cryosubstitution for studies of bacterial DNA distribution. J Bacteriol 169:2055–2062

Hobot JA (1989) Lowicryls and low-temperature embedding for colloidal gold methods. In: Hayat MA (ed) Colloidal gold: principles, methods and applications. vol 2. Academic Press, San Diego, pp 75–115

Hobot JA, Carlemalm E, Villiger W, Kellenberger E (1984) Periplasmic gel: New concept resulting from the reinvestigation of bacterial cell envelope ultrastructure by new methods. J Bacteriol 160:143–152

Humbel BM (1984) Gefriersubstitution - Ein Weg zur Verbesserung der morphologischen und zytologischen Untersuchungen biologischer Proben im Elektronenmiskroskop. Doktorthesis 7609 ETH Zürich, Switzerland

Humbel B, Müller M (1986) Freeze-substitution and low temperature embedding. In: Müller M, Becker RP, Boyde A, Wolosewick JJ (eds) The science of biological specimen preparation. SEM, AMF O'Hare, Chicago, pp 175–183

Humbel B, Marti T, Müller M (1983) Improved structural preservation of combining freeze substitution and low temperature embedding. In: Pfefferkorn G (ed) Beitr Elektronenmikroskop Direktabb Oberfl, vol 16. Antwerpen, pp 585–594

Humbel B, Schwarz H (1989) Freeze-substitution for immunochemistry. In: Verkleij AJ, Leunissen JLM (eds) Immuno-gold labeling in cell biology. CRC Press, Boca Raton, Florida pp 115–134

Hunziker EB, Schenk R (1984) Cartilage ultrastructure after high pressure freezing, freeze substitution and low temperature embedding. II. Intercellular matrix ultrastructure preservation of proteoglycans in their native state. J Cell Biol 98:277–282

Hunziker EB, Hermann W, Schenk RK, Müller M, Moor H (1984) Cartilage ultrastructure after high pressure freezing, freeze-substitution, and low temperature embedding. Chondrocyte ultrastructure-implications for the theories of mineralization and vascular invasion. J Cell Biol 98:267–276

Inoue K, Kurosumi K, Deng ZP (1982) An improvement of the device for rapid freezing by use of liquid propane and the application of immunocytochemistry to the resin section of rapid frozen, substitution-fixed anterior pituitary gland. J Electron. Miscrosc 31:93–97

Kauzmann W (1957) Some factors in the interpretation of protein denaturation. Adv Prot Chem 14:1–61

Kellenberger E (1987) The response of biological macromolecules and supramolecular structures to the physics of specimen cryopreparation. In: Steinbrecht RA, Zierold K (eds) Cryotechniques in biological electron microscopy. Springer, Berlin Heidelberg New York, pp 35–63

Kellenberger E, Villiger W, Carlemalm E (1986a) The influence of surface relief of thin sections of embedded unstrained material on image quality. J Micron Microsc Acta 17:331–348

Kellenberger E, Carlemalm E, Villiger W (1986b). Physics of specimen preparation and observation of specimens that involve cryoprocedures. In: Müller M, Becker RP, Boyles A, Wolosewick JJ (eds) Science of biological specimen preparation. SEM Inc, AMF O'Hare, Chicago, pp 1–20

Kellenberger E, Durrenberger M, Villiger W, Carlemalm E, Wurtz M (1987) The efficiency of immunolabel on lowicryl sections compared to theoretical predictions. J Histochem Cytochem 35:959–969

Keller GA, Tokuyasu KT, Dutton AH, Singer SJ (1984) An improved procedure for immunoelectron microscopy: ultrathin plastic embedding of immuno-labeled ultrathin frozen sections. Proc Natl Acad Sci USA 81:5744–5797

Kraehenbuhl JP, Jamieson JD (1972) Solid-phase conjugation of ferritin to Fab fragments of immunoglobulin G for use in antigen localization in thin sections. Proc Natl Acad Sci USA 69:1771–1775

Kraehenbuhl JP, Racine L, Jamieson DJ (1977) Immunocytochemical localization of secretory proteins in bovine pancreatic exocrine cells. J Cell Biol 72:406–414

Kraehenbuhl JP, Weibel ER, Papermaster DS (1978). Quantitative immunocytochemistry at the electron microscope. In: Knapp W, Holufer K, Wick W (eds) Immunofluorescence and related staining techniques. North Holland, Amsterdam, pp 245–253

Kraehenbuhl JP, Racine L, Griffiths GW (1980) Attempts to quantitate immunocytochemistry at the electron microscope level. Histochem J 12:317–332

Lethias C, Hartmann DJ, Masmejean M, Ravazzola M, Sabbagh I, Ville G, Herbage D (1987) Ultrastructural immunolocalization of elastic fibers in rat blood vessels using the protein A-gold technique. J Histochem Cytochem 35:15–21

MacKenzie AP (1972) Freezing, freeze-drying, and freeze-substitution. Scanning Electron Microsc 2:273–286

Mar H, White TN (1988) Colloidal gold immunostaining on deplasticized ultra-thin sections. J Histochem Cytochem 36:1387–1395

McClean JD, Singer SJ (1970) A general method for the specific staining of intracellular antigens with ferritin-antibody complexes. Proc Natl Acad Sci USA 65:122–128

McDowall AW, Chang JJ, Freeman R, Lepault J, Walter CA, Dubochet J (1983) Electron microscopy of frozen hydrated sections of vitreous ice and vitrified biological samples. J Microsc 131:1–9

Meissner DH, Schwarz H (1990) Improved cryoprotection and freeze-substitution of embryonic quail retina: a TEM study on ultrastructural preservation. J Electron Microsc Tech 14:348–356

Millonig G (1961) Modified procedure for lead staining of thin section. J Biophys Biochem Cytol 11:736–739

Moor H (1987) Theory and practice of high pressure freezing. In: Steinbrecht RA, Zierold K (eds) Cryotechniques in biological electron microscopy. Springer, Berlin Heidelberg New York, pp 175–191

Moore PJ, Staehelin LA (1988) Immunogold localization of the cell-wall-matrix polysaccharides rhamnogalacturonan I and xyloglucan during cell expansion and cytokinesis in *Trifolium pratense* L; implication for secretory pathways. Planta 174:433–445

Mowary J, Chesner J, Spangenberger S, Hixon DC (1989) Rapid low molecular weight polyethylene glycol embedding protocol for immunocytochemistry. J Histochem Cytochem 37:1549–1552

Müller M, Moor H (1984) Cryofixation of thick specimens by high pressure freezing. In: Revel JP, Barnard T, Haggis GH (eds) The science of biological speciment preparation. SEM Inc, AMF O'Hare, Chicago, pp 131–138

Müller M, Meister N, Moor H (1980) Freezing in a propane jet and its application in freeze-fracturing. Mikroskopie 36:129–140

Murata F, Suzuki S, Tsuyama S, Suganuma T, Imada T, Furihata C (1985) Application of rapid freezing followed by freeze-substitition acrolein fixation for histochemical studies of the rat stomach. Histochem J 17:967–980

Mutasa HCF, Pearson EC (1988) Use of light microscopic immunotechniques in selecting preparation conditions and immunoprobes for ultrastructural immunolabelling of lactoferrin. Histochem J 20:558–566

Nagele RG, Kosciuk MC, Wang SC, Spero DA, Lee H (1985) A method for preparing quick-frozen, freeze-substituted cells for transmission electron microscopy and immunocytochemistry. J Microsc 139:291–301

Newman GR (1987) Letter to the editor. Histochem J 19:118–120

Newman GR, Hobot JA (1987) Modern acrylics for post-embedding immunostaining tech niques. J Histochem Cytochem 35:971–981

Newman GR, Jasani B (1984) Post-embedding immunoenzyme techniques. In: Polak JM, Varndell M (eds) Immunolabelling for electron microscopy. Elsevier, Amsterdam, pp 53–70

Newman GR, Jasani B, Williams ED (1983) A simple post-embedding system for the rapid demonstration of tissue antigens under the electron microscope. Histochem J 15:543–555

Nicolas M-T (1989) Immuno-gold labeling after rapid-freezing fixation and freeze-substitution: application to the detection of luciferase in dinoflagellates In: Verkleij AJ, Leunissen JLM (eds) Immuno-gold labeling in cell biology. CRC Press, Boca Raton, Florida, pp 277–290

Nicolas M-T, Nicolas G, Johnson CH, Bassot J-M, Hastings JW (1987a) Characterization of the bioluminescent organelles in *Goryaulax polyedra* (dinoflagellates) after fast-freeze fixation and antiluciferase immunogold staining. J Cell Biol 105:723–735

Nicolas M-T, Sweeney BM, Hastings JW (1987b) The ultrastructural localization of luciferase in three bioluminescent dinoflagellates, two species of pyrocystes, and noctiluca using antiluciferase and immunogold labelling. J Cell Sci 87:187–196

Nicolas M-T, Bassot J-M, Nicolas G (1989) Immunogold labeling of luciferase in the luminous bacterium *Vibrio harveyi* after fast-freeze fixation and different freeze-substitution and embedding procedures. J Histochem Cytochem 37:663–674

Northcote DH, Davey R, Lay J (1989) Use of antisera to localize callose, xylan and arabinogalactan in the cell-plate, primary and secondary walls of plant cells. Planta 178:353–366

Orci L, Halban P, Amherdt M, Ravazzola M, Vassalli JD, Perrelet A (1984) A clathrin-coated, Golgi related compartment of the insulin secreting cell accumulates proinsulin in the presence of monensin. Cell 39:39–47

Ottersen OP (1987) Postembedding light- and electron microscopic immunocytochemistry of amino acids: description of a new model system allowing identical conditions for specificity testing and tissue processing. Exp Brain Res 69:167–174

Ottersen OP (1989) Quantitative electron microscopic immunocytochemistry of neuroactive amino acids. Anat Embryol 180:1–15

Papermaster DS, Schneider BG, Krachenbuhl JP (1978) Immunocytochemical localization of opsin in outer segments and Golgi zones of frog photoreceptor cells. An electron microscopic analysis of cross-linked albumin embedded-retinas. J Cell Biol 77:196–203

Pease DC (1966) The preservation of unfixed cytological detail by dehydration with "inert" agents. J Ultrastruct Res 14:356–378

Peterson RG, Pease DC (1970) Features of the fine structure of myelin embedded in water-containing aldehyde resins. Proc 7th Int Congr Electron Microscopy, Grenoble, France pp 409–410

Petsko GA (1975) Protein crystallography at sub-zero temperatures: cryo-protective mother liquose for protein crystals. J Mol Biol 96:381–392

Plattner H, Bachmann L (1982) Cryofixation: a tool in biological ultrastructural research Int Rev Cytol 79:237–304

Posthuma G, SlotJW, GeuzeHJ (1987) The usefulness of the immunogold technique in quantitation of a soluble protein in ultrathin sections. J Histochem 35:405–410

Richards CL Jr, Anderson TF, Hance RT (1942) A microtome sectioning technique for electron microscopy illustrated with sections of striated muscle. Proc Soc Exp Biol Med 51:148–152

Ring PKM, Johanson V (1987) Immunoelectron microscopic demonstration of thyroglobulin and thyroid hormones in rat thyroid gland. J Histochem Cytochem 35:1095–1104

Roos N, Morgan AJ (1990) Cryopreparation of thin biological specimens for electron microscopy: methods and application. Microscopy Handbook 21. Royal Microscopical Society. Oxford University Press

Roth J (1989) Postembedding labeling on lowicryl K4M tissue sections: detection and modification of cellular components. Methods Cell Biol 31:513–550

Roth J, Berger EG (1982) Immunocytochemical localization of galactocyl transferase in Hella cells: codistribution with thiamino pyrophosphatase in *trans* Golgi cisternae J Cell Biol 92:223–229

Roth J, Bendayan M, Orci L (1978) Ultrastructural localization of intracellular antigens by the use of protein A-gold complexes. J Histochem Cytochem 26:1074–1081

Roth J, Bendayan M, Carlemalm E, Villiger W, Garavito M (1981) Enhancement of structural preservation and immunocytochemical staining in low temperature embedded pancratic tissue. J Histochem Cytochem 29:663–671

Roth J, Taatjes DJ, Tokuyasu KT (1990) Contrasting of Lowicryl K4M thin sections. Histochemistry 95:123–136

Schwarz H, Humbel BM (1989) Influence of fixatives and embedding media on immunolabelling of freeze-substituted cells. Scanning Microsc Suppl 3:57–64

Simon GT, Thomas JA, Chorneyko KA, Carlemalm E (1987) J Electron Microsc Tech 6:317–324

Sjöstrand FS (1976) The problem of preserving molecular structure of cellular comonents with electron microscopic analysis. J Ultrastruct Res 55:271–282

Small JV (1984) Polyvinylalcohol, a water soluble resin suitable for electron microscope immunocytochemistry. Proc 8th Eur Congr Electron Microsc, Budapest 3:1799–1800

Small JV, Fürst DO, De Mey J (1986) Localization of filamin in smooth muscle. J Cell Biol 102:210–220

Spaur RC, Moriarty GC (1977) Improvements in glysol methacrylate 1. Its use as an embedding medium for electron microscopic studies. J Histochem Cytochem 25:163–174

Spurr AR (1969) A low-viscosity epoxy resin embedding medium for electron microscopy. J Ultrastruct Res 26:31–43

Stierhof YD, Schwarz H (1991) Yield of immunolabel compared to resin sections and thawed cryosections. In: Hyat MA (ed) Colloidal gold: principles, methods and applications. vol 3. Academic Press, San Diego pp 87–115

Stierhof YD, Schwarz H, Frank H (1987) Transverse sectioning of plastic embedded immunolabeled cryosections: morphology and permeability to protein A-colloidal gold complexes. J Ultrastruct Mol Struct Res 97:187–199

Studer D, Michel M, Müller M (1989) High pressure freezing comes of age. Scanning Microsc Suppl 3:253–269

Swords KMM, Staehelin LA (1989) Analysis of extensin structure in plant cell walls. In: Linskens HF, Jackson JF (eds) Modern methods of plant analysis. vol 10. Plant Fibers. Springer, Berlin Heidelberg New York, pp 219–231

Templeman KH, Wira CR (1986) An improved specimen carrier for large scale embedment with lowicryl. J Electron Microsc Tech 4:73–74

Thorens B, Roth J, Norman AG, Perrelet A, Orci L (1982) Immunocytochemical localization of the vitamin-D-dependent calcium binding protein in chick deodenum. J Cell Biol 94:115–1221

Tokuyasu KT (1973) A technique for ultracryotomy of cell suspensions and tissues. J Cell Biol 57:551–565

Tokuyasu KT (1978) A study of positive staining of ultrathin frozen sections. J Ultrastruct Res 63:287–307

Tokuyasu KT (1989) Use of poly(vinylpyrrolidone) and poly(vinyl alcohol) for cryoultramicrotomy. J Histochem 21:163–171

Trahair JF, Neutra MR, Gordon JI (1989) Use of transgenic mice to study the routing of secretory proteins in intestinal epithelial cells: analysis of human growth hormone compartmentalization as a function of cell type and differentiation. J Cell Biol 109:3231–3242

Valentino KL, Cruonrine DA, Ruchardt LF (1985) Lowicryl K4M embedding of brain tissue for immunogold electron microscopy. J Histochem Cytochem 33:969–973

van Harrevald A, Cromwell J, Malhorta SK (1965) A study of extracellular space in central nervous tissue by freeze-substitution. J Cell Biol 25:117–137

van Tuinen E, Reizman H (1987) Immunolocalization of glyceroldehyde-3-phosphate dehydrogenase, hexokinase, and carboxypeptidase Y in yeast cells at the ultrastructural level. J Histochem Cytochem 35:327–333

Venetie R, Blumenink JG, Verkleij AJ (1981) Propane jet-freezing: a valid ultra-rapid freezing method for the preservation of temperature liquid phases. J Microsc 123:287–292

Verkleij AJ, Humbel B, Studer D, Müller M (1985) "Lipidic particle" systems as visualized by thin-section electron microscopy. Biochem Biophys Adv 812:591–594

Viale G, Dell'Orto P, Braidotti P, Coggi G (1985) Ultrastructural localization of intracellular immunoglobulin in Epon-embedded human lymph nodes. J Histochem Cytochem 33:400–406

Villiger W (1991) Lowicryl resins In: Hayat MA (ed) Colloidal gold principles, methods and applications. vol 3. Academic Press, San Diego, pp 59–71

Voorhout WF, Leunissen-Bijvelt JJM, van der Krift ThP, Verkleij AJ (1989a) The application of cryo-ultramicrotomy and freeze-substitution in immuno-gold labelling of hybrid proteins in Escherichia coli. A comparison. Scanning Microsc Suppl 3:47–56

Voorhout W, Leunissen-Bijvelt J, Leunissen J, Tommassen J, Verkleij A (1989b) Immuno-gold labeling of Escherichia coli cell envelope components. In: Verkleij AJ, Leunissen JLM (eds) Immuno-gold labeling in cell biology. CRC Press, Boca Raton, Florida pp 292–304

Weibull C, Christiansson A (1986) Extraction of protein and membrane lipids during low temperature embedding of biological material in electron microscopy. J Microsc 142:79–86

Weibull C, Christiansson A, Carlemalm E (1983a) Extraction of membrane lipids during fixation, dehydration and embedding of Acholeplasma laidawii cells for electron microscopy. J Microsc 129:201–207

Weibull C, Villiger W, Carlemalm E (1983b) Extraction of lipids during freeze-substitution of Acholeplasma laidawii for electron microscopy. J Microsc 134:213–216

White JF, Hughes JL, Kumaratilake JS, Fanning JC, Gibson MA, Krishnan R, Cleary EG (1988) Post-embedding methods for immunolocalization of elastin and related components in tissues. J Histochem Cytochem 36:1543–1551

Wolosewick JJ (1980) The application of polyethyleneglycol (PEG) to electron microscopy. J Cell Biol 86:675–681

Wolosewick JJ (1984) Cell fine structure and protein antigenicity after polyethylene glycol processing. In: Barnard T, Revel JP, Hagg G (eds) The science of biological specimen preparation for microscopy and microanalysis. SEM Inc, AMF O'Hare, Chicago, pp 83–97

Wright R, Rine J (1989) Transmission electron microscopy and immunocytochemical studies of yeast: analysis of HMG-CoA reductase overproduction by electron microscopy. In: Tartakoff AM (ed) Methods in cell biology. vol 32. Academic Press, San Diego, pp 473–512

Wynford-Thomas D, Jasani B, Newman GR (1986) Immunohistochemical localization of cell surface receptor using a novel method permitting simple, rapid and reliable LM/EM correlation. Histochem J 18:387–396

Yokota S, Tsuji H, Kato K (1985) Localization of cathepsin D in rat liver Immunocytochemical study using post-embedding immunoenzyme and protein A gold technique. Histochemistry 82:141–149

Yokota S, Tsuji H, Kato K (1986) Immunocytochemical localization of cathepsin B in rat kidney. II. Electron microscopy using the protein A-gold technique. J Histochem Cytochem 34:899–907

Chapter 5

Cryo and Replica Techniques for Immunolabelling

Two different kinds of cryo techniques for immunolabelling will be discussed in this chapter. The first, the **Tokuyasu cryo-sectioning method**, has now become a widely used method for immunolabelling at both the light- and electron microscopy level. The second category – **freeze fracture and replica methods** – deals with a group of techniques that are primarily used in a small number of laboratories that have specialized in these approaches.

In order to give as much theoretical background as possible the Tokuyasu thawed cryo-section method, it is essential to describe also the hydrated cryo-section method, an approach for investigating high resolution structure of thin sections of unfixed, vitrified material. Although this method is still in its infancy, it seems logical to assume that it will become an important technique for the analysis of fine structure in the years to come.

5.1 Cryo-Sectioning Techniques

5.1.1 Historical Perspectives

The origins of ultra thin frozen sectioning go back as far as 1952, when Fernándes-Morán showed that unembedded fixed or fresh material could be sectioned thin enough for examination in the electron microscope (Fernándes-Morán 1950). He used a modified Spencer microtome in a −35 °C cold room, cut the sections with glass knives and floated the sections onto glycerol.This was, therefore, the beginning of the "thawed frozen section technique". These early attempts at cryo-sectioning were considered disappointing however, and interest switched to sectioning of methacrylate-embedded material.

Interest in cryo-sectioning was rekindled in France in the 1960s, when Bernhard and his group used the technique to enable various cytochemical tests to be performed on thawed ultra-thin sections of fixed frozen tissue (Leduc et al. 1967). In an attempt to overcome the freezing damage reported by Fernándes-Morán, the French group embedded material in gelatin which was then partially dehydrated in 50% glycerol and frozen onto tissue holders by immersion in liquid nitrogen. Sectioning was carried out on a modified Porter-Blum MT-1 microtome in a commercial freezer at −35°C. The sections, cut with glass knives, were floated onto a 40% dimethylsulphoxide (DMSO) solution, removed with plastic loops onto distilled water then placed on Formvar/carbon coated grids, air dried and negatively stained using various heavy metal solutions.

Among those who were impressed with the results achieved by Bernhard was Christensen, who was studying cells that secreted steroid hormones. After a brief visit with Bernhard at Villejuif in 1966, Christensen felt it might be possible to follow the intracellular pathways of steroid synthesis and secretion in these cells at the electron microscope level, using autoradiography on ultra-thin frozen sections. However, in order to maintain these diffusible substances in place, it would clearly be necessary to cut at a much colder temperature than Bernhard was using, and the tissue could not be allowed to come in contact with water or any other solvent during the procedure (which precluded fixation or cryo-protection). Christensen devised a cryo-sectioning bowl that allowed ultra-thin frozen sectioning on a standard Sorvall MT-2 ultramicrotome. The first results, from section fragments cut near liquid nitrogen temperature, were described by Christensen in 1969. The Sorvall Company became interested in the cryo-sectioning bowl and produced it as an accessory to the MT-2 ultramicrotome in 1970. The main description of the method was published in 1971 (Christensen 1971). Although occasional sections exhibited reasonable morphology, the majority showed serious ice crystal damage, because of the lack of fixation and cryo-protection.

A paper that was to have a striking impact on the use of ultra-thin frozen sections, stimulating broad interest in this approach, was that of Tokuyasu (1973). He utilized the cutting method and device of Christensen (1971) and, like Bernhard, used aldehyde fixation, cryo-protection and negative staining. In addition to his skill and insight in the use of these procedures, Tokuyasu introduced two innovations that brought about a dramatic improvement in tissue preservation and section quality, namely, the use of sucrose as a cryo-protectant, and a novel method for picking up the sections, stretching them in the process, and applying them to grids. In this 1973 paper, Tokuyasu showed that it was possible to cryo-section, and then contrast by negative stain, a wide variety of aldehyde-fixed cells and tissues. The fine-structural preservation in these preparations was remarkable. The thick layer of negative stain used in these studies not only gave excellent contrast but served as a scaffolding that protected the sections against the severe surface tension effects of air-drying. Except for the hydrated cryo-approach (Dubochet et al. 1988; see below), air-drying is a prerequisite for observing the sections in the electron microscope. The high degree of contrast provided by such negative staining, however, made it incompatible with antibody labelling studies. The immuno-markers ferritin, and to a lesser extent, colloidal gold, were difficult, if not impossible, to visualize. Initially, a compromise was reached using lower concentrations of stain. This was far from satisfactory since not only could structures not be clearly visualized, but air-drying artefacts became a serious problem (Tokuyasu and Singer 1976).

The introduction of methyl cellulose to protect thawed frozen sections against surface tension damage caused by air-drying was an essential step in the development of the technique (Tokuyasu 1978). This subtle procedure, together with the initial aldehyde fixation, determines the structural preservation as well as the contrast of the sections. The latter can vary from positive contrast (Tokuyasu 1978), where membranes appear dense on a lighter background, to partial negative contrast, where membranes are negatively contrasted but other cell structures, such as the nucleus, show positive contrast (see Griffiths et al. 1982). The later introduction of colloidal gold for this approach (Slot and Geuze 1981), which is much more dense than ferritin, enabled the methyl cellulose method to be modified to give higher membrane contrast (Griffiths et al. 1982, 1983, 1984). More recent innovations involve embedding the labelled cryo-sections in plastic (Epon or LR white) (Keller et al. 1984) or the use of polyvinyl alcohol in the embedding mixture (Tokuyasu 1986).

Two excellent methods for double-labelling have been introduced for the method, namely the combination of ferritin and imposil (a rod-shaped iron-dextran compound) by Singer's group (Geiger et al. 1981) and the use of different sizes of colloidal gold (Slot and Geuze 1981; Slot and Geuze 1985).

A recent development was the improvement in the quality of glass knives used for sectioning (Griffiths et al. 1983) based on principles outlined 20 years earlier by Tokuyasu and Okamara (1959). Tungsten coating of glass knives, first suggested by Roberts (1975), also significantly improved section quality (Griffiths et al. 1984). In the last 2–3 years, however, a new generation of diamond knives has become available from a number of commercial sources that are at least as good as glass knives.

Finally, it is essential to mention that the recent surge of interest in cryo-sectioning techniques is to a large part the result of significant technological advances in the commercial manufacture of cryo-microtomes. For a recent review see Sitte et al. (1989).

In the next section of this chapter we shall look in more detail at the processes of freezing and cryo-sectioning. From a theoretical point of view it also becomes essential to discuss in some detail the hydrated cryo-section method for examining unfixed, vitrified material. In this method, unlike the Tokuyasu thawed cryo-section method, the sections must remain vitrified (below −150°C) during the whole procedure. Although not directly relevant for immunocyto-chemistry, this method serves as an important **reference** technique, both for assessing the effects of freezing on biological systems and for deciding whether or not a structure one obtains after fixation is a real one.

5.1.2 Theory of Freezing

Freezing is the process of solidifying liquid by cooling. In all living organisms liquid water makes up the bulk of the body weight (typically ≥ 60–70%) and for most living processes it is essential that this water remains in the liquid state. The process of freezing the water in a organism can obviously have significant physico-chemical effects on other molecular constituents of cells. These effects are complex, and it is beyond the scope of this chapter to delve into all of the

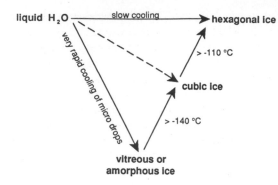

Fig. 1. Theoretical diagram to show the relationship between liquid water and the three forms of ice found at normal pressures. The *dashed line* indicates the theoretical possibility to freeze directly to the cubic ice form at intermediate freezing rates (from J. Dubochet, pers. commun.)

complexities, but only to give a simple general outline of the processes involved. For more comprehensive reviews the reader is referred to articles by Nei (1976), Franks (1977), Mazur (1984) and Dubochet et al (1988).

Let us first consider the solidification or freezing of pure water (Fig. 1). When water is cooled under normal pressures, one of three different physical states of solid water may be obtained depending on the rate of cooling. When cooling relatively slowly, **hexagonal ice** crystals form. This is the ice of glaciers, frozen ponds and snow. When small volumes of water (in the form of minute droplets or suspended films) are cooled more rapidly, an intermediate form of ice known as **cubic ice** may form. Hexagonal and cubic ice have crystalline structures that give characteristic X-ray and electron-diffraction patterns. When small volumes of water are cooled very rapidly (cooling rates $\geq 10^5 °C/s$) the **vitreous** or **amorphous** state may be attained. This is defined as solid water that is devoid of extended crystalline structure. It may help to think of the water molecules being "arrested" in the positions they occupied in the liquid at the moment of cooling. There is, apparently, however, still some debate amongst physicists about the actual structure of amorphous ice (Dubochet, et al. 1988; and pers. commun.).

In practice, vitrification of liquid water is very difficult because of the high cooling rate required. The vitrified state was first unambiguously demonstrated only a few years ago (Brüggeller and Mayer 1980; Dubochet et al. 1981). Since this state is devoid of crystalline structure, it is perhaps not appropriate to refer to it as "ice", a term which connotates crystallinity. Nevertheless, the term vitreous ice has recently gained wide usage.

At the next level of complexity one should consider the solidification of a simple solution, that is a mixture of water and solutes. When a solution is frozen, it is cooled in such a way that the water forms either hexagonal or cubic ice crystals. In this process, following the initial nucleation step, ordered microcrystal structures grow by the addition of water molecules from the relatively disordered liquid state. The molecules of solute are excluded from these crystals. As freezing continues, the concentration of the solutes in the liquid phase increases until the residual mixture becomes solid without crystallizing (the latter may be referred to as the **eutectoid** between the ice crystals).

If we now consider "freezing" of biological material, it becomes immediately evident that the situation is more complex. The first obstacle is the presence of semi-permeable membranes inside cells. Aside from the plasma membrane, there are membranes that enclose different intracellular compartments [namely, the endoplasmic reticulum, the Golgi complex, mitochondria, peroxisomes (microbodies), lysosomes, endosomes], in addition to the multitude of carrier vesicles believed to be involved in the various traffic between most of these compartments. It is possible that water and ions may be heterogeneously distributed among these different compartments: certainly their protein and to some extent their lipid compositions differ. A second complication is the idea that a significant proportion of cellular water is intimately associated with molecular surfaces, so-called hydration or bound water. This water is considered to behave differently from the rest of the water (so-called bulk water) upon freezing because it is not available for ice crystal growth. Clegg (1982) has estimated that in an "average" cell the filaments, membranes and other structures may provide a total surface area of ~750000 μm^2. According to this author's calculations, a monolayer or bilayer of water molecules bound to this amount of surface could represent as much as 10–15% of the cell's total water. The structure and properties of this "bio-water" are poorly understood and are currently the subject of heated debate (Dubochet, pers. commun.).

In order to keep this description at a simple level, I shall try to illustrate the effects of solidification of water on cells by describing what happens at three hypothetical different freezing rates.

- At slow freezing rates, hexagonal ice crystals form extracellularly, concentrating salt in the remaining liquid. Intracellular water moves rapidly out of the cell by osmosis in order to maintain osmotic equilibrium. The result is a shrinkage of the cytoplasm and a corresponding increase in the concentration of solutes.
- At intermediate freezing rates, hexagonal ice crystals form extracellularly and, before water has time to diffuse out, the nucleation of ice crystals begins intracellularly. This results in large, intracellular hexagonal ice crystals in the cytoplasm, possibly destroying cellular structures. Contrary to common belief, direct measurements of the sizes of hexagonal ice crystals inside cells indicate that their size is in the range of many microns (up to 100 μm^3), and their shapes are very irregular, comparable to the roots of a tree. Hence, no point within a crystal is very far from its surface (Dubochet et al. 1988, and pers. commun.). It is a common mistake to equate **profiles** of extensions of ice crystals (or the space left behind after the preparations have been, for example, freeze-substituted) with profiles of the **whole** ice crystals.
- At fast freezing rates cubic ice crystals may be formed. These are relatively small ($\approx$ 10 nm^3) when compared to hexagonal ice crystals. At very high freezing rates, vitreous solid water is formed both intracellularly and extracellularly. Since the cooling time required to vitrify water is necessarily very brief (10^{-4} sec, Dubochet et al. 1988) it is unlikely that significant redistribution of ions, molecules and macromolecular assemblies would occur

and, for all practical purposes, they should be arrested in their in vivo hydrated state (Dubochet et al. 1988). For this reason, the vitreous state has been termed the "solidified in vivo state" (Franks 1977). Clearly, this state should always be the goal when cells or tissues are solidified by cooling for structural studies whether for cryo-sections or for freeze substitution. Nevertheless, it seems evident that acceptable fine-structural preservation in the latter approach can also exist under conditions where vitrification has not occurred. Note that in freeze substitution devitrification must occur before substitution begins. The most dramatic demonstration that the vitrified state is compatible with maintenance of life comes from the experiments of Rall and Fahey (1985) on whole mouse embryos. These authors found that a complex mixture of solvents was suited for vitrification of these embryos when cooled in liquid nitrogen. The mortality rate of the embryos due to the effects of the solvent mixture alone, in the absence of vitrification and thawing, was about 15%. Remarkably, however, when embryos were immersed in solvents and then vitrified and thawed relatively rapidly, there was no significant increase in the rate of mortality. In other words, there was no increase in mortality as a result of the vitrification step itself.

5.1.2.1 Freezing in Practice

Except for a very thin surface layer, vitrification of living cells or tissues is extremely difficult to achieve because of the high rates of cooling required. It should be emphasized that the term "vitrification" is often used rather loosely in this context and in many, if not most studies, is it incorrectly used to mean freezing with no evidence of ice-crystal damage at the EM level. Even though the freeze-fracture literature contains many claims that "good freezing" has been achieved within an extensive depth of rapidly frozen, fresh tissue pieces, actual observation of frozen hydrated specimens has not supported this idea (Dubochet et al. 1988). Studies using the hydrated cryo-section technique indicate that, with one exception (see below), in all techniques for rapid cooling of fresh, hydrated tissues, vitrification is, at most, only possible down to 10–20 µm, even using the helium or liquid nitrogen cooled metal-mirror technique (van Harrevald and Crowell 1964; van Harrevald et al. 1974; Heuser et al. 1979; Escaig 1982, 1984; Plattner and Bachmann 1982; Robards and Sleytr 1985). This would indicate that, without some intracellular dehydration (either by "poor" freezing or by other physical means) one or, at most, two fully hydrated whole cell layers may be the limit for vitrification by any cooling method (except by high pressure freezing, see below). In the freeze-fracture technique, where one looks at a metal replica of a fractured, "frozen" tissue block, it is likely that the structural damage due to hexagonal ice crystals is easily recognizable. It is extremely unlikely, however, that the effect of the intermediate, cubic ice crystals, which are also capable of causing structural alterations in hydrated cryo-sections, would be detectable at the high resolution level (McDowall et al.

1984, Dubochet et al. 1988) since the techniques for rapid freezing are technically demanding, detailed descriptions are outside the scope of this chapter. For comprehensive reviews, see Gilkey and Staehelin (1986), Knoll et al. (1987), Sitte et al. (1987) and Roos and Morgan (1990). The latter is especially recommended for a general review of cryo preparation techniques.

The exception to the "10–20 µm rule" of vitrification, appears to be the **high-pressure freezing** method. Since the potential of the latter approach has only very recently become widely appreciated, it seems pertinent to discuss the principles in a little more detail. As described in two recent comprehensive reviews by Studer et al. (1989) and Dahl and Staehelin (1989), the main idea behind this approach is that the high pressure prevents the volume increase which occurs when water freezes at normal pressures. The effective advantage of the approach is that the cooling rate required for vitrification is significantly reduced (from a maximum of about $-10^6 °C/s$ down to about $-100 °C/s$). From theoretical calculations it was estimated that the optimal pressure for this effect is about 2100 bar pressure. A prototype machine for high pressure freezing of samples was designed by the group of Moor (1987) and is now made commercially by Balzers. In this machine, the specimen is sandwiched between two metal plates and subjected to 2100 bar pressure just prior to (less than 10 ms before) and during the application (≈ 0.5 s) of a pressurized stream of liquid nitrogen. In many cases, such as for plant tissues, I-hexadecane has been used to replace the water in the specimen immediately before freezing. This solvent is insoluble in water and believed to be osmotically inert; for a discussion see Studer et al. (1989). The available evidence now strongly suggests that a piece of tissue up to about 600 µm in one dimension can be effectively frozen by this approach (Hunziker et al. 1984a,b; Craig et al. 1987; Craig and Staehelin 1988). In all these published reports the evidence is indirect, coming from the appearance of the tissue after freeze substitution. Recently, direct evidence for vitrification of these large blocks has been obtained from electron diffracto-grams of hydrated cryo-sections (M. Müller, University of Zürich, pers. commun.). The approach is clearly still in its infancy, and the effects of the brief exposure to the high pressure on cells has not been fully studied. Nevertheless, the recently published (as well as not yet published) data are remarkable especially for more "difficult" tissues such as insect embryos (H. Schwarz, University of Tübingen, pers. commun.) and plant tissues (Craig et al. 1987; Craig and Staehelin 1988). Although no data on immunolabelling have yet appeared in the literature, there seems to be no question that, when combined with freeze substitution, this approach will in future become significant for immunocytochemistry.

5.1.2.2 Cryo-Protection

With the possible exception of high-pressure freezing, the only certain way to vitrify a complete block of tissue is by the use of cryoprotectants. These are compounds which alter the properties of water in such a way that they reduce the ability of its molecules to "nucleate" or "seed" ice crystals, the first step in ice crystal formation. Many of the well-known cryoprotectants such as polyvinyl pyrrolidine (PVP) are incapable of passing through membranes of living cells. Others, notably glycerol and dimethylsulfoxide (DMSO), are freely permeable and can suppress crystal growth in living cells, facilitating cooling of living cells to the solid state in such a fashion that the cells survive after thawing (see Mazur 1970, 1984). For the same reason, glycerol has been used extensively in the freeze-fracture technique. Paradoxically, even though under certain conditions DMSO and glycerol are compatible with cell viability, these two compounds may at least partially destroy or modify cell organization (Nei 1976). The implication is that living cells can, in some way, repair the "damage". For a discussion of cryo-protectant-induced artefacts in freeze fracture see Plattner and Bachmann (1982).

It is fortunate that in the Tokuyasu thawed cryo-section technique the use of fixation, followed by sucrose infusion, makes the process of solidification extremely simple, both in theory and practice. This is because, after fixation, cells become permeable to sucrose. As long as the concentration of the sucrose is high enough (see below) vitrification of the complete block of tissue is easily accomplished using liquid nitrogen, a relatively ineffective coolant.

5.1.3 The Knife

The quality of the knife edge is of paramount importance for the quality of cryo-sections as well as resin-embedded specimens. This point cannot be over-emphasized. For cryo-sections it is even more critical because, in the absence of an embedding medium, the section is more vulnerable to damage by a poor knife. A second argument is that, if one can guarantee the sharpness and evenness of the knife, trouble-shooting is more simple; any difficulty in sectioning can then be attributed to other factors such as unstable temperature in the cryo-chamber or poor freezing of the block.

5.1.3.1 Diamond Knives

As for routine plastic sectioning, diamond knives can be used routinely for cryo-sectioning (Chang et al. 1983; McDowall et al. 1983, 1989). Cutting dry cryo-sections is a little more difficult than cutting plastic sections on a water trough, primarily because the water in the latter approach reduces friction during the sectioning process. During sectioning static electricity is generated which often hinders further sectioning (see p. 159 below). Until recently, routine diamond knives were much more of a problem than glass knives in this respect, and sections tended to stick more avidly to diamond than to glass. A related problem was that the regular cooling and warming cycles within the cryo chamber tended to crack the adhesive that holds the diamond to the knife holder. However, new diamond knives, specially prepared for cryo-sectioning, are now commercially available: our laboratory, as well as that of Drs. J. Slot and H. Geuze (University of Utrecht, pers. commun.) has had considerable success recently using a series of these knives made by the Diatome company . Evidently, other companies are now also producing successful cryo diamond knives as well. Considering both the cost of glass and the time and effort it takes to make good glass knives, we would now recommend the use of diamond knives, as it is possible to cut thinner sections routinely on such a diamond knife than on the average quality glass knife. In our experience these diamond knives cut well at a clearance angle of 6° and should be slowly warmed up to room temperature after use, cleaned with a jet of distilled water (while still wet) and dried with a jet of nitrogen.

5.1.3.2 Glass Knives

Despite the availability of diamond knives, glass knives offer a valid alternative for cryo-sectioning and are actually still preferred by many experienced workers. Their widespread use for cryo-sectioning in the last 15 years revived an interest in the processes involved in making them. The introduction of diamond knives for plastic sectioning had, understandably, led to a decline in interest in the glass knife, which was relegated to a subsidiary role of rough trimming plastic blocks. Even in many laboratories where only glass knives are used for plastic sectioning, the quality of the knives is often questionable. The most serious problem is a basic lack of understanding of the principles involved in making a good glass knife, despite the fact that this has been known for a long time (Tokuyasu and Okamara 1959).

In conjunction with Tokuyasu, we have previously described the principles as well as practical details for making good glass knives, using the principles initially outlined by Tokuyasu and Okamara (1959), (see Griffiths et al. 1983). Only the major points will be repeated here.

Theory of Knife-Making. The theory behind the method for making glass knives is very simple. In order to make 45° knives, or knives approaching this angle, a perfect "square" (described for simplicity's sake in two dimensions) of glass with orthogonal edges, is cut into two equilateral right-angled pieces by breaking the glass, as close as possible through the "corners". The closer the break goes to the two vertices, the closer the knife angle approaches 45°, and consequently the sharper the knife (Tokuyasu and Okamara 1959). When a glass knife and its complementary piece are examined (see Fig. 2), the most striking feature is a curved line extending from a point just beneath one corner towards the opposite side of the cutting face. This line, which in three dimensions is a protuberance or hillock, may be the result of unequal forces being applied to the two "triangles" during the fracture process.

Depending on whether the fracture goes to the left or the right of a vertex, a knife may have a left or a right-curving hillock (Fig. 2). In routine glass-knife making, a general misconception is that the part of the knife edge nearest to the

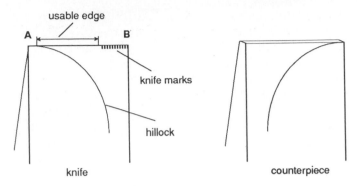

Fig. 2. Diagram of the knife (*left*) and its counter or complementary piece. In this example the knife is referred to as a "left" one since the hillock, a slight elevation in the glass surface, curves to the right (forwards *A*). (Conversely, on the counter-piece the curve goes in the opposite direction). The sharpest part of the knife is on the opposite side (*B*). Note, however, that side B is more likely to have knife marks (alternatively, the edge becomes smoother towards *A*). We advise starting sectioning just to the right of *A* and during sectioning to move gradually (if necessary) towards the centre. Eventually one moves into a region with increasing knife marks. The part of the knife from which one usually gets the best sections is indicated. Note that during the knife-breaking when the fracture goes through the opposite side of the theoretical centre plane (which would split the glass exactly down the middle) the knife would be a "right" knife and *A* would be on the right and *B* (the part with knife marks) on the left

line of the hillock (marked in Fig. 2) is the sharpest part of the knife. In fact, because of the presence of this hillock, the actual knife angle may be greater at this point than on the opposite side (B in Fig. 2). This misconception has developed because, almost invariably with knives made according to the commercial manufacturer's instructions, many nicks or knife marks are present on the side marked B in Fig. 2. Often, these marks will extend about two-thirds the length of the edge. The important point is that,

as the fracture plane approaches the vertex, the extent of the knife marks "originating" from corner B decrease and the edge becomes smoother.

In an ideal knife (with a counterpiece no more than ~ 0.1 mm thickness) the knife marks will not be visible in a stereo dissecting microscope using fibre optic illumination. In these "ideal" knives, the best and thinnest sections are usually obtained from the middle parts of the knife where a compromise is reached between sharpness and smoothness. A summary of the appearance of good and bad knives and, in particular, their counterpieces with which the knife edge is judged, is shown in Fig. 3.

It is often emphasized that the best knives are made when the glass is fractured relatively slowly. Perhaps even more important both for making squares or knives is that once the fracture starts it should proceed as evenly and smoothly as possible (Ward 1977; Algy Persson, pers. commun.). This helps to reduce the "interference lines" which are evident on the face of newly fractured glass. Too fast a break may produce chatter, and an uneven knife edge and should be avoided. Ward has recommended a fracture time of less than 1 s for

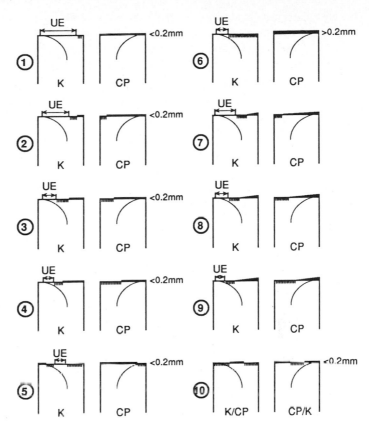

Counterpiece (CP) indicated by a thick line and real knife edge (K)
by a thin line. Knife marks are indicated by ▬ .
Usable region indicated by ' UE '.
N.B.: Knives of the type 4-6 and 8-10 are not used in practice because
the length of the usable edge (UE) is too short.

Fig. 3. Diagrams of the different kinds of knives and counter-pieces that are routinely obtained in practice. The counterpieces (*CP*) are indicated by a *thick line* and real knife edge (*K*) by a *thin line*. Knife marks are indicated by the lines perpendicular to the knife edge. Usable region indicated by *UE*. N.B. Knives of the type *4–6* and *8–10* are not used in practice because the length of the usable edge (*UE*) is too short

this process. Dr. K. Tokuyasu (pers. commun.) has always stressed the importance of a smooth, even fracturing process, which we find is easier to attain when the fracturing time is at least 2 s. A more careful study of this process would be very beneficial. It is important to note that, unless the knife meets the conditions described in the previous part (and in our publication, Griffiths et al. 1983), i.e. that the knife is an "optimal" one, tungsten coating serves no purpose: a bad knife remains a bad knife!

Even following these recommendations, not all of the knives are useful. In our experience, out of ten "ideal" knives (according to the criteria of having been broken very close to the vertex, and therefore having a complementary corner of less than about 0.2 mm (see Fig. 3), in practice about eight will be very good and the rest variable. The reasons for this variability is not clear. The situation has improved considerably with availability of the latest model of LKB knife breaker (now Cambridge Instruments), however.

5.1.3.3 The LKB Glass Knife Breaker

The LKB glass knife breaker designed by Algy Persson for the manufacturer in 1962 is at present the only machine capable of producing glass knives of sufficient quality for thin cryo-sectioning. A few years ago, the company produced a newer model, the Knifemaker II, which, though not ideal, is a definite improvement over the old machine, and is recommended. The latest model (now made by Cambridge Instruments) is again improved. An alternative possibility is to carry out modifications to the old model, as recently described by Stang (1987). Another device of similar concept is now also commercially available from Glass Ultra Micro (Box 2100, S 103 13 Stockholm, Sweden).

5.1.3.4 A Simple Modification for Use with the LKB Knife Maker

Recently, Paul Webster (Yale University) has introduced a method for making glass knives that both simplifies the procedure and increases the chances of getting "ideal knives". Since this elegant method is both unpublished and involves the older model of the LKB knife maker that is a standard feature of most EM labs all over the world, I will give the details (courtesy of Paul Webster, pers. commun.). The idea behind this method is to use a "see-saw" principle for making the balance breaks in the glass, first advocated by Tokuyasu and Okamura (1959). This method has been developed so as to require a minimal modification to the LKB knife maker II. As no engineering is required, the knife maker can be easily returned to its normal operation if required. All the part numbers referred to below come from the terminology of the LKB operating instructions supplied with each machine. A diagram outlining the principles is shown in Fig 4.

The Method

- Remove the angle setting plate (20) together with the guide plate (21), screws (40) and locking screw (17) from the cover plate (14) of the knife maker.
- Tape down the support plate (16) to stop it touching the glass. If a more permanent modification is required then this support plate can be removed by lifting off the cover plate and releasing the clips which hold the support plate.
- Tape down the arresting studs (29 and 29a) and loosen the locking screw (23) for the front glass holder (22).

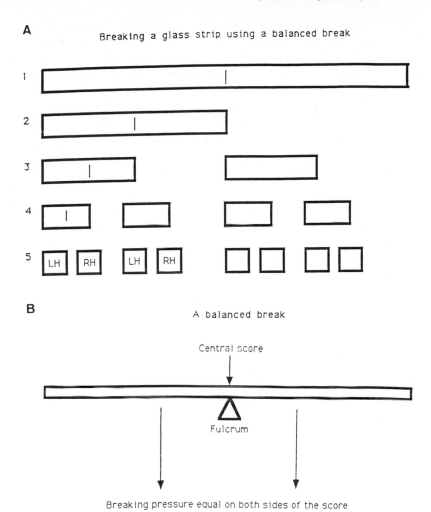

A   Breaking a glass strip using a balanced break

B   A balanced break

Central score

Fulcrum

Breaking pressure equal on both sides of the score

Fig. 4A. Illustration of the principle of making a balanced break. It is important to note that if everything is done reproducibly, each right half (*RH*) should be identical, as is each left half (*LH*). However, LH and RH are likely to be of slightly different dimensions. It is recommended to make a large number of RH and LH before starting to break knives. Choosing one half first, each time positioned in the knife maker in the same orientation (with respect to the other half) one starts breaking knives. Usually the frequency of obtaining good knives is low at first as one needs to constantly adjust the positioning of the fracture relative to the two vertices (one "aims" for the vertices). In **B**, the "see-saw" principle is shown for balancing the glass rod in order to break it exactly in the centre. For the final double square, one uses the arresting studs as a stop. Courtesy of Dr. Paul Webster, Yale University

- Adjust the front glass holder to its most forward position (<4) and reclamp with the locking screw.
- Balance the clean glass strip on the breaking pins (26) at the equilibrium point. The only part of the knifemaker touching the glass strip should be the breaking pins.
- Bring down the rear glass holder (28) to clamp the glass strip in place and adjust the clamping pressure to two divisions on the scale of the rear glass holder by releasing the locking screw (9) and pushing back the spring loaded support for the rear glass holder.
- Select the rough cutting symbol on the score selector (13) by setting it in the up position. It is not necessary to change this setting for other scores.
- With scoring shaft (15) pushed in, bring down the clamping head (8) onto the glass strip and lock it in position with the locking lever (2).
- Score the glass and then break it by turning the breaking knob (18) clockwise and then reset the knob.
- Support the scoring shaft and lift the clamping head until it is locked up.
- Remove the two equal halves of glass strip and break each half by repeating step 5 to 10 and then break each quarter piece in the same way.
- Remove the tape from the arresting stub and 25 mm glass strips (29) and break each 2" length (double squares) as recommended in the LKB instructions for making squares. When removing the broken squares from the machine keep each set in the same orientation as when they were broken. Do this for all the double squares.
- When all the double squares have been broken readjust the knife maker by setting the front glass holder to setting 7 on the scale.
- Place a left hand square in the apparatus so that one of the newly broken corners is set between the pegs of the front glass holder. It is important when repeating this step to use the same orientation for all the left hand squares. The other corner is similarly placed between the pegs of the rear glass holder.
- With the score selector on the rough break setting bring down the clamping head onto the glass square and then score and break.
- Remove the two knives for evaluation and adjust the position of the glass holders as required. The best knives are obtained when the break passes as close as possible to the two corners of the glass square. The glass holders should be adjusted to maximize the possibility of the break passing through these corners.
- The left hand squares are broken by repeating steps 13 to 16 until usable knives are obtained. Although it is possible to occasionally obtain good knives from the right hand squares, do not adjust the glass holders for these squares, as they are not all the same size.

Once the knife breaker has been modified in this way the optimal position of the glass holders has been determined, knife making becomes a simple task with no need to further adjust the settings on the apparatus other than those mentioned.

5.1.3.5 Tungsten Coating

When "ideal" knives are obtained the edge appears to "dull" over a period of days, although this is a difficult point to prove. If knives are immediately, i.e. within a few hours, coated with a thin layer of tungsten, however, they maintain their sharpness indefinitely (at least for 2 years). The use of tungsten was originally proposed by Roberts (1975), who coated glass knives with a variety of metals and empirically found that tungsten coating facilitated sectioning while all other metals tested had no noticeable effect on the quality of sectioning. For the past few years we, and others, have used this method routinely (Griffiths et al. 1984). The precise role of tungsten is not clear but, according to Peters (1984, and pers. commun.), it is unlikely that such a low amount of tungsten($\approx$ 1 nm) would uniformly cover the entire surface of the glass. The more likely situation, he believes, is for the metal to form small crystals or microdomains. These may act as "ball-bearings" on the glass surface which facilitate the flow of the sections over the glass cutting surface. The effect in practice is striking: in contrast to the cutting edge of an uncoated knife the cutting surface of a tungsten coated knife allows the cutting of many more sections. Accumulation of section debris or ice, which normally result in increased friction and dulling of the knife edge, has no noticeable effect on these knives. Further, good knives may be rinsed with distilled water after use, dried with a jet of nitrogen gas and used on subsequent days.

5.1.3.6 Procedure for Shadowing with Tungsten

The knives are placed in the vacuum evaporator such that both the front and the back of the knives can be equally coated (Fig. 5). The v-shaped tungsten wire is placed about 7 cm above the knives. We use one filament of the $\approx$ 1 mm, three filament helix of tungsten sold by Balzers. We have empirically found that the best results are obtained when the current is set so that the tungsten wire would

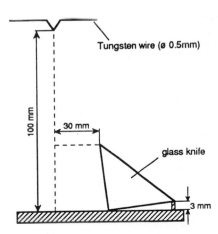

Fig. 5. Diagram to show how tungsten is coated on the freshly fractured knife

burn out in about 3 min at a voltage of 30 V. Using this setting, the actual shadowing is done for only 30 s. A simple rule of thumb is that the time of shadowing should be roughly one-sixth of the time required for the filament to burn out. This thin layer of tungsten is enough to produce a very light grey colour on filter paper strips. Stang and Johansen (1987) have recently described the design of a convenient round holder that facilitates the ease with which many knives can be mounted in the evaporator.

The knives are conveniently stored in glass or plastic containers on strips of double-sided tape. We routinely store each knife next to its complementary piece in order to facilitate the evaluation of the quality of the knife-edge (Fig. 2).

5.1.4 The Microtome

All the cryo-microtomes commercially available are now capable of cutting thin cryo-sections reproducibly and with ease. By far the most important attribute of a good cryo-ultramicrotome is temperature stability, which is a measure of how well the cryochamber is insulated. The cryochamber, which holds both the specimen and the knife at the required temperature of sectioning, is mechanically coupled to the microtome body, which is at room temperature. It is imperative that heat transfer between these two compartments be kept at a minimum otherwise the block or the arm of the microtome containing the block will contract or expand independently of the microtome-controlled mechanical or thermal advance. In the ideal machine, a 100 nm advance setting would always give 100 nm sections (assuming the knife is optimal). Such precision, however, cannot be expected of any commercial microtome. It should be said, however, that the latest models by the three manufactures, LKB, Reichert (now combined as Cambridge Instruments and recently taken over by Leica) and RMC Inc (formerly Sorvall-Dupont) are vastly improved when compared to most of their predecessors, and all are capable of cutting thin or semi-thin sections in a reproducible manner (all these are used on a routine basis in our laboratory and all can be recommended).

A fibre-optic illumination system, which is not supplied with all of the current cryo-microtomes, can sometimes facilitate the cryo-sectioning process since it eases the identification of section interference colours, as well as making it easier to judge the quality of the knife and its position relative to the block face. For the latter purpose, the "back"-lighting system present on some of the latest cryo-microtomes is also very suitable.

All of the commercial cryo-microtomes can be easily and quickly modified for conventional plastic sectioning. For more details of these instruments see Sitte (1984), Robards and Sleytr (1985), Menco (1986), and Sitte et al. (1987).

5.1.5 Cryo-Sectioning Theory

A perfect section is a three-dimensional segment of the block that has the same dimension as the block face. It is made of two flat planes and an edge of finite thickness, the section thickness. However, in practice ultra thin sections from any kind of material, be it plastic or blocks of frozen tissue, are never faithful segments of the block. This statement has profound implications for the evaluation of fine structure.

It may be surprising for a beginner to discover that the sectioning process is very complex and far from being understood. If sections are not faithful images of the block face, it is important to know why this is the case. If, for example, the "errors" are systematic ones, we might be able to correct for them in our final interpretation of the image. The answer, in short, is that some of the errors are systematic but others are not. In this respect, the use of electron diffraction to unambiguously assess the physical state of the embedding medium, water, in cryo-sectioning has provided much useful information (see Dubochet et al. 1988).

Before we discuss the actual mechanism of section formation, the artefacts or "errors" in cryo-sectioning as we presently understand them will be described in order of importance (see also Zierold 1987).

5.1.5.1 Artefacts of Cryo-Sectioning

• *Compression.* The most serious unavoidable artefact is a compression of the section in the direction of cutting which reduces the height of the section. Dubochet and colleagues have made a systematic study of this phenomenon after cryo-sectioning (Chang et al. 1983) and found that, whereas the length of the section is reduced, on average, by 30–50%, the width is not measurably altered. This will increase the **thickness** of the section in proportion to the amount of compression. Compression of a similar magnitude is also evident during plastic sectioning if the sections are cut, as in cryo-sectioning, on a dry knife (Griffiths et al. 1984a). In the latter study the compression was not evident when the Epon sections were floated on a water surface. This may not always be the case, however (see Weibel 1979).

• *Crevasses.* These are a series of cracks going deep into the sections, as their name suggests. They are mostly parallel to the knife edge. Their width is in the 100 nm range and they appear mainly on the last formed surface of the section. Crevasses, unlike compression, are to some extent avoidable: whereas they are always present in samples where water has crystallized (either hexagonal or cubic crystals), they are far less common when vitrified samples are sectioned with a sharp knife (Dubochet and McDowall 1984; Chang et al. 1983).

• *Chatter.* Chatter, a common artefact of plastic sectioning, may also occur during cryo-sectioning. It is a periodic variation of section thickness in the cutting direction (with a wavelength of > 0.5 mm). Chatter is generally attributable to either mechanical vibrations in the microtome or to using too large a clearance angle for the knife. Hagler (pers. commun.) has pointed out that chatter during cryo-sectioning may also be introduced when the microtome is operated manually or by the habit of resting one's hand on the microtome to manipulate the sections during the sectioning process. When cutting plastic sections on water, it is widely accepted that one should not touch the microtome during the sectioning process since this may produce vibrations and thus chatter. The precise events that occur during the sectioning process are poorly understood. For a discussion the reader is referred to Leunissen et al. (1984), Sitte et al. (1987) and Dubochet et al. (1988).

● *Surface Deformation.* A perfect section will have two flat planes on the top and bottom surface. Any deviation from this ideal will result in surface deformations. Crevasses are one example of surface deformations, knife marks are another. In cryo-sections, fracture, as in freeze fracture, is an additional possible cause of deformation. Since dry sections are manipulated, usually with a hair, this physical manipulation may also be a source of cracks in the sections.

● *Knife Marks.* The surfaces of an ideal knife are two perfect planes. Any breaks in these planes result in knife marks in the section. These are areas which coincide with the discontinuity in the knife edge where the section is "scraped" (i.e. is damaged) rather than sectioned. According to the model proposed by Dubochet et al. (1988; see their Fig. 21), knife marks are always on the surface of the section farthest from the block face and are formed during the cutting of the preceding section. For more details see also Richter et al. (1991) J Microscopy 163:19–28.

5.1.5.2 The Process of Cryo-Sectioning

Consider a knife sectioning ice cream. When the ice cream has been left in the deep freeze it becomes harder and more force is required to cut a section, even with a sharp knife. What tends to happen in this case is that the knife cuts the superficial layer of the material and at a certain depth the ice cream "fractures" along the plane started by the knife. Whereas the "cut" surface is rather smooth, the fractured surface is always rough. When ice cream is "sectioned" at warmer temperature the knife can "glide" through the block without the fracturing phenomenon but now it becomes difficult to section thin slices.

Leunissen et al (1984) have improved our understanding of the sectioning process by looking at replicas of the surface of cryo-sections which are similar to replicas made in the freeze fracture method. They showed, for example, that the cutting speed could influence the outcome of the sectioning process. Using high speeds (90 mm/s), especially combined with warmer temperatures (−70°C or 203 K), the surface of cryo-sections resembled typical freeze-fracture images. At low cutting speeds (0.1 mm/s) fracture images were only seen in a few cases at −120°C (153 K) or colder. Dubochet and colleagues have further extended this work by using the hydrated cryo-section method. This is a very sensitive method in that a vitrified block, and sections thereof, will remain vitrified only if the temperature remains below the re-crystallization temperature ($-135\,^{\circ}$C or 138 K for pure water; higher for biological specimens). Hence they could address the question – is cryo-sectioning possible without "melting", i.e. raising the temperature. The answer to this question is "yes": cryo-sections of vitrified blocks, that remained vitrified, have been routinely obtained. This also agreed with the study of Karp et al (1982) that showed that solidified toluene could be cryo-sectioned at a temperature only $1\,^{\circ}$ below its melting point: in other words, the amount of heat generated during sectioning was not enough to raise the temperature of the section by $1\,^{\circ}$. These observations are also supported by theoretical calculations showing that the amount of energy dissipated at the knife edge during sectioning and converted completely to heat should not significantly raise the temperature of the section or block (Hodson and Marshall 1972). In support of Leunissen's work, Dubochet and colleagues have now found that slow cutting speeds (< 1 mm/s) are a prerequisite for cryo-sections to remain completely vitrified. At higher sectioning speeds small hexagonal ice crystals were observed on one surface of the cryo-section, presumably due to "melting" or a temperature rise during sectioning.

For the Tokuyasu technique, the presence of high sucrose concentrations results in tissue blocks which are much softer than the non-cryoprotected blocks prepared for hydrated cryo-sections, or for X-ray microanalysis. The effect of the sucrose, as already mentioned, is to increase the **plasticity** of the block, thus giving it properties which are more "liquid like". Blocks infiltrated with PVP-sucrose (see below) are even more plastic. This plasticity should ensure that the sectioning process is more of a "cutting" rather than "fracturing" process. Furthermore, using high sucrose concentrations (> 1.0 M), ribbons of sections are often obtainable: in this case the ends of the sections are "glued" together. This phenomenon is probably not due to a melting of the ends of the sections, as suggested by Appleton (1974), since we were able to show that even ribbons of sections produced after fixation and sucrose infusion were vitreous when examined in the cold stage of the EM below $-140\,^{\circ}$C (Griffiths et al. 1984). If the temperature had been raised at anytime above about $-120\,^{\circ}$C (≈ 150 K), ice crystals would have been detected.

The model suggested by Dubochet and McDowall (1984) (see also Zierold 1987) is recommended as a guide in order to appreciate what may be happening during the sectioning process. It is not really clear how the sectioning process leads to the compression that invariably occurs during sectioning, but it appears likely that friction building up at the surface of the knife plays an important role. Thinner sections are usually more compressed than thicker ones: this may be due to the fact that a larger percentage of the total mass of the section is involved in interactions with the surface of the knife in the case of a thin section (Dubochet and McDowall 1984). When the section is allowed to float on a liquid, the forces that caused the compression can be relaxed and the effects of the compression reversed. Hence, Epon sections are compressed when cut on a dry knife, but if a water trough is present at the knife edge, little compression is seen (Griffiths et al. 1984). The surface of a dry knife definitely offers a certain amount of friction to the surface of the section. In our experience, this is greatest with a conventional diamond knife and least with a tungsten coated glass knife or the new generation of "cryo"-diamond knives available commercially (see above). The effect of this friction is to further increase section compression, as well as to hinder the cutting of subsequent sections.

The compression phenomenon can probably not be avoided in hydrated cryo-sections. Until now, attempts to reduce the extent of compression of sections to below about 30%, have been unsuccessful. Even very low angle diamond knives (20–30°), which, from theoretical considerations, might be expected to reduce compression gave no obvious advantage (McDowall, pers. commun.). The recently introduced anti-static device may, however, improve this situation (see below). As mentioned above, thawing of the cryo-protected sections in the Tokuyasu technique may allow the effects of compression, to be reversed and the original dimensions of the sections as well as the cellular and subcellular details to be restored. One cannot assume, however, that compression is always quantitatively overcome during thawing. This must, in part, depend on the plasticity of the material being sectioned. As shown in Fig. 13, latex beads taken up by phagocytosis by macrophages are always compressed in our cryo-sections, even after picked up on a sucrose drop. It is also possible that the stretching during thawing may overcompensate for the effects of compression.

The ideal theoretical situation for cutting cryo-sections would be to use a collecting liquid (as with plastic sections) that remains liquid at the cutting temperatures. One of the problems with this approach is that suitable liquids such as freon or ethane have a very low surface tension (unlike water). Sections will rather sink than float, and cannot be recovered. Until now, these attempts have always failed (Hodson and Marshall 1972).

In conclusion four factors should be taken into consideration for successful cryo-sectioning:

Speed of Cutting. The slowest cutting speeds are recommended in order to avoid fracturing and possibly devitrification of the section surface. This statement is based on theoretical considerations. In practice, however, it is often possible to obtain acceptable sections for the Tokuyasu technique with relatively high cutting speeds. This is especially true if the blocks are relatively soft. This is the case, for example, when the machines are beginning to run out of liquid nitrogen. There is a period during which the temperature is raising where it is still possible to cut good sections (mechanically) at relatively fast speeds.

Temperature. The colder the block the harder it becomes and, given suitable plasticity (see below) the easier it should be to cut thinner sections. For hydrated cryo-sections, where high concentrations of sucrose have not been used, temperature is a more critical parameter: one should aim to stay below $-135\,^{\circ}C$ (138 K), which is the devitrificaton temperature for pure water. At higher temperature (starting at $-100\,^{\circ}C$ or 173 K), water will start to evaporate (by sublimation). For fixed, sucrose-infused tissues, the temperature chosen depends more on the section thickness required and on the sucrose concentration used to infuse. Using 2–2.3 M sucrose we normally cut thin sections around −100°C (173 K) and 0.5 μm sections at about −60°C (213 K). Some workers prefer to have a temperature difference between the block (−100°C) and the knife (−90°C). While this may facilitate sectioning, it is clearly not essential. Using the original Sorvall MT2B with FCS cryo-attachment, many of us were able to section routinely with both block and knife at the same temperature.

Plasticity of the Block. The ease with which a block sections rather than fractures is a function of its "plasticity", which depends both on its chemical composition and on the state of solid water within it. Vitrified blocks are often, but not always, relatively easy to section compared with crystalline blocks. Furthermore, the presence of large, hexagonal ice crystals usually makes blocks impossible to section. The extent to which a block can be vitrified depends on both its water content and the speed of cooling (cooling rate). Replacing even a part of the water with sucrose as is done for the Tokuyasu technique, for example, increases the chances of vitrification. It also makes blocks softer and more plastic. The higher the sucrose concentration the softer the block. At temperatures below about −120°C (153 K) high sucrose concentration-infused blocks usually become brittle. At lower sucrose concentrations of sucrose, the plasticity can also be greatly affected by the amount of protein and lipid in the tissue (see Tokuyasu 1980). At the highest sucrose concentrations (2.3 M) the plasticity is, in general, more a function of the sucrose itself, so that different tissues tend to section similarly. The recent PVP-sucrose method of Tokuyasu (1986 1989), however, may be used to significantly improve the plasticity of the tissue block, if necessary (see below).

The Knife

Four factors must be considered here:

- *The sharpness or knife angle*. In our experience knives with angles "approaching" 45° appear to section most smoothly. Up to now there have been no practical possibilities for making glass knives with angles less than 45°, despite claims to the contrary by some of the companies manufacturing knife breakers. Although not generally commercially available, low angle diamond knives have been tested over the years, in the hope that they would reduce the compression phenomenon (Dubochet et al. 1988). The general conclusion was that the lower angle did not improve matters (but see below). For plastic sectioning, however, distinct improvements have been seen by Jésior (1989). This author has suggested that the reason for the lack of improvement in cryo-sectioning is due to the friction that occurs when sectioning in the absence of liquid. According to Mr. H. Gnägi from the Diatome company (pers. commun.), the use of an anti-static device, providing a negative charge in the cryo-chamber, has recently looked promising in combination with these low angle knives (see below).

- *The quality of the edge.* The presence of knife marks or discontinuities in the edge are clearly to be avoided if possible. With diamond knives, one must rely on the manufacturer for the quality. Here I should point out that cryo-sections, especially hydrated cryo-sections, are a more sensitive indicator of knife marks than plastic sections. In our experience, even new diamond knives have caused some knife marks. With glass knives the situation is similar. When very sharp knives are made, some are as relatively free of knife marks as are good diamond knives, others are full of them. The reasons for this are far from clear: it cannot, at present, be attributed to any obvious factor such as the speed the glass is broken for example. Furthermore, tungsten coating has no apparent effect on the presence of knife marks. Cutting cryo-sections would be greatly facilitated if conditions could be found that reproducibly give edges free of knife marks.

- *The surface characteristics of the knife.* The amount of friction the sections are exposed to is a function of the "surface quality" of the knife, a phenomenon about which we understand little, except to say that it clearly varies from one knife to the next.

- *The clearance angle.* Sectioned with a $0°$ clearance angle is obviously not possible. However, as the angle becomes larger, the chance of "chatter" increases, that is sections have periodic regions of uneven thickness. Alternatively, chatter may also occur when the specimen touches the surface of the knife at small clearance angle. A compromise is usually obtained empirically. In practice, clearance angles between $4°$ and $8°$, as for plastic sectioning, are found acceptable for cryo-sectioning.

5.1.5.3 Sectioning in Practice

After inserting the knife and specimen into the cryo-chamber it is advisable to wait for 5–10 min for their temperature to equilibrate with that of the cryo-chamber.

The knife must then be advanced towards the specimen manually. We recommend starting the sectioning near the edge of the knife which is over the curve (towards A in Fig. 3), that is the part with the least knife marks. It is also useful to rotate the block before advancing so that the narrowest, and hopefully the straightest edge meets the knife edge first: this reduces the area/length of knife edge used, and may also reduce the effects of friction.

Extra care must be taken when the knife is advanced towards the mounted specimen. Ideally one would like to go as close as possible, without touching the block. Cutting (or, more likely, fracturing) a very thick section (> 1–$2\,\mu m$) from the block face should be avoided at all costs. A useful tip is to advance the knife as close as possible using the binocular microscope and then, setting the automatic advance of the fine-feed mechanism to cut a thick section (up to 0.5 μm). The specimen is then advanced, either mechanically or manually, as fast as possible, until contact is made. At this point the mechanical advance feed may be reduced to zero.

It is important to ensure that the block is not advancing thermally due to a temperature difference between the block and the knife and/or chamber.

Once it is clear that no sections are cutting with the advance setting on zero, the process of sectioning can begin. For beginners, a "Matterhorn" type block is recommended, that is, one which tapers from a larger base to give a very small block face. With such blocks the chances that a very thick section will fracture deep into the block face are reduced. Although the initial sections are very small, with time, the size of the sections obtained will increase.

For an experienced operator a few sections are enough to evaluate the quality of the knife, the block, as well as the temperature stability of the chamber. Once a few complete semi-thin sections have been obtained from the block, ultra-thin sections for electron microscopy should be the goal. The presence of any clear interference colours is an important sign that the surface of the block and section is relatively smooth (polished). For every knife, at any particular temperature, there exists a critical section thickness below which sections will be scraped rather than cut. With a good glass knife this point is reached, in our experience, with sections that give gold interference colours. These are probably 50–80 nm, plus 30–50% due to compression (Griffiths et al. 1984). With very small blocks thinner sections can be cut (Tokuyasu 1974 1989). Sections giving gold, red, purple or blue interference colours are acceptable for routine labelling studies. Those giving greenish hues or gold-green, which can easily be mistaken for true gold sections, are generally too thick. The consistency of the section, as it comes off the block, is a more useful guide to thickness than are interference colours. Very thin sections are fluid or cellophane like and come off the block as do tissue-paper sheets from a dispenser box whereas thicker sections are more rigid; to continue the paper analogy, medium-thick sections would come off like "kitchen-roll" paper, and thick and very thick sections as sheets of writing paper or card, respectively.

The sections, once cut, should be manipulated with an eyelash tip so that they are positioned in small groups without sections lying on top of each other.

The buildup of **static electricity** within the chamber is often a hindrance to sectioning and to the ease with which the sections can be manipulated. Whereas this varies from day to day and cannot be completely avoided, a few suggestions can be made to reduce its effects. The simplest is to use a commercially made anti-static device which emits a small electric current that allows the accumulated ions to escape ("zerostat"). This is especially useful when sections jump from the knife onto the block face. In this case, the block face is moved as far above the knife as possible and slowly irradiated with the "gun" by pressing the trigger slowly in one direction only, avoiding the "click" at the end. If it does not work in one direction (+ charge), try the other (- charge). Other possibilities are devices that can be inserted into the cryo-chamber, such as polonium rods (available in the USA only) that emit low levels of beta radiation.

Recently, a new low voltage anti-static device has been introduced and is available from the Diatome company, Biel Switzerland. Mr. H. Gnägi (Diatome Co) recently gave us a demonstration of this device, which can fit into any cryo chamber. The improvement in the ease with which serial sections were obtained with this device was remarkable. It is envisaged that for diamond knives, to which the sections have tended to adhere strongly when compared to glass knives, this device will offer even more advantages. According to Mr. Gnägi (pers. commun.), when used in combination with this anti-static machine, the low knife-angle diamond knives (20–35°) do, in fact, appear to reduce compression. He has suggested that the reason that an improvement was not seen in the past with low angled knives was that an increased amount of static electricity was produced which caused a relatively higher friction on the knife edge. For more details see Michel et al. (1991) J Microscopy 166:43–56.

5.1.6 The Hydrated Cryo-Sectioning Technique

In practice the hydrated cryo-section technique is much more difficult than the thawed frozen section technique. The major problems can be briefly outlined. For details and examples see Chang et al. (1981, 1983); McDowall et al. (1983, 1984, 1989); Dubochet et al. (1988).

- Since the cooling step is usually done on fresh (unfixed) tissue, rapid-freezing methods must be used. As mentioned, there are severe limitations on the volume of tissue that can be vitrified.
- Dry sectioning and all subsequent steps must be carried out below $-140\,°C$ (130 K) to avoid devitrification.
- The sections must be transferred from the knife to the grid in the cryochamber and then pressed onto the grid. This is a technically demanding step. The grids are then transferred, usually under liquid nitrogen, to the cold stage of the electron microscope where they are examined at temperatures below $-140\,°C$.

Even after the desired vitreous state is obtained there are a number of other difficulties with hydrated cryo-sections. The first is the unavoidable presence of cutting artefacts, the major one being compression. This is usually in the magnitude of 30–50% in the direction of cutting (Chang et al. 1983). Less serious but still problematical are knife marks, crevasses and other surface deformations (Chang et al. 1983). The second major difficulty is the low inherent amplitude contrast in the unstained sections. This technique therefore utilizes phase-contrast information as opposed to amplitude contrast information in conventional EM specimens.

Dubochet and colleagues have made a number of observations which are of fundamental importance to any cryo-sectioning technique including the thawed frozen section method (see Dubochet et al. 1988). The main points of relevance will be listed here:

- In the EM the state of solid water can be unambiguously identified using electron diffraction. It is the **only** approach whereby this is possible.
- From the optical density of micrographs taken under standard operating conditions of the EM, the **mass thickness** of structures in sections can be estimated:
  Mass thickness = density × section thickness = mass per unit area of section. From this formula, the section thickness can be estimated if the density of material in the section (i.e. grams per cubic centimetre) is known, or conversely, the density can be estimated if section thickness is known (Dubochet et al. 1983).
- Using mass-thickness measurements, the amount of water in frozen sections can be directly estimated by comparing sections before and after freeze-drying (done by raising the temperature of the cold stage to between −80° to −110°C (163–193 K) (Dubochet et al. 1983).

5.1.6.1 Use of the Hydrated Cryo-Section Technique to Study the Tokuyasu Technique

In a combined effort with Dubochet and McDowall, we have used the hydrated cryo-section technique to study the major physical parameters involved in the Tokuyasu technique, namely "freezing", sectioning, thawing and drying of the sections (Griffiths et al. 1984). The main points of this study will be described here.

A spherical tissue culture cell grown in suspension was used as a model for this study. Cell pellets, up to 1 mm³ in volume, were fixed with glutaraldehyde, infused with sucrose, frozen in liquid nitrogen and cryo-sectioned. The sections were then observed in the cryo-state in the EM and compared with sections from the same block that were thawed and dried. The main conclusions drawn are listed below:

- Using 1·6 − 2·3 M sucrose, the cells were routinely vitrified in liquid nitrogen. When 1 M sucrose was used, the cells were **either** vitreous (the majority) or they had cubic ice intracellularly while hexagonal ice was always present extracellularly. Although not measured, it is reasonable to assume that the formation of hexagonal ice extracellularly resulted in an outward movement of water from the cells (and shrinkage). When 0.6 M sucrose was used, hexagonal ice was present both intra- and extracellularly. Hence with high sucrose concentrations, vitrification can be expected throughout relatively large blocks (> 1 mm³) by using liquid nitrogen only, despite the fact that this coolant is known to be one of the least efficient (see Roos and Morgan 1990). When a lower concentration of sucrose must be used, more efficient coolants are recommended. I should emphasize, however, that these results refer directly to one cultured cell line, CHO (Chinese Hamster Ovary) and are not necessarily applicable to tissues with different water contents. It should also be noted that the hardness of the material is only one of the relevant criteria for sectioning: other factors, in particular the plasticity of the block, must also be considered (see below).

- We observed an average compression of 42% in the direction of cutting during sectioning. Upon thawing, however, the cell profiles completely regained their circularity. A similar phenomenon was also observed when the same cells were embedded in Epon and sectioned either on a dry knife (resulting in 36% average compression) or with a liquid trough (no compression).
- The average thickness of our routine hydrated cryo-sections was 110 nm as determined by mass-thickness measurements, but thinner sections can certainly be cut without difficulty, down to about 50 nm. In special cases, sections as thin as 20–30 nm have been cut, (see Tokuyasu 1974). When sections were thawed, washed and simply air-dried, the originally 110-nm-thick sections shrank to give an average thickness of about 30 nm. The density of material in the section increased correspondingly. As described later, this air-drying "collapse" is accompanied by severe structural damage. Under conditions of "protected" drying using the methyl cellulose technique the collapse was to a large extent prevented (see below).
- As discussed in more detail in Chapter 3, there is evidence in the literature both for (e.g. Pentilla et al. 1974) and against (e.g. Bone and Denton 1971) the theory that sucrose is free to enter cells after fixation. Mass-thickness measurements enabled us to show directly that the concentration of sucrose in the cell interior after vitrification was the same as that in the extracellular space, that is, that the sucrose was freely accessible to the cytoplasm (for more details see Griffiths et al. 1984).

5.1.7 Tokuyasu Thawed Cryo-Section Technique

The thawed frozen section technique (Tokuyasu 1973, 1978; Tokuyasu and Singer 1976) offers a number of advantages over all other methods for high resolution immunolabelling of ultra-thin tissue sections. These can be listed as follows:

- It is the most sensitive post-sectioning technique for immunolabelling since the initial aldehyde fixation is the only potential chemical denaturation step for antigens. Freezing and thawing, which could be envisaged to be harmful, appear not to significantly affect antigenicity in this technique.
- In cryo-sections, fine structural preservation down to the molecular level is as good as, if not better, than all other available techniques for sectioning chemically fixed cells and tissues. Further, the possibilities for contrasting structures, especially to show high resolution detail, are greater than for any other method.
- The entire procedure, from fixation to photographic documentation, can be routinely completed in one working day.

The principle steps in the procedure are (see Fig. 6 for a summary):

- Chemical fixation.
- Embedding cells or tissues.
- Cryoprotection and freezing.
- Cryo-sectioning.
- Retrieval, thawing and transfer of sections to grids.
- Antibody labelling and marking.
- Contrasting with heavy metal stains and drying of sections.

5.1.7.1 Chemical Fixation

Since fixation has already been discussed at length in Chapter 2, only a few relevant points will be made in this chapter. Chemical fixation here essentially means aldehyde-fixation since, until now, no other cross-linking agents have yet been used for this technique. In the future, however, other reagents may need to be tested for situations where aldehydes are not satisfactory. For example, an imidate can be used to partially block amino groups on proteins in order to modify subsequent glutaraldehyde cross-linking (Tokuyasu and Singer 1976; Geiger et al. 1981).

In addition to stabilizing cell structure, aldehyde fixation has an additional important role in the Tokuyasu cryo-section technique. Unfixed cells are impermeable to sucrose, the most widely used cryo-protectant in this technique, whereas fixed cells are permeable to sucrose (Griffiths et al 1984). Infiltration with cryo-protectant is essential for vitrification of relatively large tissue pieces.

Using the Tokuyasu technique for the first time, or applying it to an unfamiliar tissue, it is advisable to fix the tissue initially under conditions that have given good morphological preservation with conventional plastic embedding. In other words, if 2% glutaraldehyde for 1 h gives good morphological results in plastic sections, it is reasonable to assume that this protocol will give adequate cross-linking for frozen sections. At this stage, obtaining frozen sections with good structural preservation should be the major objective: preservation of antigenicity and conditions for labelling can be considered at a later stage. The presence of sucrose will ensure that the tissue pieces will be vitrified upon plunging into liquid nitrogen. Specimens with large extracellular spaces may need to be additionally supported by embedding with compounds such as gelatin or agarose, before infusion with sucrose (see below).

For optimal antibody labelling, fluorescence microscopy (or other sensitive light microscope techniques) may be used to assess the effects of fixation on the antigen, as well as to give some indication of the signal which is to be expected from the immunolabel. **Conditions should first be used that are most likely to give a positive signal at the light microscope level**[1], such as thick frozen sections, 0.5–

---

[1] These statements assume that the preparation for light microscopy does not result in loss af antigen. This is an important consideration for small molecules.

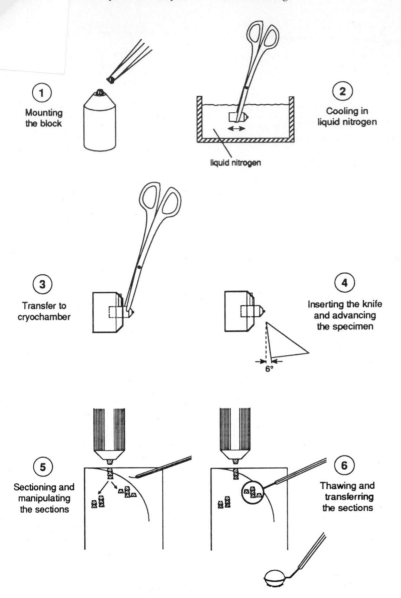

Fig. 6. Schematic diagram of all the practical steps involved in the Tokuyasu procedure

5 μm (sectioned on a cryostat or ultracryomicrotome) or by the use of permeabilized tissue culture cells. **Failure to obtain a significant signal under these conditions makes it pointless to even attempt localization at the electron microscopic level.** In this way, one can establish the range of antibody dilutions most useful for electron microscopy and obtain a rough idea of the distributions of label over the cells or tissues being examined. This is also an ideal way to test

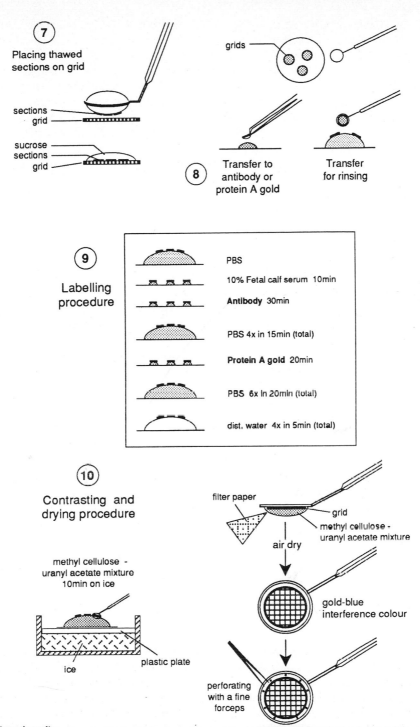

7

Placing thawed
sections on grid

sections
grid

sucrose
sections
grid

grids

8   Transfer to
antibody or
protein A gold

Transfer
for rinsing

9

Labelling
procedure

PBS

10% Fetal calf serum 10min

Antibody 30min

PBS 4x in 15min (total)

Protein A gold 20min

PBS 6x in 20min (total)

dist. water 4x in 5min (total)

10

Contrasting and
drying procedure

methyl cellulose -
uranyl acetate mixture
10min on ice

ice   plastic plate

filter paper

grid

methyl cellulose -
uranyl acetate mixture

air dry

gold-blue
interference colour

perforating
with a fine
forceps

Fig. 6 (continued)

the effects of different fixatives on the labelling signal. Additionally, **unfixed**, fresh frozen 5–20 μm sections may also be tested in this way, if necessary. In general, the highest fixative concentration consistent with an acceptable signal at the light microscopy level, should be used for EM. One should again note, however, that in general the degree of cross-linking required for frozen sections is far less than that required for plastic sections. Even a few minutes in 4–8% formaldehyde may be adequate (although not ideal) for giving reasonable fine structure preservation of cultured cells provided the cells are not subsequently washed with buffer (i.e. they can be infiltrated directly with 2.1 M sucrose containing 1% formaldehyde and the fixation allowed to continue during the infusion period).

5.1.7.2 Embedding Cells or Tissues Before Freezing

Often, organelle pellets/suspensions, cells or tissues may need to be embedded in a suitable support medium in order to keep cells together during fixation, and cryoprotection, to facilitate trimming and mounting the block in a preferred orientation, or simply to keep the tissue together after thawing the sections. This is especially useful for tissues which have large extracellular spaces, such as lung. It should be noted that all the compounds used for facilitating embedding are high molecular weight compounds that will not pass through plasma membranes and, therefore, can only play an extracellular role. In practice three kinds of compounds have been used – gelatin, fibrinogen and agar compounds. These will be described in some detail. Note that this approach can also be used to embed material in plastic resins such as Lowicryl.

Gelatin. Fixed tissues can be left to infiltrate with a 10% aqueous solution of gelatin at the lowest temperatures at which the solution is liquid ($\approx 30°C$) for up to 15 min. Note that the time required for infiltration needs to be sufficient only for access to all large **extracellular** spaces. The tissue pieces are then placed on ice to solidify the gelatin. This can either be done by first taking the tissue pieces out of the bulk gelatin and placing them in a suitable container (e.g. a petri dish) or by leaving them in the bulk gelatin. In the latter case, after cooling, tissue pieces must be trimmed with a scalpel or razor blade prior to the fixation step. Fixation is necessary here to cross-link the gelatin to the tissue and either glutaraldehyde or formaldehyde may be used. As for fixation in general, the smaller the pieces, the more effective will be the cross-linking. Preferably they should be less than 1 mm^3.

For suspensions of fixed material such as cultured cells or organelles prepared after homogenization the warm gelatin is added to a pellet of the material. It is better not to resuspend the preparation in the gelatin but simply to centrifuge the gelatin onto the specimen (this can usually be done for 1–2 min at room temperature, before the gelatin sets, Fig. 7A). This is conveniently done in an Eppendorf microfuge and, after centrifugation, the tube is placed on ice and after a few minutes, fixative is added. After sufficient time for cross-linking, the

A.

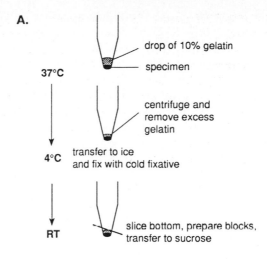

drop of 10% gelatin
specimen

37°C

centrifuge and
remove excess
gelatin

4°C   transfer to ice
and fix with cold fixative

RT   slice bottom, prepare blocks,
transfer to sucrose

B.

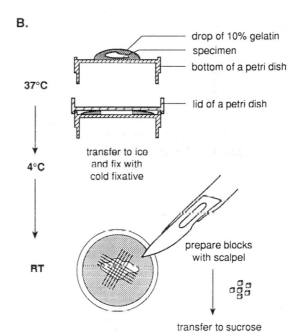

drop of 10% gelatin
specimen
bottom of a petri dish

37°C

lid of a petri dish

transfer to ice
4°C   and fix with
cold fixative

RT   prepare blocks
with scalpel

transfer to sucrose

Fig. 7A,B. Practical details of
methods to embed cell pellets or
pieces of tissues in a gel prior to
cryo-sectioning

tube is sliced with a razor-blade or scalpel and the gelled material can be
trimmed into smaller pieces and, ideally, into optimal shapes for sectioning. It is
emphasized that it is worth taking extra care to prepare the block since it greatly
facilitates the ease with which sections, and especially ribbons of sections, can
be subsequently cut. The blocks are infused with sucrose before sectioning (see
below).

A useful alternative to centrifugation has been suggested by Dr. Jan Slot
(pers. commun., Fig. 7B). In this approach the suspension of material and

gelatin is poured onto the bottom (or top) of a 3 cm plastic cell culture petri dish (most brands of petri dishes have a rim on one or both parts). A second petri dish top is now placed on top of the drop, thus making a sandwich containing both gelatin and the material to be sectioned. This is now placed on ice for a few moments and the two lids forced apart. The thin layer of gelled material will stay with one surface of plastic. This can now be fixed, and subsequently, can be accurately trimmed for sectioning. The advantage of this approach is that the top and bottom of this thin layer is perfectly flat and parallel, which facilitates subsequent sectioning.

When cells need to be sectioned while still attached to their supporting matrix, they should be grown on glass microscope slides dipped in a solution of 1% gelatin to give a thin film (0.5 mm or less) which, fixed in aldehyde and rinsed, is dried and sterilized by UV and placed in the petri dish of nutrient medium. The gelatin plus cells can be sliced off the glass in small pieces using a scalpel or a razor blade. In this way, the damage to the cells is minimized and the pieces of monolayer can be mounted for sectioning after infusion with sucrose. The bulk of the sucrose is remove with filter paper and the monolayer is lifted with a pin or one point of a tweezers into an elongated form before freezing. A second embedding in gelatin may be required. This kind of approach was successfully used by Chen and Singer (1982), who prepared a "Swiss-roll" of cells with gelatin for cryo-sectioning. An alternative modification was introduced by Langanger and De Mey (1988) whereby the cells can be cut in parallel to the substratum. This reference gives a detailed protocol.

After freezing the tissue in liquid nitrogen or other coolant, the specimen holder is rapidly transferred into the pre-cooled chamber of the microtome. A hemostat is most convenient for this purpose. Blocks can be conveniently and indefinitely stored under liquid nitrogen, even after repeated sectioning.

Fibrinogen-Thrombin. A recent innovation by the group of Dr. Ivan Raska in Prague (pers. commun.) is to take advantage of the blood clotting protein fibrinogen for use as an embedding matrix. This is cheap and commercially available (e.g. Human Fibrinogen Type I from Sigma). A 1% solution of this protein is made up in phosphate buffered saline (or other suitable buffer) containing calcium and magnesium and used in the same way as gelatin above. For the gelation a drop of thrombin solution (e.g. the stock solutions sold by Sigma) is added. Within 1 min the material will form a hardened gel which is suitable for trimming in the absence of further cross-linking. This is an advantage over the use of gelatin. With the petri dish lid method the sandwich is made as soon as possible by rapidly adding and mixing the thrombin and fibrinogen.

Agarose. Pellets of cells/organelles can be suspended in small volumes (50–200 µl) of low melting point agarose (2% in PBS or water) at temperatures just above the melting point (35°C). If necessary the material can be centrifuged briefly and the tube placed on ice: there is no need to post-fix this pellet and suitable-sized pieces can be cut (≈ 1 mm^3). This approach is used by Schwarz (pers. commun.) for embedding cells in Lowicryl resins.

5.1.7.3 Cryo-Protection

The purpose of freezing in the Tokuyasu technique is simply to solidify the block so that it can be sectioned. As already pointed out (Chap. 2), this is not cryo-fixation per se since the tissue has already been chemically fixed and the sections are subsequently thawed. Since this freezing step is an important part of the technique, however, it is essential to understand as much as possible about the phenomena involved in freezing as well as the subsequent physical steps of sectioning, thawing and drying of the sections.

In principle, the freezing or solidification step in the thawed frozen section technique is extremely simple. The sucrose-infused, fixed tissue pieces are mounted on a specimen holder and frozen by plunging it into liquid nitrogen. However, in practice, there are a number of additional subtleties to consider.

In addition to controlling the quality of freezing, the concentration of sucrose used also affects the plasticity and, therefore, the cutting qualities of the block. In general, as the sucrose concentration increases the block becomes softer. In addition, the plasticity of the block is determined by the tissue composition and diminishes with increasing sucrose concentrations. At concentration of 2.1–2.3 M sucrose, the sucrose alone appears to determine the plasticity (Tokuyasu 1984). By this reasoning, any tissue infiltrated with 2.1–2.3 M sucrose should section in the same fashion, and at a similar optimal temperature, for any given thickness (for thin sections, usually −80 to 110°C (163 to 193 K)). With a few exceptions, this is what has been observed. Since tissue blocks infused with the highest concentrations of sucrose are relatively soft, thin sections can only be cut at low temperatures (as the temperature decreases, the block becomes harder). Below about −120 °C, however, blocks infused with high sucrose concentrations become very brittle and difficult to section. At any temperature, infusion with lower concentrations of sucrose will lead to harder (but less plastic) blocks. Obviously, a lower sucrose concentration may necessitate higher cooling rates: this is especially true when one approaches 1 M or below. If for any reason sucrose, which was empirically found to have the best sectioning properties, cannot be used, other sugars, such as glucose or raffinose may be employed (Tokuyasu 1973). Alternatively, high molecular weight compounds (non-penetrating cryo-protectants) such as polyvinyl pyrrolidone (PVP) or dextran can serve the purpose of binding the extracellular water and reducing its ability to form ice crystals (Tokuyasu 1986). In fact, the recent suggestion to use a mixture of PVP and sucrose is highly recommended and will now be described.

The recent innovation by Tokuyasu (1986, 1989) has the capacity to significantly improve the plasticity of tissue blocks. The method involves infusion of the tissue with a mixture of 1.8 M sucrose and 20% (w/v) polyvinylpyrrolidone (PVP; MW 10000). To make 100 ml of the mixture, prepare a paste consisting of 20 g of PVP, 4 ml of 1.1 M Na_2CO_3 in a buffer such as 0.1 M phosphate (Na_2HPO_4) and 80 ml of 2.3 M sucrose prepared in the same buffer. Cover it lightly and leave it at room temperature overnight so that minute air bubbles in the paste can escape into the air, leaving behind a clear solution. Air bubbles can also be removed by centrifugation or sonication. A solution prepared in this way will be nearly neutral in pH, but if desirable, the pH may be further adjusted with 1 N NaOH. The time required for a thorough infusion of the solution into the specimen ($\approx$ 1 mm^3) is usually at least 2 h. Specimen blocks infused with this mixture can be frozen in liquid nitrogen. Prepared in such a manner, the frozen blocks can often be sectioned much more smoothly than those frozen after infusion with sucrose alone in the range of $-60\,^{\circ}$C (213 K) to $-80\,^{\circ}$C (193 K) for semi-thin sections and in the range of $-80\,^{\circ}$C (193 K) to $-120\,^{\circ}$C (153 K) for ultra-thin sections (Tokuyasu 1986, 1989).

As a rough guide, it can be assumed that the time required for the sucrose to penetrate throughout the block will be approximately the same as the time required for the minimal fixation of that tissue. For most tissue blocks smaller than 1 mm^3, 15–60 min will be sufficient by the use of 2.1–2.3 M sucrose. The tissues can also be left overnight (or longer) in these infusion solutions, at 4 °C. It is easy to detect regions which have not been infused properly because, upon freezing, the presence of hexagonal ice crystals will render these regions opaque. Furthermore, these regions have a powdery consistency and tend to crumble during sectioning. Beginners can see this phenomenon for themselves by mounting a fixed tissue block onto the specimen holder and freezing it in liquid nitrogen in the absence of sucrose infusion. Such preparations are impossible to section.

If sucrose diffuses very slowly or not at all into the tissue of interest, it may be necessary to try glycerol or dimethyl sulphoxide (DMSO) (at concentrations up to 50% wt/vol), both of which can penetrate even unfixed membranes. The DMSO has the additional advantage of having a very low viscosity.

5.1.7.4 Mounting Freezing and Sectioning

On all commercial cryo-microtomes, the tissue piece to be sectioned is mounted on a small specimen stub (Fig. 8), usually made of either copper or aluminium . The size and shape of the tissue block on the specimen holder is crucial for the ease and quality of the sectioning process. As with resin embedded specimens, the sectioning is facilitated by a small and regular block face. The upper and lower surfaces of the block, (parallel to the knife) should be straight and parallel to each other. This increases the chances of obtaining ribbons of sections and reduces the possibility of the sections curling back over the knife edge. Circular section profiles will often curl upwards. It is sometimes easier to control the size and shape of the block face before freezing. If this is not possible, block trimming in the microtome itself with a cooled scalpel, the edge of a glass knife, or with the trimming devices present in some ultramictrotomes may be necessary.

The specimen holder may be modified for different purposes. If, for example, cross-sections of a rod-shaped object such as muscle fibres are required, holders as shown in Fig. 8 may be used. Suspensions of cultured cells or isolated organelles can be prepared for sectioning in a number of ways. The easiest

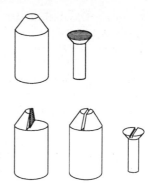

Fig. 8. Different kinds of specimen stubs for cryo-
sectioning

method is to centrifuge the cells (or organelles) and to fix the pellet in situ.
Suitably shaped pieces of the pellet can be infused with sucrose and mounted on
the specimen holder. If formaldehyde is used as a fixative it is often important,
in our experience, not to disturb the pellet once it has come into contact with the
fixative. Otherwise the pellet is more likely to disperse in the sucrose (to which
1% formaldehyde may be added). This approach has the advantage over
embedding in gelatin in that cell surface antigens within the entire section are
completely exposed to the antibody reagents. This increases the sensitivity for
labelling of such antigens.

Once the specimen is suitably mounted, it is simply plunged into liquid
nitrogen and, after bubbling has stopped, inserted into its position in the
microtome (Fig. 6). Extra specimens can also be stored in the cryo-chamber. It
should be noted that a small amount of sucrose (or sucrose-PVP) left on the
bottom of the block is sufficient to adhere it to the metal stub. It is undesirable
to have sucrose covering the whole specimen.

The specimens, once frozen, can be kept indefinitely in liquid nitrogen. They
can also be repeatedly sectioned and stored without any problem. For further
details of cryo-sectioning freezing and cryo-protection see sections 5.1.2 and
5.1.5.

5.1.7.5 Section Thawing and Retrieval

An important question is what happens to the sections upon thawing them on
the sucrose drop?. The gross changes of stretching overcoming the effect of
section compression have already been described. What happens, however, to
the water in the blocks which is vitrified in the cryochamber? Can this
recrystallize as the temperature is raised above the theoretical devitrification
temperature? The evidence suggests it does not. According to MacKenzie
(1977), vitrified solutions of pure sucrose above 60% (1.7 M) may not pass
through a crystalline phase upon warming. It is likely that, in practice, the
transition from vitrified water to the liquid state is too rapid to present any
serious problems. This statement is also supported by preliminary experiments

that we made to obtain a rough idea of the effects of freezing (vitrification) and thawing on immunolabelling. Cultured cells grown on cover slips were fixed for indirect immunofluorescence microscopy. Two different conditions were compared. One set of cells was prepared normally (i.e. permeabilized with Triton X-100 and labelled). The other set was infused with sucrose, vitrified in liquid nitrogen and thawed in buffer at room temperature before permeabilization and labelling. No qualitative differences in the amount of fluorescence was seen between the two methods, for a number of different antigens tested (G. Griffiths and B. Burke, unpubl. results).

Groups, or small series of sections are picked up using a loop containing 2.3 M sucrose in PBS (Tokuyasu 1973). The surface of the sucrose is then brought into contact with the sections (Fig. 6). It is important that the sucrose remains fluid long enough for the sections to be able to stretch. Sections that have attached to already solidified sucrose will not stretch when the sucrose is subsequently melted. It is important, therefore, to avoid pressing down on the surface of the knife with the loop, as this will increase the rate of cooling of the sucrose. An indication that this has happened is a layer of frost on the area of the glass knife which has been warmed by the sucrose.

The loop containing the sucrose is made of any pliable metal, e.g. copper or platinum wire, and may have a diameter between 1 mm (small loop) or 2 mm (large loop). The small loop has the advantage that, when the sucrose it carries is thawed and touches the grid, the sections on the surface of the sucrose drop will always be somewhere near the centre of the grid. The disadvantage of small loops is that there is less time for the operator to bring the loop into the chamber and touch the sections before the sucrose solidifies. In this respect the colder temperatures for cutting are also a disadvantage. A large loop allows significantly more time for retrieval of sections. When such loops, together with the sections, are brought into contact with the surface of coated and glow discharged grids, it is possible that the sections may be found on the peripheral regions of the grid. This makes visualization difficult or impossible. This problem can be overcome by using a stereo microscope, using light reflected from above, to locate the position of the sections while still on the loop. This procedure is easier to learn using large, semi-thin sections [0.5–2 μm].

Sections picked up on the drop of sucrose have one surface on the sucrose drop and the other exposed to the air. The latter surface will stick avidly and immediately to the surface of a coated EM grid or to a microscope slide (Fig. 6–7). We routinely use 100-mesh hexagonal copper or copper-paladium grids coated first with 1% Formvar, and then with carbon. Nickel grids can also be used but since they tend to magnetize they are more likely to give problems. These are then ionized by glow discharging (see Dubochet et al. 1982). This process makes the surface of the grid hydrophilic, which has two benefits. First, the sucrose containing the sections spreads more evenly over the grid. Second, hydrophilicity of the grid surface may facilitate the final embedment step with methyl cellulose (see below).

The grids, plus sections can now be washed and stored until enough are available for immunolabelling (Fig. 6–8). It is often useful to check one grid for structural preservation prior to labelling the rest of the grids. This is simply done by rinsing the grid by floating it on the surface of distilled water for a few minutes, then embedding it with a methyl cellulose-uranyl acetate mixture.

The grids may be stored on the surface of cold PBS, or on a gel consisting of a mixture of agarose and gelatin (see Tokuyasu 1980) or on 2% gelatin on ice (J. Slot,pers. commun.). In the latter case, the solid gelatin is warmed to room temperature to make it liquid just before labelling. In this way, the first step in the immunolabelling procedure, the saturation of non-specific protein binding sites, is already completed. In our laboratory we now routinely store grids on 1–5% newborn or fetal calf serum in PBS on ice for up to 1–2 h, which is the time it takes to prepare all the sections required. The use of newborn calf serum or fetal calf serum was found to reduce background significantly in immuno-blotting experiments by colleagues at the EMBL. If necessary this storage may be extended up to 24 h at 4°C. A better way of long term storage is to float the grids on 2.3 M sucrose and store them at −20°C (Tokuyasu, pers. commun.). Before labelling they are allowed to equilibriate with room temperature.

5.1.7.6 Antibody Labelling of Thawed Frozen Sections

It is important to master the essential technical skills of sectioning and contrasting of the sections before attempting any labelling experiments. The subsequent labelling technique presents no technical difficulties whatsoever: it is simply a matter of transferring grids from one drop of solution to another. The most important point in labelling however, is the quality of antibodies and marker molecules. Without high quality reagents labelling studies make little sense. Since we will discuss antibodies and marker reagents in detail in Chapters 6 and 8 respectively, these subjects will not be fully discussed here, except to point out those aspects that are directly relevant to labelling of thawed frozen sections. It is again necessary to mention fixation briefly here. Cells or tissues should be cross-linked as extensively as is compatible with labelling. A negative result may be due to too much, or too little, fixation (see Chap. 3). When the labelling is affected by too extensive cross-linking, one must ask whether this is due to a direct effect on the antigen or to a decreased accessibility of the antigen in the sections. The former can be best studied in vitro, if the purified antigen can be immobilized in some way, such as on a solid support system. When the cross-linking is affecting the accessibility of the antibody to the antigen it is sometimes possible to obtain an indication of this by labelling very thin sections. In this case it is usually possible to find areas where the structure(s) of interest is "opened", probably during thawing (the thinner the sections the lower the chances for the cross-links to hold the structures together). If cross-linking is really hindering accessibility one should get more intense labelling of these very thin sections.

Even if the antibody is able to react efficiently with the aldehyde-fixed antigen in the section there are other possibilities for negative labelling results. The first is that the antigen has been washed out of sections due to insufficient cross-linking. The second is that the antigen is present in concentrations too low to give a sufficient labelling signal (see Chap. 11). Both these problems can present major obstacles for immunocytochemistry and must be addressed for each individual case. The basic method for labelling thawed frozen sections is summarized in Fig. 6–9. The details of using secondary labels such as colloidal gold are described in Chapter 7.

5.1.7.7 Contrasting and Drying

In practice, it is possible to obtain well cross-linked sections of high quality that are also satisfactorily immunolabelled. What is perhaps surprising is that the care taken at the above steps will have been completely in vain if equal care is not taken at the final contrasting and drying steps. The potential for destroying fine structure at this step cannot be emphasized enough. The best way to illustrate this point is to cut thin cryo-sections of tissue, and after washing them with water, simply to air-dry them. Two things may be surprising at first sight. First, as mentioned, the destruction of the fine structure and second, the relatively high contrast in these unstained sections. The destruction is a direct effect of surface tension at the air-water interface when the sections are dried: as already mentioned our mass thickness measurement have shown that sections of cultured cells cut to a average thickness of 110 nm collapse to a thickness between 30 and 50 nm after washing and air drying (Griffiths et al. 1984).

It is to be expected that the latter phenomenon will be affected by the number of cross-links and therefore protein concentration, in the section: relatively dense tissues, such as muscle, would not be expected to collapse as much. The relatively high contrast seen in dried, unstained sections can be explained by their high mass per unit area (which is related to their density: in the case of cultured cells about 1 g/ml, that is, approaching that of pure protein (1.3–1.6 g/ml). In the above study when the approximately 80–110 nm sections were thawed and dried/protected with methyl cellulose they were no thinner than 70 nm. This might even be an underestimation (Griffiths et al. 1984).

The development of a simple procedure to prevent the collapse phenomenon upon air-drying was a critical step in the development of the cryo section technique for immunolabelling (Tokuyasu 1978). In Tokuyasu's first publication of thawed frozen sections (1973), excellent fine structure preservation was obtained by simply embedding and drying the thawed frozen sections in a thick layer of stain, mostly uranyl acetate or phosphotungstic acid, that is, the sections were simply negatively stained. In this case the high concentration of stain at least partly protected the sections against the effects of drying. Using such heavy metals, however, there is little chance to see gold particles and almost none to see anything with less contrast, such as ferritin. Tokuyasu's idea for contrasting after immunolabelling was to use low concentrations of heavy metal stains mixed with a chemically inert organic polymer that, upon drying, would provide a protective scaffold to the sections. In the original publication (Tokuyasu 1978) a number of compounds were tried that resulted in hydrophilic– (water-miscible) as well as hydrophobic (requiring prior dehydration of the sections) embedment. Other modifications were introduced later (Tokuyasu 1980, 1984, 1986, 1989; Griffiths et al. 1983, 1984; Griffiths 1984; Keller et al. 1984). For routine contrasting, methyl cellulose has proved the most useful compound in our laboratory. Recent innovations by Tokuyasu (1986, 1989) involving the use of polyvinyl alcohol and by Keller et al. (1984), using resins, also give excellent results, however. In these procedures, which will be described in detail below, the general aim is to provide a thickness of embedding material that is similar to the thickness of the section (Tokuyasu 1989).

After the last rinsing step following protein A-gold labelling the sections must be rinsed with distilled water since the presence of any phosphate will precipitate uranyl ions (Fig. 6–9). Tokuyasu (1978) has recommended a "pre-stain" with neutral pH uranyl acetate oxalate prior to contrasting. He empirically found that if, in the absence of this treatment, sections were put directly on aqueous uranyl acetate (pH ~4), artifactual blebbing and vesiculation of membranes was possible: this phenomenon was avoided by a pre-staining with neutral pH uranyl oxalate. It is emphasized that, at no stage during the labelling procedure should the face of the grid opposite to that which contains the sections come in contact with any of the solutions. Even tiny drops of PBS on the "back" of the grid lead to significant contamination of the grid. The practical details of the embedment procedure are shown in Fig. 6–10.

"Positive-Negative" Staining (Griffiths et al. 1982). A 2% solution of low viscosity (25 centipoise) methyl cellulose is made by mixing the powder with ice cold triple-distilled water. The powder takes many hours to dissolve fully. Note that methyl cellulose is more soluble at low temperatures. The solution is best left in the refrigerator for 1–2 days, with periodic or continuous stirring. The solution is then centrifuged (at 100000 g at 4°C) and stored in the refrigerator without disturbing the insoluble pellet. This solution is stable for 4–6 weeks. Nine parts of this is mixed with one part of a 3% solution of uranyl acetate, just before use. After one or two changes to remove the water the grids are left to float on this mixture on ice for about 10 min. They are then taken out using loops slightly larger than grids (3–3.5 mm) and the excess fluid removed and dried by careful blotting, without touching the part of the grid containing the sections directly on the filter paper. Enough excess solution should be removed so that the final interference colour of the dried film is between gold and blue. Often, a rainbow of colours is obtained, which may be acceptable.

This method, which uses a relatively high concentration of stain gives a mixture of positive and negative contrast. The nucleus and microtubules, for example, are almost always positively stained, whereas membranes usually give a characteristic strong negative contrast. It is an excellent technique for visualizing membranes. These are recognizable even in oblique sections through membrane profiles since there is sufficient contrast between the background, high stain regions and the unstained membranes.

Assuming fixation is adequate, the thickness of the embedment determines the overall fine structural appearance as well as the contrast. The thicker the

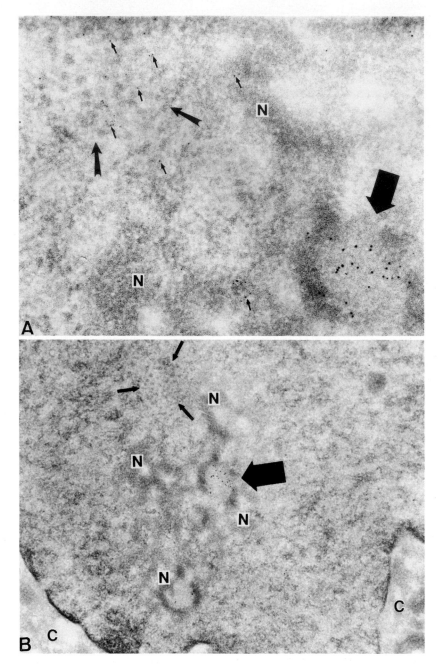

Fig. 9 A, B

film, the better the fine structure because there is more protection against air drying collapse. The effective contrast signal decreases, however, as the thickness increases, because the "noise" due to uranyl ions present in the methyl cellulose mixture increases. Conversely, as the methyl cellulose layer decreases in thickness there is less protection against air drying and significant collapse is inevitable: for more stable structures, however, this combination may offer the best contrast, as the stain dries down onto the surface of the structures in the section. The best practical advice is to experiment with the thickness of embedment and to examine every part of every grid. Note, also, that the films tend to be thicker near the bars of the grid.

If, after drying and examination, the film of methyl cellulose is too thick, it can be dissolved on the surface of cold water for a few minutes. The process can be repeated without apparent effect on structures or labelling. For examples of this method see Fig. 9–13.

Positive Staining

● *Polyethyleneglycol-methyl cellulose.* A procedure introduced by Tokuyasu is to embed the sections in a thick layer of a mixture of low concentration (0.2%) methyl cellulose with a high concentration (2%) of a low molecular weight polyethylene glycol (PEG –400); plus a low concentration (0.02%) of uranyl-acetate. The idea here is that the low molecular weight PEG should be able to penetrate much smaller spaces in the section (that would otherwise collapse during drying) than does methyl cellulose. The sections are left on this mixture, on ice, for about 5–10 min. During this time, the low concentration of uranyl-acetate will bind to some cell components to give positive staining. The background concentration of stain with this method is insignificant compared to the positive-negative staining method so that negative staining effects are not usually observed. When sections are sufficiently thin and the embedding film thick enough, a very subtle contrast is obtained with this technique so that even ferritin or very small (3 nm) gold particles can be clearly seen. With the positive-negative staining method the contrast is usually too high, unless the sections are extremely thin. For typical examples of this procedure see Tokuyasu (1989).

◀———————————————————   ————   —

Fig. 9A,B. Cryo-section of Hella cells showing parts of the nucleus. The cells were fixed in 8% paraformaldehyde in 0.2 M PIPES (pH 7) for 1 h, infused with 2.1 M sucrose, frozen and sectioned on a Reichert cryomicrotome. The sections were double-labelled with (1) human auto-immune antibody that recognizes RNA polymerase I followed by 10 nm gold conjugated to antishuman IgG and (2) a rabbit antibody that recognizes specifically capped RNA followed by 5 nm gold conjugated to rabbit IgG. The labelling with the two primary antibodies and the two gold conjugates was done concomitantly (in other words, the labelling was done in two steps only). The large gold particles associate predominantly with the less dense fibrillar centres (*large arrows*). The small gold particles (*small arrows* in **A**) are mostly associated with the coiled body (*intermediate-sized arrows*). **A** ×77000; **B** ×27000. *N* nucleolus; *C* cytoplasm. (Courtesy of Dr. Ivan Raska et al. 1989 Biol of the Cell, 65, 79–82 Czechoslovakian Academy of Science, Prague. For more details see the review by Raska et al. 1990)

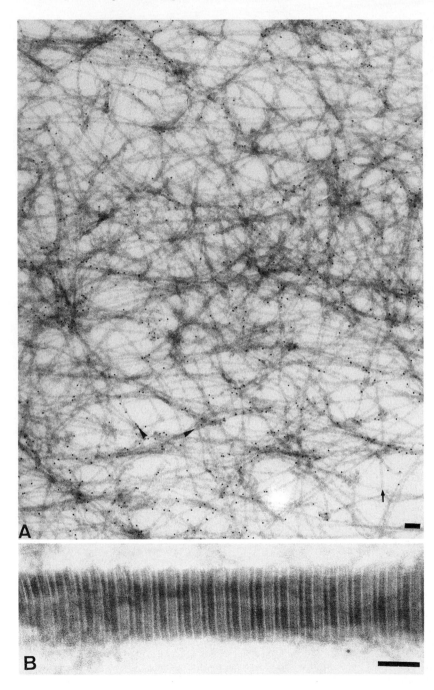

Fig. 10 A, B

• *Polyvinyl alcohol.* A recent innovation of Tokuyasu's involves the use of polyvinyl alcohol (PVA). For this a solution of 3% polyvinyl alcohol (MW 10000) is made in water. This is best made by sonication to avoid air bubbles. Mix nine parts of this with one part of 3% aqueous uranyl acetate. After the final water rinse place the grids for 10 min at room temperature on this solution. The uranyl acetate concentration may be varied from about 0.2–0.5%: at the higher concentrations more negative contrast is obtained. Because of the lower viscosity of the embedding mixture, one should use either relatively small loops (ca. 3 mm in diameter) or, simply keep the grid in an anti-capillarity forceps while removing the excess fluid: in our experience it is more difficult to control the thickness of the film with this procedure. It is important to have relatively thin sections and relatively thick films for optimal positive-staining effects. For more details of this approach, including the use of osmium, lead citrate and uranyl acetate see Tokuyasu (1989) and Fig. 14.

• *Plastic embedding of thawed cryo-sections.* An alternative method to using water-soluble methyl cellulose compounds to protect thawed cryo-sections against air drying is to embed them in a thin film of epoxy or acrylic resins which are then polymerized as for conventional electron microscopy (Keller et al. 1984). The results are both striking and indistinguishable from conventional osmicated plastic sections. The following details have been provided by Dr Gilbert Keller (pers. commun., see Fig. 15 for an example).

Procedure: After immunolabelling, the grids are placed on drops of 0.5 to 2% osmium tetroxide (with or without 1% potassium ferrocyanide), in e.g. 0.1 M cacodylate buffer pH 7.2 for 10 min. They are rinsed in water and post-stained for 5–20 min on aqueous 0.5% uranyl acetate in either water or barbital-acetate buffer pH 5.2 containing 5% sucrose. The grids are then sequentially dehydrated in ethanol: e.g. 2 min each in 40, 50, 60, 80, 90, 95 and 100%. The grids should sink in the latter solution. An alternative possibility to increase contrast is to stain the grids in a 2% solution of uranyl acetate in 50% ethanol for 10–15 min. The grids are then infused with plastic. Originally, a 2% solution of Epon 812 in ethanol was used (pure Epon was too viscous) (Keller et al. 1984). Subsequently, the acrylic resin LR white (see Chap. 4) was used because it is hydrophilic and easier to use. The grids are put into the pure monomer for 30 min and then excess liquid is removed. This is the critical step in the method. In

◀ ─────────────────────────────────

Fig. 10A, B. Cryo-section of the collagen matrix of a proliferating zone of the growth cone of a 19-day-old chick embryo tibia. The tissue was fixed in 2.5% glutaraldehyde in 0.2 M HEPES buffer for 1 h at room temperature before sucrose infusion and freezing. The sections were labelled with an antibody against type II collagen and protein A gold (9 nm). Note that no digestion with proteases or hyaluronidase was performed as are generally used in other methods, in order to allow penetration of antibodies. The labelled type II collagen fibres (arrowheads can be distinguished from unlabelled type X collagen (*arrow*). In **B** high resolution details are shown of a type I collagen fibre. Sections post-embedded in uranyl acetate-methyl cellulose. **A** ×43000; **B** ×128000. *Bars* 100 nm. (Courtesy of Dr. Carlo Tacchetti, Instituto Scientifico per lo Studio e la Cura dei Tumori, Genoa, Italy)

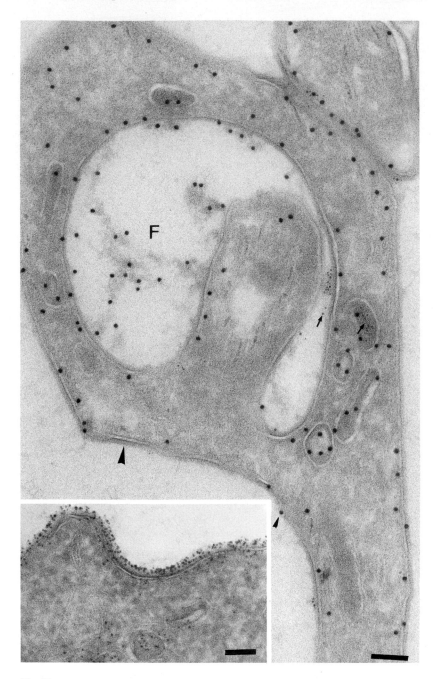

Fig. 11

the original publication, the grids were dried between two pieces of hardened blotting paper (Whatman No. 50). Tokuyasu (pers. commun.) recommends drawing the grid with a forceps along the surface of filter paper. A recent procedure from Dr. Keller is to place the grids on the filter paper (with sections upwards) and wait until the bulk of the embedding medium has been adsorbed (which takes a few minutes). The grids are then placed on fresh filter paper and polymerized at 60 °C for 1–2 h. According to Keller (pers. commun.), for a quick check the grids can also be polymerized in 5 min in a microwave oven; the results of this procedure are, however, less satisfactory. The amount of excess liquid that has to be removed is determined empirically, as for the methyl cellulose or PVA methods. Too thick a film makes it difficult to visualize the structures and the gold or ferritin particles, whereas too thin films lead to the characteristic air-drying collapse phenomenon. In our experience it is not always easy to obtain optimal thicknesses of films in a reproducible fashion with this method.

The grids should preferably be polymerized in a vacuum oven (oxygen should not be present for the LR white polymerization). Alternatively, the grids could be polymerized at room temperature using UV light. Dr. Keller recommends preparing ≈ 5 ml aliquots of a freshly opened bottle of LR white resin, keeping these at −20 °C and thawing and Millipore filtering it before use. There is a possibility to further contrast the grids with lead citrate, uranyl acetate, or if ferritin is used as the marker, with 0.04% bismuth nitrate (Keller et al. 1984).

An important contribution of this modification, I believe, is to show convincingly that, despite the freezing, sectioning and labelling, the fine-structure preservation is at least as good as that of conventional Epon sections, until the drying stage. It also offers another contrasting alternative for those who prefer conventional images without having the detrimental effect of osmication, dehydration, infiltration and polymerization prior to the immuno-labelling. However, since the adverse effects of these treatments on structural preservation are unavoidable, the preservation of fine structure at the high-resolution level faces the same problems as with conventional plastic sections.

● *Negative contrast using ammonium molybdate.* A very useful method for delineating membrane structures is to use the negative stain ammonium

Fig. 11. Cryo-section of *Trypanosoma brucei* that had been incubated for 1 h at 37 °C with bovine transferrin conjugated to 5 nm gold (*arrows*) before fixation in 0.5% glutaraldehyde in 200 mM PIPES (pH 7.0) for 30 min. The tissue was infused with 2.1 M sucrose and sectioned on a Sorvall MT2B/FTS with a tungsten-coated glass knife at −90 °C. The section was labelled with an antibody against the variable surface glycoprotein (VSG) followed by 9 nm protein A gold (*arrowhead*). Post-embedding was in methyl cellulose-uranyl acetate. The internalized transfer-rin conjugate co-localizes with the VSG in endocytic vesicles and in the flagellar pocket. The antiserum used against the VSG had relatively low titre since large areas of the cell surface where the VSG coat was evident by its electron density (*large arrowhead*) appeared unlabelled. The normal appearance of densely labelled VSG with a high affinity anti-VSG is shown in the inset. ×117000; (inset ×90000). *Bars* 100 nm. (Courtesy of Dr. Paul Webster ILRAD; Nairobi, Kenya; presently in Dept of Cell Biology, Yale University, New Haven, USA)

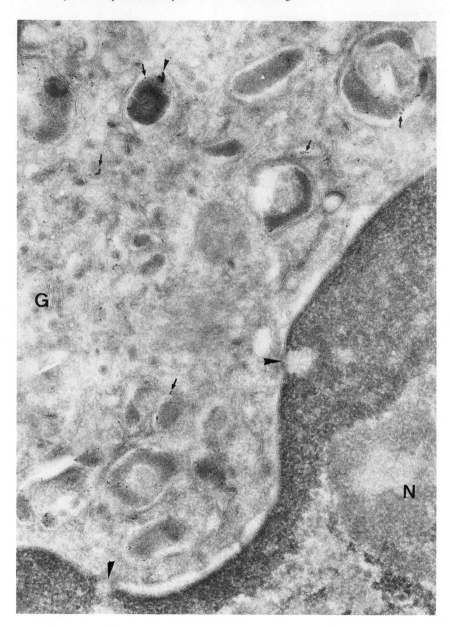

Fig. 12

molybdate (Tokuyasu, pers. commun.). After the final water rinse, the grids are placed on a freshly made mixture of 1 % ammonium molybdate (nine parts) plus 2% methyl cellulose (25 cp) (one part) for 10 min on ice. It is important to have relatively thin sections and relatively thick films for a good negative stain effect. For the drying step, the same procedure is used for the positive negative contrast above. An alternative approach is to put the grids on the above mixture for 10 min and then on 2% methyl cellulose, without stain, for 10–15 s, before removing the excess liquid in the usual way. The latter method has proven more reproducible in our recent studies. For an example of this method see Fig. 16.

It is recommended that anyone starting with the thawed frozen section technique should try all the above techniques as well as experimenting with other possibilities (e.g. the use of osmium tetroxide and lead citrate; see Tokuyasu 1989). The potential rewards at this stage with respect to contrasting and preserving fine structure, even at high resolution, are enormous as, on the other hand, are the possibilities for complete destruction of structure.

5.1.7.8 Visualization of Cryo-Sections

A few points should be discussed about the visualization of the sections in the electron microscope. It is obviously important, after the effort put into the earlier stages, that as much useful information as possible be obtained from the sections. It should be noted that unlike plastic sections, which deteriorate with time due to plastic "flow", cryo-sections appear to be indefinitely stable. If necessary they can be put on water (cold water in the case of methyl cellulose) and the polymer can be dissolved and the sections re-embedded.

● *Contrast.* In order to obtain the highest contrast on the sections, a small objective aperture (<20 30 µm) and relatively low accelerating voltages (i.e. no higher than 80 kV) are recommended.

● *Beam Sensitivity.* The methyl cellulose is very beam-sensitive, in other words the effect of the electron beam is to cause mass loss. This effect is easily noticeable as the part of the section that has been exposed to the beam becomes

◀ ───

Fig. 12. Example of cryo-section contrasted with methylcellulose, uranyl acetate mixture. Primary cultures of mouse peritoneal macrophages were removed from the culture dish with 5 mM EDTA, fixed for 5 min in 4% formaldehyde in 200 mM HEPES pH 7.4, followed by 4 h in 8% formaldehyde in the same buffer. The cells had internalized a 16 nm gold-BSA conjugate for 4 h followed by an overnight chase into late endocytic compartments (*small arrowhead*). Just before fixation a 5 nm gold-BSA conjugate was taken up for 8 min followed by a 22 min chase. This marker (arrows) fills the prelysosomal compartment (late-endosome) and can be found together with the large gold. The membranes in these cells are invariably positively contrasted by the methyl cellulose procedure since the sections is sufficiently thin. *Large arrowheads* indicate nuclear pores. *G* Golgi complex; *N* nucleolus. ×28000. [From a study by the author with H Horstmann (EMBL) S Rabinowitz and S Gordon (University of Oxford). See J Cell Biol (1992) 116:95–112.]

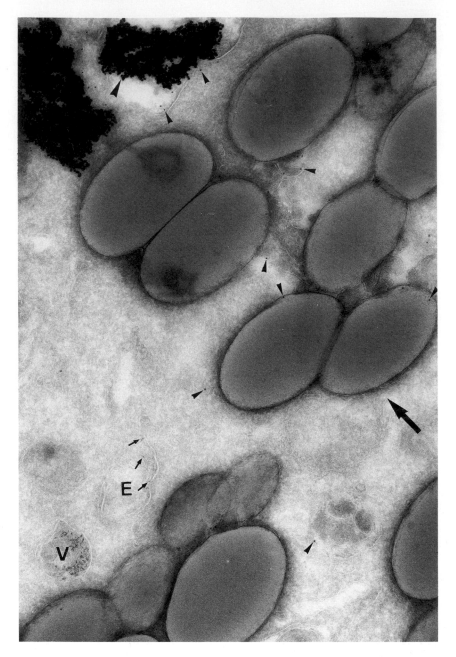

Fig. 13

more transparent. This is often useful since it facilitates the recognition of those areas that have already been looked at. One must be careful to illuminate evenly the area to be photographed to avoid the presence of annoying density differences across the negative. This is often seen when first examining an area at high magnification, but then recording an image at a lower magnification. For high resolution work, the effect of the electron dose on the specimen itself, which has not been systematically studied with thawed and dried frozen sections, may become detrimental. In this case, "low-dose" conditions may have to be used.

● *Section Thickness.* It has been shown for unstained hydrated cryo-sections that, within a broad range, the optical density of the micrograph is proportional to the section thickness (Dubochet et al. 1982). Although not systematically studied for thawed, stained cryo-sections, it is intuitively clear that the greater the thickness the denser the section appears (under standard conditions). Electron density, evaluated subjectively, is the best indicator of section thickness and, with experience, it becomes easy to select those sections on the grid that are thin enough to be useful. When one wants to measure the thickness of the dried section, a number of approaches may be used. The easiest is to look for folds in the section: the width of the fold is taken to be twice the section thickness(Small 1968; see Griffiths et al. 1983). For frozen sections it is important that the methyl cellulose is thick enough to protect the fold against a drying collapse in the same way it protects cell structures. Alternatively, the grids containing the sections could be embedded in Epon and sectioned transversely.

An additional method to estimate thickness is to take advantage of the fact that immunogold particles are, for many structures, primarily on the top and bottom surfaces of the section (Stierhof and Schwarz 1989). The depth of penetration of each organelle depends primarily on the protein concentration and on the amount of cross-linking. The position of gold particles with respect

Fig. 13. Cryo-section through mouse peritoneal macrophages prepared as in Fig. 11. The large gold-BSA (*large arrowhead*) was chased overnight as in Fig. 11. The living cells were treated with a 30 min pulse of latex beads then with wheatgerm agglutinin (WGA, 25 µg/ml) for 6 h before fixation. BSA-gold 5 nm (*small arrows*) was added to the culture medium for 8 min followed by a 22-min chase just before fixation. This WGA treatment blocks the delivery of material from the endosome carrier vesicle (*V*), which originates from the early endosome (*E*); note the small amount of residual 5 nm gold which has not chased from the carrier vesicles into the prelysosomal compartment/late endosome (cf. Fig. 12). The latter have large amounts of large gold as well as the latex beads taken up by phagocytosis. The section was labelled with monoclonal antibodies against LAMP 2, a lysosomal membrane glycoprotein (kindly provided by Dr. Tom August), followed by rabbit anti-mouse and 9 nm gold protein A (*small arrowheads*). This labels the membrane of the prelysosomal compartment. The large arrow indicates the direction of sectioning: this is evident in this preparation since the latex beads, initially spherical have become compressed during sectioning. Unlike the endosome vesicle (*V*), which has presumably regained its circular appearance after section stretching, the latex beads remain compressed (they are approximately twice the diameter in one direction that they are in the other) (see legend to Fig. 12)

Fig. 14. Example of polyvinyl alcohol embedded cryo-section showing high resolution details. Chicken pectoralis muscle, fixed in a mixture of 4% formaldehyde and 0.5% glutaraldehyde in 0.1 M phosphate buffer for 1 h at room temperature. After infusing with a 10% polyvinylpyrrolidone-2.07 M sucrose mixture, it was frozen in liquid N_2 and sectioned at $-125\,°C$. The section was first stained with 2% uranyl acetate for 10 min, washed briefly with water and adsorption-stained with a 0.006 M lead citrate-3% polyvinyl alcohol mixture. Periodicities of 14 nm (sets of four *dots*) are recognized either as fine cross-striations across several thick filaments or *dark spots* on individual thick filaments. They are believed to represent regularly arranged head regions of myosin molecules. For more details see Tokuyasu (1989). (Micrograph courtesy of Dr. K. Tokuyasu, Dept. of Biology, University of California, San Diego, California, USA)

to the thickness of the section can be visualized in three dimensions by taking a tilt series (or stereo pair of photographs) using the goniometer stage of the electron microscope. The images can be "fused" to give a stereo image using a stereo viewer. One should use the highest tilt angle that the observer can fuse together into one image: this decreases with increasing magnification as well as with section thickness. The recommended maximal tilt angles in relation to magnification and section thickness can be estimated from plots given by Hudson and Makin (1970). For a general description of the use and interpretation of stereo images see Howell and Boyde (1972) and Nemanic (1974).

5.2 Freeze-Fracture and Replica Labelling Methods

During the past few years a number of techniques have been introduced which combine freeze-fracture approaches with immunolabelling. I have already noted some general criticisms of the freeze-fracture approach as it is conventionally used, especially when combined with freeze-etching (see Chap. 3). Most of these criticisms apply also to the immunolabelling approaches – especially when high resolution information is the goal. Nevertheless, a few recent publications make it clear that these approaches offer a valid alternative to those of cryo-sectioning and the new resins in some cases. Since I have no personal experience with these techniques, and since they have been adequately described in a number of recent publications and reviews (see below), only the theoretical background will be discussed here. For comprehensive reviews of this area see Pinto da Silva (1984, 1987, 1989); Boonstra et al. (1987) and Severs (1989). Three basic kinds of freeze-fracture labelling approaches have been carried out, which have been referred to as "fracture-label", "label fracture" and "fracture-flip".

5.2.1 Fracture-Label Method

The protocol here is as follows (Pinto de Silva et al. 1981a-c).

- Fix in glutaraldehyde.
- Treat with glycerol as cryoprotectant.
- Freeze in liquid nitrogen.
- Fracture with either a knife (as with thin conventional freeze-fracture methods) or with a glass pestle (freeze-crushing).
- Thaw in the presence of glutaraldehyde, then wash.
- Immunolabelling, e.g. antibody followed by gold.

After this step at least two different approaches are possible. First, the suspension of fragments may be processed for thin section EM after embedding

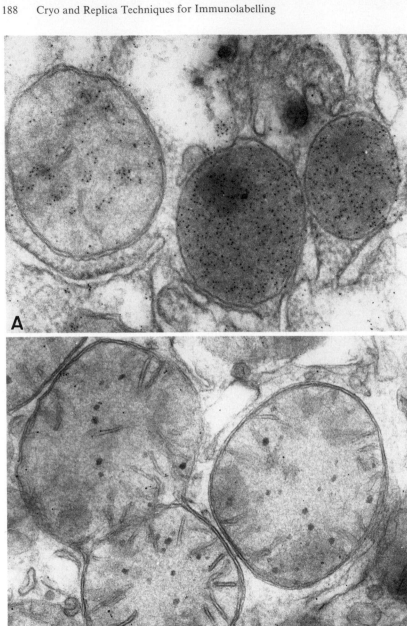

Fig. 15A, B

in plastic; second they can be critically-point-dried, shadowed with platinum-carbon as for routine freeze-fracture methods and the replicas observed in transmission EM.

Epon embedding has recently been used for a number of elegant studies. The most striking of these were the studies by Torrisi and Bonatti (1985; Torrisi et al. 1987) on Sindbis virus. They showed that the viral spike membrane glycoproteins partitioned, as expected, with the exoplasmic faces of the endoplasmic reticulum and Golgi complex membranes . When the membrane of these organelles was cleaved, the immunolabel associated with spike proteins stayed with the luminal, rather than the cytoplasmic leaflet of the membrane bilayer. Note that here the inner luminal leaflet corresponds to the outer leaflet of the plasma membrane. However, at the plasma membrane the opposite was true:

◄ ──

Fig. 15A,B. Example of LR white embedding of immunolabelled cryo-section. Since many of the details of this preparation have not previously been published, a detailed protocol is given here.

Specimen preparation: Livers from Sprague-Dawley rats were fixed by portal perfusion with 3% paraformaldehyde, 0.25% glutaraldehyde in 0.1 M cacodylate buffer. After 5 min of perfusion, small blocks were cut and immersed in the same fixation solution for 2 hr. All the fixation solutions are stored in aliquots of $-20\,^{\circ}C$, thawed and millipore filtered (0.22 μm) before use. Blocks were infused in 2.3 M sucrose for 30 min, rapidly frozen and gold-coloured sections were cut on a Diatome Diamond knife in a Reichert Ultracut E microtome equipped with the FC4E cryokit. The sections were labelled with affinity purified rabbit antibody against rat prenyl transferase (10 μg/ml). In normal rat liver, the immunolabelling can be seen in close apposition to the ER membranes (where HMG-CoA reductase is also located) but also in the mitochondria, close to the cristae. In mevinolin treated animals, the labelling becomes predominantly peroxisomal. Fig. B shows peroxisomes and mitochondrial labelling obtained with an affinity purified rabbit antibody against SCP2 (sterol carrier protein 2, see Keller et al. 1989) was used at a concentration of 5 μg/ml for 10 min. The immunolabelling is seen in three different cellular compartments: i) the cytosol, ii) the mitochondria, iii) the peroxisomes. The cytosolic labelling is thought to represent the protein en route to the mitochondria and the peroxisomes.

The gold particles of 6–8 nm diameter were prepared by sodium borohydride reduction and coupled to an affinity purified guinea pig antibody against rabbit IgG. This reagent was applied to the sections for 45 min.

After immunolabelling and rinsing the grids by floating in a beaker containing PBS for 10–15 min, the grids were floated onto droplets of 1% ferrocyanide reduced osmium for 15 min, thawed and rinsed in distilled water for 2×2 min. The grids are then "*en block* stained" by floating on 2% uranyl acetate for 10–15 min (a stock 4% aqueous- or saturated uranyl acetate solution can be conveniently kept in the refrigerator and mixed with the same volume of ethanol, just prior to use). The grids were dehydrated in increasing concentrations of ethanol and usually sink at the 100% ethanol step. If they do not sink they should be gently pushed beneath the surface. After a few minutes in pure ethanol, the grids are transferred to pure resin (3 changes) and left for at least 30 min (they can be left overnight if convenient). Note that if the LR white has started polymerizing, the results will be poor. The grids are then placed, sections up, on a disk of hardened Whatman no. 50 filter paper.

The excess of monomer diffuses in the paper and when the diameter of diffusion circle does not expand any more, the grids are transferred to a fresh filter paper in a (plastic or glass) petri dish and placed either in the microwave oven (5 min on roast beef!) or, better, in an oven at $60\,^{\circ}C$ for 1–2 h. The microwave should only be used if you are in a hurry or to check just one grid, as it often makes holes in the Formvar film. The sections can be counterstained by a short treatment (30 s-1 min) in lead citrate, but this treatment is usually avoided since it imparts a grainy appearance to the membranes. (Micrographs and information kindly provided by Dr. Gilbert Keller, Genentech Inc., San Fransisco, California, USA

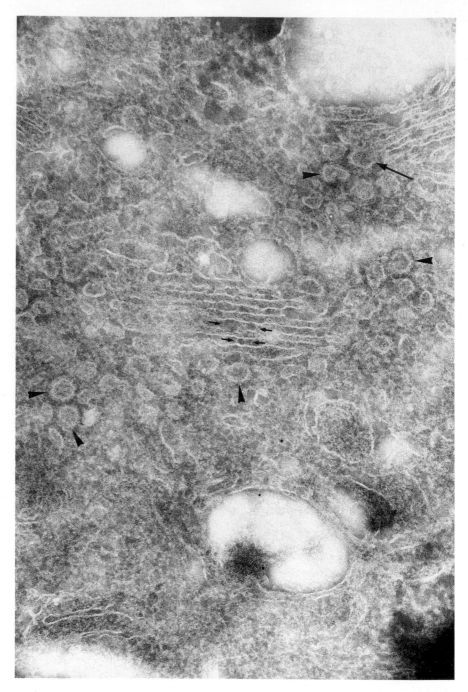

Fig. 16. Cryo-section of an NRK cell contrasted with ammonium molybdate. Using this procedure "coats" (*arrowheads*) on the vesicles in the vicinity of the Golgi stack become more prominent as do the strands of fibrous material (*arrows*) that appear to connect the Golgi cisternae. ×120000

the label associated with the spikes now stayed with the cytoplasmic rather than with the exoplasmic leaflet (Torrisi and Bonatti 1985). This was taken as evidence for an association of the spikes with the underlying nucleocapsids, a hypothesis which is also supported by other approaches (Whitt et al. 1989). In a later publication, Torrisi et al. (1987) showed that the spikes of this virus were present in the inner nuclear envelope, providing convincing evidence that the whole of the inner nuclear envelope was in functional continuity with the outer membrane (to which the ribosomes are bound). (See also studies by Pinto da Silva et al. 1981a,b; Torrisi and Pinto da Silva 1982 for other examples.)

The second approach for "fracture label" is to use replicas, as for conventional freeze-fracture studies (Pinto da Silva et al. 1981b). The advantage of this method is that it enables the label to be seen over a large surface, in particular of membranes. A potential difficulty with this approach, however, is to interpret the images. This is especially true for those of us who do not routinely examine freeze fracture images. In addition, the way in which the fracture cleaves the components of the membrane may not always be reproducible, nor interpretable. As for all freeze-fracture studies, for example, there is no generally accepted notion of how an intra-membranous particle (IMP) is formed during the fracture procedure. For examples of the applications of this approach see the references in Pinto da Silva (1987).

5.2.2 Label-Fracture Method

The idea here is first to label the outer surface of a cell with an electron dense gold marker and then to carry out conventional freeze fracture (Pinto da Silva and Kan 1984). For this, cells may be fixed before or after labelling and a cryo-protectant may be used before freezing. The cells are fractured without etching and the fracture faces are replicated using standard procedures (e.g. platinum-carbon). Unlike routine freeze-fracture, however, the replicas are not treated with acid, alkali or detergents which are normally used to remove the biological material. Instead they are visualized directly, using transmission EM where a superimpositioned image is obtained of the replica metal, as well as the more-electron-dense gold marker. For examples of this approach, see Pinto da Silva (1987); Boonstra et al. (1985a,b). Noteworthy also was the study by van Berkel et al. (1986), who used this method to label a chloroplast membrane protein. Significantly, the density of gold labelling obtained by this technique agreed with the estimates obtained after pre-embedding labelling and Epon sectioning.

The main criticism of this approach in my opinion is the difficulty in relating the image of the replica to the underlying structures (but see Pinto da Silva 1987). The use of double- or even multiple-labelling strategies would be very beneficial if the position of one marker could then be correlated to a reference marker, whose position has already been established. Technical problems associated with this technique are discussed by Boonstra et al. (1987).

5.2.3 Fracture-Flip

The latest of the freeze-fracture labelling approaches is the "fracture flip" method (Forsman and Pinto da Silva 1988; Fujimoto and Pinto da Silva 1988). Following freeze-fracture the membrane halves are stabilized simply by carbon evaporation. After thawing and washing the replicas, with their exoplasmic membranes attached, they are picked up on formvar coated grids. These are then turned upside-down, exposing the membrane surface away from the carbon for immunocytochemical labelling followed by metal shadowing.

5.3 Other Replica Techniques for Labelling

Aside from the freeze-fracture type of replica methods, two other approaches must be discussed within this general category of replica/labelling techniques. Both are primarily applicable only to cells in culture. The first is an approach for looking at the outside surface of the cell while the second is a way of visualizing the cytoplasmic surface of the plasma membranes. Both these approaches can be combined with immunogold labelling. Although most of these methods are best suited for transmission EM, the advent of very high resolution scanning EM's offers an alternative possibility for visualization (see Müller et al. 1989).

5.3.1 Surface Replica Methods

The idea here is to first label the material of interest with a probe followed by a gold marker (the protocols are essentially the same as for labelling sections). Suitable material may be whole particles (e.g. viruses) on grids, whole- or detergent-treated cells growing on coverslips. Subsequently, the sample must be dried, either by critical-point drying or by rapid freezing followed by freeze-drying. Both approaches have potential artefacts associated with them, especially shrinkage. However, these problems can be reduced if the cells are fixed before the drying step with aldehyde. The material is then shadowed (coated) with a thin film of a heavy metal: simultaneous evaporation of platinum and carbon, often followed by another coat of carbon, is most widely used for this purpose. For particles such as viruses on grids, the material may be viewed directly after this stage. For cells on coverslips the replica must be

——————————————————————————————▶

Fig. 17A, B. Surface replicas of cultured human skin fibroblasts incubated with **A** low density lipoprotein-gold (12 nm) conjugate for 30 min at 4°C. The gold particles are localized almost exclusively in clusters. ×32 500. **B** A polyclonal rabbit anti-human fibronectin antibody followed by gold- (14 nm) labelled protein A. The gold label is exclusively associated with a fibrillar network on the plasma membrane surface. ×17000. (Courtesy of Dr. Horst Robenek, Medizinische Cytobiologie, University of Munster, FRG)

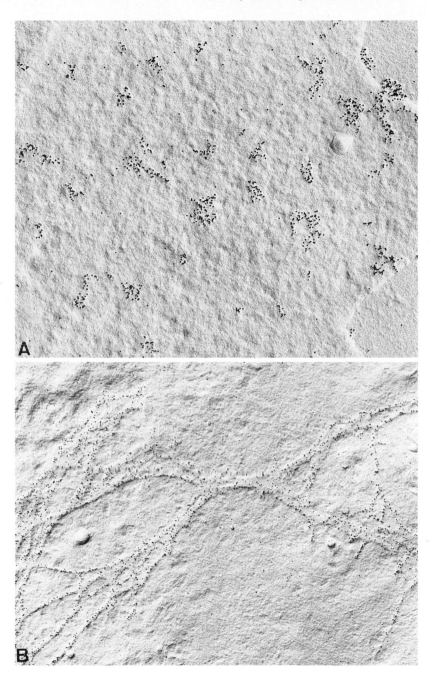

Fig. 17A, B

Fig. 18. Surface replica of a mouse peritoneal macrophage incubated with acetylated low density lipoprotein-gold (40 nm) conjugate for 1 h at 4 °C. The gold complexes preferentially bind in the intermediate region of the plasma membrane surface which surrounds the nucleus. ×62000. (Courtesy of Dr. Horst Robenek, Medizinische Cytobiologie, University of Munster, Germany)

separated from the coverslips. If the material being shadowed is suitable and thin enough to be visualized in transmission EM, e.g. cytoskeleton, the structures plus replica are examined after placing the complex on a grid. If the preparation is thicker, such as whole cells, it is necessary to use solvents, acids or bleach to dissolve the biological material leaving only the replica to be picked up and examined on an EM grid.

Whereas the labelling methods are simple, the drying and shadowing procedures are technically more complex and, ideally, should be learnt from the laboratories that specialize in these procedures. For recent, detailed reviews see Robenek (1989a, b), Hohenberg (1989) and Nermut and Nicol (1989). For example of surface replica labelling see Figs. 17 and 18.

The replica method can be considered complementary to sectioning methods in that they allow large areas of surface to be seen in one view. This is especially true for membranes. When a shadowed membrane has also been immunolabelled, patterns of surface labelling may often be appreciated (e.g. a label that follows an underlying cytoskeletal organization) that would be difficult to see in sections through that membrane. For excellent examples of the use of this approach for immunolabelling, see Mannweiler et al. (1982), Hopkins (1985), Nicol et al. (1987), Schmidt et al. (1987), Veltel and Robenek (1988), Robenek (1989a), Severs (1989) and Semich and Robenek (1990). For an example of double-labelling see Robenek and Severs (1984).

According to Hohenberg (1989), the detachment of the labelled and shadowed cells from the underlying glass or plastic support is a critical step that can lead to loss of gold particles. This author has empirically worked out and described in this reference conditions that minimize such losses.

5.3.1.1 Approaches for Visualizing and Labelling the Cytoplasmic Surfaces of Cultured Cells

Of the various techniques that have been developed for this purpose two main approaches have been most successful (Robenek 1989a). The first involves placing a polylysine-coated coverslip over a cell monolayer and allowing it to adhere. The coverslip is then separated, cleaving the cell and exposing predominantly the upper (dorsal) cytoplasmic surface leaving the remainder of the cells bound to the surface on which the cells were growing (Aggeler and Werb 1982; Rogalski et al. 1984). The second is the "lysis squirting method" (Clarke et al. 1975; Avnur and Geiger 1981; Nermut 1982). In this the cells are first exposed to a hypotonic buffer and then the bulk of the cell material is sheared away by squirting with a syringe (see Nermut and Nicol 1989). After either procedure the cells can be fixed with a suitable aldehyde, immunolabelled, dehydrated, critical-point dried and replicated with platinum carbon (see Robenek 1989a; Nermut and Nicol 1989; Mannweiler et al. 1989 for details). This method, both with or without immunolabelling, has proven to be an excellent approach for looking at associations of cytoskeletal elements with the plasma membrane and for looking at clathrin coated pits (see Fig. 19 as well

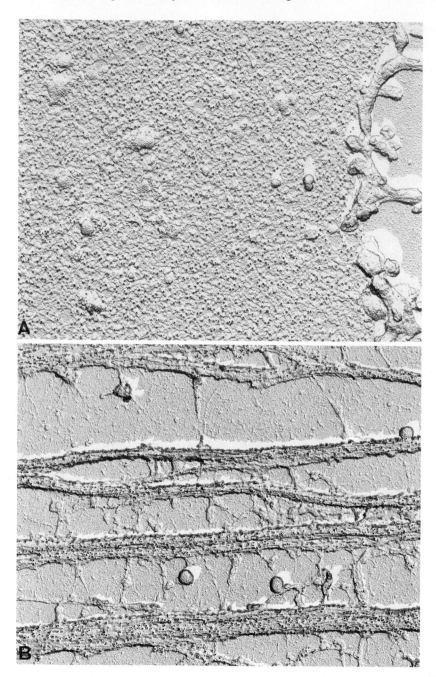

Fig. 19 A, B

Fig. 19A,B. Replicas of the cytoplasmic surfaces of ventral plasma membranes of cultured human skin fibroblasts after lysis-squirting and labelling. **A** With polyclonal antibodies against p 36, a member of the Ca^{12}/lipid binding proteins that is a substrate for the tyrosine kinase encoded by the SVC oncogene (see Semich et al. 1989). **B** With polyclonal anti-actin antibodies; both followed by protein A-gold (14 nm). p 36 is uniformly distributed over the entire surface of the ventral plasma membrane. Coated pits are not labelled. Actin is localized over stress fibers. Coated pits are free of gold label. **A** ×26000; **B** ×26000. (Courtesy of Dr. Horst Robenek, Medizinische Cytobiologie, University of Münster, Germany)

◄───

as the references in the above reviews). For examples of immunolabelling with this approach see Rutter et al. (1985), Hohenberg et al. (1985), Nicol et al. (1987) and Semich et al. (1989). The power of this approach can be extended by pre-labelling the upper surface of the cells with one marker and gold before tearing off the surface with the polylysine-coverslip method. After fixation, a second cytoplasmic antigen can be labelled with a second size of gold. In this way the distribution of one antigen on the (outer) cell surface can be correlated directly with another on the cytoplasmic surface on large areas of replica (see Rutter et al. 1988; Robenek 1989a; Mannweiler et al. 1989 for more details).

5.3.2 Replicas of Labelled Sections

As recently pointed out to me by H. Schwarz, (University of Tübingen, pers. commun.), useful technical information can be made available by coating the surface of gold-labelled, cryo- or plastic sections with a metal such as platinum. As shown in Fig. 12 of Chapter 4, even in Lowicryl sections the label is sometimes able to find its way to the bottom side of the section (that is, between the supporting film and the section); this label is consequently not shadowed. This argues against the general applicability of the double-labelling method of Bendayan (see Chap. 7) where the two surfaces of the section (without a supporting film) are labelled with different antibodies and distinguishable gold markers. An alternative approach suggested by Schwarz (pers. commun.) might be to replicate the two sides of double-labelled grid in different directions. This would enable the gold on both surfaces to be more easily distinguished using the two different directions of shadow.

References

Aggeler J, Werb Z (1982) Initial events during phagocytosis by macrophages viewed from outside and inside the cell: membrane-particle interactions and clathrin. J Cell Biol 94:613

Appleton TC (1974) A cryostat approach to ultrathin "dry" frozen section for electron microscopy: a morphological and X-ray analytical study. J Microsc 100:49–74

Avnur Z, Geiger B (1981) Substrate-attached membranes of cultured cells. Isolation and characterization of ventral cell membranes and the associated cytoskeleton. J Mol Biol 153:361–369

Bone G, Denton EJ (1971) The osmotic effect of electron microscope fixatives. J Cell Biol 49:571–576

Boonstra J, van Belzen N, van Maurik P, Hage WJ, Blok FJ, Wiegant FAC, Verkleij AJ (1985a) Immunocytochemical demonstrations of cytoplasmic and cell-surface EGF-receptors in A431 cells using cryo-ultramicrotomy, surface replication, freeze etching and label fracture. J Microsc (Oxford) 140:119–129

Boonstra J, van Maurik P, Defize LHK, de Laat SW, Leunissen JLM, Verkleij AJ (1985) Visualization of epidermal growth factor receptor in cryo-sections of cultured A431 cells by immuno-gold labeling. Eur J Cell Biol 36:209–216

Boonstra J, van Maurik P, Verklij AJ (1987) Immunogold labelling of cryo-sections and cryo-fractures. In: Steinbrecht RA, Zierold K (eds) Cryotechniques in biological electron microscopy. Springer, Berlin Heidelberg New York

Brüggeller P, Mayer E (1980) Complete vitrification in pure liquid water and dilute aqueous solutions. Nature 288:569–571

Chang J-J, McDowall AW, Lepault J, Freeman R, Walter CA, Dubochet J (1981) Freezing, sectioning and observation artefacts of frozen hydrated sections for electron microscopy J Microsc 132:109–123

Chang J-J, McDowell AW, Lepault J, Freeman R, Walter CA, Dubochet J (1983) Freezing, sectioning and observation artefacts of frozen hydrated sections for electron microscopy J Microsc 132:109–123

Chen WT, Singer SJ (1982) Immunoelectron microscopic studies of the sites of cell substratum and cell-cell contact in cultured fibroblasts. J Cell Biol 95:205–222

Christensen AK (1969) A way to prepare frozen thin sections of fresh tissue for electron microscopy. In: Roth LJ, Stumpf WE (eds) Autoradiography of diffusible substances. Academic Press Inc, New York, 349

Christensen AK (1971) Frozen thin sections of fresh tissue for electron microscopy, with a description of pancreas and liver. J Cell Biol 51:772–804

Christensen AK, Komorowski TE, Wilson B, Syan-Fu M, Stevens RW (1985) The distribution of serum albumin in the rat testis. Studied by electron microscopic immunocytochemistry on ultrathin frozen sections. Endocrinology 116:1983–1996

Clarke M, Schatten G, Mazia D, Spudich JA (1975) Visualization of actin fibers associated with the cell membrane in amoebae of *Dictyostelium discoideum.* Proc Natl Acad Sci USA 72:1758

Clegg JS (1982) Alternative views on the role of water in cell function In: Franks F, Mathias S(eds) Biophysics of water. Wiley, Chichester pp 365–383

Costello MJ, Corless JM (1977) The direct measurement of temperature changes within freeze-fracture specimens during rapid quenching in liquid coolants. J Microsc 112:17–37

Craig S, Staehelin LA (1988) High pressure freezing of intact plant tissues. Evaluation and characterization of novel features of the endoplasmic reticulum and associated membrane systems. Eur J Cell Biol 46:80–93

Craig S, Gilkey JC, Staehelin LA (1987) Improved specimen support cups and auxilliary devices for the Balzers high pressure freezing apparatus. J Microsc 48:103–106

Dahl R, Staehelin LA (1989) High-pressure freezing for the preservation of biological structure: theory and practice. J Electron Microsc Tech 13:165–174

Dubochet, J, McDowall A W (1984) Frozen hydrated sections In: Revel JP, Barnard T, Haggis GH (eds) Science of biological specimen preparation. SEM Inc, AMF O'Hare, Chicago, pp 147–152

Dubochet J, Booy FP, Freeman R, Jones AV, Walter CA (1981) Low temperature electron microscopy. Ann Rev Biophys Biochem 10:133–149

Dubochet J, Lepault J, Freeman R, Berriman JA, Homo JC (1982) Electron microscopy of frozen water and aqueous solutions. J Microsc 128:219–237

Dubochet J, McDowall AW, Menge B, Schmid EN, Lickfield KG (1983) Electron microscopy of frozen-bacteria. J Bacteriol 155:381–390

Dubochet J, Adrian M, Chang J-J, Homo J-C, Lepault J, McDowall AW, Schultz P (1988) Cryo-electron microscopy of vitrified specimens. Q Rev Biophys 21:129–228

Escaig J (1982) New instruments which facilitate rapid freezing at 83K. J Microsc 126:221–226

Escaig J (1984) Control of different parameters for optimal freezing conditions. In: Revel JP, Barnard T, Haggis GH (eds) The science of biological specimen preparation. SEM Inc, AMF O'Hare, Chicago, pp 117–122

Fernándes-Morán (1952) Application of the ultrathin freezing-sectioning technique to the study of cell structures with the electronmicroscope. Ark Fys 4:471–491

Forsman CA, Pinto da Silva P (1988) Fracture-flip: new high-resolution images of cell surfaces after carbon stabilization of freeze-fractured membranes. J Cell Sci 90:531–539

Franks F (1977) Biological freezing and cryofixation. J Microsc 111:3–16

Fujimoto K, Pinto da Silva P (1988) Macromolecular dynamics of the cell surface during the formation of coated pits is revealed by fracture-flip. J Cell Sci 91:161

Geiger B, Dutton AH, Tokuyasu KT, Singer SJ (1981) Immunoelectron microscopic studies of membrane-microfilament interactions: distribution of α-actin in tropomyosin, and vinculin in intestinal brick border and chicken gizzard smooth muscle cells. J Cell Biol 91:614–628

Gelderblom H, Kocks C, L'Age-Stehr J, Reupke H (1985) Comparative immunoelectron microscopy with monoclonal antibodies on yellow fever virus-infected cells: pre-embedding labelling versus immunocytomicrotomy. J Virol Methods 10:225–239

Gilkey JC, Staehelin LA (1986) Advances in ultrarapid freezing for the preservation of cellular ultrastructure. J Electron Microsc Tech 3:177–210

Green J, Griffiths G, Louvard D, Quinn P, Warren G (1981) Passage of viral membrane proteins through the Golgi complex. J Mol Biol 152:663–698

Griffiths G (1984) Selective contrast for electron microscopy using thawed frozen sections and immunocytochemistry In: Revel JP, Barnard T, Haggis GH (eds) Proc Congr Specimen Preparation. Traverse City, Michigan, SEM Inc, Chicago, pp 153–159

Griffiths G, Brands R, Burke B, Louvard D, Warren G (1982) Viral membrane proteins acquire galactosyl in *trans* Golgi cisternae during intracellular transport. J Cell Biol 95:781–792

Griffiths G, Simons K, Warren G, Tokuyasu KT (1983) Immunoelectron microscopy using thin, frozen sections: applications to studies of the intracellular transport of Semliki Forest virus spike glycoproteins. Methods Enzymol 96:466–484

Griffiths G, McDowall AW, Back R, Dubochet J (1984a) On the preparation of cryo-sections for immunocytochemistry. J Ultrastruc Res 89:65–78

Heuser JE, Reese TS, Dennis MJ, Jan Y, Jan L, Evans L (1979) Synaptic vesicle exocytosis captured by quick freezing and correlated with quanal transmitter release. J Cell Biol 81:275–300

Hodson S, Marshall J (1972) Ultramicrotomy – a technique for cutting ultrathin sections of unfixed frozen biological tissues for electron microscopy. J Microsc 95:459–465

Hohenberg H (1989) Replica preparation techniques in immuno-gold cytochemistry In: Verkleij AJ, Leunissen JLM (eds) Immuno-gold labeling in cell biology. CRC Press Inc, Boca Raton, Florida, pp 157–178

Hohenberg H, Bohn W, Rutter G, Mannweiler K (1985) Plasma membrane antigens detected by replica techniques. In: Müller M, Becker RP, Boyde A, Wolosewick JJ (eds) Science of biological specimen preparation. SEM Inc, Chicago, pp 235–244

Hopkins CR (1985) The appearance and internalization of transferrin receptors at the margins of spreading human tumor cells. Cell 40:199–207

Howell PGT, Boyde A (1972) Comparison of various methods for reducing measurements from stereo-pair scanning electron micrographs to "Real 3-D Data". In: Scanning electron microscopy. Proc 5th Annu Scanning Electron Microsc Symp, IIT Research Institute, Chicago, p 234

Hudson B, Makin MJ (1970) The optimum tilt angle for electron stereo-microscopy. J Sci Instrum 3:311–323

Hunziker EB, Hermann W, Schenk RK, Müller M, Moor H (1984a) Cartilage ultrastructure after high pressure freezing, freeze substitution, and low temperature embedding. I. Chondrocyte ultrastructure – implications for the theories of mineralization and vascular invasion. J Cell Biol 98:267–276

Hunziker EB, Hermann W, Schenk RK, Müller M, Moor H (1984b) Cartilage ultrastructure after high pressure freezing, freeze substitution, and low temperature embedding. II.

Intercellular matrix ultrasturcture – preservation of proteoglycans in their native state. J Cell Biol 98:277–282

Jésior J-C (1989) Use of low-angle diamond knives leads to improved ultrastructural preservation of ultrathin sections. Scanning Microsc Suppl 3:147–153

Karp RD, Silcox JC, Somlyo AV (1982) Cryoultramicrotomy: evidence against melting and the use of a low temperature cement for specimen orientation. J Microsc 125:345–352

Keller GA, Tokuyasu KT, Dutton AH, Singer SJ (1984) An improved procedure for immunolabelling: ultrathin plastic embedding of immunolabelled ultrathin frozen sections. Proc Natl Acad Sci USA 81:5744–5747

Keller GA, Scallen TJ, Clarke D, Maher PA, Krisans SK, Singer SJ (1989) Subcellular localization of sterol carrier protein-2 in rat hepatoytes: its primary localization to peroxisomes. J Cell Biol 108:1353–1361

Knoll G, Verkleij AJ, Plattner H (1987) Cryofixation of dynamic processes in cells and organelles In: Steinbrecht RA, Zierold K (eds) Cryotechniques in biological electron microscopy. Springer, Berlin Heidelberg New York, pp 258–271

Langanger G, De Mey J (1988) Ultrathin cryosections in the plane of cell monolayers: evaluation of their potential for antibody localization studies of the cytoskeleton. J Electron Microsc Tech 8:391–399

Lepault J, Booy FP, Dubochet J (1983) Electron microscopy of frozen biological suspensions. J Microsc 129:89–102

Leunissen JLM, Elbers PF, Leunissen-Bijvelt JJM, Verkleij AJ (1984) An evaluation of the cryosectioning of fixed and cryoprotected rat liver. Ultramicroscopy 12:345–352

MacKenzie AP (1977) Non-equilibrium freezing behaviour of aqueous systems. Philos Trans R Soc Lond B278:167–189

Mannweiler K, Hohenberg G, Bohn W, Rutter G (1982) Protein-A gold particles as markers in replica immunocytochemistry: high resolution electron microscope investigations of plasma membrane surfaces. J Microsc 126:145–156

Mannweiler K, Bohn W, Rutter G (1989) Labeling of viral antigens in replica techniques In: Verkleij AJ, Leunissen LJM (eds) Immuno-gold labeling in cell biology. CRC Press Inc, Boca Raton, Florida pp 317–329

Mayer E, Brüggeller P (1982) Vitrification of pure liquid water by high pressure jet freezing. Nature 298:715–718

Mazur P (1970) Cryobiology: the freezing of biological systems. Science 168:939–949

Mazur P (1984) Freezing of living cells: mechanisms and implications. Am J Physiol 247(16) C125–C142

McDowall AW, Chang J-J, Freeman R, Lepault J, Walter CA, Dubochet J (1983) Electron microscopy of frozen hydrated sections of vitreous ice and vitrified biological samples. J Microsc 131:1–9

McDowall A, Gruenberg J, Römisch K, Griffiths G (1989) The structure of organelles of the endocytic pathway in hydrated cryosections of cultured cells. Eur J Cell Biol 49:281–294

McDowall AW, Hofmann W, Lepault J, Adrian M, Dubochet J (1984) Cryo-electron microscopy of vitrified insect flight muscle. J Molec Biol 178:105–111

Menco BPM (1986) A survey of ultra-rapid cryofixation methods with particular reference to freeze-fracturing, freeze-etching, and freeze-substitution. J Elec Micro Techn 4:177–240

Moor H (1987) Theory and practice of high pressure freezing. In: Steinbrecht RA, Zierold K (eds) Cryo-techniques in biological electron microscopy. Springer, Berlin Heidelberg New York, pp 175–191

Müller M, Walther P, Hermann R, Schwarb P (1989) SEM immunocytochemistry with small (5 to 15 nm) colloidal gold markers. In: Verklei AJ, Leunissen JLM (eds) Immuno-gold labeling in cell biology. CRC Press Inc, Boca Raton, Florida, pp 199–216

Nei T (1976) Review of the freezing techniques and their theories In: Yamada E, Mizuhima M (eds) Recent progress in EM of cells and tissues. Georg Thieme Publ, Stuttgart pp 213–243

Nemanic M K (1974) Preparation of stereo slides from electron micrograph stereopairs. In: Hayat MA (ed) Principles and techniques of scanning electron microscopy: biological applications, vol 1. Van Nostrand Reinhold Co pp 135–148

Nermut MV (1982) The cell monolayer technque in membrane research: A review. Eur J Cell Bio 28:160-172

Nermut MV, Nicol A (1989) Colloidal gold immunoreplica method. In: Hayat MA (ed) Colloidal gold: principles, methods and applications, vol 1, Academic Press, New York pp 349-374

Nicol A, Nermut MV, Doeinck A, Robenek H, Wiegand C, Jockusch BM (1987) Labeling of structural elements at the ventral plasma membrane of fibroblasts with the immunogold technique. J Histochem Cytochem 35:499-506

Parton RG, Prydz K, Bomsel M, Simons K, Griffiths G (1989) Meeting of the apical and basolateral pathways of the Madin-Darby canine kidney cells in late endosomes. J Cell Biol 109:3529-3272

Pentilla A, Kalimo H, Trump BF (1974) Influence of glutaraldehyde and/or osmium tetroxide on cell volume, ion content, mechanical stability, and membrane permeability of Ehrlich ascites tumor cells. J Cell Biol 63:197

Peters K R (1984) Continuous ultra thin metal films. In: Revel JP, Barnard T, Haggis GH (eds) Science of biological specimen preparation.SEM Inc, AMF O'Hare, Chicago, pp 221-223

Pinto da Silva P (1984) Freeze-fracture cytochemistry. In: Polak J, Varndell I (eds) Immunolabelling for electron microscopy. Elsevier, Amsterdam, pp 179-188

Pinto da Silva P (1987) Molecular cytochemistry of freeze-fractured cells: freeze-etching, fracture-label, fracture-permeation, and label-fracture. Adv Cell Biol I:157-190

Pinto da Silva P (1989) Visual thinking of biological membranes: from freeze-etching to label-fracture. In: Verklei AJ, Leunissen JLM (eds) Immuno-gold labeling in cell biology. CRC Press Inc, Boca Raton, Florida, pp 179-198

Pinto da Silva P, Kan F K (1984) Label facture: A method for high resolution labelling of cell surfaces. J Cel Biol 99: 1156-1161

Pinto da Silva P, Kachar B, Torrisi MR, Brown C, Parkison C (1981a) Freeze-fracture cytochemistry: replicas of critical point-dried cells and tissues after "fracture label". Science 213:230-233

Pinto da Silva P, Parkison C, Dwyer N (1981b) Fracture-label: cytochemistry of freeze-fracture faces in the erythrocyte membrane. Proc Natl Acad Sci USA 78:343-347

Pinto da Silva P, Parkison C, Dwyer N (1981c) Freeze-fracture cytochemistry, II: Thin sections of cells and tissues after labelling of fracture faces. J Histochem Cytochem 29: 917-928

Pinto da Silva P, Douglas SD, Branton D (1982) Freeze-fracture cytochemistry: partition of glycophorin in freeze-fractured erythrocyte membranes. J Cell Biol 93:463-469

Plattner H, Bachmann L (1982) Cryofixation of biological materials for electron microscopy by the methods of spray-, sandwich-, cryogen-jet- and sandwich-cryogen-jet-freezing: a comparison of techniques. In: Revel JP, Barnard T, Haggis GH (eds) Science of biological specimen preparation, vol 2, Chicago, SEM Inc, AMF O'Hare, Chicago, pp 139-146

Posthuma G, Slot JW, Geuze HJ (1984) Immunocytochemical assays of amylase are chymotrypanogen in rat pancrease secretory granules. J Histochem Cytochem 32:1028-1032

Posthuma G, Slot JW, Geuze HJ (1985) The validity of quantitative immuoelectron microscopy on ultrathin sections as judged by a model study. Proc Ro Microsc Soc 20 1MS

Rall WF, Fahy GM (1985) Ice-free preservation of mouse embryos at -196°C by vitrification. Nature 313:573-575

Raska I, Ochs RL, Salamin-Michel L (1990) Immunocytochemistry of the cell nucleus. Electron Microsc Rev 3:301-353

Robards AW, Sleytr UB (1985) Low temperature methods in biological electron microscopy. In: Glauert AM (ed) Practical methods in electron microscopy, vol 10, Elsevier, Amsterdam, pp 309-324

Robenek H (1989a) Distribution and mobility or receptors in the plasma membrane. In: Sek WH (ed) Freeze-fracture studies of membranes. CRC Press, Boca Raton, Florida, pp 61-86

Robenek H (1989b) Topography and internalization of cell surface receptors as analyzed by affinity- and immunolabeling combined with surface replication and ultrathin sectioning techniques. In: Plattner H (ed)Electron microscopy of subcellular dynamics. CRC Press, Boca Raton, Florida, pp 141-163

Robenek H, Severs NJ (1984) Double labeling of lipoprotein receptors in fibroblast cell surface replicas. J Ultrastruct Res 87:149-158

Roberts IM (1975) Tungsten coating – a method of improving glass microtome for cutting ultrathin frozen sections. J Microsc 103:113–119

Rogalski AA, Bergmann JE, Singer SJ (1984) Associations of elements of the Golgi apparatus with microtubules. J Cell Biol 99:1092–1100

Roos N, Morgan AJ (1990) Cryopreparation of thin biological specimens for electron microscopy: methods and applications. Microscopy Handbooks of the Royal Microscopical society 21 Oxford University Press

Rutter G, Bohn W, Hohenberg H, Mannweiler K (1985) Preparation of apical plasma membranes from cells grown on coverslips. Electron microscopic investigations of the protoplasmic surface. Eur J Cell Biol 39:443

Rutter G, Bohn W, Hohenberg H, Mannweiler K (1988) Demonstration of antigens at both sides of plasma membranes in one coincident electron microscopic image. A double immunogold replica study of virus infected cells. J Histochem Cytochem 36:1015

Schmidt G, Robenek H, Harrach B, Glössl J, Nolte V, Hörmann H, Richter H, Kresse H (1987) Interaction of small dermatan sulfate proteoglycan from fibroblasts with fibronectin. J Cell Biol 104:1683–1691

Semich R, Robenek H (1990) Organization of the cytoskeleton and the focal contacts of bovine aortic endothelial cells cultured on type I and III collagen. J Histochem Cytochem 38:59–67

Semich R, Gerke V, Robenek H, Weber K (1989) The p36 substrate of pp60[src] kinase is located at the cytoplasmic surface of the plasma membrane of fibroblasts; an immunoelectron microscopic analysis. Euro J Cell Biol 50:313–323

Severs N J (1989) Freeze-fracture cytochemistry: review of methods. J Electron Microsc Tech 13:175–203

Singer SJ, Tokuyasu KT, Dulton AH, Chen WT (1982) High resolution immunoelectron microscopy of cell and tissue ultrastructure. In: Griffith J (ed) Electron microscopy in biology, vol 2, Wiley, New York, pp 55–106

Sitte H (1984) Equipment for cryofixation, cryoultramicrotomy and cryosubstitution in biomedical TEM routine. Zeiss MEM 3:25–30

Sitte H, Edelmann L, Neumann K (1987) Cryofixation without pretreatment at ambient pressure. In: Steinbrecht RA, Zierold K (eds) Cryotechniques in biological electron microscopy. Springer, Berlin Heidelberg New York, pp 87–113

Sitte H, Neumann K, Edelmann L (1989) Cryosectioning according to Tokuyasu vs rapid-freezing, freeze-substitution and resin embedding. In: Verkleij AJ, Leunissen JLM (eds) Immuno-gold labeling in cell biology. CRC Press, Boca Raton, Florida, pp 63–96

Slot JW, Geuze HJ (1981) Sizing of protein A colloidal gold probes for immunoelectron microscopy. J Cell Biol 90:533–536

Slot JW, Geuze HJ (1982) Ultracryotomy of polyacrylamide embedded tissue for immunoelectron microscopy. Biol Cell 44:325–328

Slot JW, Geuze HJ (1983) Immunoelectron exploration of the Golgi complex. J Histochem Cytochem 31:1049–1056

Slot JW, Geuze HJ (1983) The use of protein A-colloidal gold complexes as immunolabels in ultrathin frozen sections. In: Giello AC (ed) Immunohistochemistry. IBRO Handbook series. Wiley, Chichester, England, pp 323–332

Slot JW, Geuze HJ (1985) A novel method to make gold probes for multiple labelling cytochemistry. Eur J Cell Biol 38:87–93

Small JV (1968) Measurement of section thickness. In: Bocciarelli DS (ed) Proceedings of the fourth European Congress on Electron Microscopy, vol 1 Tipografia Poliglotta Vaticana, Roma, pp 609–610

Stang E (1987) Modification of the LKB 7800 series knifemaker for symmetrical breaking of cryo knives. J Microsc 149:77–79

Stang E, Johansen BV (1987) Improved glass knives for ultramicrotomy: A modified procedure for tungsten coating the knife edge. J Ultrastr Res, Mol Struct 98:328

Stierhof Y-D, Schwarz H (1989) Labeling properties of sucrose-infiltrated cryosections. Scanning Microsc 3:35–46

Stillinger FH (1980) Water revisited. Science 201:451–457

Studer D, Michel M, Müller M (1989) High pressure freezing comes of age. Scanning Microsc 3:253–269

Tokuyasu KT (1973) A technique for ultracryotomy of cell suspensions and tissues. J Cell Biol 57:551–565

Tokuyasu KT (1974) Some applications of cryoultramicrotomy. In: Sanders JV, Goodchild DJ (eds) Proc 8th Int Congr Electron Microscopy. Australian Acad Sci Canbera 2, pp 34–35

Tokuyasu KT (1976) Membranes as observed in frozen sections. J Ultrastruct Res 55:281–287

Tokuyasu KT (1978) A study of positive staining of ultrathin frozen sections. J Ultrastruct Res 63:287–307

Tokuyasu KT (1980) Immunochemistry of ultrathin frozen sections. Histochem J 12:381–403

Tokuyasu KT (1983) Present state of immunocryoultramicrotomy. J Histochem Cytochem 31:164–167

Tokuyasu KT (1984) Immuno-cryoultramicrotomy. In: Polak JM, Varndell IM (eds) Immunolabelling for electron microscopy. Elsevier, Amsterdam, pp 71–82

Tokuyasu KT (1986) Application of cryoultramicrotomy to immunocytochemistry. J Microsc 143:139–149

Tokuyasu KT (1989) Use of poly(vinylpyrrolidone) and poly(vinyl alcohol) for cryoultramicrotomy. Histochem J 21:163–171

Tokuyasu KT, Okamara S (1959) Glass knives for electron microscopy J Biophys Biochem Cytol 6:305

Tokuyasu KT, Singer JS (1976) Improved procedures for immunoferritin labelling of ultrathin frozen sections. J Cell Biol 71:894–906

Tokuyasu KT, Dutton AH, Geiger B, Singer SJ (1981) Ultrastructure of chicken cardiac muscle as studied by double-immunolabelling in electron microscopy. Proc Natl Acad Sci USA 78:7619–7623

Torrisi MR, Bonatti S (1985) Immunocytochemical study of the partition and distribution of Sindbis virus glycoprotein in freeze-fractured membranes of infected baby hamster kidney cells. J Cell Biol 101:1300–1306

Torrisi MR, Pinto da Silva P (1982) T lymphocyte heterogeneity: WGA labelling of transmembrane glycoproteins. Proc Natl Acad Sci USA 79:5671–5674

Torrisi MR, Lotti LV, Pavan A, Migliaccio G, Bonatti S (1987) Free diffusion to and from the inner nuclear membrane of newly synthesised plasma membrane glycoproteins. J Cell Biol 104:733–737

van Berkel J, Steup M, Völker W, Robenek H, Flügge UI (1986) Polypeptides of the chloroplast envelope membranes as visualized by immunochemical techniques. J Histochem Cytochem 34:577–583

Van Harrevald A, Crowell J (1964) Electron microscopy after rapid freezing on a metal surface and substitution fixation. Anat Rec 149: 381–386

Van Harrevald A, Trubatch J, Steiner J (1974) Rapid freezing and electron microscopy for the arrest of physiological processes. J Microsc 100:189–198

Veltel D, Robenek H (1988) Immunogold surface replica study on the distribution of acetylcholine receptors in cultured rat myotubes. J Histochem Cytochem 36:1295–1303

Ward RT (1977) A method for breaking glass knives slowly. Stain Technol 52:116–117

Weibel ER (1979) Stereological methods, 1. Practical methods for biological morphometry. Academic Press, New York

Whitt MA, Chong L, Rose JK (1989) Glycoprotein cytoplasmic domain sequences required for rescue of a vesicular stomatitis virus glycoprotein mutant. J Virol, 63:3569–3578

Zierold K (1987) Cryoultramicrotomy. In: Steinbrecht RA, Zierold K (eds) Cryotechniques in biological electron microscopy. Springer, berlin Heidelberg New York, pp 132–148

Chapter 6

Elementary Immunology

BRIAN BURKE

To obtain meaningful results from immunocytochemical studies requires the use and production of well-characterized antibodies of defined specificities. The aim in this chapter is to briefly summarize our current understanding of the immune system as much as it concerns the production of specific antibodies for experimental purposes. For a more complete discussion of the immune system and cellular immunology the following reviews and texts are recommended (Paul 1984; Roitt et al. 1985; Harlow and Lane 1988). Detailed practical aspects can be found in Mayer and Walker (1987) and Hudson and Hay (1989). I will then discuss in more detail strategies which may be used for the production and purification of antibodies against defined antigens. Finally, I will point out the various methods available for characterization of antibodies, their respective antigens, and precisely what information can be obtained from these methods.

6.1 Basic Immunology

6.1.1 The Immune Response

The role of the immune system is to recognize and subsequently destroy foreign or abnormal agents, be they viruses, toxins, bacteria or tumour cells. Traditionally, immunologists have subdivided the immune system, or more precisely, the immune response into two broad classes. The first is the humoral or antibody-mediated response which involves the induction of antibodies by a foreign antigen. These antibodies circulate in the blood stream and specificially bind to the inducing antigen. Formation of an antibody-antigen complex may have a number of consequences. For instance, if the antigen is a virus particle, binding of antibody may neutralize it, i.e. render it non-infective. Binding of antibody may also facilitate phagocytosis and destruction of the antigen by phagocytic cells such as macrophages. A final consequence is that a complex of serum proteins, collectively known as the complement system, may become activated, which will lead to the lysis of cells perceived as being foreign, e.g. invading bacteria or virally infected cells. The second class of the immune response is known as cell-mediated and involves the induction, again in an antigen-specific manner, of cells capable of destroying host cells bearing foreign antigens, e.g. virally infected cells. It is this class of the immune response which is typically involved in the rejection of tissue and organ transplants.

6.1.2 Cells of the Immune System

The cells primarily involved in conferring immunity are lymphocytes. These are to be found in blood and lymph and also in various lymphoid tissues which include the bone marrow, thymus, spleen and lymph nodes. There are also other accessory cells such as macrophages, whose behaviour, including phagocytosis, may be modulated, directly or indirectly. Just as the immune response falls into two broad classes, so do the effector cells, the lymphocytes. In higher vertebrates lymphocytes originate in the bone marrow (and in the liver in fetuses) from multipotent progenitor cells. Many of these lymphocytes subsequently mature in the thymus and are thus known as **T cells** (or thymus-derived lymphocytes). The others which mature independent of the thymus are known as **B cells**. The "B" in this case, refers to Bursa-derived, since in birds these cells mature in a specialized organ known as the Bursa of Fabricius. In mammals this maturation appears to occur primarily in bone marrow. Studies involving animals which are genetically or experimentally deficient in one or the other of these cell types have shown that it is the T cells which are solely responsible for cell-mediated immunity, while it is B cells which produce antibodies. The two branches of the immune system do not operate totally independently. In the majority of cases, for a B cell to become activated to produce antibody requires the cooperation of certain subsets of T cells, most notably helper T cells. It is true, however, that some antigens, invariably polymeric, are T-independent, being able to elicit an antibody response in the absence of T cell help.

The thymus, bone marrow and bursa (in birds only) are referred to as primary lymphoid tissues. Following maturation, the lymphocytes migrate, via the bloodstream, to secondary or peripheral lymphoid tissues. These include the spleen, lymph nodes, Peyer's patches in the small intestine, appendix and tonsils. It is in these secondary lymphoid tissues that T and B cells may be first challenged with foreign antigens.

When an animal is exposed to a foreign antigen such as a virus, the immune system rapidly responds to produce antibodies which can specifically bind to the antigen. The most extraordinary aspect of this, however, is that there appears to be almost no limit to the number of antigens which an animal's immune system can recognize and, just as importantly, distinguish between. Certainly the number runs to several million. It is only very recently that our understanding of the mechanisms underlying the very diverse nature of the immune response has become clear at all.

One of the keys to this understanding is the concept of clonal selection, formulated more than 30 years ago (Burnet 1959). In short, this hypothesis, subsequently verified, predicted that during their maturation, lymphocytes become committed to recognize and respond to single specific antigens without ever having been exposed to those same antigens. Thus the immune system of an animal never exposed to foreign antigens will nevertheless contain millions of lymphocyte clones or families (both B cells and T cells), each being dedicated to responding to a single specific antigen.

Exposure of an animal to a particular antigen does not cause a general mobilisation of the immune system, but instead results in the stimulation and proliferation of only a limited number of B and T cells, each clone will produce a single class of antibody of predetermined specificity. Thus while an individual may be capable of expressing millions of different antibodies, challenge with a single antigen results in the appearance in the blood stream of only 10–100 antibody types, each resulting from the stimulation of a specific B cell clone. On first exposure to an antigen, the immune response takes several days to develop, as judged by the appearance of antibodies in the serum, and reaches a peak after about 2 weeks. This is termed the **primary response**. The immune system, however, retains a memory of that first exposure, since the proliferation and expansion of the various B cell clones gives rise not only to so called **memory cells**. Memory cells do not themselves secrete antibody, but if re-exposed to the original antigen they are able to undergo rapid proliferation and differentiation to yield mature antibody secreting plasma cells. As a consequence, if the individual is later challenged with the same antigen, the response occurs very much more rapidly, is more intense and is of longer duration. This is known as a **secondary immune response**.

6.2 The Structure of Antibodies

Antibodies or **immunoglobulins** (reviewed by Jeske and Capra 1984) represent the γ globulin class of serum proteins and account for about one fifth of the total serum protein. In mammals five distinct antibody classes have been recognized. These are known as immunoglobulin (Ig) G, IgA, IgM, IgD and IgE. They all have common features but may be distinguished by their biological activities (see below and Table 1).

All antibodies are related by general structural features. The simplest, IgG for example, are Y shaped molecules with two identical antigen binding sites, one being located at the tip of each arm of the Y. They are thus bivalent and, as a consequence, will form large aggregates with multi-valent antigens. Most protein antigens, it turns out, are effectively multi-valent, since they almost invariably possess several unique antigenic determinants or epitopes against which antibodies can be raised. This is discussed in greater detail below.

The basic structure of an immunoglobulin molecule is shown in Fig. 1. It consists of two identical heavy chains and two identical light chains all linked by intra-chain disulphide bonds. Proteolytic cleavage of the molecule with papain yields three large fragments. Two of these, termed Fab fragments, are identical, each containing a single antigen binding site. As can be seen from Fig. 1, they

Table 1. Properties of Immunoglobulins

Property	IgG	IgA	IgM	IgD	IgE
Usual form	Monomer	Monomer Dimer Hexamer	Pentamer	Monomer	Monomer
Formula	$k_2\gamma_2$ $l_2\gamma_2$	$(k_2a_2)_n$ $(l_2a_2)_n$	$(k_2\mu_2)_5$ $(l_2d_2$	k_2d_2 l_2e_2	k_2e_2
MW (kDa)	150	160	950	175	190
Sedimentation coefficient	6.6S	7–14S	19S	7S	8S
Serum concentration	8–16 mg/ml	1–4 mg/ml	0.5–2 mg/ml	0–0.4 ng/ml	0.01–0.4 µg/ml
% of total Ig in serum	75–78	7–15	5–10	0.3	0.003
Half life days	23	6	5	3	2.5

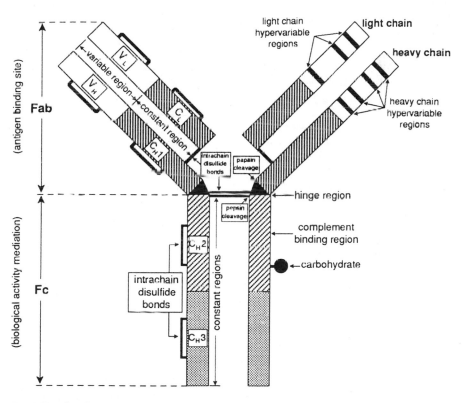

Fig. 1. Details of structure of an IgG molecule. Both heavy chains are linked by disulphide bonds in the hinge region. Similarly each light chain is disulphide bonded to each heavy chain. As indicated in the figure, proteolytic digestion may give rise either to a divalent (Fab')$_2$ fragment or to two monovalent Fab fragments depending upon whether cleavage occurs proximal or distal to the inter-heavy chain disulphide bond

each contain a single intact light chain and in the amino terminal half a heavy chain. The third fragment, known as the Fc fragment (Fc, since it is easily crystallized) is a dimer of the carboxyl terminal half of each heavy chain. It is this portion of the molecule which confers the different biological activities of each immunoglobulin class. If another proteolytic enzyme, pepsin, is used, a slightly different pattern of fragmentation is seen. In this case only a single fragment termed Fab′$_2$ is generated which is bivalent. However, treatment of this fragment with ß mercaptoethanol, a reducing agent, yields two monovalent Fab′ (pronounced Fab prime) fragments. From Fig. 1 it will be seen that the Fab′ fragment is slightly larger than Fab. After pepsin digestion, no Fc fragment is seen, since this part of the molecule is completely degraded.

Both the heavy and light chains are folded into a series of homologous domains with very similar tertiary structures, each domain being stabilized by an intra-chain disulphide bond. This structure is frequently referred to as the immunoglobulin fold and is a diagnostic feature of proteins beloning to the immunoglobulin super gene family (e.g. major histocompatibility complex class I and II antigens, Thy-1, ß2 microglobulin, T cell antigen receptor, secretory component.).

The immunoglobulin light chains contain approximately 220 amino acid residues and are folded into two domains. Two distinct types of light chain have been identified. These are known as λ and ϰ and are found in all five classes of immunoglobulins. However, a single antibody molecule may contain only one or the other type of light chain, but not both. There are five types of heavy chain γ, μ, α, δ and ε, corresponding to each of the five immunoglobulin classes, G, M, A, D and E respectively. The heavy chains contain between about 440 and 570 amino acid residues and are folded into either four (γ, α, δ) or five (μ, ε) domains, depending upon the immunoglobulin class. In contrast to the light chains, immunoglobulin heavy chains are glycosylated. The extent of glycosylation and the location of oligosaccharides varies considerably between the five immunoglobulin classes, with IgGs being the least glycosylated and IgEs being the most heavily glycosylated. As a general rule oligosaccharides are restricted to the region of the molecule on the carboxy terminal side of the disulphide bridge which links the heavy and light chains (i.e. ∼ the Fc region). They are never found in the antibody binding domain. As with the light chains, a single antibody molecule must have identical heavy chains. The heavy chains contain an additional structural feature known as the "hinge" region. This is a flexible segment of the polypeptide chain which lies between the Fc region and the two N terminal domains within the Fab portion. It is particularly sensitive to proteolysis and allows variation in the angle between the two antigen binding sites. This improves the ability of the antibody to bind two antigens.

6.3 The Biological Functions of Immunoglobulins

6.3.1 IgG

IgGs are the most abundant antibodies in serum, accounting for about 75% of the total immunoglobulin and are produced as the major component of the secondary immune response. They conform to the basic immunoglobulin scheme consisting of two heavy chains and two light chains. The Fc region of these molecules contain complement binding sites and are thus able to activate the complement system. In addition to this, they may bind to specific Fc receptors on phagocytic cells, thereby enhancing phagocytosis of antigens. Four subclasses of IgG have been recognized in both humans (subclass 1, 2, 3, 4) and mice (1, 2a, 2b, 3). They can be distinguished serologically and exhibit different affinities with various Fc binding proteins (e.g. Fc receptors, protein A). Because of their abundance, solubility properties and amenity to both chemical and proteolytic modification, IgGs are the most useful immunoglobulins for immunocytochemical work.

6.3.2 IgM

IgM (Fig. 2) is the major component of the primary immune response. In serum it accounts for about 5–10% of the total immunoglobulin. It is found in two quite different forms – secreted and membrane associated. The secreted form is a pentamer of the basic immunoglobulin unit containing 10 μ heavy chains and 10 light chains (either γ or $\varkappa$). As a consequence secreted IgM has 10 antigen

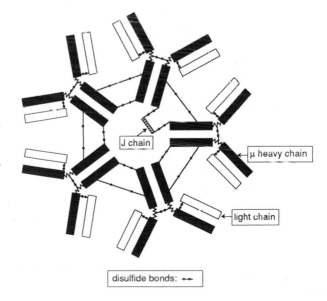

Fig. 2. Subunit structure of a circulating IgM molecule. While the overall scheme is similar to that of the IgG molecule, IgMs possess a third disulphide bonded subunit, a J chain. In addition, they exhibit additional inter-heavy chain disuphide bonds which covalently link the individual H_2L_2 heterotetramers

binding sites. As can be seen in Fig. 2, each of the IgM monomers (containing two heavy chains and two light chains) are linked by disulphide bridges. A single copy of a third polypeptide chain, the joining or J chain (M_r20000 Da), is also found in each pentamer. It is linked to two of the IgM monomers by disulphide bonds and probably nucleates assembly of the pentamer. This polypeptide also associates with IgA, although it is not a member of the immunoglobulin family. The molecular weight of a secreted IgM is about 950000 D as compared with about 150000 Da for IgG.

The membrane bound form of IgM is monomeric having only two antigen binding sites and no J chain. Each heavy chain contains a C terminal hydrophobic membrane anchor segment and arises through an alternate splicing mechanism. Membrane bound IgM functions as the major antigen receptor on B lymphocytes.

6.3.3 IgA

IgA is the second most abundant class of immunoglobulin after IgG accounting for about 10–15% of total serum Ig. It is, however, the predominant immunoglobulin of extra-vascular secretions such as tear drops. In these fluids it is found as a polymeric form containing two IgA monomers, a J chain and a fourth glycopeptide known as secretory component (SC). SC is not of lymphocyte origin, but is synthesised as a membrane bound precursor in the epithelial cells across which the IgA is transcytosed. It functions as an IgA receptor in these cells.

6.3.4 IgD and IgE

These two immunoglobulin classes are extremely minor components of serum. Both are monomeric. IgD is found in a membrane associated form and functions as an antigen receptor on B cells. IgE, on the other hand, binds to Fc receptors on mast cells, and in the presence of antigen results in the release from these cells of histamine and other pharmacologically active compounds. It is thus the immunoglobulin responsible for allergy and anaphylaxis.

6.3.5 Generation of Antibody Diversity

The domains of immunoglobulin heavy and light chains fall into two categories. The first are constant or C region domains whose structure remains unchanged between antibodies within a given class. The second are variable or V region domains and it is these which form the antigen binding site. As the name implies, these domains vary between antibodies secreted by different B cell clones resulting in different specificities and affinities. Every heavy and light chain has a single V domain located at the extreme N terminus. Thus, each

GERM LINE DNA

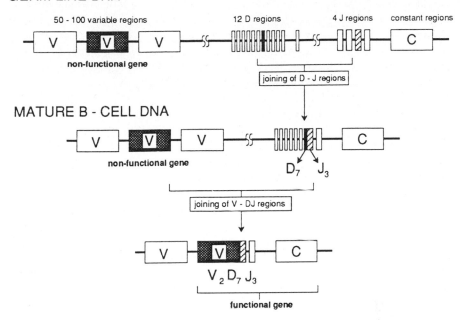

MATURE B - CELL

primary response :

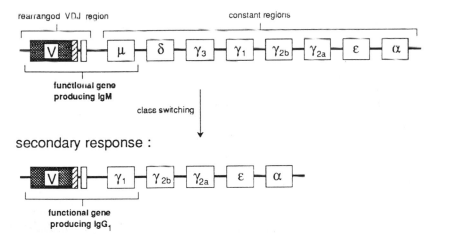

secondary response :

Fig. 3. Primary rearrangements of an immunoglobulin heavy chain gene to produce first a functional μ heavy chain followed by class switching to produce a functional γ1 chain gene during a secondary response. Removal of additional 3′ J segments occurs post transcriptionally. A more detailed explaination of these rearrangements is found in the text

antigen binding site is formed by two V domains, one from the light chain and one from the heavy chain.

It has been estimated that the antibody repertoire of a mouse may be as high as 10^9, yet it is clear that mice do not have 10^9 genes coding for antibodies of different specificities. How then can such diversity be generated? It turns out that immunoglobulin heavy and light chains are coded for by several gene segments which undergo rearrangement at the genomic level during lymphocyte maturation (reviewed by Tonegawa 1983). In the case of kappa light chains there is a single gene coding for the C region on chromosome 6 plus a large set of non-identical, although interchangeable, V region genes (proximal or 5' to the C region gene). Each V region gene codes for only about 90% of the V domain the rest being coded for by a separate short segment of DNA called the J gene segment. For the kappa light chain there are four J segments which reside in a cluster very close to the 5' end of the C gene (see Fig. 3). During lymphocyte maturation, the entire sequence between a chosen (presumably at random) V gene and a J segment is deleted such that the V gene now lies next to the J segment. The entire V-J-C unit can then be transcribed. RNA splicing is employed to remove any additional J segments. The heavy chain genes, on chromosome 12, undergo similar but more complex rearrangement since the V domains are coded for not only by V and J genes, but also by a D or diversity gene segment.

Thus, even if there are only 1000 V-J or V-D-J combinations for heavy and light chains, the fact that any light chain should be able to assemble with any heavy chain will give rise to 10^6 possible antibody molecules. In practice the number of possible V-D-J combinations for heavy chains may in fact be closer to 10000. Finally, the V regions are found to be particularly prone to somatic mutation, a property which will also greatly enhance antibody diversity.

Since lymphocytes are diploid they must contain two sets of heavy and light chain genes. The fact that they produce only a single species of antibody means that one set of genes (either paternal or maternal) for heavy and light chains must be inactive in any given lymphocyte. This is known as **allelic exclusion**. Similarly, in any given B cell, only one set of light chain genes, either λ or ϰ, which are found on separate chromosomes (16 and 6 respectively in mouse) are active.

6.4 The Nature of Antigenicity

In this final section on basic immunology we will discuss very briefly the mechanisms by which an antigen generates an immune response (for a review see Berzofsky and Berkover 1984). We will also consider what features of an antigen are recognized by the immune system. For the sake of simplicity these discussions will be restricted mainly to soluble protein antigens.

6.4.1 Antigen Size – Carriers and Haptens

Almost without exception, molecules with a molecular weight of less than about 5000 Da are unable to elicit an immune response when injected to an animal. However, it may nevertheless be possible to obtain antibodies against such a molecule if it is first cross-linked to a larger molecule prior to injection. In this case the large molecule, bovine serum albumin for instance, is referred to as a **carrier**, while the small molecule, e.g. dinitrophenol, is termed a hapten. A hapten is thus defined as a small molecule which cannot itself stimulate antibody synthesis, but which will combine with an antibody once formed.

6.4.2 Antigen Presentation

Generation of an immune response requires the interaction of several distinct cell types, both with each other and with the stimulating antigen (reviewed by Roitt et al. 1985). The antigen, however, must first be processed by antigen presenting cells (APC) of which the best known examples are macrophages. This process involves endocytosis and partial proteolysis, followed by exposure of the fragments on the presenting cell surface, apparently in association with class II major compatibility complex, or MHC glycoproteins. It is in this processed state that the antigen is able to stimulate helper T cells (T_H). These antigen specific T_H cells then interact, in the presence of antigen, with antigen specific B cells causing their proliferation and maturation. While the T_H and B cells are antigen specific, they do not recognize the same antigenic sites or **epitopes** on the molecule. This is best understood if one considers immunization with a hapten-carrier complex. In such a situation it can be experimentally demonstrated that the T_H cells are specific for the carrier while the B cells are specific for the hapten.

6.4.3 Proteins as Antigens
(see Lerner 1982; Berzofsky and Berkover 1984 for reviews)

Any antigenic determinant on a **native** protein must be situated at the surface, not in the interior, since it must be accessible to antibody. Constraints imposed by the dimensions of the antigen binding sites of antibodies limits the size of an antigenic determinant to six to eight amino acid residues. Studies on the antigenicity of native proteins such as myoglobin, lysozyme and influenza virus haemagglutinin (HA) have indicated that many antigenic states may be dependent upon secondary, tertiary and even quaternary structure since they may be composed of discontinuous segments of the polypeptide chains(s) brought into close proximity by folding. Often they are found to incorporate loops or turns in the polypeptide chain. As an example of this discontinuity, Atassi and his colleagues have mapped the amino acids forming antigenic sites on the surface of hen egg white lysozyme (see Atassi and Lee 1978). It is clear from a comparison of the amino acid sequence and of the crystallographic structure that each of the sites are highly conformation-dependent. For instance one of the sites which they mapped is formed by Lys 116, Asn 113, Arg 114, Phe 34 and Lys 33. The preponderance of charged or polar amino acids is also a consistant feature. Out of 16 amino acids mapped to three sites in lysozyme, nine are either lysine or arginine (reviewed by Lerner 1982).

A rather dramatic practical demonstration of the importance of conformation in the structure of antigenic sites has been provided by Crumpton (1974), who demonstrated that antibodies raised against native myoglobin fail to react with apomyoglobin. However, antibodies against the apo molecule favoured this conformation so strongly that they would trap myoglobin in the apo conformation (the two forms being in equilibrium) causing it to lose its haem group. On a very simplistic level one can view the antibody as "squeezing" the haem out of the myoglobin.

These studies have also shown that native proteins possess only a very limited number of sites against which antibodies are preferentially made during an immune response. In the case of the influenzy HA most antibodies are found to be directed against regions near the top of the molecule which vary between serotypes. It would appear, therefore, that many regions of a native molecule are not immunogenic. That is not to say, however, that antibodies cannot be made against these regions. Lerner and his colleagues have shown that short (as few as 10–20 amino acids) synthetic peptides derived from the HA sequence can be used to elicit antibodies against intact (although not necessarily native) HA. They have reported antibodies against about 75% of the HA1 sequence which will react with the full length molecule. This is a far broader spectrum of antisera than could be elicited by intact native HA. It is clear then, that while antigenic sites in native molecules are frequently discontinuous in nature, it is nevertheless possible to obtain antibodies against sites formed from a short segment of the polypeptide chain. In this way one can generate antibodies against regions of protein normally ignored by the immune system. This strategy for obtaining antibodies will be discussed further in Section 6.5.4.

6.4.4 The Interaction of Antibodies with Their Antigens
(see Berzofsky and Berkover 1984)

Since the binding of an antibody to its antigen is, in principle always reversible, it may be expressed in terms of an affinity constant, K_A. This is defined by the following equation:

$$K_A = \frac{[Ab.Ag]}{[Ab][Ag]} ,$$

where [Ab], [Ag] and [Ab.Ag] are the concentrations, at equilibrium, of antibody, antigen and complex, respectively. The antigen in this case is monovalent, having only a single copy of any given epitope per molecule. It may of course have several different epitopes. **Affinity** is thus a measure of the precision of fit between an antibody and its complementary epitope or antigenic site. Values for K_A are most frequently found in the range of 10^7-10^9-M^{-1} although they may be as high as 10^{12} M^{-1}. Since antibodies have a minimum of two antigen binding sites, it is intuitively obvious that when an antibody binds to a polyvalent antigen (i.e. one containing identical, repeating epitopes), binding of the first site may increase the likelihood of the second site being able to bind. This will be reflected in much tighter binding of the antibody as a whole and is referred to as the **avidity**. It is important to appreciate that affinity is a property only of a single antigen binding site whereas avidity is dependent upon the number of binding sites. Thus an IgM with antigen binding sites of the same affinity as those on an IgG will have a much greater avidity for a polyvalent antigen since it has five times the number of binding sites. It is worth pointing out, however, that in practice, IgMs generally have lower affinities than IgGs.

Another term which is frequently referred to as a charactaristic of an antibody preparation (usually just serum, in fact) is **titre**. This is literally a value for how much the antibody can be diluted before its association with antigen can no longer be detected, the method of detection being irrelevant. Clearly, the titre of any given antiserum will depend on both the affinities and avidities of the various antibodies in the serum and also on their concentrations. It is a particularly useful value for comparison of antisera. For instance, for immunocytochemistry it would be preferable to work with a high titre rather than a low titre antiserum.

Since many different lymphocyte clones may respond to a single antigen the immune response is said to be polyclonal. It is exceedingly rare to observe a monoclonal response. Even a single epitope may stimulate several lymphocyte clones. The result of this is that an immune serum will contain a broad spectrum of antibodies directed against a single stimulating antigen. These antibodies will, in all likelihood, exhibit a variety of affinities (and avidities) and be directed against a variety of epitopes on the antigen. It is obvious, then, that if the antigen possesses three or more epitopes it must form extensive aggregates with the antibodies at a certain defined relative concentration. This relative concentration is known as the **equivalence point**. Aggregates are not formed under conditions of either antibody or antigen excess. The formation of these aggregates or immunoprecipitates is the basis of a number of analytical techniques. Some of these are discussed in the following sections.

6.5 Practical Aspects of Immunology

6.5.1 Preparation of Antibodies

The most important first step in preparing an antibody is actually to ask oneself two very simple questions, namely:

- "Do I know what the antigen is?"
- "Can I purify it?"

The answers to these two questions are critical since they determine the long-term strategies which one has to adopt. The simplest situation of course is where the answer to both of these questions is "yes" and the purified, well-characterized antigen is available. In this case, one only has to decide whether to make polyclonal or monoclonal antibodies (below). This will depend to a large extent on the ultimate use to which these antibodies are to be put.

The next level of difficulty is reached when an antibody is required against a component which can be identified, but cannot be purified. Examples of such a component could include the precursor form of a protein which has only a transient existence, a prohormone for instance (see below), or perhaps a very low abundance enzyme. The strategy adopted in this case will depend to a large extent upon what is known about the antigen. If sequence information is available, e.g. in the case of the prohormone, it may be possible to elicit either polyclonal or monoclonal antibodies against the molecule by immunizing with a synethetic peptide (see below). If no such information is available, one will almost certainly be forced to adopt the monoclonal approach, which is described below. It is worth bearing in mind that inability to purify a protein for the purpose of immunization is not as common as could be imagined. Proteins which cannot be purified by conventional techniques can often be obtained in microgram quantities as bands excised from polyacrylamide gels following electrophoresis. In many cases sufficient material can be obtained in this fashion for a successful immunization.

The most complex situation arises when an antibody is required against a component about which essentially nothing is known. The best one can do in this case is to formulate ideas regarding what properties are to be expected of the component. The next step is then simply to identify antibodies which recognize an antigen having those properties following immunization with material (usually a complex mixture of proteins) in which the antigen should be found. Generally, it will be possible to produce only monoclonal antibodies against such an antigen although occasionally a polyclonal approach may be feasible. A typical example of this situation would be the generation of organelle specific antibodies, for instance, against the Golgi complex. The type of assumptions one would make about the antigen in this case are that it is (1) confined to the Golgi complex and (2) it is a protein, probably an intregral membrane protein. A more detailed discussion of organelle specific antibodies is given below.

6.5.2 Production of Polyclonal Antibodies

In short, this involves immunization (by injection) of a suitable animal (e.g. rabbit, mouse, guinea pig) with a chosen antigen, adminstration of further doses of antigen to induce a secondary response, and finally collection of immune serum. Immunisation schemes merely differ in the form and quantity in which the antigen is adminstered, and the sites (e.g. intramuscular verus sub-cutaneous) and timing of each administration. It is important to appreciate that an antiserum obtained from an animal immunised using conventional techniques is, in most cases, a good reflection of the antigen which was originally injected. Consequently, if a complex mixture of proteins is injected, then in all likelihood the antiserum will contain a complex mixture of antibodies directed against many, if not all of these proteins. In most situations such an antiserum will not be particularly useful for immunoctochemistry since there will always be ambiguity regarding the origin of any observable signal, i.e. is it due to antibodies against component X or component Y, or against both, or perhaps neither? Polyclonal antisera against complex antigens should therefore only be employed if there is a reliable method available for purifying the useful antibodies or alternatively of removing those which are unwanted. Such methods are discussed below.

Usually, acellular antigens are initially administered in combination with an **adjuvant**, the overall effect of which is to potentiate the immune response. The most commonly used is Freund's adjuvant, consisting of a mixture of mineral oils which may or may not contain a killed mycobacterium. The former is referred to as complete adjuvant while the latter is known as incomplete. The antigen and this type of adjuvant are usually combined in equal volumes as an emulsion just prior to injection. Initial administration of the antigen should be carried out in complete adjuvant, while subsequent doses should be in incomplete. In this way a large secondary immune response against the mycobacterium can be avoided. A typical immunization regime is shown in figure 5. We routinely use the protocol for producing rabbit antibodies against mouse IgG. It is, however, suitable for any good, readily available immunogen.

If only very small amounts of antigen are available or if it is only weakly immunogenic, a modified immunization regime may be used. In such cases, good response can be obtained if the animal (a rabbit) is initially immunized by injection of the antigen directly into the **popliteal lymph node**, located behind the knee joint. Accurate delivery of the antigen in this way requires minor surgery to expose the lymph node. However, this is a relatively simple procedure to perform and is explained in detail by Sigel et al. (1983). Apart from its efficacy in producing a good immune response with only a few micrograms of antigen,

Table 2. Immunization regime for producing rabbit anti-mouse IgG

Day 1	100–500 µg IgG in CFA[a]	Multiple intra-dermal sites on back
Day 21	100–500 µg IgG in FA[b]	2–3 subcutaneous sites
Day 42	20 µg IgG in PBS[c]	Intramuscular
Day 43	100–200 µg IgG in PBS	Intravenous
Day 52	Bleed via marginal ear vein	

[a] Complete Freunds adjuvant.
[b] Incomplete Freunds adjuvant.
[c] Phosphate buffered saline.

delivery of a very complex mixture of antigens directly into the lymph node seems to result in a response against only a relatively small number of the more immunogenic components of the mixture. This has proved particularly useful in generating polyclonal antisera against various intracellular organelles. This is discussed in more detail below.

A further refinement which can only be used to produce high titre antisera against proteins which are only weakly immunogenic is described by Louvard et al. (1982). They were able to obtain antibodies against mammalian clathrin, a notoriously poor immunogen, by first chemically cross-linking it to keyhole limpet haemocyanin (KLH), followed by direct injection into the popliteal lymph node as an emulsion with complete adjuvant. Presumably the immunogenicity of the clathrin is considerably enhanced when it is cross-linked to KLH which is itself a very powerful immunogen.

Polyclonal antibodies are almost invariably obtained in serum from the blood of immunized (or autoimmune) animals, the serum being the liquid remaining after the blood has been allowed to clot. It corresponds, therefore, to plasma depleted of clotting components (fibrin etc.) There are techniques available for producing polyclonal antibodies in ascites fluid (Tung 1983). However, these are not very commonly used. Antisera are usually extremely stable and may frequently be stored, under sterile conditions at 4°C, for months or even years with only marginal losses in titre. While many antisera may be used for immunocytochemical studies without further purification, it is frequently found that they give unacceptably high background or non-specific labelling due to the presence of unwanted antibody species. Another problem may be that the specific signal is very weak due to a low concentration (or titre) of the specific antibdoy. These problems are frequently overcome by purification, or at least, partial purification, of the antibody of interest. The strategies available for this range from a very crude purification of the immunoglobulin fraction from the serum to affinity purification of the specific antibody. The relevant methods are discussed below.

6.5.3 Monoclonal Antibodies

Because of the polyclonal nature of the immune response, immunization of an animal with a single protein antigen, or indeed several, will give rise to an antiserum containing a heterogeneous mixture of antibodies of varying affinities and directed against a variety of epitopes. Furthermore, because of the huge antibody repetoire of any single individual, it is extremely unlikely that any two animals will produce precisely the same spectrum of antibodies following exposure to the same antigen or antigen mixture. For many purposes this is not a problem. However, for others, it is highly desirable to be able to produce a monospecific antibody of defined affinity and directed against a single epitope.

Ideally, one would like to be able to select a single B lymphocyte, secreting an antibody of interest, and allow it to grow up as a clone of cells in culture. From the culture medium one could then collect antibody at will. Such antibody would, by definition, be monoclonal since it is produced by a single clone of antibody-secreting cells. In practice, B cells have only a very limited life span in culture, and as a consequence, production of a monoclonal antibody requires a means of immortilizing B cells. The methodology for doing this was provided in 1975 by Köhler and Milstein. They showed that B lymphocytes, if induced to fuse with a continuously growing myeloma cell (a myeloma is a lymphoid tumour cell), acquired the myeloma's ability to grow indefinitely in culture while continuing to secrete antibody. This new cell line, with properties of both the myeloma and the B cell, is known as a hybridoma.

The cells employed for producing hybridomas are usually of mouse origin (although rat and human hybridomas have been produced), B cells being most commonly obtained from the spleen of a suitable donor mouse. Since it is not feasible to identify and select useful B cells at this stage, the total spleen cell population is employed in the fusion with the myeloma cells. Use of selective medium then ensures that only hybridomas are able to grow (below). In order to maximize the chances of obtaining antibodies of the desired specificity, the spleen cell population should be enriched for B cells specific for the antigen of interest. This may be achieved by immunizing the donor mouse with the antigen during the weeks prior to collection of the spleen cells. As an alternative to this in vivo immunization, in vitro immunization is is also possible. In this case, spleen cells are collected from an unimmunized animal and cultured, under defined conditions, in the presence of antigen, resulting in the proliferation of antigen specific B cells. There are several potential advantages of in vitro immunization versus in vivo immunization. (1) It requires only 4–5 days instead of several weeks. (2) Much smaller (1/10) amounts of antigen can be used (3) In vitro immunization may negate some problems of immunological tolerance or suppression. The only obvious disadvantage of the technique is that it tends to give rise to a preponderance of monoclonal IgMs which are generally less desirable than IgGs for immunocytochemical work. The theory and practice of in vivo immunization for the purpose of producing monoclonal antibodies has been extensively reviewed by Reading (1982).

Several mouse myeloma cell lines have been employed for monoclonal antibody production, most being ultimately derived from the MOPC 21 plasmacytoma. This cell line was derived from the Balb/c mouse strain. The most commonly used myelomas are listed in Table 3. Some of these myelomas secrete their own endogenous IgG or k light chain (of MOPC 21 origin). Consequently, they have been mainly superceded by the other non-secreting derivatives. All of these myelomas have been chosen because of their inability to grow in selective medium containing Hypoxanthine, Aminopterin and Thymidine (HAT medium).

Each of the myelomas listed are deficient in the enzyme hypoxanthine guanine phosphoribosyl transferase (HGPRT) and is consequently resistant to the drug 8 azaguanine. HGPRT is a key enzyme in the salvage of purines and will utilize exogenously supplied hypoxanthine to produce inosinic acid, the common purine nucleotide precursor. Thus, a cell blocked in purine biosynthe-

Table 3. Commonly used myeloma lines for hybridoma production

Mouse myelomas	Endogenous immunoglobulin chains expressed
P3-X63-Ag8 (Köhler and Milstein 1975)	Gamma 1, kappa
NS1/1-Ag4-1 (Köhler et al. 1976)	Kappa
X63-Ag8.653 (Kearney et al. 1979)	None
Sp2/0-Ag14 (Shulman et al. 1978)	None
F0 (de St. Groth and Scheidegger 1980)	None
NS0/1 (Galfre and Milstein 1981)	None
FOX-NY (Taggart and Samloff 1984)	None

sis will nevertheless be able to survive if supplied with hypoxanthine, provided that the HGPRT is functional. Inhibition of de novo purine biosynthesis may be achieved with the drug aminopterin. Aminopterin is a potent inhibitor the enzyme dihydrofolate reductase (DHFR) which catalyzes the reduction of dihydrofolate to terahydrofolate (FH_4). FH_4, in turn, is an essential cofactor in the biosynthesis of both purines and thymidine. To survive in the presence of aminopterin, a normal cell needs to be provided with both a source of purines (in the form of hypoxanthine) and with thymidine, which can be utilized via the enzyme thymidine kinase (TK). Hence, a normal cell will survive in HAT medium while a cell deficient in HGPRT (or TK, in fact) will not.

● *Fusion* between spleen cells and myelomas is most frequently achieved using polyethylene glycol as fusogen. Following fusion, the cells are transferred into HAT medium which selects for the hybrid cells. The unfused myelomas are killed by the HAT medium while the spleen cells have only very limited growth potential in culture. The hybridomas, however, continue to grow since the spleen cell genome provides functional HGPRT while the myeloma provides the ability to survive indefinitely in culture.

● *Identification* of hybridomas secreting useful antibody is achieved by plating the cells, immediately after fusion, in 100 or more small (100–1000 μl) cultures. In this way, each culture should initially contain only a limited number (<10) of viable hybrid cells which may then grow up as individual clones. Once sufficient cells have grown up in each culture, the media must be tested for the presence of the desired antibody. The **testing procedure** will depend to a large extent upon the nature of the antigen, although typically some form of **radioimmunoassay**, **immunoblotting** or **immunofluorescence microscopy** is frequently used. Whatever the method chosen, it must be fast and reliable since the time between antibody becoming detectable in each culture supernatant and the cells becoming overgrown and dying may be only a day or two. Having identified a culture containing antibody secreting cells, the cells of interest must be separated from other non-secreting cells in the culture or from cells secreting irrelevant antibodies. This is done by growing up individual clones of cells from the culture and testing each for specific antibdoy production. Such monoclonal growth of the cells can be achieved be plating them in micro cultures at very low density so that on average there is only one single cell per culture. This is known as **limiting dilution cloning**. An alternative method is to culture the cells at **low density in agarose**. The consistency of the agarose is such that while the cells are allowed to grow normally, they remain in contact after division. Thus, each cell initially present in the culture will give rise to a clone which can be identified in the light microscope as a ball of cells suspended in the agarose. Each clone can then be picked out, transferred to a larger individual culture and tested for specific antibody production. Irrespective of the cloning methods chosen, it should be carried out at least twice to ensure the monoclonality of the cells.

Having obtained specific hybridomas, there are basically two methods for obtaining antibody. The first involves simply harvesting culture medium which for many purposes may be used directly without further purification steps. The disadvantage of this is that the antibody concentration is relatively low (10–20 µg/ml at best). The alternative is to grow the hybridoma as an ascites tumour in the pertioneal cavity of a syngeneic mouse (usually Balb/c). The ascites fluid so formed may contain the monoclonal antibody at a concentration of 10 mg/ml. The disadvantage here is that it will be contaminated with the mouse's own endogenous antibodies. However, for most purposes this is not a problem.

It is clear from this discussion of monoclonal antibody production that the enormous potential of the technique lies in the fact that it can be used to produce monospecific antibodies against antigens from complex mixtures. In principle, monoclonal antibodies could be obtained against any protein in the cell without ever having to purify it. What is required, however, is a method of identifying hybidoma clones secreting the desired antibody.

6.5.4 Anti-Peptide Antibodies

As discussed above, chemically synthezised peptides consisting of sequences exposed on essentially any part of the surface of a protein can elicit antibodies directed against the native protein molecule. For cell and molecular biologists there are several consequences of this. Firstly, the methodology may be used to obtain antibodies, polyclonal or monoclonal, against proteins whose existence have only been predicted from nucleic acid sequence data (reviewed by Lerner 1982). Secondly, it may be used to generate specific antibodies against defined regions of a protein molecule, e.g. polyoma virus large T antigen and influenza virus haemagglutinin (reviewed by Lerner 1982). Tooze et al. (1987) have recently described an anti-peptide antibody specific for a prohormone, proopiomelancortin (POMC), the precursor of ACTH, β-endorphin and other bioactive peptides. Since conventionally raised antisera invariably recognize both POMC and one or more of its cleavage products, they synthezised a peptide containing eight amino acids which span the endoproteolytic cleavage site in POMC between ACTH and β-endorphin. Polyclonal antibodies raised against this peptide were found to be specific for POMC and did not recognize either of the two product peptides. The reason for this is that endoproteolytic processing of the POMC molecule results in elimination of the epitope recognized by the anti-peptide antibodies. This very simple example shows that with careful strategies, highly specific immunological probes can be elicited using appropriate synthetic peptides.

The peptide, being a small molecule, is routinely cross-linked to a larger carrier molecule such as KLH or BSA prior to injection. Subsequently, it need be treated no differently to any other immunogen. A suitable method for coupling peptides to carrier molecules, and also a typical immunization regime are described in detail by Kreis (1986).

6.5.5 Use of Molecular Biology Approaches to Express Immunoglobulin Molecules in Cells

A recent innovation in antibody production has come from the ability to clone and express immunoglobulin genes in non-lymphoid eukaryotic cells and prokaryotic systems. The immunoglobulin molecule is a complex disulphide linked heterotetramer, requiring associated proteins for efficient assembly in the ER. For this reason, it was a long held view that expression of isolated immunoglobulin chains in heterologous systems would be unlikely to result in reagents with useful binding affinities for their original antigens. It is now clear that this is an overly pessimistic view, and useful reagents may be generated by a variety of recombinant techniques.

Perhaps the simplest method of producing antibodies by expression in a heterologous system is to micro-inject mRNA isolated from the parent hybridoma into target cells of interest (Burke et al. 1984; Rosa et al. 1989). Simultaneous expression of heavy and light chain genes even in non-lymphoid cells can result in the secretion of functional immunoglobulin (Cattaneo and Neuberger 1987). Subsequently, it has been shown that removal of the signal sequences from cloned heavy and light chains can result in the expression of assembled functional immunoglobulin in the cytoplasm of the target cells (Biocca, Neuberger and Cattaneo 1990). The addition of a nuclear localization signal to the 5′ ends of these truncated constructs results in the transport of intact immunoglobulin into the nucleus (Biocca, Neuberger and Cattaneo 1990). These new techniques will undoubtedly have a powerful impact, but they are not all appropriate for generating immunocytochemical reagents.

A major obstacle to the efficient production of recombinant immunoglobulins is the fact that they are constructed from two chains encoded by separate genes. X-ray crystallographic data on solved antigen-antibody complexes confirms the involvement of variable region residues from both the light and heavy chains in the combining site (Amit et al. 1986). However, there is other evidence that the contribution of the heavy chain may predominate for many antibodies:

- isolated light chains have little or no antigen affinity (Utsumi and Karush 1964);
- isolated heavy chains may have antigen affinity only two orders of magnitude below intact immunoglobulin (Rockey 1967);
- antibodies of different specificity have very different heavy chains but may use very similar light chains (Hong and Nisonoff 1966).

Two approaches have been taken to the generation of secreted, truncated, immunoglobulin-like reagents. Winter and his colleagues have relied on the residual binding affinity of heavy chains alone, expressed in *E. coli* as a truncated single domain molecule (Ward et al. 1989). Lerner and his colleagues have established a system for random combination of a truncated light chain and a truncated heavy chain by joining individual heavy and light chain libraries, derived by PCR from splenic RNA from immunized animals (Huse et al. 1989). Both of these approaches are in their early stages, but it is clear that rapid developments may be expected. In the years to come this type of technology will undoubtedly provide a wide range of low molecular weight immunocytochemical reagents, with the potential for further improving the resolution limit defined by the size of an intact immunoglobulin molecule.

6.5.6 Purification of Antibodies

Several procedures are available for the purification of both monoclonal and polyclonal antibodies and are described briefly in this section (for a review see Johnstone and Thorpe 1982). Each of these methods yields a somewhat different preparation in terms of purity. Consequently, the use to which the antibodies are to be put, and also their source, should determine which method is most appropriate.

6.5.6.1 Preparation of γ Globulin Fraction

This may be achieved simply by ammonium sulphate precipitation (40% saturation) of serum. This yields only a very crude antibody preparation. However, the technique is useful as a source of material for further ion exchange purification of IgG. Ammonium sulphate precipitation may also be employed as a convenient method for the concentration of monoclonal antibodies in hybridoma culture supernatants.

6.5.6.2 Purification of IgG

There are two commonly used methods for the purification of IgGs. The first is based upon the charge properties of IgG, while the second utilizes the specific interaction between certain immunoglobulin Fc domains and the two immuno-

Table 4. Immunoglobulin G binding to protein A versus protein G. (Harlow and Lane 1988)

Species	Protein A	Protein G
Human	++++	++++
Horse	++	++++
Cow	++	++++
Pig	+++	+++
Sheep	+/−	++
Goat	−	++
Rabbit	++++	+++
Chicken	−	+
Hamster	+	++
Guinea pig	++++	++
Rat	+/−	++
Mouse	++	++

Protein A/G affinities for mouse IgG subclasses

Antibody		
IgG1	+	++++
IgG2a	++++	++++
IgG2b	+++	+++
IgG3	++	+++

Note that this table is primarily relevant for immunochemical experiments. For immunocyto-chemistry see the table in Chapter 9.

globulin binding proteins from *Staphylococcus aureaus*, protein A and protein G. In an immune serum, IgG specific for the immunizing antigen usually accounts for only about 10% of the total IgG.

- Purification of an IgG fraction from serum is most conveniently achieved by ion exchange chromatography of ammonium sulphate precipitated γ globulins, using a DEAE cellulose column. Most IgG remains unbound to the column at low ionic strength (10 mM Na phosphate buffer) and can be collected as the flow through. The remainder can then be preferentially eluted using 50–70 mM NaCl. Above this concentration IgA is eluted. Monoclonal IgG from ascites fluid can also be purified in this manner. Because the monoclonal antibody will be at a very high concentration (maybe > 10 mg/ml) and completely homogeneous, eluting from the column at a defined ionic strength, it can be obtained in a highly enriched form.
- The second method also involves immobilization of the IgG on a solid matrix under one set of defined conditions followed by elution under a second set of conditions. In this case the matrix consists of a material such as Sepharose 4B (agarose beads) with covalently attached protein A or protein G. These two proteins exhibit a high affinity interaction with the Fc domains of certain IgG subclasses (Table 2: their binding differs somewhat between

species) and also with IgMs from some species. Binding of IgG to protein A occurs at roughly neutral or slightly alkaline pH, while elution takes place at low pH. The high specificity of this interaction provides a single step purification of IgG directly from serum. It is, also however, invaluable for use on an analytical scale for immunoprecipitations (below). The other important use of this method is the purification of monoclonal IgGs (of appropriate subclass) from hybridoma culture supernatants. The conventional ammonium sulphate/ion exchange method is not particularly convenient for this purpose since the concentration of monoclonal antibody is usually very low. This necessitates having to handle large volumes of tissue culture medium. It is much simpler, therefore, just to pass the medium once over a small (1 mg protein A will bind >5 mg IgG) protein A-Sepharose column and then to elute the bound IgG at low pH. In our experience, contamination with bovine IgG from fetal calf serum in the medium is minimal. Because a 1 ml column can bind monoclonal IgG from about 500–1000 ml of culture supernatant, the antibody is recovered highly concentrated.

Since these two techniques are based on general properties of IgG neither are capable, therefore, of providing an antigen-specific IgG subfraction from serum.

6.6 Affinity Purification

The methods which we have so far described for the purification of antibodies are based on bulk properties of immunoglobulins and do not, therefore, discriminate between antibodies of different specificities. Purification of antibodies (from serum) which recognize a particular antigen requires a scheme which utilizes their specific interaction. This is referred to as affinity purification. Given the availability of antigen, affinity purifications are extremely simple to perform and will yield a pure antibody preparation in a single step. It is a particularly valuable procedure for the concentration of specific antibodies. For instance, very low titre antisera may be useless if used directly for immunocytochemical purposes. However, the antibodies which they contain, once concentrated and recovered by affinity purification, may turn out to be extremely valuable immunological reagents.

The entire process involves binding of specific antibody to a solid matrix (the affinity matrix) under conditions where no other antibodies will bind. Following a washing step to remove unbound or only loosely bound antibodies, the conditions are adjusted such that the specific antibodies are eluted from the matrix. The procedure is very similar to that employed in the purification of IgG using protein A. The difference lies in the fact that the specific interaction that is employed for affinity purification involves the Fab portion of the antibody molecule which is antigen specific, and not the Fc portion which is antibody class specific.

The affinity matrix should consist of purified antigen covalently coupled to a solid, insoluble support. Chemically activated supports, ready for coupling with antigen, are commerically available. The most commonly used include cyanogen bromide activated Sepharose 4B (Pharmacia), glutaraldehyde activated Ultrogel (LKB) and Affi-Gel (Bio Rad). In each case, coupling with antigen is a relatively simple process. The various procedures involved in the preparation of an affinity matrix are described in detail by Johnstone and Thorpe (1982) and by Ternynck and Avrameas (1976).

Binding of specific antibody to the affinity matrix is most conveniently achieved using a small chromatgraphy column through which the serum can be slowly passed. The washing and elution steps may then be performed simply by changing the buffers in the column reservoir. The conditions most frequently employed for eluting specific antibodies from the affinity matrix are those of low pH, usually between 2 and 3 or other mildly chaotropic agents. Immediately following elution, these agents may either be neutralized or dialyzed away. Because of the high affinities of many antibodies for their respective antigens it may sometimes be necessary to use rather more harsh conditions to effect the elution (e.g. 6–9 M urea or even 1% SDS). Under such circumstances there is always the danger of permanently damaging the antibody. Further practical details may be found in Johnstone and Thorpe (1982).

It is clear from the foregoing discussion that affinity purification of antibodies on a preparative scale requires a source of purified antigen for the preparation of the affinity matrix. Such an antigen preparation may not always be available, and in these cases the technique may only be of limited use. However, affinity purifications can frequently be carried out on a microscale using antigens which can only be separated by SDS gel electrophoresis. This procedure is adapted from the immunoblotting technique which is described below. In short, electrophoretically separated proteins may be transferred (blotted) onto a sheet of chemically activated paper or, more commonly, onto a nitrocellulose filter, where they are immobilized, thereby forming a replica of the original gel. Individual strips of nitrocellulose containing a single protein band may then be excised from the filter and used directly as an affinity matrix. In this way it is possible to obtain sufficient affinity purified antibody for several immunocytochemical experiments or for a small number (e.g. one to five) of immunoblots or immunoprecipitations (below). These procedures are described in full by Olmsted (1981), Burke et al. (1982) and Smith and Fisher (1984).

Frequently, it may not be possible to produce affinity-purified antibody even using this microscale procedure. However, it may nevertheless be feasible to produce a useful antibody using a crude antigen preparation for the affinity matrix. Alternatively, specific antibody enrichment may be achieved by removing unwanted or interfering antibodies from an antiserum by affinity procedures. This is best explained by a relevant example. Louvard et al. (1982) were able to obtain antibodies specific for the Golgi complex of rodent cells using precisely these techniques. They initially immunized rabbits with rat liver Golgi membranes via the popliteal lymph nodes and obtained an antiserum which recognized, in addition to the Golgi complex, many cellular membranes and also secreted proteins such as fibronectin. They subsequently prepared affinity matrices using sub-cellular membrane fractions discarded during the original rat liver Golgi membrane preparation and also using rat plasma which contains proteins secreted by the liver. These affinity matrices were then used to absorb from the antiserum all unwanted reactivities as determined by an immunofluoresence screening assay. The remaining antibodies which did not bind to the affinity matrices were found to be specific for the Golgi complex, recognizing a single protein of 135 kDa. Similar procedures also yielded antibodies specific for the endoplasmic reticulum (Louvard et al. 1982)

6.7 Characterization of Antibodies

Having produced an antibody, one is invariably faced with the prospect of demonstrating its specificity against the antigen of interest. Furthermore, following immunization with a complex mixture of antigens, it may even be necessary to identify the antigen against which the antibody is directed. This is a frequently encountered situation when raising monoclonal antibodies. Even in the simplest case where a single pure antigen has been used to raise polyclonal antibodies, it is important to be able to verify that the antibody does not show any irrelevant **cross reactivities**. This may occur when the antigen of interest may, by chance, share a common epitope with an apparently unrelated protein. Such cross reactivities can be extremely misleading in immunocytochemical experiments although they may, at one level, shed light on the relationship between certain proteins.

6.8 Precipitin Techniques

The simplest methods for demonstrating antibody specificity rely on the co-precipitation of an antibody with its antigen at their equivalence point (see above). These methods are generally only applicable to polyclonal antibodies since monoclonal antibodies alone (i.e. in the absence of a second precipitating antibody), recognizing only one epitope per antigen molecule, tend not to form precipitable immune complexes. By far the most straightforward of these methods is double immunodiffusion (Ouchterlony and Nilsson 1986). In this technique, antibody and antigen are allowed to encounter each other by diffusion through an agarose gel, having originally been placed in separate wells about 1 cm apart. If the antigen is recognized by the antibody, an immunoprecipitin band appears between the two wells at the precise location where the concentrations of the two components are at equivalence. Clearly, adjusting the original concentration either of the antibody or of the antigen will result in a shift or even in the disappearance of the precipitin band. This very simple technique is most suitable for demonstrating a reaction against a **purified antigen**. For complex mixtures it is less useful since it provides little information regarding which antigens in the mixture are recognized by the antibody (or antibodies). Some of these disadvantages can be overcome to a certain extent by combining the precipitin methodology with agarose electrophoresis. Such techniques, and their suitability for quantitative analysis, are described in detail by Johnstone and Thorpe (1982). These procedures have, however, been generally superceded by more powerful methods involving polyacrylamide gel electrophoresis and by approaches where detection of antigen-antibody binding utilizes more sensitive enzymic and radiochemical techniques. These are discussed in the following sections.

6.9 Immunoblotting

Also known as Western blotting (Burnette 1981), this technique allows detection of protein antigens separated by SDS polyacrylamide gel electrophoresis. Because the technique resolves proteins in terms of their sub-unit molecular weight it may be employed to determine the specificities of antibodies against antigens in complex mixtures, e.g. in total cell lysates, for instance. For this reason it is particularly useful in the initial characterization of monoclonal antibodies where specificities may be largely unknown. In the case of polyclonal antisera raised against purified antigens, the technique may be employed as a final check of specificity and also as a means to determine existence of any cross reacting proteins in cells and tissues etc. on which the antibody is to be used. Such cross reactivities can lead to extremely ambiguous interpretations of immunocytochemical experiments if they go undetected. Obviously, as an analytical method in its own right, immunoblotting may be used to follow the fate of protein antigens in a large variety of different biochemical and cell biological procedures.

The great advantage of the immunoblotting technique is that it is quick and relatively simple to perform, requiring few manipulations. It is carried out in four main stages. (1) Fractionation of the antigen mixture on an SDS gel. (2) Electrophoretic transfer of the fractionated proteins onto a filter, usually nitrocellulose, thereby producing a replica of the gel. (3) Labeling of the filter with specific antibodies. (4) Detection of bound antibodies.

Each of these steps is essentially straight forward. Electrophoretic fractionation of proteins can be performed conventionally using either one-dimensional (Laemmli 1970) or, where necessary, two-dimensional gels (O'Farrel 1975). The transfer or blotting step is described by Towbin et al. (1979) and by Burnette

(1981), and requires the use of an electrophoretic transfer chamber. These are commercially available from several sources including Hoeffer and Bio Rad. Labelling of the filter with antibody requires first saturating all non specific protein binding sites by soaking the filter in a concentrated protein solution, e.g. BSA at 20 mg/ml. Incubation with a suitable diluted antibody is then performed in the same solution.

Detection of specifically bound antibody is invariably accomplished using either a second antibody, an anti-IgG, or protein A, each conjugated to a specific marker. The most commonly used markers are enzymes (peroxidase, alkaline phosphates, Hawkes et al. 1982; Dao 1985), [125]I (e.g. Burnette 1981), and colloidal gold (Brada and Roth 1984). The enzymic markers are used in conjunction with substrates which yield insoluble coloured products giving rise to coloured bands on the white nitrocellulose. Enzymic detection has the advantage of being extremely fast, but is generally non-quantitative. [125]I is detected autoradiographically (or sometimes by γ counting) and is useful for quantitation. It may, however, be rather slow since autoradiographic exposures can last from a few hours to several days. With colloidal gold, the colour develops directly on the filter during incubation with the labelled reagent. When combined with a silver enhancement step, its sensitivity is as high as the other three methods. It is not, however, easily amenable to quantitation. Kits for gold labelling are available from Janssen Pharmaceuticals. Enzyme-linked antibodies and protein A are available from a large number of companies, including Tago, Sigma, Miles, Vector Labs, Dako and Cappel.

The one major drawback of the immunoblot technique is that antigens are denatured prior to reaction with antibodies. As a consequence many antibodies fail to recognize their antigens following blotting, despite the fact that some renaturation probably does take place on the filter. In a few cases this may be an insurmountable problem, for instance, where an epitope is formed by the interaction of two non-identical subunits. Such an epitope could never be detected on conventional blots. In these situations immunoprecipitation, described below, may be the method of choice. Further discussions of the immunoblot technique are provided by Gershoni and Palade (1982).

6.10 Immunoprecipitation

Immunoprecipitation essentially involves the immobilization of a soluble (or solubilized) antigen, often from a complex mixture on a solid support (e.g. protein A Sepharose) using a specific antibody. Usually the antigen is radioactively labelled and the precipitated material may be analyzed by counting radioactivity or more commonly by autoradiography/fluorography following gel electrophoresis. Thus, like immunoblotting, the immunoprecipitation technique allows one to identify protein antigens by molecular weight (and/or iso-electric point). The major differences between the two techniques stem from the fact that the antibody-antigen interaction takes place after electrophoresis in the case of immunoblotting, but before electrophoresis in the case of immunoprecipitation. There are two consequences of this.

- Immunoblotting will only detect the **total population** of a given antigen, whereas immunoprecipitation, depending upon the labelling conditions chosen, may be used to detect a selected subset of an antigen population. For instance, if a protein antigen is to be labeled by metabolic incorporation of, say, ^{35}S-met, a labelling period of short duration (e.g. 10 min) will allow one to detect only newly synthesized protein. A long labelling period, on the other hand, (e.g. 10 h) or chemical iodination with ^{125}I, will, like immunoblotting, allow detection of the entire population of antigen molecules.
- While immunoprecipitations are rather tedious to perform, they may work in situations where immunoblots fail since many antibodies will not recognize their respective antigens if the antigens have been denatured in SDS. This tends to be more noticeable when using monoclonal antibodies rather than polyclonal antisera, owing to the heterogeneity of specificities and affinities in the latter.

A prerequisite for a "clean" precipitation is that the antigen under investigation must be rendered completely soluble under conditions where it will still associate with its specific antibody. Since the solubility of proteins varies immensely, depending, among other things, upon their function and subcellular (or tissue) localization, methods employed for their solubilization are equally diverse. Ideally, one should attempt to employ conditions which are just sufficient to solubilize an antigen, breaking all its interactions with other proteins, without completely denaturing it. In many cases, such as with membrane proteins or other highly insoluble proteins it may not be possible to tread this rather fine line.

Following the precipitation step, the immunoprecipitate, consisting of a solid matrix (protein A-Sepharose) with attached immunecomplexes, must be washed free of non-specifically absorbed material. The washing regime used will depend to a large extent upon both the protein being precipitated and also upon the initial solubilization method. Precipitated antigen can be subsequently released from the matrix with SDS and analyzed by electrophoresis. Because there are a huge number of immunoprecipitation protocols in the literature, many tailored for specific proteins, we have included three protocols in the Appendix which we have found generally useful. The applicability of these protocols in various situations is discussed.

- Protocol A involves the harshest solubilization conditions. However, if the antigen is still recognized after denaturation in SDS, this method routinely gives the cleanest precipitates with minimal background.
- Protocol B is particularly useful for membrane proteins. It employs the detergent Triton X-114 which undergoes a cloud point precipitation at 30°C yielding a detergent phase into which membrane proteins and other

hydrophobic proteins preferentially partition and an aqueous phase which contains hydrophilic proteins.

● Protocol C is most suitable for freely soluble proteins and some membrane proteins.

Following the washing step the immunoprecipitates may be analyzed by electrophoresis.

6.11 Immunoassay

In the final section of this chapter on immunology, we will focus very briefly on the topic of immunoassays. In short, these use the specific antibody-antigen interaction to provide quantitative information regarding either the antigen or the antibody. Immunoassays can be extremely sensitive and may vary greatly in complexity. However, we shall consider only the simplest form, which is the primary binding assay. Because of its simplicity, it is particularly useful, for instance, for testing antisera from immunized animals and for screening hybridoma tissue culture supernatants for secretion of specific antibody. These assays are usually performed using 96-well plastic microtitration plates and are described in detail by Johnstone and Thorpe (1982). For identification of antibody directed against a specific antigen, the microtitration plate is first coated with antigen. This can usually be achieved by simply filling each well with an antigen solution and incubating for an hour or two, during which time a small amount of the antigen will become tightly, albeit non-specifically, absorbed to the plastic surface. The plate is then washed free of unbound antigen and all remaining protein binding sites are saturated by incubation with a concentrated solution of an irrelevant protein such as haemoglobin or bovine serum albumin. Each of the wells is next incubated with a sample (50–100 µl) of the solution, e.g. hybridoma supernatant, under test. After a washing step, wells containing specifically bound antibody are identified using a labelled second antibody, e.g. rabbit anti-mouse IgG, or protein A. If the label is radioactive (e.g. ^{125}I), the assay is referred to as a radioimmunoassay (or RIA) and analysis is performed by counting the radioactivity in each well. Alternatively, the label may be an enzyme in which case the assay is referred to as an enzyme linked immunosorbent assay (or ELISA for short) and analysis is performed spectrophotometrically employing enzyme substrates which yield coloured products. With only minor modifications, this technique can be adapted for estimation of antigen concentrations (Johnstone and Thorpe 1982). Finally, the microtitration plate may for some purposes be replaced simply by a sheet of nitrocellulose on which an antigen solution can be spotted. The whole sheet is then processed with first and second antibodies just as it would be in a Western blot, with detection being either radiometric or enzymic. This is frequently referred to as a dot-blot (Glenney et al. 1982; Hawkes et al. 1982). It is especially useful in estimations of antigen concentration.

Appendix

Stock Solutions and Materials

- Cells grown in 6.0 cm petri dishes.
- Minimal essential medium without methionine (MEM-met).
- Labelling medium: MEM-met, 10% fetal calf serum (dialyzed), 1.5 mg/l methionine (for long-term labelling only e.g. > 1 h), 35 µCi/ml ^{35}S-met.
- Chase medium: MEM, 10% FCS, 0.15 mg/ml methionine (unlabelled).
- Phosphate buffered saline (Dulbecco).
- 20% (w/v) Triton X-100.
- 50 mM Triethanolamine (TEA) pH 7.4, 100 mM NaCl, 0.4% SDS, 40 µg/ml PMSF (TNS).
- 50 mM Triethanolamine (TEA) pH 7.4, 100 mM NaCl, 40 µg/ml PMSF, (TN).
- 50 mM Tris.Hcl pH 7.4 (Tris).
- CLAP: 10 mg/ml chymostatin, leupeptin, antipain, pepstatin in DMSO.
- 50 mM Triethanolamine (TEA) pH 7.4, 100 mM NaCl, 0.5% (w/v) TX-114, 40 µg/ml PMSF, CLAP 1:1000, (TNTX-114).
- 50 mM Triethanolamine (TEA) pH 7.4, 100 mM NaCl, 0.5% (w/v) TX-100, 40 µg/ml PMSF, (TNTX-100).

Labelling of Cells

1. Wash cells once briefly in PBS
2. Wash cells in MEM-met (37°C) ×2.
3. Drain off medium and add 1.5 ml labelling medium. Incubate for desired time.
4. Add 3 ml chase medium (optional).
5. Place cells on ice, remove medium (remember it is radioactive!)
6. Wash 2×with PBS.
7. Lyse cells (below).

Cell Lysis and Immunoprecipitation

Protocol A
1. Drain all liquid from monolayer.
2. Add 0.5 ml TNS, suspend cells using P1000 and transfer to an Eppendorf tube. The cells are solubilized very rapidly releasing their DNA.
3. Sonicate or pass through a 21G hypodermic needle to shear the DNA.
4. Add 20% TX-100 to 2%.
5. Add CLAP 1:1000.

6. Centrifuge, 10 min, Eppendorf. Discard pellet.
7. Add specific antibody (or non-immune): 1–10 µl serum or 1 µg affinity purified antibody.
8. Add 20 µl protein A-Sepharose (50% slurry in PBS).
9. Rotate overnight in cold room (or 3 h at RT).
10. Wash Sepharose in TNSX 5. Rotate 10 min between washes.
11. Wash 2× with Tris.
12. Drain pellet, add SDS gel sample buffer.
13. Run on gel.

Protocol B
1. Drain all liquid from monolayer.
2. Suspend and lyze cells in 1 ml Triton X-114.
3. Centrifuge 5 min, Eppendorf. Discard nuclear pellet.
4. Perform TX-114 cloud point precipitation as described
5. Suspend final TX-114 pellet in TN to original volume.
6. Add antibody and protein A-Sepharose (above). Rotate overnight 4°C.
7. Wash Sepharose 5× in TNTX-100.
8. Wash 2× in Tris.
9. Drain, add SDS gel sample buffer.

Protocol C
1. Drain monolayer.
2. Lyse cells in 0.5–1 ml TNTX-100.
3. Centrifuge 10 min at 13000 g (Eppendorf). Discard nuclear pellet.
4. Add antibodies, protein A-Sepharose and rotate overnight in the cold.
5. Wash 5×, in TNTX-100.
6. Wash 2×, in Tris.
7. Drain and add sample buffer.
8. Run gel.

For all the protocols, a pre-adsorption step may be included immediately prior to the antibody incubation. This step is sometimes useful for the elimination of "background" bands, especially in samples which are not easily solubilized, and is carried out as follows.

1. To solubilized cell extract add 1 µg non-immune IgG and 20 µl protein A-Sepharose.
2. Rotate 1–3 h RT.
3. Centrifuge, 10 min Airfuge, 22 psi (or 15–20 min, Eppendorf). Discard pellet.
4. Add specific antibody and protein A to supernatant.
5. Continue according to method A, B or C.

References

Amit AG, Mariuzza RA, Phillips SN, Poljak RJ (1986) Three-dimensional structure of an antigen-antibody complex at 2.8 Å resolution. Science 233:747–753

Atassi MZ, Lee C-L (1978) The precise and entire antigenic structure of native lysozyme. Biochem J 171:429–434

Berzofsky JA, Berkover IJ (1984) Antigen-antibody interactions. In: Paul WE (ed) Fundamental immunology. Raven Press, New York, pp 595–644

Biocca S, Neuberger MS, Cattaneo A (1990) Expression and targeting of intracellular antibodies in mammalian cells. EMBO J 9:101–108

Brada D, Roth J (1984) "Golden Blot" -Detection of polyclonal and monoclonal antibodies bound to antigens on nitrocellulose by protein A-gold complexes. Anal Biochem 142: 79–83

Burke B, Griffiths G, Reggio H, Louvard D, Warren G (1982) A monoclonal antibody against a 135 k Golgi membrane protein. EMBO J 1:1621–1628

Burke B, Warren G (1984) Microinjection of mRNA coding for an anti-Golgi antibody inhibits intracellular transport of a viral membrane protein. Cell 36:847–856

Burnet FM (1959) The clonal selection theory Of aquired immunity. Vanderbilt University Press, Nashville, TN

Burnette WN (1981) "Western blotting": electrophoretic transfer of proteins from sodium dodecyl sulphate-polyacrylamide gels to unmodified nitrocellulose and radiographic detection with antibody and radioiodinated protein. A Anal Biochem 112:195–203

Cattaneo A, Neuberger MS (1987) Polymeric immunoglobulin M is secreted by transfectants of non-lymphoid cells in the absence of immunoglobulin J chain. EMBO J 6:2753–2758

Crumpton MJ (1974) Protein antigens: the molecular bases of antigenicity and immunogenicity. In: Sela M (ed) The antigens. vol 2, Academic Press, New York, pp 1–79

Dao MY (1985) An improved method of antigen detection on nitrocellulose: in situ staininng of alkaline phosphatase conjugated antibody. J Immunol Methods 82:225–231

de Groth F, Scheidegger D (1980) Production of monoclonal antibodies: strategy and tactics. J Immunol Meth 35:1–21

Galfre G, Milstein C (1981) Preparation of monoclonal antibodies: stategies and procedures. Methods Enzymol 73:3–46

Gershoni JM, Palade GE (1982) electrophoretic transfer of proteins from sodium dodecyl sulphate-polyacrylamide gels to a positively charged membrane filter. Anal Biochem 124:396–405

Glenney JR , Glenney P, Weber K (1982) erythroid spectrin, brain fodrin and intestinal brush border proteins (TW260/240) are related molecules containing a common calmodulin-binding subunit bound to a variant cell type-specific subunit. Proc Natl Acad Sci USA 79:4002–4005

Harlow E, Lane D (1988) In: Antibodies: a laboratory manual. Cold Spring Harbor Laboratory Press, New York, pp 726

Hawkes R, Niday E, Gordon J (1982) A dot immunobinding assay for monoclonal and other antibodies. Anal Biochem 119:142–147

Hong R, Nisonoff A (1966) Heterogeneity in the complementation of polypeptide subunits of a purified antibody isolated from an individual rabbit. J Immunol 96:622–628

Hudson L, Hay FC (1989) Practical immunology, 3rd edn. Blackwell, Oxford

Huse WD, Sastry L, Iverson SA, Kang AS, Alting-Mees M, Burton DR, Benkovic SJ, Lerner RA (1989) Generation of a large combinatorial library of the immunoglobulin repertoire in phage λ. Science 246:1275–1281

Jeske DJ, Capra JD (1984) Immunoglubulins: structure and function. In: Paul WE (ed) Fundamental immunology. Raven Press, New York, pp 131–166

Johnstone A, Thorpe R (1981) In: Immunochemistry in practice. Blackwell, Oxford, UK, pp 306

Kearney JF, Radbruch A, Liesegang B, Rajewsky K (1979) A new mouse myeloma cell line that has lost immunoglobulin expression but permits the construction of antibodt secreting hybrid cell lines. J Immunol 123:1548–1550

Koehler G, Milstein C (1975) Continuous cultures of fused cells secreting antibody of predefined specificity. Nature 256:495–497

Koehler G, Howe SC, Milstein C (1976) Fusion between immunoglobulin-secreting and nonsecreting myeloma cell lines. Eur J Immunol 6:292–295

Kreis TE (1986) Microinjected antibodies against the cytoplasmic domain of vesicular stomatitis virus glycoprotein block transport to the cell surface. EMBO J 5:931–941

Kurstak E (1971) The immunoperoxidase technique: localization of viral antigens in cells. Methods Virol 5:423–429

Lerner R (1982) Tapping the immunological repertoire to produce antibodies of predetermined specificity. Nature 299:592–596

Laemmli UK (1970) Cleavage of structural proteins during assembly of the head of bacteriophage T4. Nature 227:680–685

Louvard D, Reggio H, Warren G (1982) Antibodies to the Golgi apparatus and the rough endoplasmic reticulum. J Cell Biol 92:92–107

Mayer RJ, Walker JH (1987) Immunochemical methods in cell and molecular biology. Academic Press, London

O'Farrell P H (1975) High resolution two-dimensional electrophoresis of proteins. J Biol Chem 250:4007–4021

Olmsted J (1981) Affinity purification of antibodies from diazotized paper blots of heterogeneous protein samples. J Biol Chem 256:11955–11957

Ouchterlony O, Nilsson LA (1986) Immunodiffusion and immunoelectrophoresis. In: Weir DM, Herzenberg LA, Blackwell C (eds) Handbook of experimental immunology, 4e Sect 32 Blackwell, Oxford, vol 1, pp 19–44

Ouchterlony O, Nilsson LA in D M Weir, ed Handbook of Experimental Immunology, 3rd edition (Blackwell, Oxford) chapter 19

Paul WE (1984) Fundamental Immunology. Raven Press, New York

Reading CL (1982) Theory and methods for immunization in culture and monoclonal antibody production. J Immunol Methods 53:261–291

Rockey JH (1967) Equine antikapten antibody. The subunits and fragments of anti-beta-lactoside antibody. J Exp Med 125:249–275

Roitt IM, Brostoff J, Male D (1985) Immunology; 2nd edn., Blackwell, Oxford p 260

Rosa P, Weiss U, Pepperkok R, Ansorge W, Niehrs Ch, Stelzer EHK, Huttner WB (1989) An antibody against secretogranin I (chromogranin B) is packaged into secretory granules. J Cell Biol 109:17–34

Shulman M, Wilde CD, Koehler G (1978) A better cell line for making hybridomas secreting specific antibodies. Nature 276:269–270

Sigal MB, Sinha YN, Van der Laan WP (1983) Production of antibodies by innoculation into lymph nodes. Methods Enzymol. 93:3–12

Smith DE, Fisher PA (1984) Identification, developmental regulation, and response to heat shock of two antigenically related forms of a major nuclear envelope protein in Drosophilla embryos. J Cell Biol 99:20–28

Taggart RT, Samloff IM (1984) Stable antibody producing murine hybridomas. Science 219:1051–1054

Ternynck T, Avrameas S (1976) Polymerization and immobilization of proteins using ethylchloroformate and glutaraldehyde. Scand J Immunol Suppl 3:29–35

Tonegawa S (1983) Somatic generation of antibody diversity. Nature 302:575–581

Tooze J, Hollinshead M, Frank R, Burke B (1987) An antibody for an endoproteolytic cleavage site provides evidence that pro-opiomelanocortin is packaged into secretory granules in AtT-20 cells before its cleavage. J Cell Biol 105:155–162

Towbin H, Staehelin T, Gordon J (1979) Electrophoretic transfer of proteins from polyacrylamide gels to nitrocellulose sheets: procedures and some applications. Proc Natl Acad Sci USA 76:4350–4354

Tung A (1983) Production of large amounts of antibodies, non-specific immunoglobulins, and other serum proteins in ascitic fluid in individual mice and guinea pigs. Methods Enzymol 93:12–23

Utsumi S, Karush F (1964) The subunits of purified rabbit antibody. Biochemistry 3:1329–1338

Ward ES, Güssow D, Griffiths AD, Jones PT, Winter G (1989) Binding activities of a repertoire of single immunoglobulin variable domains secreted from *Escherichia coli*. Nature 341:544–546

Chapter 7

Labelling Reactions for Immunocytochemistry

In the preceding chapter we have dealt with the theoretical principles and practices of antibody production and characterization. This chapter deals with the next stage – how to apply these antibodies on thin sections in order to label antigens at the electron microscope level. Since characterization of the labelling reaction depends on the immunochemical characterization of the antibodies, some overlap with the preceding chapter is unavoidable.

7.1 Historical Perspectives

Most, if not all, of the basic principles involved in the application and characterization of labelling reactions used for electron microscopy originate from light microscopic studies. The reason for this is simple: whereas satisfactory EM techniques have become generally available only during the last decade, reliable light microscope techniques have been extensively applied for almost 50 years. A useful way to understand the principles involved in labelling is to look back at the historical developments of the techniques.

A key step in the development of antibody labelling methods originated with the finding by Marrack (1934) that covalent addition of dye molecules to antibodies did not necessarily affect their ability to bind antigen. In a simple, four-paragraph letter to **Nature,** Marrack (without actually showing any data!) described his observations that bacteria treated with dye-conjugated anti-bacterial antibodies gave a specific coloured reaction in the test tube. The precise experiment was as follows: Marrack mixed typhoid antibodies stained with a dye with an undyed cholera antiserum. When he added typhoid bacteria to this mixture a red agglutinated precipitate was observed: in the presence of cholera bacteria, however, the agglutinated bacteria were not coloured. The significance of this finding cannot be overstated. It may appear trivial today but it should be emphasized that very little was known about antibody molecules at that time; even their protein nature had not been universally accepted.

The person who justifiably became known as the father of immunocyto-chemistry, however, was Albert Coons, who soon realized the implication of Marrack's findings. In their first paper on the subject in 1941, Coons et al., used antibodies against *Pneumococcus* bacteria that were directly coupled to a fluorescent compound, β-anthryl-carbido isocyanate. In what were essentially whole mounts, the bacteria were shown to give a blue fluorescence reaction in the fluorescence microscope after treatment with a specific antibody but not after treatment with a non-specific control. The cytochemical basis for the use of isocyanates as fluorescent components was due to the work of Creech and Jones (1941a;b), who were co-authors with Coons on the first two papers (Coons et al. 1941, 1942).

It was evident to Coons that, since antibodies could be coupled to fluorescent dyes without losing their biological activity, two different approaches were possible for labelling antigens, namely **direct** and **indirect**. In the former the tag was attached to the specific antibody itself whereas in the latter it was bound to a general purpose immunoglobulin. Although most of the initial work was done using the direct approach, it later became apparent that the indirect method offered at least two major advantages. First, the same second antibody conjugate could be used to label a family of (unconjugated) primary antibodies. Second, since more than one second antibody molecule could bind a single molecule of primary antibody the signal became amplified.

In 1950, Coons' group introduced the idea of using sections of tissue to localize microbial antigens (Coons and Kaplan 1950; Kaplan et al. 1950; Coons et al. 1950). In these studies they described two approaches for preparing the sections. The first was to use cryostat sections of freshly frozen tissues that were subsequently fixed with methanol prior to labelling. The second method, which worked for more robust antigens, was to use paraffin embedding, after formaldehyde fixation. Sections prepared using this method had to be rehydrated before labelling. This was done after removal of the wax with xylene followed by treatment with a descending series of alcohol. During this period they also described the synthesis and use of fluorescein isocyanate as a fluorescent tag, and, equally important, the optimal conditions for coupling it to proteins (Coons and Kaplan 1950). The choice of this compound was made for two reasons. First, the yellow-green fluorescence that it emitted is apparently rare in mammalian tissues and second, the human retina is extremely sensitive to this wavelength.

The concept of non-specific staining was introduced in the Coons and Kaplan (1950) paper when they found a fluorescent reaction with their bacterial antiserum in control, uninfected tissue. They empirically found that this reaction could be abolished when they mixed their specific antibody-fluorescein conjugate with a dried suspension of ground liver tissue. A variation of this procedure is used (or should be used) in essentially every immunocytochemical experiment in use today (see below). From this observation it became clear that contaminants in their conjugated antisera were responsible for much of this reaction (a non-specific "sticking" reaction). This was especially a problem for the direct labelling methods, since it made it essential to purify the antibody and maintain its specificity properties after conjugating it to the fluorescein. In later years, it became apparent that specific antibodies themselves could also bind in an apparently non-specific manner to tissue sections.

In the 1950 papers they also listed three criteria which they considered essential to be fulfilled in order to have proof of the specificity of the labelling reactions. These were:

- An excess of unconjugated antiserum should be able to compete with the fluorescinated antibody.
- Pretreatment of the conjugated antibody with antigen should block the reaction.
- Their antibody against pathogenic microbial antigens should not react with tissues from normal, uninfected animals.

The essence of these principles do, of course, remain valid even today.

In later studies, Coons, and others, found repeatedly that the indirect labelling methods increased the sensitivity of the labelling reactions. Indirect methods detected significantly smaller amounts of antigen when compared to the direct methods. The main reason for this is that more than one molecule of second antibody can bind to the first antibody. This finding was the basis for the introduction of the sandwich techniques where multiple layers were used in order to amplify small levels of primary antibody binding (reviewed by Sternberger 1979).

In the early years of immunocytochemistry all the studies were performed using antibodies against microbial pathogens, that is, antigens not intrinsic to the tissues of higher animals. The obvious advantages of these antigens is that antibodies of high titre were easily obtained and antigen-negative control preparations were readily available. In 1951, Marshall became the first to apply an immunohistochemical technique for the study of native tissue antigens. He made antibodies in rabbits against a commercial preparation of the pituitary, adrenocorticotropic hormone (ACTH). In the initial part of this study, it was found that the antiserum not only stained the pituitary but gave a significant reaction with plasma. After treating the antibody-containing serum with a mixture of porcine serum and porcine kidney powder, however, this non-specific staining was effectively eliminated. This idea of removing unwanted antibody contaminants in antisera by adsorption is still an important tool in immunocytochemistry. An additional advantage with the use of a hormone as an antigen was that its biological activity could easily be assayed. Accordingly, Marshall showed that this specific antibody conjugate could precipitate active hormone from a purified preparation of hormone. As Coons (1956) later pointed out, this was fortunate in Marshall's case since many antibodies will also block the bioactivity of proteins.

In addition to the use of fresh frozen cryostat sections, Marshall also used cryo-sections of freeze-dried tissues. A serious problem with this and similar approaches which the early workers became aware of was the possibility of loss of antigen during the labelling reactions. This led to the need to improve specimen preparation methods. In combination with cryostat sections, fixation conditions were devised which helped preserve antigenicity. The earlier fixation methods involved the use of ethanol, methanol or acetone (Coons 1956). Subsequently, formaldehyde was introduced for use with frozen sections, first used by Coons et al. (1942) in combination with paraffin embedding and labelling.

Conditions for immunoreaction were also worked out during this period, conditions that to a large extent are still used today. With respect to the time of labelling, Coons (1956) considered 10–30 min to be optimal while longer times had no apparent advantage. For washing procedures, it also became evident that relatively brief periods, around 10 min, with a coupled charge, were sufficient. These times for labelling and washing agree well with the optimal times empirically determined for cryo-sections (Painter et al. 1973; Tokuyasu and Singer 1976).

With respect to the dilution of antibody a potentially important finding was that made by Bateman et al., in 1941. During this period there was a lot of interest in an optical method used to measure the thickness of monolayers of molecules which were non-specifically absorbed onto a thin film of barium stereate. The principle of the method was that, using interference microscopy, the thickness of the absorbed film could be detected as a change in the angle of refraction of light passing through the upper and lower surfaces of the film (Blodjett and Langmuir 1937). Using this technique, a number of groups had shown that, after pre-absorbing antigen onto the film (which gave a measurable thickness) a layer of specific antibody could be bound and the thickness of this additional layer could also be estimated. Observations that such bound antibody could still bind to a second layer of antigen was, in fact, one of the lines of evidence showing that antibodies were multi-valent (Porter and Pappenheimer 1939; Rothen and Landsteiner 1942). In retrospect, it is now clear that this technique had given reliable estimates of the gross dimensions of a number of molecules.

The key observations of Bateman et al. (1941) were as follows: when concentrated solutions of antiserum was allowed to bind pre-absorbed antigen the thickness of the film was estimated to be 150 Å (the technique was considered to be accurate to about 3 Å). When, however, a 300-fold dilution of the antiserum was used the thickness of the film was only 55 Å. Although not shown directly in the study, it was clear that significantly more antibody must have bound in the former than in the latter case. The interpretation of the authors was that at high concentrations, the elongated antibody molecules packed on the antigen film with their long axis perpendicular to the film. Their estimate of 15 nm is certainly in reasonable agreement with the maximum length of IgG molecules determined in later studies. At lower dilutions, they argued, the antibody molecules covered the antigen by having their long axis tilted away from the perpendicular. The significance of this is that, once bound in this fashion at low dilution, these antibody molecules would block the addition of further antibody molecules (i.e. steric hindrance). Provided one can extrapolate from the simple situation where one has a uniform, totally accessible surface to the typical sections used for immunocytochemistry, the concept which emerges is as follows. When labelling sections containing high concentrations of antigen, the use of a relatively high concentrations of antibody for short incubation times may actually give a significantly greater signal than longer incubations using relatively low concentrations. In other words, the length of the incubation time cannot necessarily compensate for too high a dilution of the antibody. Further, they found that at high concentrations the antibody covered the surface of the antigen in 5 min whereas it took significantly longer, 30 min, for the diluted serum to do the same. I will argue below that these observations, which may not be so significant for light microscopic immunolabelling, can have fundamental importance for labelling studies at the EM level.

In the 1956 review Coons discussed the sensitivity of the immunofluorescent technique. Based on the assumptions about the size of his labelled bacteria and the amount of protein contained in them, he predicted that one should be able to detect 10 µg antigen/cm^3 of tissue, although he believed the practical limitations at the time was probably more in the range of 100 µg/cm^3.

The next important milestone in the development of immunocytochemistry occurred around 1960 with the introduction of ferritin as an electron-dense marker for electron microscopic labelling (Singer 1959; Singer and Schick 1961).This showed remarkable foresight and it took another decade before specimen preparation techniques had been developed which enabled the full potential of this marker to be utilized. Indeed, it was in the Singer laboratory that both the BSA embedding (Farrant and McLean 1969; McLean and Singer 1970) and later the cryo-section method (Tokuyasu 1973; Painter et al. 1973; Tokuyasu and Singer 1976) were developed in order to label intracellular antigens on sections at the EM level. In parallel to those developments, the use of enzymes such as horseradish peroxidase was introduced as an alternative antibody marker, both for light- and electron microscopy (Avrameas and Uriel 1966; Nakane and Pierce 1967). The attraction of this approach was that the enzyme, conjugated mostly to a secondary antibody, would catalytically amplify the reaction and, according to dogma (see comments below), increase the sensitivity of the method. This latter approach, which was to dominate EM immunocytochemistry for the next 15–20 years, led to a multitude of different recipes, including the use of multiple layers (see Sternberger 1979). I have argued elsewhere (see Chap. 1) that, for labelling studies at the EM level, this latter approach will probably be made redundant by the availability of quantitative, particulate markers, primarily colloidal gold.

The introduction of colloidal gold and its conjugation to protein A and antibodies represented one of the last recent significant advancements with respect to immunolabelling methods at the EM level (Faulk and Taylor 1971; Romano and Romano 1977; Roth et al. 1978). It should be emphasized, however, that this innovation was not really a conceptual innovation but a practical one: gold particles are just more easy to visualize than ferritin and other electron-dense particulate markers.

7.2 Labelling in Practice

In this part I shall discuss the current appreciation of the conditions which are required to label sections of biological material with antibodies. The goal here is strictly to visualize antigens at the EM level. In Chapter 2 I have explained why I believe that only the use of particulate markers can suffice for this purpose. For reasons explained in Chapter 4, for most purposes the use of either thawed cryo-sections or of the "new generation" methacrylate resins (primarily Lowicryls and LR resins) is recommended.

7.2.1 Purity of Antibody: Use of Light Microscopy

In general, the use of purified antibodies is preferred over simple antisera for EM immunocytochemistry. There is no "hard and fast" rule in this, however. The two most important factors to be considered are first:

- does the antibody preparation give a positive signal on sections,
- can its specificity be shown by an independent immunochemical method.

I emphasize again that, concerning the first point, the antibody should be tested under conditions that are most likely to give a positive signal. For most purposes, the easiest test is at the light microscopy level, either with whole cultured cells or with thick, cryostat sections. This approach enables the same fixation protocols that will be employed for the subsequent EM localization to be tested. By comparing cryostat sections of unfixed with fixed tissue, the effect of the fixatives can be directly assessed: note, however, that many antigens may be lost during the washing steps in the case of unfixed tissues (for discussion as well as practical details, see Larsson 1988). Irrespective of the system that is used to visualize or detect the antibody, e.g. fluorescence, horseradish peroxidase or gold, it is important to stress that the quality of, and the conditions for optimally using the "detecting" (usually secondary) antibody (or protein A) should be first worked out using a characterized primary antibody. At the light microscope level the quality of the antibody labelling (as well as the optimal concentrations – see below) can be assessed. Both at the light- and especially at the electron-microscope level, some knowledge of the biological system is essential. An antibody against a membrane protein should not label the nuclear matrix; similarly, tissues or cell types, which from independent immunochemical assays are known not to contain the antigen, should not bind antibody. It should be pointed out, however, that for novel antigens these generalizations may be more difficult to apply. Immunochemical assays may give false negatives whereas the antibody may recognize the fixed antigen in a section.

When, or before, a positive signal is obtained at the light microscope level, it is important to obtain proof of labelling specificity. If the antigen is a protein, for instance, the antibody should recognize a molecule of appropriate molecular weight shown by immunoblotting or immunoprecipitation analysis.

In addition, if it is an enzyme or hormone, the antibody should bind a molecule with the appropriate biological activity. Further, these assays should be shown for the same tissue or cell type that is used for immunocytochemistry. This reduces the possibility of obtaining spurious immunocytochemical results. By a combination of light microscopic labelling and the use of an immunochemical test, the experimenter can decide whether the antibody preparation is good enough to test at the EM level. It goes without saying that if the original antiserum does not show any signs of specific labelling by light microscopic immunocytochemistry there is usually no point in proceeding to the EM level. If, however, the antiserum gives a satisfactory reaction at the light microscopic level, and appears specific by immunoblotting or immunoprecipitation, it can be tested directly on thin sections for EM. When there is evidence of background labelling at either the LM or EM levels, then different purification protocols will have to be applied (see Chap. 6). Often, such background problems first appear at the EM level.

The recommendation for testing antibodies by light microscopy prior to testing them at the EM level is important for a number of reasons. First, it enables useless antibodies to be screened out and potentially good antibodies to be recognized. Second, it enables satisfactory fixation conditions to be worked out and, third, in the case of a positive signal, it helps to assess whether further purifications of the antiserum may be necessary. Finally, it provides the experimenter with a much better overview of the cellular distribution of the antigen. Such an analysis may also be facilitated by the use of double-labelling procedures at the light microscope level. Often, the use of many useful second (reference) markers such as the DAPI (4,6 Diamino-2-phenylindole) or Höchst dye (No 33258) stains for DNA as well as many histological dyes are not possible at the EM level.

7.2.2 Class of Antibody; Hybrid Antibodies

In practice, the great majority of immunocytochemical experiments are carried out using the bivalent IgGs and there is no doubt that these are the preferred reagents (see Chap. 6). However, since the primary response of the animal to antigen usually involves antibodies of the IgM class, one is often obliged to use these less desirable, bulky pentameric molecules. With the advent of the in vitro immunization approaches (see Chap. 6), the need to do immunocytochemistry with IgMs is becoming more common. It is difficult at this stage to describe any general principles for immunolabelling using these reagents. While they are often more difficult to handle than IgGs (they tend to give higher background labelling, for example) this is not always the case. For an example of labelling a cryo-section with an in vitro prepared IgM see Fig. 1.

For some special problems it may be an advantage to use hybrid antibodies, that is, artificial combinations of two Fabs of different specificity. For a recent discussion of the state of the art and potential uses see Larsson (1988).

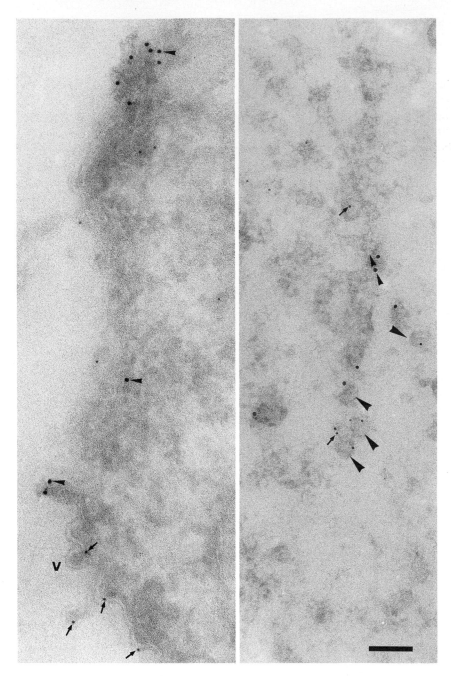

Fig. 1

7.2.3 Antibody Concentrations

The optimal concentrations of antibody for EM immunocytochemistry is far more critical than for light microscopy. The reason for this is simple but little appreciated. A thin section is really only a minute slice through a cell. Note again, that in the case of resin sections and even in most parts of cryo-sections, since the labelling is mostly restricted to the section surface, the effective section thickness, that is – that part of the section in which the antigen is able to recognize the antibody is significantly less than the entire section thickness. In accordance with this, all the available calculations indicate that, for most antigens, (even under optimal conditions using cryo-sections) not more than $\approx$ 10–15% total antigens present in the section can be labelled. The latter values represent the upper limit using cryo-sections under conditions where significant or even complete penetration of the label is expected. An example would be the outer surface of the plasma membrane in sections through a pellet of unembedded cells (see Chap. 11 for a more detailed discussion). In order, therefore, to have the chance to visualize as much of the antigen as possible in this relatively small sample it is imperative that the antibody concentration be optimal, in other words, it should give the highest signal to-noise-ratio. In practice, this means using relatively high concentrations of antibodies in order to obtain as high a signal as possible over "specific" structures without obtaining unacceptable levels of labelling over structures that should not contain the antigen. This statement may sound heretical to many in the field of immunohistochemistry, where one is often proud of using antibody dilutions which may be 100000 times or more! At such dilutions the total number of antibody molecules is vanishingly small. The rationale in such light microscopic

◄ ——

Fig. 1. Example of labelling of cryo-sections with an anti-idiotypic IgM antibody prepared by an in vitro immunization protocol using mouse lymphocytes. The goal of this project was to show that the nucleocapsid of Semliki Forest virus interacted with the cytoplasmic domain of the E_1 spanning membrane glycoprotein of the virus, a process which initiates the budding of the virus through the plasma membrane (for more details see Fig. 1, Chap. 10). An (idiotypic) antibody was made against (and recognized) a peptide present in the cytoplasmic tail of E_1. Anti-idiotypic antibodies were then made using in vitro protocol that recognized the nucleocapsid (the anti-anti idiotypic antibody made against the latter antibody recognized the initial immunogen, the E_1 tail). For details of this study, see Vaux et al. 1988. The figure here shows complementary EM data from BHK cells infected with SFV that were fixed with 4% paraformaldehyde for 1 h. The sections were labelled with the anti-idiotypic IgM antibody followed by a commercial preparation of a rabbit anti mouse IgM (Cappel) and then protein A gold(9 nm, small arrowheads). After a protein A block (Geuze et al 1981) a second label was applied, namely an affinity-purified anti capsid (rabbit) antibody followed by protein A gold (6 nm – *arrows*). The data collectively suggest that the anti-idiotypic antibody does not recognize all the nucleocapsids, but a sub-set of the population. In the micrograph on the *left*, some electron dense regions directly beneath, and close to, the plasma membrane are labelled (*v* budding virion). The micrograph on the *right* is from a very thin section that was contrasted with a thin layer of ammonium molyldate-methyl cellulose (see Chap. 5). Under this condition some of the individual nucleocapsids (large arrowheads) can be better resolved. ×125000; *bar* 100 nm. (Work done in collaboration with David Vaux, EMBL). In a recent report the specificity of this anti idiotypic antibody has been questioned (see Suomalainen and Garoff (1992) J Virol 66:5106–5109)

labelling experiments is to amplify this relatively small signal to an enormous extent using various "sandwich" procedures and enzymes such as horseradish peroxidase. Clearly, however, any non-specific labelling will also be amplified to the same extent. In particular, there may be a danger of selecting relatively minor, unwanted antibody species that are of high affinity. I have already referred to the classic paper by Bateman et al. (1941), which showed that using a more dilute antibody solution for a longer period may not necessarily give the same level of specific binding as a high concentration of antibody for a relatively short period. Although a systematic study of this phenomenon has not been made, empirical observations on cryo-sections from a number of groups tend to agree with this statement: the highest specific labelling densities are usually obtained at relatively high **concentrations of antibody**. The theoretical reasons for this observation are far from clear, especially if a significant proportion of antibodies would be capable of binding antigen in a bivalent fashion as in the case of IgG on sections. If such bivalent binding occurred, it would be expected to be more prevalent at lower antibody concentration: as the concentration of antibody increased, the Fabs would compete with each other for binding sites and more antibodies would bind by a single arm only. This, then, would effectively reduce the avidity of binding, and with increasing concentration there would be an increasing tendency for the bound antibody to be washed off during the rinsing steps. But what is the evidence suggesting that a significant proportion of antibodies can bind to antigens on sections in a bivalent fashion (as the schematic diagrams often show in immunocytochemistry reviews)? While it is hard to document direct evidence either for or against this phenomenon, from theoretical considerations it would appear that bivalent-binding would be the exception rather than the rule. On the surface of a section the antigens are literally fixed in position. Once an antibody has bound an antigen with one of its Fabs, it has the possibility to bind a second antigen (which is usually on a different molecule) only if the latter is positioned within a well-defined zone close to the first antigen. Whereas the angle between the two Fabs can vary depending on the distance between binding sites, the extent of this variability is limited. Once bound by one Fab it is likely that, for the second Fab to bind, the second antigen must be positioned precisely with respect to the first. In addition, its orientation and accessibility in three-dimensional space must be critical for the second arm to have a possibility of binding. It appears intuitively difficult to imagine that this situation occurs often, even for antigens that are packed at very high density. Perhaps the best evidence that at least some bound antibodies on sections have free binding sites comes from the gold labelled antigen detection (GLAD) method of Larsson (1979). In this method after the first antibody step the sections are treated with colloidal gold particles conjugated to pure antigen. If no free antigen-binding sites were available on the antibody the antigen-gold complexes would obviously not be able to bind. For a discussion of the use of this, and related approaches, the reader is again referred to the book by Larsson (1988).

The idea that most antibody molecules bind to antigens on the section by a single arm rather than two is also more compatible with respect to the empirical observations that relatively high concentrations of antibody are, in fact, required in order to give the highest signal-to-noise ratios. In this scenario the degree of binding would strictly be a function of the affinity of the antibody for the fixed antigen. This theory predicts that Fab fragments should bind as strongly as do bivalent antibodies on sections. Clearly, experimental data are required to prove or disprove this idea.

The recommendation to use the highest antibody concentration compatible with the highest signal to noise ratio should be taken with caution in the case of either monoclonal antibodies or when using polyclonal antibodies against small peptides. These antibodies are made against a linear sequence of amino acids present in the antigen so there is always the danger that unrelated antigens may share some sequence homology, purely by chance. An excellent example is given by the study of Nigg et al. (1982), who made an antibody against a hexapeptide present in the transforming protein $p60^{src}$ of Rous sarcoma virus. Although not strictly a monoclonal antibody, since the size of the peptide was about the size of one epitope, the antibody could be effectively considered as a monoclonal. At low concentrations (1–3 µg/ml) this antibody gave the expected distribution of $p60^{src}$ by immunofluorescence in transformed but not in untransformed cells. At higher concentrations (10–20 µg/ml), however, the antibody recognized unrelated cytoskeletal (and other) proteins in both transformed and untransformed cells. These immunofluorescence data were fully supported by immunoblotting results showing the monospecificity of the antibody at low (0.12 µg/ml) but not at high (3 µg/ml) where additional bands appeared that co-migrated with known cytoskeletal proteins. Significantly, all the reactions by both fluorescence and blotting could be abolished by pre-incubation with the free peptide (at concentrations of 200 and 90 µg/ml, respectively). In the case of ß tubulin, one of the proteins with which the antibody cross-reacted, sequence information was available. From this sequence the authors could identify one tetrapeptide that shared structural homology with four of the six amino acids in the $pp60^{src}$ peptide and was clearly a good candidate for the cross-reaction. As the authors pointed out, cytoskeletal proteins are usually present in relatively high concentrations in cells and unwanted cross-reactions with this class of protein are not uncommon, even with monoclonal antibodies.

Obviously, cross-reaction phenomena,which can in principle occur for any antibody, are more likely to occur at high antibody concentrations. This is especially true for monoclonal antibodies, since any potential for spurious cross-reaction will be shared by all the antibody molecules. In contrast, polyclonal sera are likely to consist of mixtures of antibody species of different specificities. Further, these cross-reactions may not always be evident by independent immunochemical tests (see below). Nevertheless, as stated above, empirical observations indicate that one needs relatively high concentrations of antibody for the highest signal to noise ratio on sections. Below this optimal concentration the signal drops while above it the levels of background binding increase. In our experience, as a rule of thumb, the optimal concentration is generally about a factor of 10 higher than those which give an optimal immunofluorescence signal on permeabilized tissue culture cells. As a rough

Table 1. Empirically determined concentration range for using antibodies for labelling cryo-sections[a]

Antibody[b]	Antibody protein concentration	Dilution factor
Antiserum	$\approx$ 50 mg/ml	1:5 – 1:200
IgG Fraction	1–10 mg/ml	1:5 – 1:100
Affinity purified antibody	100–500 µg/ml	1:1 – 1:50
Monoclonal-culture supernatant	10 µg/ml	1:1 – 1:10
Monoclonal-ascites fluid	50 mg/ml	1:20 – 1:1000

[a] This is intended as a rough guide only.
[b] All antibody solutions should be centrifuged just before the labelling for 1 min at $\approx$ 10 000 g to remove aggregates.

guide, I have given our empirically determined range of concentrations for different types of antibody preparations in Table 1.

Our (limited) experience with Lowicryl sections, as well as the experience of other groups working more extensively with this resin, tends to agree with the above recommendations.

When testing a new primary antibody at the EM level (it makes sense to use a secondary labelling reagent, e.g. protein A gold) whose conditions of reactivity are known. This is best done independently using a well characterized primary antibody.

Finally, when encountering a negative result at the EM level the possibility that the concentration of antigen may be below the detection threshold should be considered (see Chap. 11). The antigen concentration sufficient to give a positive immunocytochemical result at the light microscopy level, using whole cells or thick cryo-sections, may not necessarily suffice to give an acceptable signal with thin sections at the EM level.

7.2.4 Time, Temperature and Repeated Applications of Antibody

In our experience, a well as that of numerous colleagues in the field, the initial recommendation of Coons still holds with respect to the optimal time of incubation with antibodies. At the optimal concentration, a 5–30 min incubation is sufficient and only small gain can be expected by increasing the incubation period. Although there are few quantitative studies on this point, it appears likely that on the thin sections used for immunocytochemistry the majority of the antibody molecules will bind in the first few minutes.

Similarly, for the temperature of incubation, room temperature appears to be satisfactory for most purposes. Antibodies will, however, bind antigen at temperatures from 4°C to about 40°C if a need should arise for a low or high temperature of incubation.

A significant observation was made by Gu et al. (1983), who showed that a repeated application of the primary, polyclonal antibody gave a higher signal for immunolabelling when compared to a single application. The authors suggested that the low affinity antibody molecules were washed off after each labelling step and the procedure, in effect, selected for the higher affinity antibodies. Similar data were shown subsequently by Scopsi and Larsson (1985).

Recent data from light microscopic studies suggest that the use of microwaves may significantly speed up the rate of immunolabelling reactions. Most, if not all, of these studies have dealt with thick cryostat or paraffin sections and it seems likely that the microwave radiation facilitates penetration of the reagents (see, for example, Suurmeijer et al. 1990 as well as the whole issue of the Histochemical Journal 6/7 1990). It remains to be seen whether this method will have any advantage for labelling thin sections.

7.2.4.1 A Quantitative Analysis of the Times of Labelling and the Concentration of Antibody in a Model System

In order to look at these two parameters in more detail, we decided to quantitate two well-characterized antigens in a model system with which we are very familiar using cryo-sections. These are the cation-independent mannose 6-phosphate receptor (MPR), which is involved in delivering newly synthesized, phosphorylated lysosomal enzymes from the Golgi complex to the endocytic pathway of animal cells and the lgp 120, a membrane protein first characterized by Lewis et al. (1985), which is found in both the prelysosomal compartment (PLC or late endosomes) and lysosomes (for details see Griffiths et al. 1988, 1990). Both these molecules are present in high concentrations in the PLC and in small amounts on the plasma membrane as well as other locations. Normal rat kidney cells were allowed to internalize 16 nm BSA gold particles conjugated with a passenger protein, bovine serum albumin (BSA), for 4 h followed by an overnight "chase" in medium free of gold. Under this condition this marker distributes between the PLC and the lysosomes. The PLC is a complex structure that is, by definition, MPR and lgp positive, while the lysosomes are small, spherical electron dense vesicles whose membranes are lgp positive but MPR-negative.

Thawed cryo-sections of these cells were labelled with antisera against the two membrane proteins at four different concentrations and for two different incubation times. The antibody concentrations used covered the range that would normally be expected for these antisera. The concentration referred to as 1 was in fact the optimally determined concentration.

In the microscope the cells were systematically sampled: the position of the grid was moved in one direction until the next PLC structure was evident (labelled with either MPR or lgp and 9 nm protein A gold as well as the 16 nm BSA gold). At the lowest magnification ($\times 13\ 000$) in which both the structure and the label were clearly visible, a micrograph was taken. At this magnification

Table 2. Quantitation of labelling of two membrane proteins on cryosections. Effect of incubation time and concentration of antibody

Labelling over PLC (gold/μm2)

Dilution factor	MPR 30 min	MPR 3 h	lgp 30 min	lgp 3 h
0.1	43.8 ± 3.2 n=18	39.9 ± 2.8 n=16	24.5 ± 2.9 n=10	54.1 ± 6.0 n=10
1[a]	53.5 ± 5.1 n=18	70.4 ± 3.8 n=17	32.0 ± 2.5 n=10	27.3 ± 2.0 n=15
10	4.6 ± 0.7 n=16	15.0 ± 1.3 n=15	19.4 ± 2.3 n=10	22.6 ± 3.3 n=14
100	2.5 ± 0.7 n=18	4.3 ± 0.8 n=15	3.7 ± 0.7 n=12	7.6 ± 1.3 n=11

Labelling on plasma membrane (gold/μm)

Dilution factor	MPR 30 min	MPR 3 h	lgp 30 min	lgp 3 h
0.1	0.239 ± 0.033 n=18	0.295 ± 0.028 n=16	0.164 ± 0.019 n=10	0.441 ± 0.052 n=10
1[a]	0.098 ± 0.016 n=18	0.086 ± 0.012 n=17	0.150 ± 0.026 n=10	0.056 ± 0.009 n=15
10	0.008 ± 0.004 n=16	0.022 ± 0.005 n=15	0.086 ± 0.017 n=10	0.060 ± 0.012 n=14
100	0.017 ± 0.004 n=18	0.009 ± 0.004 n=15	0.005 ± 0.003 n=11	0.026 ± 0.012 n=11

Background labelling over nucleus (gold/μm2)

Dilution factor	MPR 30 min	MPR 3 h	lgp 30 min	lgp 3 h
0.1	1.04 ± 0.18 n=14	1.49 ± 0.23 n=14	1.34 ± 0.17 n=8	3.331 ± 0.36 n=10
1[a]	0.52 ± 0.08 n=15	0.62 ± 0.09 n=13	0.86 ± 0.13 n=10	0.27 ± 0.05 n=11
10	0.02 ± 0.01 n=14	0.08 ± 0.02 n=11	0.21 ± 0.04 n=9	0.20 ± 0.05 n=8
100	0.03 ± 0.02 n=12	0.04 ± 0.01 n=15	0.03 ± 0.01 n=10	0.08 ±2000.02 n=10

Values: means and SEM.

n: Number of negatives counted.

[a] 1 here means the previously determined optimal concentration (1:150 for the rabbit anti chicken MPR antiserum and 1:30 for the rabbit anti rat lgp antiserum). Thus, for MPR, 0.1 means 1:15 and 10 means 1:1500 etc.

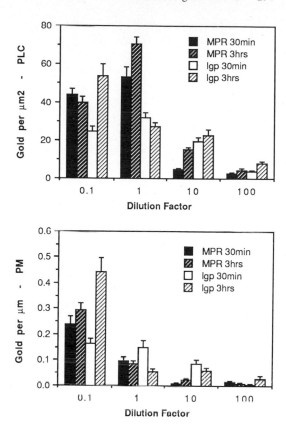

Fig. 2. Graphic display data from Table 2 to show the labelling over the PLC and plasma membrane (PM) for MPR and Lgp

a part of the plasma membrane is always included in the picture: this represented our unbiased sample for quantitating the label over the plasma membrane. The negatives were enlarged and the number of gold particles relating to the area of the PLC or to the linear length of plasma membrane were determined by point and intersection counting (see Chap. 11 for details).

The results are listed in Table 2 and the main points from this table are summarized graphically in Fig. 2. Note that whereas the highest level of labelling tends to be obtained at the highest concentration of antisera, this condition also gives the highest level of background labelling. Since the two antigens are spanning membrane proteins, it is reasonable to assume that all gold particles associated with the nuclear matrix are not specific and the labelling over the nuclear matrix was our reference for background labelling. Interestingly, for the anti lgp (but not for the anti MPR), it was possible to dilute the antiserum a factor of 10 beyond the routinely used 1:30 dilution with only slight loss of signal and, after the 30-min labelling time, this was accompanied by a significant decrease in background labelling. Overall, the data strongly suggest that increasing the incubation time from 30 min to 3 h has little effect on either signal or noise. The data also tend to support the major points made above with respect to incubation time and concentration, namely; (1) that outside of a rather narrow window of antibody concentration the signal to noise ratio decreases significantly, and (2) for most purposes a 30-min incubation time is not only sufficient but cannot be improved upon by longer times. More such data are needed, however, before general rules can be established.

7.2.5 Avoiding Non-Specific Labelling

By non-specific labelling I am simply referring here to a "sticking" phenomena where immunoreagents adsorb to structures not containing antigen: the more insidious problem of spurious cross-reactions where the antibody shows affinity for a molecule(s) not related to the antigen is a separate problem (see Chap. 6 for discussion).

Provided the primary antibody as well as the reagents for detecting the antibody are specific and used at the optimal concentrations, two additional factors are critical to prevent non-specific labelling. The first is the need to use a reagent that blocks non-specific labelling and the second involves the use of judicious washing steps. The need for these processes stems from the fact that all proteins have a tendency to adsorb non-specifically onto surfaces almost irrespective of the nature of that surface. This binding may be caused by ionic, hydrophobic or other (mostly) weak interactions. The obligatory first step in order to avoid non-specific sticking of antibodies to sections is to treat the sections with a general purpose protein solution. The idea here is to block any sites on the section which have an affinity for proteins. For this purpose, a range of proteins have been used, including gelatin (Tokuyasu and Singer 1976; see Behnke 1986 for a discussion), bovine serum albumin (McLean and Singer 1970; Kraehenbühl and Jamieson 1974), haemoglobin (H. Reggio, pers. commun.), ovalbumin (Roth 1989) and even non-fat dried milk (Duchamel and Johnson 1985). In recent years the use of non-immune sera has also become widespread. For many years we have been routinely using 1–10% newborn calf serum (or fetal calf serum) following observations by colleagues in our laboratory that these reagents greatly reduce background labelling on Western Blots, when compared to gelatin solutions. It should be noted that these reagents are virtually free of IgG and what little they contain does not significantly bind protein A. A similar rationale was introduced earlier by Geuze et al. (1981), who used normal goat serum. The use of such sera makes good sense since they present a complex mixture of proteins, lipids and salts and are more likely to bind all the "sticky" sites on sections than is a solution of a single protein. In this respect it should be noted that commercial gelatin solutions are also predominantly mixtures of different peptides; this is especially true for those with low "Bloom" number. This is an indication of the strength of the gels produced and is indicated by the manufacturers. Behnke (1986) recommended gelatin with Bloom numbers between 60 and 100. A recent innovation is to use gelatin purified from the skin of cold water fish at a final concentration of up to 0.5–1% (Birrell et al. 1987). In our group, this reagent, which has the advantage of remaining liquid even at relatively high concentration on ice, has significantly reduced background labelling when compared with either newborn calf serum or normal gelatin in immunolabelling cryo-sections of yeast cells.

Dr. T. Johnson (University of Fort Collins, Colorado, pers. commun.) has recently pointed out a striking observation his group has made on immunolabelling of freeze-fracture replicas, namely that the relatively large proteins normally used to block background labelling may also sterically block the ability of antibody to recognize the antigen. Their solution was to use either a trypic digest of collagen or hydrolyzed soy peptone which were then size-fractionated (e.g using an Amicon PM10 filter) in order to remove the high molecular weight products (e.g. above 500 MW). These small molecular weight products kept background labelling low while enabling the specific signal to be maintained.

It is advisable to pretreat sections for at least 5 min with whatever reagent that is used to reduce non-specific labelling as well as to have the latter present, in excess, in the primary and secondary antibody, (or protein A gold) mixtures. Even in the presence of an excess of a blocking reagent, such as serum, it is to be expected that antibody molecules may still bind with weak affinity to sites on the section other than the antigens. These weak interactions can usually be removed by the washing or rinsing steps.

The term washing or rinsing often connotes the need for an intense physical action, such as immersing the grids in a jet of water or PBS. In practice, however, all that may be needed is to float the grids on four to six different (≈ 0.1–0.5 ml) drops of buffer (such as PBS) for a few min (total period 10–15 min). It should not be forgotten that the size of the sample is minute and simple diffusion alone almost always suffices to remove the non-specifically bound antibody. In the case of especially "sticky" antibodies, various modifications can be tried such as incubating the antibody as well as the washing steps in high salt solutions (up to 0.5 KCl or NaCl) or incubating at different pHs (Grube 1980; Craig and Goodchild 1982). Other methods include the addition of weak solution of non-ionic detergents such as Tween 20 (Craig and Goodchild 1982; Roth 1989) in the antibody gold preparations and/or washing solutions. As pointed out by Larsson (1988), however, there is always the danger that these treatments may lower or alter the affinity of the antibody for the antigen. After the incubation with gold-conjugated antibody, or protein A, a rinsing step is even more important than after the antibody incubation. On sections, the time for rinsing at this step must usually be at least 20 min. Following empirically determined procedures used by the laboratory of J. Slot and H. Geuze (University of Utrecht, pers. commun.), we routinely use 30 min with six different drops of PBS in order to remove the non-specifically bound gold particles. It should be noted that when gold conjugates are used above their optimal concentrations, they bind non-specifically and, in our experience, it may be impossible to remove this bound gold, even after a 24-h period of rinsing (our own unpublished data).

For pre-immune labelling studies of whole cells, thick non-embedded sections or whole mounts, extra precautions may be needed to reduce non-specific binding of primary antibody and gold conjugates. When compared to labelling the surface of thin sections, the total surface area available for both specific and non-specific binding in such thick preparations is significantly larger (for a discussion see Behnke et al. 1986).

Finally, some parts of some tissues may be especially sticky to antibodies and gold reagents. An example is given by erythrocytes which often bind gold conjugates non-specifically. For those experiments where such non-specific labelling is unacceptable, procedures will have to be worked out empirically on a case by case basis in order to block these non-specific reactions.

7.2.6 Evaluation and Control of Labelling Reactions

An antibody prepared against component X labels only structure "A" in a tissue "T". How do we assess the validity of this result: in other words is it reasonable to conclude that, within tissue T, the antigen is predominantly or exclusively in structure A? In immunocytochemistry this is not always a trivial question. There are two interrelated aspects to the answer, an immunological aspect and a biological aspect, and both must be considered carefully in order to avoid pitfalls.

The immunological part has already been dealt with at some length in the preceding chapter. How was the antigen prepared, how was the antibody prepared and how were both characterized? These are key questions. Even when these steps are undertaken in "textbook" fashion, however, there is still no way of completely eliminating the possibility of obtaining spurious cross-reactions. Even monoclonal antibodies can give problems in this respect. A monoclonal antibody against the cell surface antigen Thy-1, for example, was also able to bind significantly to an epitope present in an otherwise totally unrelated intermediate filament protein in fixed cells (Dulbecco et al. 1981). Note also the observations of Nigg et al. (1982) discussed above.

We will discuss in more detail below why there appears to be no immunological control in immunocytochemistry that can, in theory, completely eliminate the possibility for error. Nevertheless, in practice, these controls usually work and are important in that they enable the experimenter to build up confidence that the interpretations are correct.

Larsson (1988) has recently classified these immunocytochemical controls as either "first level", that is, directly involved with the specificity of the antibody-antigen interaction and the "second level" controls which he defines as all other factors that can affect the labelling reaction.

7.2.7 The Immunocytochemical Controls

7.2.7.1 First Level Controls

- Absorption of antibody with purified antigen should eliminate the labelling reaction. Larsson (1988) has additionally divided this category of control into "liquid-phase" adsorption, in which the antigen and antibody solutions are simply mixed in a tube, and "solid-phase" adsorption, whereby the antigen is adsorbed onto a solid support such as sepharose. This author has discussed in detail the potential pitfalls with these approaches. The first is when the antigen with or without bound antibody shows affinity for any sites in the section which would increase non-specific labelling. This problem is avoided by the use of solid-phase adsorption. The second general problem is the presence of "contaminating" antibodies, that is, antibodies against minor components of the original immunogen. Neither type of adsorption protocols can get rid of this problem. This is especially difficult if the contaminating species does not reveal itself by the classical immunochemical tests such as immunoblotting or immunoprecipitation. As pointed out by Larsson (1988) "a successful adsorption control in no way implies identity between the molecule used for adsorption but merely implies a similarity between an epitope present on the molecule used for adsorption and the molecule stained in the tissue". For a detailed discussion as well as other examples of possible pitfalls, this reference is highly recommended.

From a theoretical point of view, when chemically synthesized peptides are used as immunogens, an adsorption control would seem to be an essential one for any immunocytochemical experiment. In this case, the antigens (but not necessarily the precise epitopes) are both defined at the molecular level and available in very pure form. It should be noted, however, that free peptides exist in many different conformations that are in a dynamic equilibrium with each other, and the relevant structure (the one that best mimics the antigenic site on the protein) may only represent a fraction of a percent of the total peptide (Dyson et al. 1988a,b). In this case, very high concentrations of peptide would be required to inhibit the reaction. This appears, in fact, to be an increasingly common observation in immunochemical studies.

• For some antibodies against defined peptide antigens an interesting approach to confirm the specificity of an immunocytochemical reaction is to use chemicals that can specifically alter defined amino acids. This approach was first used by Larsson (1979), who used, for example, cyanogen bromide to specifically modify methionine residues in order to distinguish between two functionally distinct enkephalin-like molecules that differ only in having either methionine or leucine in their carboxy-termini (see also Jirikowsky 1985).

7.2.7.2 Second Level Controls

This kind of control deals with all non-antibody antigen interactions of any of the reagents used for labelling with the tissue (Larsson 1988). In this case, the two following protocols could be taken in order to see if the observed labelling is really due to the antibody-antigen interaction.

- Substituting a pre-immune (or non-immune) serum for the specific antibody should eliminate the labelling. Although this is a widely used procedure, when true, it tells us simply that the reaction in the specific serum has something to do with the immunization but it does not necessarily constitute proof of specificity. The same can be said for controls where the antibody step is omitted from the labelling reaction. Again, on this point I can do no better than quote Larsson (1988), who stated that "in the case of polyclonal sera, differences between animals, presence of contaminating antibodies and differences in antibodies between successive bleedings of the same animal make this control nearly useless".

- Deletion of specific steps of labelling reaction. Removal of a horseradish peroxidase-conjugated antibody, for example, allows one to determine whether any endogenous peroxidase is present that could account for a reaction observed in sections.

7.2.7.3 Independent Methods for Establishing Specificity

As pointed out in Chapter 6, immunoblotting of SDS denatured antigens or immunoprecipitation of native antigens are the most commonly accepted methods today which are used as independent immunochemical methods for establishing the specificity of the reaction. While the strength of the approaches has been discussed in that chapter, the reader is again referred to the excellent treatise of Larsson (1988) on this point. This author has pointed out many weaknesses of these approaches as they are generally used for immunocytochemical experiments (see, for example, Fig. 6 from this reference). The first problem relates to the fact that the antigen must not only recognize the antibody after homogenization but it must also be resolved on a gel system (most commonly polyacrylamide). The second problem is that in immunocytochemistry the antigen has been chemically fixed and processed in a manner that is quite different to the above approach. A striking example in this context comes from the elegant work of Milstein et al. (1983) who studied a monoclonal antibody against serotonin (5-hydroxytryptamine) which gave a good immunohistochemical signal for this antigen in sections of fomaldehyde-fixed tissues. In solution, this antibody also recognized the closely related molecule dopamine, in fact with an even higher affinity in solution than for the original antigen. In sections of formaldehyde-fixed tissues, however, parts of the brain known to contain significant amounts of dopamine (but not serotonin) gave a negative immunohistochemical signal with this antibody. The results strongly suggested that the best antigen was a condensation product of formaldehyde with serotonin (which the authors argued has more reaction sites for formaldehyde than dopamine).

For some antigens that are available in pure form, it is possible to study the effect of fixation/processing steps on the antigen in situations that mimic (but are never identical to) conditions on, for example, the section surface. Some of the simpler blotting techniques have been discussed in Chapter 3. A more sophisticated procedure, first used by Brandtzaeg (1972), is to embed the antigen in a protein gel that can be treated in the same way as tissue, then sectioned and labelled. The strength of this approach was shown in the study by Schipper and Tilders (1983), who embedded different concentrations of antigens into a gelatin gel which was then be fixed in identical fashion to the tissue of interest. This gel can then be sectioned. In the original article the authors used cryostat sections for light microscopic evaluation. Since the antigens are cross-linked to a protein matrix and to each other, this system gives an in vitro model that more closely resembles the situation in cells and tissues than do the more classical immunochemical procedures. Obviously, the effects of the fixatives on the antigen can also be directly tested, both with respect to its ability to keep the protein in the gel and, directly, on antibody-antigen interactions (see Schipper and Tilders 1983 for examples of this). In their paper the authors compared serotonin (5-hydroxytryptamine) and 14 closely related indole derivatives with respect to their ability to bind to an antibody raised against serotonin. The results were quantified with respect to the concentration of these compounds that was necessary to block the fluorescence signal obtained with the anti-serotonin antibody, as determined fluorimetrically. The latter approaches could also be compared with in vitro binding studies (i.e. in solution). The results obtained are both elegant and striking with respect to their ability to determine the specificity of the antibody, that is, to identify precisely the parts of the molecule that are recognized by the antibody. By using either cryosections or resin embedded sections of such gels combined with immunogold procedures, it is clear that this kind of approach is potentially very powerful for future EM immunocytochemical studies (see Chap. 11 for similar approaches by Slot, Ottersen and co-workers for quantitative EM purposes).

One of the most difficult classes of antigens with respect to establishing specificity of an immunocytochemical reaction involves the small peptide hormones that often share considerable sequence homology. For a detailed discussion of this area the reader is referred yet again to the excellent book by Larsson (1988), who has had considerable experience in this area.

7.2.8 The Biological Controls

When one has taken reasonable precautions in making and characterizing antibodies for use in immunocytochemistry, there is no question in my opinion that the ultimate controls for the reaction involve considerations of the biological system in question. At the electron microscope level, the assessment of the labelling reaction is far easier than it is at the light microscopical level, since it is straightforward to resolve the individual structures involved: spots and blobs at the light microscope level are more likely to become identifiable organelles at the EM level.

The first biological consideration is the nature of the antigen, and any knowledge in this respect helps to asses the validity of the immunocytochemical labelling. Is the antigen a membrane protein, for example; is it soluble or insoluble? is it a glycoprotein? does it associate with nuclear fractions, and so on? A cytosolic protein is unlikely to be found in the extracellular space (and vice versa). Cells and tissues which are known not to contain the antigen should not label. Treatments that remove, re-locate or reduce the amounts of antigen should have a corresponding effect on the amount of labelling. In this respect the quantitative factor can also be a powerful judge of the specificity of labelling at the EM level, whereas it is unlikely to play such a role at the light microscope level. In chapter 11 I have outlined proposals for obtaining rough estimates of the amount of antigen in a tissue and the amount of label one should expect. Note that the presence of too much label should be viewed with as much (if not more) suspicion as too little label.

7.2.9 Problems in Labelling

In Table 3, I have listed some of the most common reasons for a lack of a specific signal in immunocytochemistry as well as potential solutions/suggestions to those problems.

Perhaps the more common situation than that mentioned above is to have a "specific" signal combined with background "noise". In Table 4 I have listed the most common reasons for these

Table 3. Common resons for a lack of signal in EM immunocytochemical experiments

Fault	Solutions
1. "Poor" antibody	Alternative Immunizations etc.
2. Too little antigen	Look for alternative systems in which express higher amounts of antigen are expressed
3. Antigen removed (a) or destroyed (b)	Increase cross-linking (a) or in the case of (b) use alternative specimen preparation. Alternatively, compare fixed and unfixed antigen at light microscopy level
4. Problems with gold conjugates, high background aggregates etc.	Make new batch of gold conjugate. Use alternative markers. Test with characterized first antibody
5. Antigen inaccessible on section	Use treatments to try to 'unblock' the antigen.[a] A more realistic approach is some kind of in vitro preparation or even cells that have been "badly" fixed on purpose (e.g. fixed in hypoosmotic conditions) Alternatively in the case of cryo sections very thin sections of lightly fixed cells can sometimes be useful

[a] This approach has been used with some success for light microscope immunocytochemistry, especially by using protease treatment for highly resistant antigens such as keratins (see Battifora and Kipinski 1986 for references). To my knowledge, however, this approach has not been successfully used at the EM level and, therefore, remains hypothetical.

Table 4. Common reasons for having signal plus "background[a]" in EM immunocytochemical labelling

Fault	Potential solutions
1. Antibody cross-reacts with unwanted antigen(s) in addition to the desired antigen	Purification/adsorption using purified antigen (or contaminating bands). In the case of monoclonal antibodies which show a real cross-reaction the only solution is to make a new antibody
2. Antigen displaced during specimen preparation	Alternative fixation procedures
3. Too high concentration of antibody or gold preparation	Use more dilute solutions
4. Aggregates in antibody or gold	Purification/alternative storage. N.B. antibody solutions should always be centrifuged for 1 min at $\approx$ 10,000 g before use
5. Particularly sticky structures in tissue	Check other tissues. Use of high or low salt, weak detergent solutions, different blocking reagents
6. Biological premise wrong! i.e. "background" is not really background	Look for an independent biochemical/ immunocytochemical proof

[a] Background is defined as labelling over structures which the experimenter is convinced should not contain significant antigen.

It goes without saying that the word "background" is a subjective one. The experimenter must assess what is considered to be a "background structure" (free of antigen) as well as the levels of "background" labelling which is acceptable in an average experiment.

With "good" antibodies and the gold preparations presently available, both with cryo-sections and the new generation plastic sections, the level of background labelling can be kept extremely low without too much difficulty (less than 0.1–1 gold per Øm^2 over antigen–"negative" structures). In all cases the assessment of the signal to noise ratio is a critical parameter for evaluating the immunocytochemical result (see Chap. 11).

7.2.10 Accessibility Problems on Sections

A surface of a section potentially exposes epitopes all over the section surface. The proportion of those epitopes that are completely free to bind antibody depends on many complex factors. Two different factors can contribute to sterically hindering accessibility of antibody to the antigen. First, the density of the matrix which can be due to the biological material itself, to their cross-linking by aldehyde as well as by exogenously added embedding media such as polyacrylamide (see Chap. 11). The second factor is the density of antigen; when antigens are closely spaced the binding of one antibody molecule can be expected to sterically hinder the binding of a subsequent antibody. The studies of Posthuma et al. (see Chap. 11) show clearly that for soluble proteins such as amylase embedded in a gelatin matrix, the latter factor appears not to be a serious problem since the number of immunogold particles gives a linear relationship to antigen concentration. This is true even for very high antigen concentrations. In such a model system little interaction can be expected between molecules of the protein either with themselves or with the gelatin matrix. In many parts of cells, however, the fact that antigens can interact with other cellular components may become a significant factor in reducing their accessibility to antibodies, even on the surface of a section. This is only one manifestation of the complex potential role of the matrix in limiting access of antibody to antigen in the section. A noteworthy biological example of a structure in which the matrix can pose a serious problem is the nucleus in which the proteins and DNA are relatively densely packed and, perhaps more importantly, are involved in complex interactions with each other.

The starting point for most immunocytochemical studies of the nucleus are usually light microscopic labelling studies following detergent extraction. Often the latter give a strong positive signal whereas a negative result may be obtained with EM labelling studies of sections of the same cells after fixing (without permeabilization) and plastic embedding or cryo-sectioning. In this case, a sensible approach would be to section and label the cells after the same permeabilization protocol that is used for light microscopy. The severe extraction procedure may remove components that cover antigenic sites which can then be labelled on the section. An example of such a phenomenon in practice comes from the recent study by Waitz and Loidl (1990). These authors investigate the detailed localization of the transforming nuclear protein c-myc using immunogold labelling of LR white sections of the slime mould *Physarum*. When whole nuclei were embedded and labelled, the signal was extremely poor. In contrast, when nuclear "matrices", prepared by extracting the nuclei were extracted with a low salt- and detergent-containing buffer, were fixed and embedded, the labelling of the sections of this material was excellent In fact, this labelling could be correlated with changes in the cell cycle and verified biochemically.

Assuming the antigen is indeed abundant enough (see Chap. 11), a few other approaches might be considered if one suspects that one has accessibility problems on sections. First, one should try to prepare and label EM sections following very brief cross-linking in low concentrations of fixative, preferably formaldehyde. Second, in the case of cryo-sections, one should attempt to make very thin sections, especially of preparations that are relatively weakly cross-linked: in this case one is more likely to get penetration into the section.

When the above procedures fail, one is forced to try more drastic methods that go in the general direction of an in vitro preparation The use of detergent treatment has already been mentioned for the example of the nucleus. Alternatively, try gentle homogenization of tissue or cells, or scraping monolayers of cells with a rubber policeman, before fixation. For cells in culture, the "permeabilized cell technique", whereby a filter is placed briefly on the monolayer and then "ripped-off", can often be useful in removing cytosolic components (Simons and Virta 1987; Balch et al. 1987). The use of toxins such as streptolysin O might also be worthwhile (see Chap. 10). Finally, one might consider combining an in vitro approach with the use of enzymes to digest specific components that might hinder accessibility (proteases, nucleases etc). Such treatment could also be applied on the section (see, for example, Dell'orto et al. 1982; for a discussion see Larsson 1988, pp. 151–153). Such methods are theoretically more likely to succeed with (unembedded) cryo-sections than with plastic sections.

7.2.11 Sensitivity of Labelling

Since a thin section exposes such a small fraction of the total antigen in a cell, it is imperative that as much of those antigens as possible should be detected. In other words, the technique should be as sensitive as possible. As explained in more detail in Chapter 11, these modern techniques are reasonably sensitive and the labelling protocols can be expected to recognize up to, say 10% of the total antigen in the section. The crucial factor for EM immunocytochemistry is the ability of the primary antibody to recognize the antigen, since the secondary gold probes appear, in general, to be very efficient in finding the bound antibodies (at least on the section surface). As explained in detail in Chapter 11, for quantitative purposes the ideal situation is to have no more than one gold particle per antigen. The use of layers of antibodies mostly serves to increase the number of gold particles per "hit", not the number of hits (i.e. antigens detected) and should therefore be avoided when possible. There are, however, examples where the use of extra layers can give a significant increase in the real signal, that is in the amount of antigen detected. Hence Slot et al. (1988a) showed that on cryo-sections of gels consisting of amylase embedded in gelatin and labelled with anti-amylase there was a significant increase in the amount of labelling with a three-step (primary antibody followed by a secondary antibody followed by protein A gold) when compared to two-step labelling (primary antibody followed by protein A gold). This was assessed as a real increase in signal (not just in amplification) since clusters of gold particles, due to one primary antibody binding event, were counted as one "labelling unit" in these studies. This increase in clusters (the signal) was attributed to the greater flexibility of unconjugated secondary antibody when compared to protein A gold, to recognize the bound primary antibody in those parts of the cryo-sections that are not directly on the surface. The price to pay for this increase is, first, a significant decrease in resolution, in that the gold clusters have a broader distribution around the antigen than do the single gold particles in the two-step labelling and, second, an increase in background labelling (see Slot et al. 1988a, 1989 for discussion and Table 5). The authors' explanation for the latter observation was as follows (see Fig. 3). Protein A recognizes only the Fc portion of the bound antibody: when the primary antibody is specifically bound to

Table 5. Immunogold labelling of amylase in 100 nm-thick gelatin sections

Labelling procedure	Clusters[a]/μm^2	Gold/μm^2	Gold/cluster
	Mean and standard error of mean		
(1) 1 R α amylase 2 S α R-gold	8.8 (6.8%)[b]	16.1 (5.1%)[b]	1.8
(2) 1 R α amylase 2 prot. A-gold	7.1 (3.1%)	8.2 (3.3%)	1.1

(3) 1 R amylase
    2 S × R
    3 prot. A-gold
	17.8 (7.1%)	51.1 (5.9%)	2.9

Cryo-sections were made from 10% gelatin blocks, containing 1 mg/ml rat pancreas amylase and were fixed in 2% glutaraldehyde.

[a] Clusters mean either single gold particles or a distinct group in close contiguity. Note that this parameter is a measure of the real signal to noise ratio.

[b] Non-specifically bound label, expressed as percentage of specific lael, was measured in section parts that did not contain amylase.

(R = rabbit; S = swine).

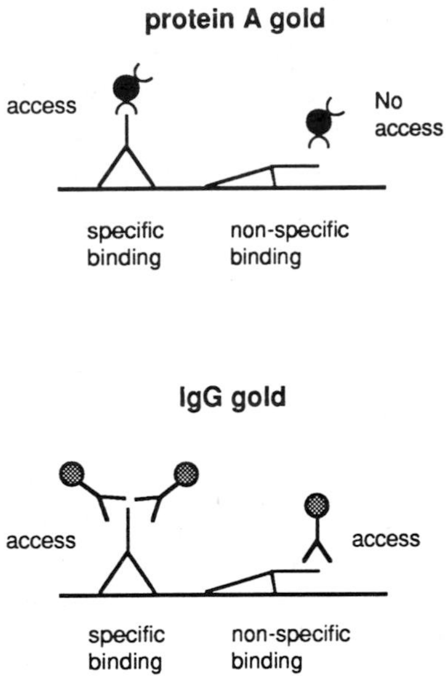

protein A gold

IgG gold

Fig. 3. Diagram to explain how IgG gold complexes may lead to higher background labelling when compared to protein A gold (Idea from Slot, pers. commun.)

antigen, one expects that the Fc part should be well exposed for binding to protein A gold, whereas when the antibody is non-specifically bound to the section that Fc part may, on average, be less accessible to the protein A on the gold particles. In contrast, when a free second antibody is applied (in the three-step method), this is more likely to recognize some exposed parts of these non-specifically bound primary antibodies. In this case the labelling is not restricted to the Fc part of the IgG.

For the same reason that a three-step protocol may give a higher density of specific labelling than a two-step one, the indirect labelling protocols can be expected to be more sensitive than direct ones. Clearly, free antibodies would be expected to have more access to antigens in the section than antibodies conjugated to, say, bulky colloidal gold particles: this point has, however, to my knowledge not been carefully studied.

7.3 Approaches for Single Labelling

As already discussed above there are two kinds of approaches for single labelling, namely direct and indirect (Fig. 4). For most purposes and for reasons given above, the indirect method is preferred. The only potential reason for preferring the direct labelling would be to improve the resolution (see below). For conventional labelling on sections, using gold particles in the range 5–9 nm it appears unlikely that, when coupled directly to antibodies, the resolution of the technique is significantly improved over a two step, indirect method using protein A gold as a detector. In order to gain a significant benefit it seems likely that Fab fragments combined with 1–3-nm gold particles would be required (Baschong and Wrigley 1990; see below). Note also the possibility for some high resolution studies of visualizing the antibody (or Fab) by its electron density, that is, without marker (see below).

7.4 Approaches for Double Labelling

7.4.1 Double Direct Labelling

This is a rarely used approach in which two primary antibodies are coupled directly to two different markers. They can be added in one step to sections (Fig. 4). Perhaps the greatest theoretical problem with this method is the possibility that the antibody bound to marker may have less affinity for antigen when compared to free (unconjugated) antibody. As for direct approaches for single labelling this method is tedious since each primary antibody has to be conjugated to the marker. For an example combined with a pre-embedding approach for visualizing α actinin and myosin see Langanger et al. (1986).

Approaches for single labelling

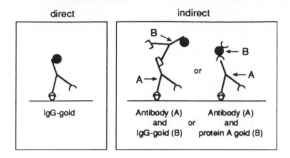

direct | indirect

IgG-gold

Antibody (A)
and   or
IgG-gold (B)

Antibody (A)
and
protein A gold (B)

Approaches for double labelling

double direct - simultaneous

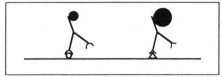

double indirect - simultaneous

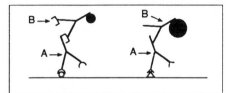

double indirect - sequential

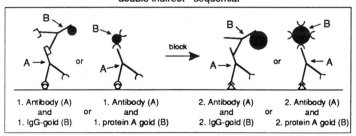

1. Antibody (A)
and   or
1. IgG-gold (B)

1. Antibody (A)
and
1. protein A gold (B)

block

2. Antibody (A)
and   or
2. IgG-gold (B)

2. Antibody (A)
and
2. protein A gold (B)

Fig. 4. Schematic diagram to show the different kinds of single- and double-labelling protocols

1. Antibodies applied simultaneously

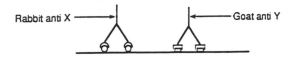

Rabbit anti X ——→   ←—— Goat anti Y

2. Second antibodies
(cross-adsorbed against each other)
applied simultaneously

Sheep anti rabbit - ——→   ←—— Chicken anti goat -
imposil   ferritin

Rabbit anti X ——→   ←—— Goat anti Y

Fig. 5. Schematic illustration of the double-labelling scheme used by the Singer laboratory (Geiger et al. 1981). Note that for simplicity both Fab's on the antibodies are shown to bind antigen. This is unlikely in practice (see p. 246)

7.4.2 Double Indirect – Simultaneous

Perhaps the most sophisticated approach for double (or triple) labelling is to use a simultaneous double indirect procedure, that is two (or more) antibody marker pairs such that any cross-labelling reaction between the two pairs is not possible (Fig. 5). The first use of such an approach for electron microscopy again came from the Singer laboratory. The conditions for this procedure were first worked out by Dutton et al. (1979), which also introduced the use of imposil (see Chap. 8) as an alternative marker to ferritin. Subsequently, in a classical double-labelling study of smooth muscle cells by Geiger et al. (1981) the first antibody pair was a rabbit anti vinculin antibody visualized with an imposil conjugated goat anti-rabbit and the second involved a guinea pig anti-tropomyosin labelled using ferritin conjugated goat anti-guinea pig. For this approach it is necessary to ensure that there is no cross-reaction between the two pairs of reagents. In practice this means that the marker-conjugated second antibodies must be cross-absorbed against the primary antibodies. In the example from Geiger et al. (1981), the goat anti-guinea pig IgG had to be passed through a column containing rabbit IgG coupled to a sepharose solid support. Any molecules in this preparation which showed affinity for the rabbit IgG were removed by this procedure. Conversely, the goat anti-rabbit IgG was passed through a guinea pig IgG column.

The advantage of this approach is that both primary antibodies can be mixed together in the first labelling reaction, and both secondary antibody conjugates can be combined in the second step. In this way, the possibility that the binding of one antibody to its antigen sterically blocks the binding of the second antibody is reduced. The main disadvantages are as follows: one is restricted to preparing antibodies in two different species such that their IgGs are not recognized by the same secondary antibody molecules and this is relatively time consuming. From a theoretical point of view one of the two markers used by Dutton et al. (1979) and Geiger et al. (1981), namely ferritin and imposil, are preferred over gold since, unlike gold, both these markers are coupled covalently to the protein. This approach has, however, also been used using two gold-antibody conjugates (Tapia et al. 1983; Doerr-Schott and Lichte 1984). In general, this approach has not been used extensively.

7.4.3 Double Indirect – Sequential

In this approach the first antibody then the first marker is applied followed by the second pair, also in two steps. In most cases the sections must be treated in such a way that the reactivity of the first pair towards the second is blocked/ destroyed without affecting the antigens to be recognized by the second antibody (Fig. 4). For all these approaches a theoretical limitation is that, for different antigens that are positioned close to each other in the section, the labelling of one antibody marker pair may block the accessibility of the second. An obvious way to check this possibility is to reverse the sequence of labelling (as well as to compare the intensity of labelling after single and double labelling).

7.4.3.1 Two Sizes of Protein A Gold

Geuze et al. (1981) introduced a very simple method for double-labelling using two sizes of gold particles, both conjugated to protein A (for later modifications see Slot and Geuze 1984). The outline of this method, shown schematically in Fig. 3, is that following the first antibody-protein A-gold pair (normal indirect labelling) the sections are treated with free-protein A, in order to block any IgG molecules that have not bound to the (first) protein A gold step. This is important because, otherwise, these could bind to the second protein A gold. The second antibody is then applied. This can bind to specific antigen as well as to any free protein A available on the surface of the first gold particles (this will increase in proportion to the size of the first gold particles). The key factor here is that these latter antibodies, once bound to protein A, should be unavailable for binding to the second protein A gold added subsequently. It should be noted that although protein A has five identical IgG binding sites, once a molecule is bound to IgG (by two of these sites) the other domains are apparently unable to bind further IgG molecules: when bound to gold, protein A is apparently unable to cross-link IgG molecules (see Chap. 9).

The advantages of this widely used approach is that it is technically simple and has proven very reliable in a large number of laboratories, including our own. The disadvantages are:

- The procedure is restricted to antibodies that have a high affinity for protein A. Most monoclonal antibodies do not bind protein A. For these it is necessary to use a three step labelling protocol (e.g. mouse IgG followed by rabbit-anti-mouse IgG followed by protein A gold). For double-labelling with two monoclonals this approach has in our hands always given spurious results and high background labelling.
- The method relies completely on the first protein A gold to stay bound during the labelling and the rinsing steps of the second antibody-gold pair. If any gold-protein A (first step) washes off (or is competed by the second protein A gold) during this latter period the second gold can bind to a proportion of the first antibody. Slot (pers. commun.) has also found evidence for the first gold label coming off its antigen and associating with the second (subsequently applied) antibody. Presumably this reflects a relatively low affinity of the protein A for the first antibody. In general, the fixative blocking step between the two antibody steps (above) is likely to reduce the magnitude of such problems.

For this approach to work, it is essential that the protein A binds only to the Fc portion of the IgG's. There are indications, however, that some antibodies may bind to protein A via their Fab domains (Young et al. 1984; Bendayan and Stephens 1984). For a discussion of this problem see Chapter 9. This phenomenon is most likely responsible for a spurious co-localization of both sizes of gold particles we have occasionally observed with this procedure. This problem is most clearly recognizable over "background" areas. For example, in localizing two membrane proteins we would routinely look at the nucleus: a false co-localization is more easily identified when the gold particles are sparsely distributed. The most likely explanation for this phenomenon is that the second primary antibody may bind to the excess (free) protein A on the surface of the first gold particles by its Fab part. In this case, its unbound Fc region is now free to bind to protein A on the second size of gold particle. We have found no satisfactory solution to this problem. In some, but not all, cases the magnitude of the problem could be reduced by fixing the sections in between the two antibody steps (see below).

In order for this approach to be successful, it is important to empirically find the concentrations of both antibodies that give low background labelling in independent single-labelling studies. In our experience, the level of background labelling is often significantly higher (for both antibody-gold combinations) than it is in single labelling studies, for reasons not entirely understood. For routine studies, it is imperative that the general distribution of both antigens be studied extensively in single labelling experiments before double-labelling studies are carried out. The possibility for recognizing spurious results can be increased by comparing the labelling patterns, as well as the amounts of label, for all four possible combinations i.e.

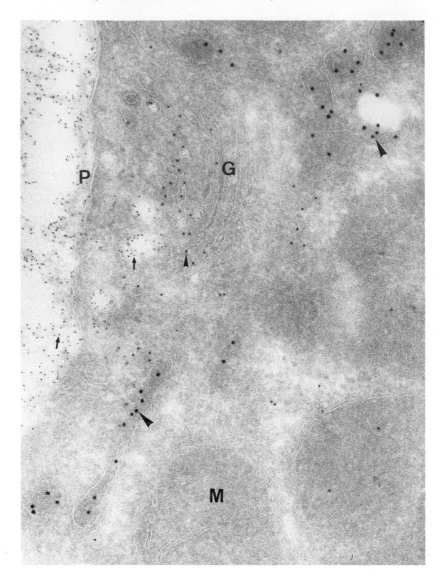

Fig. 6. Triple-labelling of rat adipose tissue in thawed cryo-section. The sections were labelled with the following protocol.

1. Affinity-purified rabbit anti insulin-regulated glucose transporter (5 μg/ml). This was followed by PBS rinses.
2. 9 nm protein A gold (OD$_{520}$ O.2).
3. Following PBS rinses, fixation in 1% glutaraldehyde in PBS for 10 min. Subsequently treatment with PBS-glycine (to quench free aldehyde groups), then PBS-BSA (0.2%) as a "blocking" step for the next labelling step.
4. Rat anti cathepsin D (affinity purified, 5 μg/ml) followed by PBS rinses.
5. 15 nm protein A-gold (OD$_{520}$ 0.3).
6. Step 3 is repeated.
7. Rat anti-albumin (affinity purified, 5 μg/ml) followed by PBS rinses.

1. Antibody 1 ⇒ large gold ⇒ antibody 2 ⇒ small gold.
2. Antibody 1 ⇒ small gold ⇒ antibody 2 ⇒ large gold.
3. Antibody 2 ⇒ large gold ⇒ antibody 1 ⇒ small gold.
4. Antibody 2 ⇒ small gold ⇒ antibody 1 ⇒ large gold.

This approach has also been successfully used for localizing three antigens (Slot and Geuze 1985, and pers. commun.). In this case, a second free-protein A step is added between the second gold step and the third specific antibody. For most purposes the use of a fixature is now recommended instead of protein A.

7.4.3.2 A Recent Improvement: the Use of a Fixative Block

A recent innovation by Slot (pers. commun.) is to use a fixative for blocking the reactivity of the first antibody marker pair towards the second pair. Thus, for example, following the first gold marker, the section is treated with either formaldehyde or, preferable, glutaraldehyde. The choice of fixative obviously depends on the sensitivity of the second antigen to the fixative: in preliminary experiments 1% glutaraldehyde appears to work well, although we have also had success with 8% formaldehyde. One reason for the success of this approach is that , in our experience, antibodies seem to be more sensitive to the effects of fixatives than antigens. This technique appears to offer a number of improvements over the protein A blocking approach. It mostly gets rid of many of the disadvantages listed above, including the spurious double-labelling phenomena that are sometimes seen (see below). This method may also work with two- or three-step labelling reactions, for example two monoclonals followed by a bridging antibody and protein A gold. Combinations of protein A gold and Ig-gold complexes should also be feasible. The method also enables triple labelling procedures to be used routinely (for an example see Fig. 6). Note that with the protein A blocking method this was a difficult procedure.

A conceptually similar approach using IgGold was introduced earlier by Wang and Larsson (1985) where the possibility of the second antibody-Ig gold combination to cross react with the first pair is eliminated by treating the resin sections with formaldehyde vapours at 80°C subsequent to the binding of the

---

8. 5 nm protein A gold (OD_{520} 0.1).
9. PBS rinses, water rinses then embedding in methyl cellulose-uranyl acetate (see Chap. 5).

Evidently, the role of the glutaraldehyde fixation between the various labelling step has been a crucial one in enabling this triple-labelling reaction to be carried out successfully.

The micrograph shows extracellular albumin (5 nm gold, *arrows*; *P* plasma membrane) as well as some in endocytic structures next to the Golgi complex (*G*). The glucose transporter (9 nm gold, *small arrowhead*) is primarily in a cisternal structure close to the Golgi stack that most likely represents the trans Golgi network/reticulum. The cathepsin D (15 nm gold, *large arrowhead*) is, in electron dense structures (prelysosomes or lysosomes), distinct from the Golgi complex as well as the endocytic structures containing albumin. M mitochondrion ×70000. (Courtesy of Dr. J W Slot, University of Utrecht, Netherlands. From a study by J.W. Slot, H.J. Geuze, S. Gigengack, G.E. Lienhard and D.E. James. See J Cell Biol)

first pair. While it seemed unlikely that this harsh procedure would be useful for cryo-sections, Larsson (pers. commun.) has pointed out that his group has, in fact, successfully used this approach, also with cryo-sections. The critical point was to protect the sections with methyl cellulose (see Chap. 5) during the formaldehyde vapour treatment. The methyl cellulose is then dissolved with cold water and the labelling of the second antigen continued.

7.4.3.3 Two-Sided Sections Method

An interesting approach for double-labelling of plastic sections was introduced by Bendayan (1982)and Bendayan and Stephens (1984). In this method, the first antibody gold marker pair are applied to one surface of the section which has been placed on a "naked" grid (without a supporting film). Subsequently that surface is dried and the other surface is labelled with a second antibody marker pair. In this method, care must obviously be taken to ensure that the section surfaces do not "see" the "wrong" antibody. This approach is unlikely to be generally useful for cryo-sections, since a supporting film on the grid is essential for such sections. A second reason is that single labelling studies of cryo-sections have indicated that both surfaces of the sections (as well as the interior) may have access to antibody and markers (Slot and Geuze 1983). As pointed out in Chapter 4 (see Fig. 12 in Chap. 4), however, this problem can also occur in Lowicryl sections. This method is theoretically attractive because it is the only method which prevents any interactions between the two sets of labelling reactions. A theoretical disadvantage is the potential loss of resolution due to the fact that the entire thickness of the section lies between the position of the two sets of markers.

7.4.4 Multiple-Labelling Using Serial-Sections

In plastic embedded sections a useful method for double and even multiple labelling for EM is to label serial sections with different antibodies. This approach is technically tedious since it demands putting adjacent sections on different grids and finding the same areas on the successive sections. However, it has been successfully used in a number of elegant studies by Orci's group (see Ravazzola and Orci 1980; Orci et al. 1985). For an example see Fig. 10 in Chapter 4. For a few sections, but not for an extensive series, this procedure is also possible using cryo-sections (Posthuma et al. 1986).

7.4.5 Avidin-Biotin Systems

From a theoretical point of view, the fact that the high affinity interaction between avidin and biotin is a non-immunological one makes this combination a very attractive one for double-labelling experiments using any of the above approaches. For example, if one of the primary antibodies is biotinylated and non-reactive towards protein A while the second antibody, a protein A binder, is not biotinylated, the two can be mixed in the first labelling reaction and a mixture of protein A gold and avidin-gold (of different sizes), added at the second labelling step should distinguish between them without any obvious reason for any cross-reaction. For reasons explained in Chapter 9, normal glycosylated chicken egg avidin is less suitable for this purpose since it is a very "sticky" molecule. The non-glycosylated *Streptococcus* avidin (streptavidin), with a significantly lower isoelectric point, should be the protein of choice in this respect. In fact, streptavidin coupled to colloidal gold is now commercially available from a number of sources. At the time of writing, this approach is only beginning to live up to its expectations. With improvements in the quality of these gold conjugates, however, it is foreseen that the method will rapidly gain in popularity. For more detail see Chapter 9.

7.4.5.1 Additional "Sandwich" Procedures

There are a large variety of multiple layer methods for single and double labelling which have been primarily developed for light microscopy including the PAP method (see Sternberger 1979), the avidin-biotin peroxidase complex (ABC) method (Hsu et al. 1981) and a whole class of "labelled antigen" detection methods developed by Larsson and his colleagues (see Larsson 1988 for a review). The latter, as mentioned, take advantage of the fact that when an antibody is bound to antigen in the section (and usually by a single arm – see above) its second arm is free to bind to labelled antigen added in a second step. The label can be colloidal gold antigen complex for example (the so-called GLAD procedure). A theoretical attraction of this approach is that the antibody which is detected is by definition specific for the antigen.

7.4.6 Resolution

The resolution of a labelling technique is a reflection of how close the position of the particulate marker is to the position of the antigen in a structure within the section. This is clearly determined by the size of the total complex that is used to bind and visualize the antigen. Imagine an antigen in a section is labelled by a rabbit antibody followed by 5 nm colloidal gold-protein A conjugate. When we ask what the resolution of this labelling is we are really interested in knowing what, on average, is the largest possible distance between the centre of the gold particle and the antigen: this distance defines the poorest resolution theoretically obtainable. Since the size of the antibody molecule is fixed, two factors can affect this "worst case" resolution. First, the size of the protein A-gold complex and second, its point of attachment to the antibody. An IgG molecule has an extended length of about 8–10 nm (see for example Amit et al. 1986). A 5 nm gold particle would be expected to have a total diameter of 7–8 nm when

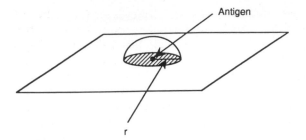

Fig. 7. The resolution is determined by the semi circular zone of radius r, where r is the size of the labelling complex (e.g. the size of the antibody plus the distance to the centre of the protein A gold). Upon projection in the electron microscope this reduces to a circle (*hatched zone*)

covered by protein A. Thus, in the worst theoretical case, where the gold attaches to the periphery of the IgG molecule, the resolution in one direction is about 15–18 nm. On the section the space occupied by antibody/gold complex could be enclosed by a hemisphere of radius $\approx$ 15–18 nm. (Fig. 7). When viewed in projection from above (as in the case of the electron microscope) this reduces to a circle of radius 15–18nm (covering an area of $\approx$ 710–1020 nm^2).

In practice, the resolution actually seen is almost always better than the theoretical values. There are at least two reasons for this. First, the visualizing gold marker is unlikely to bind to the extremity of the IgG molecule but rather, closer to its hinge region (viewed from "above") (see Baschong and Wrigley 1990). Second, most antigens will be fixed in the sections such that the antibody will have access to them only in a defined orientation. Consider, for example, antigens on the extracellular domain of the plasma membrane. When this membrane is sectioned, the access of the antibody to these antigens is restricted so that binding is predominantly to the extracellular side. In practice, the gold markers labelling such antigens are often precisely positioned in their expected sites (even with indirect labelling protocols), rather than being distributed in a wider zone on both sides of the membrane profile. It should be noted that the important factor here is the two-dimensional or projectional position rather than its precise three-dimensional position: within the focus limits of the micrograph it matters little whether the marker is 5 or 50 nm above the antigen. For an antigen whose localization is not unequivocally known, however, the theoretical arguments can be the only basis for defining the worst resolution possible.

How can the resolution be improved? The obvious possibility is to reduce the size of the labelling complex, that is antibody plus gold. Clearly, direct labelling protocols will give a higher theoretical resolution than will indirect ones. It appears unlikely, however, that direct Ig-gold complexes would offer much improvement over the indirect approaches using protein A gold. A reason for this (suggested by Slot, pers. commun.) might be that there appears to be a tendency in the latter procedure for the bound primary antibody to be rather restricted in its position above the antigen: the proposal here is that, in practice, the antibody-protein A gold complex has a higher probability of being positioned in a limited cone-shaped zone above the antigen than throughout the hemisphere as shown in Fig. 7. For the best theoretical resolution Fab fragments conjugated directly to very small gold particles must be used (Tschopp et al. 1982; Baschong and Wrigley 1990). Although gold, as well as other heavy metal particles as small as 1 nm is now available, in practice, the limit of visibility on most sections in transmission EM is about 3 nm. In combination with Fab fragments this could reduce the overall particle sizes to about 8 nm giving a "worst case" resolution of 8 nm radius (Baschong and Wrigley 1990). This, then, appears to be the practical limit to labelling antigens on sections. For high resolution studies, such as negative staining, or cryo EM, of proteins, (especially periodic objects), antibody or Fab fragments without electron-dense marker can often be visualized directly by an increase in electron density over a part of the object (see Wabl 1974; Aebi et al. 1977; Kubalek et al. 1987 and Carrascosa 1988). This will set the absolute limit of resolution for labelling reactions. In most cases, however, this approach is not feasible on sections due to the relatively high inherent density of the biological material itself and, in the case of plastic sections, of the resins.

The resolution of labelling obviously depends not only on the size of the labelling complex but also on the ability of the investigator to resolve (visualize) the structure of interest in the section. It also depends on the size and shape of the structure in relation to the magnitude of section thickness (see Slot and Geuze 1983; Slot et al. 1988a for more details). If small vesicles are being labelled that are, on average, approaching or less than the section thickness the chances of labelling antigens within the lumen of such vesicles and to visualize those vesicles clearly is very small indeed: the more clearly a vesicle is seen the more likely it is that the whole profile is not sectioned but "embedded" within the section.

An example of the effect of section thickness in cryo-sections in reducing resolution is shown in Fig. 8. Isolated mitochondria were pre-labelled with an antibody against an outer membrane protein (porin) followed by 6 nm gold-protein A. These gold particles thus define the position of the outer membrane. The mitochondria were fixed and cryo-sections were made which were then labelled with a second antibody against a 19 kDa protein that is also restricted to the outer membrane (Söllner et al 1990) followed by 9 nm gold-protein A. In the absence of an embedding medium both markers are likely to be present throughout the entire section thickness. In Fig. 8A, which shows a thin section, both markers are in their expected positions on the outer membrane. In Fig. 8B, a thicker section is shown. In the lower profile which is through the centre of the mitochondrion the membrane is cut perpendicularly and the label is in its "correct" position. In the upper profile, however, the top "cap" of the mitochondrion is revealed; i.e. only one (lower) section plane has gone through the particle and the (upper) section plane has evidently missed the particle altogether. Thus, the whole upper surface of the mitochondrion is seen. In Fig. 8C, which shows a section that is even thicker than in B, although the section plane has gone through the central region of the particle, the label is not precisely positioned but covers a broad "zone" at the periphery. For structures such as mitochondria showing a high degree of curvature, as the section thickness

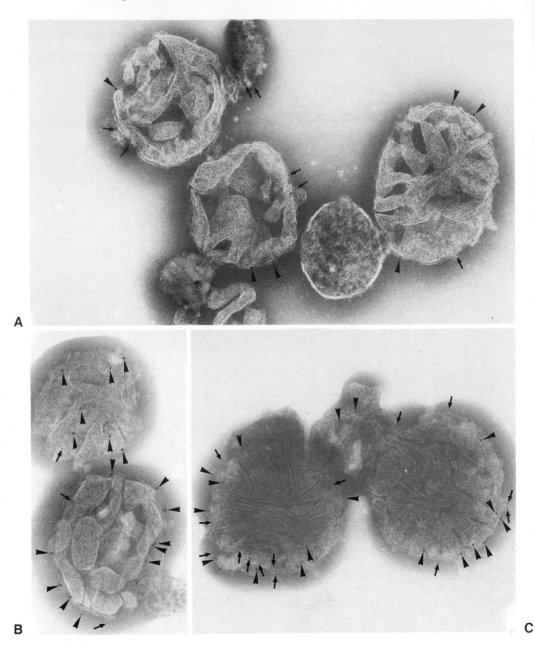

Fig. 8A–C. Neurospora mitochondria were prelabelled with porin antibodies (6 nm gold, *arrows*) and the thawed cryo-sections labelled with anti MOM 19 and 9 nm gold (*arrowheads*, see text for more details). The sections were contrasted with ammonium molyldate-methyl cellulose. (From a study done in collaboration with Dr. Walter Neupert's group, University of Munich; see Söllner et al. 1990, for details)

increases, the less will be the chance to have both section p
to the outer membrane (as in the lower profile in B) and the
resolution.

References

Aebi U, Ten Heggeler B, Onorato L, Kistler J, Showe MK (1977) New method for localizing proteins in periodic structures: Fab fragment labelling combined with image processing of electron micrographs. Proc Natl Acad Sci 74:5514–5518

Amit AG, Mariuzza RA, Phillips SEV, Poljak RJ (1986) 3-Dimensional structure of an antigen-antibody complex at 2.8-A resolution. Science 233:747–753

Avranmeas S, Uriel J (1966) Methode de marquage d'antigènes et d'anticorpes avec des enzymes; son application en immunodiffusions. CR Acad Sci 262:2543–2540

Balch WE, Wagner KR, Keller DS (1987) Reconstitution of transport of vesicular stomatitis virus G protein from the endoplasmic reticulum to the Golgi complex using a cell-free system. J Cell Biol 104:749–760

Baschong W, Wrigley NG (1990) Small colloidal gold conjugated to fab fragments or to immunoglobulin G as high-resolution labels for electron microscopy: a technical overview. J Electron Microsc Tech 14:313–323

Bateman JB, Calkins HE, Chambers LA (1941) Optical study of the reaction between transferred monolayers of Lancefield's "M" substance and various antisera. J Exp Med 71:321–341

Battifora H, Kipinski M (1986) The influence of protease digestion and duration of fixation on the immunostaining of keratins. J Histochem Cytochem 34:1095–1100

Behnke O (1986) Non-specific binding of protein-stabilized gold sols as a source of error in immunocytochemistry. Eur J Cell Biol 41:326–338

Bendayan M (1982) Double immunocytochemical labelling applying the protein A gold technique. J Histochem Cytochem 30:81–85

Bendayan M, Stephens H (1984) Double labelling cytochemistry applying the protein A-gold technique. In: Polak JM, Varndell IM (eds) Immunolabelling for electron microscopy. Elsevier, Amsterdam pp 143–154

Birrell GB, Hedberg KK, Griffith H (1987) Pitfalls of immunogold labeling: analysis by light microscopy, transmission electron microscopy, and photoelectron microscopy. J Histochem Cytochem 35:843–853

Blodjett KB, Langmuir I (1937) Built-up films of barium stearate and their optical properties. Phys Rev 51:964–982

Brandtzaeg, P (1972) Evaluation of immunofluorescence with artificial sections of selected antigenicity. Immunology 22:177–183

Carrascosa JL (1988) Immunoelectron microscopical studies on viruses. Electron Microsc Rev 1:1–16

Coons AH (1956) Histochemistry with labelled antibody. Int Rev Cytol 5:1–23

Coons AH, Kaplan MH (1950) Localisation of antigen in tissue cells. II Improvements in a method for the detection of antigen by means of fluorescent antibody. J Exp Med 91:1–13

Coons AH, Creech HJ, Jones RW (1941) Immunological properties of an antibody containing a fluorescent group. Proc Soc Exp Biol Med 47:200–202

Coons AH, Creech HJ, Jones RN, Berliner E (1942) The demonstration of pneumococcal antigen in tissues by the use of fluorescent antibody. J Immunol 45:159–170

Coons AH, Snyder JC, Cheever FS, Murray ES (1950) Localization of antigens in tissue cells. IV Antigens of rickettsial and mumps virus. J Exp Med 91:31–38

Craig S, Goodchild DJ (1982) Post-embedding immunolabelling. Some effects of tissue preparation on the antigenicity of plant proteins. Eur J Cell Biol 28:251–260

HJ, Jones RN (1941a) The conjugation of horse serum albumin with isocyanates of certain polynuclear aromatic hydrocarbons. J Am Chem Soc 63:1661–1669

eech HJ, Jones RN (1941b) Conjugates synthesized from various proteins and the isocyanates of certain aromatic polynuclear hydrocarbons. J Am Chem Soc 63:1670–1674

Dell'orto P, Viale G, Colombi R, Braidotti P, Coggi G (1982) Immunohistochemical localization of human immunoglobins and lysozyme in epoxy-embedded lymph nodes. J Histochem Cytochem 30:630–636

Doerr-Schott J, Lichte CM (1984) A double-immunostaining procedure using colloidal gold, ferritin, and peroxidase as markers for simultaneous detection of two hypophysial hormones with the electron microscope. J Histochem Cytochem 32:1159–1165

Duchamel RC, Johnson DA (1985) Use of nonfat dry milk to block nonspecific nuclear and membrane staining by avidin conjugates. J Histochem Cytochem 33:711–714

Dulbecco R, Unger M, Bologna M, Battifora H, Syka P, Okada S (1981) Cross-reactivity between Thy-1 and a component of intermediate filaments demonstrated using a monoclonal antibody. Nature 292:772–774

Dutton AH, Tokuyasu KT, Singer SJ (1979) Iron-dextran antibody conjugates: General method for simultaneous staining of two components in high-resolution immunoelectron microscopy. Proc Natl Acad Sci USA 76:3392–3396

Dyson HJ, Rance M, Houghten RA, Wright PE (1988a) Folding of immunogenic peptide fragments of proteins in water solution. I. Sequence requirements for the formation of a reverse turn. J Mol Biol 201:161–200

Dyson HJ, Rance M, Houghten RA, Wright PE, Lerner RA (1988b) Folding of immunogenic peptide fragments of proteins in water solution. II. The nascent helix. J Mol Biol 201:217

Farrant JL, McLean JD (1969) Albumins as embedding media for electron microscopy. 27th meeting of the EM Society of America pp 422–423, Hagler H, Wurzelbach P (eds) Claitor's, Baton Rouge

Faulk WP, Taylor GM (1971) An immunocolloid method for the electron microscope. Immunochemistry 8:1081–1088

Geiger B, Dutton AH, Tokuyasu KT, Singer SJ (1981) Immunoelectron microscope studies of membrane-microfilament interactions: distribution of α actinin, tropomyosin and vinculin in intestinal epethelial brush border and chicken gizzard smooth muscle cells. J Cell Biol 91:614–628

Geuze HJ, Slot JW, van der Ley P, Scheffer RCT (1981) Use of colloidal gold particles in double-labelling immunoelectron microscopy of ultrathin tissue sections. J Cell Biol 89:653–659

Griffiths G, Hoflack B, Simons K, Mellman I, Kornfeld S (1988) The mannose 6-phosphate receptor and the biogenesis of lysosomes. Cell 52:329–341

Griffiths G, Matteoni R, Back R, Hoflack B (1990) Characterization of the cation-independent mannose 6-phosphate recoptor-enriched prelysosomal compartment in NRK cells. J Cell Sci 95:441–461

Grube D (1980) Immunoreactivities of Gastrin (G-) cells. Histochemistry 66:144–167

Gu J, Islam KN, Polak JM (1983) Repeated application of first-layer antiserum improves immunofluorescence staining: a modification of the indirect immunofluorescence staining procedure. Histchem J 15:475–482

Hsu S-M, Raine L, Fanger H (1981) A comparative study of the peroxidase-antiperoxidase method and an avidin-biotin complex method for studying polypeptide hormones with radioimmunoassay antibodies. Am J Clin Pathol 75:734–742

Kaplan MH, Coons AH, Deane HW (1950) Localization of antigen in tissue cells. III Cellular distribution of pneumococcal polysaccarites Types II and III in the mouse. J Exp Med 91:15–30

Kraehenbühl JP, Jamieson JD (1974) Localization of intracellular antigens by immunoelectron microscopy. Exp Pathol 13:1–53

Kubalek E, Ralston S, Lindstron J, Unwin N (1987) Location of subunits within the acetylcholine receptor by electron image analysis of tubular torpedo marmorata. J Cell Biol 105:9–18

Jirikowski G (1985) Cyanogen bromide cleavage of methionine residues as a control method for enkephalin immunocytochemistry. Histochemistry 83:93–95

Larsson L-I (1979) Simultaneous ultrastructural demonstration of multiple peptides in endocrine cells by a novel immunocytochemical method. Nature 282:743–747

Larsson LI (1988) Immunocytochemistry: theory and practice. CRC Press, Boca Raton, Florida

Langanger G, Moeremans M, Daneels G, Sobieszek A, De Brabander M, De Mey J (1986) The molecular organization of myosin in stress fibers of cultured cells. J Cell Biol 102:200–209

Lewis V, Green SA, Marsh M, Vihko P, Helenius A, Mellman I (1985) Glycoproteins of the lysosomal membrane. J Cell Biol 100:1839–1847

Marrack J (1934) Nature of antibodies. Nature 133:292–293

Marshall JM (1951) Localization of adrenocostecote-opic hormone by histochemical and immunochemical methods. J Exp Med 94:21–29

McLean JD, Singer SJ (1970) A general method for the specific staining of intracellular antigens with ferritin-antibody conjugates. Proc Natl Acad Sci USA 65:122–128

Milstein C, Wright B, Cuello AC (1983) The discrepancy between the cross-reactivity of a monoclonal antibody to serotonin and its immunohistochemical specificty. Mol Immunol 20:113–123

Nakane PK, Pierce GB (1966) Enzyme-labeled antibodies: preparation and application for the localization of antigens. J Histochem Cytochem 14:929–931

Nakane PK, Pierce GB (1967) Enzyme-labelled antibodies for the light and electron microscopic localization of tissue antigens. J Cell Biol 33:307–315

Nigg EA, Walter G, Singer SJ (1982) On the nature of crossreactions observed with antibodies directed to defined epitopes. Proc Natl Acad Sci USA 79:5939–5943

Orci L, Ravazzola M, Amherdt M, Louvard D, Perrelet A (1985) Clathrin-immunoreactive sites in the Golgi apparatus are concentrated at the *trans* pole in polypeptide hormone-secreting cells. Proc Natl Acad Sci USA 82:5385–5389

Painter RG, Tokuyasu KT, Singer SJ (1973) Immunoferritin localization of intracellular antigens: the use of ultracryotomy to obtain ultrathin sections suitable for direct immunoferritin staining. Proc Natl Acad Sci USA 70:1649–1653

Porter EF, Pappenheimer AM (1939) Antigen-antibody reactions between layers adsorbed on built up stearate fillms. J Exp Med 69:755–765

Posthuma G, Slot JW, Geuze HJ (1986) A quantitative immuno-electronmicroscopic study of amylase and chymotrypsinogen in peri- and tele-insular cells of the rat exocrine pancreas. J Histochem Cytochem 34:203–207

Ravazzola M, Orci L (1980) Glucagon and glicentin immunoreactivity are topologically segregated in the α granule of the human pancreatic A cell. Nature 284:66–67

Romano EL, Romano M (1977) Staphylococcial protein A-coated colloidal gold particles for immunoelectron microscopic localization of ACTH on ultrathin sections. Histochemistry 60:317–320

Roth J (1989) Postembedding labeling on Lowicryl K4M tissue sections: Detection and modification of cellular components. Methods in Cell Biol 31:513–551

Roth J, Bendayan M, Orci L (1978) Ultrastructural localization of intracellular antigens by the use of protein A-gold complex. J Histochem Cytochem 26:1074–1081

Rothen A, Landsteiner K (1942) Serological reactions of proteins films and denatured proteins. J Exp Med 72:437–449

Schipper J, Tilders FJH (1983) A new technique for studying specificity of immunocytochemical procedures: specifity of secretogranin immunostaining. J Histochem Cytochem 31:12–18

Scopsi L, Larsson L-I (1985) Increased sensitivity in immunocytochemistry: Effects of double application of antibodies and of silver intensification on immunogold and peroxidase-antiperoxidase staining techniques. Histochemistry 82:321–329

Simons K, Virta H (1987) Perforated MDCK cells support intracellular transport. EMBO J 6:2241–47

Singer SJ (1959) Preparation of an electron-dense antibody conjugate. Nature 183:1523–1527

Singer SJ, Schick AF (1961) The properties of specific stains for electron microscopy prepared by the conjugation of antibody molecules with ferritin. J Biophys Biochem Cytol 9:519–537

Slot JW, Geuze HJ (1983) The use of protein A-colloidal gold (PAG) complexes as immunolabels in ultra-thin frozen sections. In: Cuello AC (ed) Immunohistochemistry. Wiley, Chichester, pp 323–346

Slot JW, Geuze HJ (1984) Gold markers for single and double immunolabelling of ultrathin cryosections. In: Polak JM, Varndell IM (eds) Immunolabelling for electron microscopy. Elsevier, Amsterdam, pp 129–142

Slot JW, Geuze HJ (1985) A new method of preparing gold probes for multiple-labeling cytochemistry. Eur J Cell Biol 38:87–93

Slot JW, Geuze HJ, Wehrkamp AJ (1988a) Localization of macromolecular components by application of the immunogold technique on cryosectioned bacteria. Methods Microbiol 20:211–236

Slot JW, Posthuma G, Chang LY, Crapo JD, Geuze HJ (1988b) Quantitative assessment of immuno-gold labelling in cryosections. In: Verkleij AJ, Leunissen JLM (eds) Immuno-gold labeling in cell biology. CRC Press, Boca Raton, Florida, pp 135–156

Slot JW, Posthuma G, Chang LY, Crapo JD, Geuze HJ (1989) Quantitative aspects of immunogold labeling in embedded and nonembedded sections. Am J Anat 185:271–281

Söllner T, Pfaller R, Griffiths G, Pfanner N, Neupert W (1990) A mitochondrial import receptor for the ADP/ATP carrier. Cell 62:107–115

Sternberger LA (1979) Immunocytochemistry, 2nd edn. Wiley, New York

Sternberger LA, Hardy PH, Cuculis JJ, Meyer HG (1979) The unlabelled antibody enzyme method for immunohistochemistry. Preparation and properties of soluble antigen-antibody complex (horseradish peroxidase-antihorseradish peroxidase) and its use in identification of sperochetes. J Histochem Cytochem 18:315–325

Suurmeijer AJH, Boon ME, Kok LP (1990) Notes on the application of microwaves in histopathy. Histochem J 22:341–346

Tapia FJ, Varndell IM, Probert L, de Mey J, Polak JM (1983) Double immunogold staining method in the ultrastructural localizations of regulatory peptides. J Histochem Cytochem 31:977–981

Tokuyasu KT (1973) A technique for ultracryotomy of cell suspensions and tissues. J Cell Biol 57:551–565

Tokuyasu KT, Singer SJ (1976) Improved procedures for immunoferritin labelling of ultrathin frozen sections. J Cell Biol 71:894–906

Tschopp, J, Podack ER, Müller-Eberhard HJ (1982) Ultrastructure of the membrane attack complex of complement: detection of the tetramolecular C9-polymerizing complex C5b-8. Proc Natl Acad Sci USA 79:7474–7478

Vaux DJT, Helenius A, Mellman I (1988) Spike-necleocapsid interaction in Semliki Forest virus reconstructed using network antibodies. Nature 336:36–42

Wabl MR (1974) Electron microscopic localization of two proteins on the surface of the 50 S ribosomal subunit of Escherichia coli using specific antibody markers. J Mol Biol 84:241–247

Wang BL, Larsson LI (1985) Simultaneous demonstration of multiple antigens by indirect immunofluorescence or immunogold staining. Histochemistry 83:47–56

Waitz W, Loidl P (1990) Cell-cycle dependent association of c-Myc protein with the nuclear matrix. Oncogene 6:29–35.

Young WW, Tamura Y, Wolock DM, Fox JW (1984) Staphylococcal protein A binding to the Fab fragments of mouse monoclonal antibodies. J Immunol 133:3163–3166

Chapter 8

Particulate Markers for Immunoelectron Microscopy

JOHN LUCOCQ

A wide range of particulate markers have been used successfully in immunoelectron microscopy on sections. In practice, however, two types of markers have
been most widely used, namely colloidal gold and ferritin. These will be
discussed at length and brief mention made of imposil.

8.1 Colloidal Gold

8.1.1 Introduction

Gold particles are now in widespread use for localizing cell components with
immunocytochemical techniques. Their popularity stems from their easy
visualization on biological specimens using a variety of microscopic methods
(see for example, Horisberger 1981) and, their ability to form stable complexes
with many of the macromolecules used in immunocytochemistry. The macromolecules may be affinity reagents such as protein A, antibodies, lectins or even
polysaccharides. Under appropriate conditions, they associate spontaneously
with gold particles without losing their specific binding properties. The result is
a gold complex that can be used in an immunocytochemical labelling sequence.
An example of such a complex is protein A-gold.

 Protein-A binds tightly to a wide range of antibody classes from different
species (see Chap. 9) and the protein A-gold complex can therefore be used to
localise an antibody already bound to an antigen in a cell or tissue section. By
means of this labelling sequence the location of gold particles reveals the
distribution of the antigen. There are many variations on this sequence in
immunocytochemistry but all depend on the preparation and use of active
complexes between gold and specific affinity vreagents.

 This part of the chapter deals with: (1) the properties and preparation of gold
particle suspensions (gold colloids); (2) the factors that influence formation of
protein-gold complexes (most affinity reagents are proteins) and (3) the
preparation of protein-gold complexes. It should be emphasized that the
preparation of most of these complexes is neither expensive nor time
consuming. Sufficient reagent for many experiments can be prepared within a
single day and, under appropriate conditions, can be stored for years. Emphasis
will be put on the basic principles. Lists of methods, which can be found in a

number of excellent reviews (e.g. Roth 1983a; Beesley 1989), are kept to a minimum.

8.1.2 Properties of Gold Colloids

8.1.2.1 Gold Particles in Microscopy –

Suspensions of finely divided gold particles, often termed gold colloids, can be produced by chemical reduction of a gold salt in aqueous solution producing a supersaturated solution of elemental gold from which crystals of metallic gold condense out. Careful manipulation of the reduction conditions enables the final size of the particles to be precisely controlled at will, yielding particles with diameters between 2 and 150 nm or more. The gold colloids most commonly used for immunocytochemistry have particle diameters between 3 and 80 nm and are red by transmitted light. Smaller particles (1.5–2.5 nm) are yellow/ brown, but larger ones, especially where the particle shape departs from spherical, are violet or blue. Red-gold colloids show a peak of absorption between 515 and 540 nm[1] with the peak moving to longer wavelengths as the particle diameter increases.

Electron Microscopy. In the transmission electron microscope the electron density of gold particles makes them an ideal marker for localization studies on ultra-thin sections of cells or tissues. In addition their particulate nature allows the labelling to be quantitated with the possibilty of estimating the amount of a particular protein molecule in a given structure (see Chap. 11).

For the preparation of protein-gold complexes used in immunoelectron microscopy, it is preferable to use gold particles with a defined size and narrow size variation. Such preparations have two main advantages: (1) they allow simultaneous double (and triple) labelling of cell components using gold particles of different sizes and, (2) they allow the conditions for protein-gold complexing to be standardized. An indication of the variability of a preparation is the coefficient of variation of the particle size (CV, the standard deviation of particle diameter/mean diameter). With current methods for preparing homogeneously sized gold particles the CV is usually less than 15% (Slot et al. 1988). In the past, non-statistical terms such as "monodisperse" and "polydisperse" have been used to describe the variation in particle size. However, these terms are better avoided because they do not give a quantitative measure of size variation.

The particle size must be selected with care. While a number of studies have shown that the smaller gold particles give a higher labelling density (Horisberger 1981; Slot and Geuze 1981; Slot and Geuze 1983; Van Bergen en Henegouwen and Leunissen 1986; Yokota 1988) it may not be possible to

---

[1] 1-nm gold particles do not show absorbance maximum in this range. Jan De Mey, EMBL, Heidelberg, pers. commun.

visualize the smallest probes on the specimens to be used. For example, on ultra-thin sections of biological specimens contrasted with uranium or lead salts, gold particles less than 4 nm in diameter are difficult to distinguish and for routine use it may be more convenient to use larger particles of 5 or 6 nm diameter.

Recently, there has been much interest in the possibility of using very small gold particles (approximately 1 nm diameter; complexes available from Janssen Pharmaceutica, Belgium) for immunolabelling studies because they should produce the highest labelling density. Such small markers are, however, difficult to detect on ultra-thin sections of biological specimens but they can be revealed by a silver enhancement procedure (Bastholm et al. 1986; Bienz et al. 1986; Namork and Heier 1989; Scopsi and Larsson 1985, see below). This technique causes the growth of the particles (see Stierhof and Schwarz 1991 for the method) to a size at which they can be visualized easily. At present, it is not clear exactly what proportion of the small gold particles are revealed by this procedure.

When double-labelling is performed, it is important that the particles of each gold complex are distinguishable in the electron microscope. In the range 3–15 nm this can usually be achieved when the larger particle diameter exceeds the smaller one by at least 60%. For example particles of 3, 5, 8 and 15 nm may be distinguished provided each colloid has a small variation in particle size. If there are any problems distinguishing the two particle populations the size distributions can be measured by electron microscopy.[2] Methods for performing double labelling experiments are given in Chapter 7.

Light microscopy. In light microscopy the particles themselves cannot be visualized directly due to their small size, but their red colour allows them to be used as a cytochemical stain (Lucocq and Roth 1985). The larger particles (e.g. 15 nm diameter and upwards) give a more intense red staining, whilst the smallest particles (<5 nm diameter) may not be visible at all.

The gold signal may be enhanced by coating gold particles with elemental silver generated by chemical reduction of silver salts (Danscher 1981; Holgate et al 1983; Hacker et al. 1988), after which even previously invisible gold labelling is revealed as a black coloured stain (Lucocq and Roth 1985). Theoretically, large increases in sensitivity can be achieved by combining high density of labelling obtained with the smallest particles and silver enhancement (see Fig. 2 in Chap. 10). A silver enhancement procedure is described in the Appendix.

Other more specialized imaging techniques such as dark field and epipolarization video enhanced contrast microscopy (Shotton 1988) have also been used to observe gold particles in the light microscope. In video enhanced contrast, the particles are visualized as diffraction limited spots (around 200 nm across)

---

[2] Particle size may be determined by stabilizing a gold colloid with excess Bovine Serum Albumin (BSA) (or any convenient protein) and adsorbing the complexes to the carbon coated plastic support film. For example, add 1 ml of gold colloid to 10 µl of 1% BSA in water. Apply coated EM grids to this solution for 5 min, rinse by floating the grids on two drops of 1 ml distilled water and drain dry on filter paper.

using a high-resolution, high-light level camera applied to a light microscope. By combining this method with either Normarski differential interference contrast (De Brabander et al. 1985, 1986, 1988) or epipolarization (Inoué et al. 1985), individual gold particles of at least 5 nm diameter have been visualized. Again, smaller gold particles may also be imaged using such methods after the particle size has been increased by a silver enhancement procedure (Hoefsmit et al. 1986).

8.1.2.2 Electrolyte Induced Aggregation

Electrolyte induced aggregation is an important property because it forms the basis for a test of successful protein-gold complexing. It is therefore discussed here in some detail.

In the absence of electrolytes gold particles in suspension are quite stable because their negative surface charges ensure that they do not stick to each other during their frequent encounters – repulsive forces always exceeding adhesive ones (see Verwey and Overbeek 1948). The surface charges are due to dissociation of salts such as $HAuCl_4$, $HAuCl_2$, $HAuOHCl$ or $HAuOH_2$ (Pauli 1949) or alternatively, reducing agents adsorbed from the reducing solution.

The addition of electrolytes to a gold colloid induces the particles to form aggregates, the colour changing from red to blue or black within seconds[3] This electrolyte-induced aggregation was observed over 100 years ago by Michael Faraday (1857) when he wrote: "Place a layer of the ruby fluid (gold colloid) in a clean white plate, dip the tip of a glass rod in a solution of common salt and touch the ruby fluid; in a few moments the fluid will become blue or violet-blue, and sometimes almost colourless ..." A key observation is that this colour change, and therefore the aggregation, can be prevented by the prior adsorption of macromolecules to the gold particles (Zsigmondy 1901; see below), an effect that can be used as a test for successful protein-gold complexing.

Electrolyte-induced aggregation is caused by cations because their positive charges neutralize the effects of the negative electrostatic charges that surround the particles, allowing the particles to adhere as they come close to each other (Verwey and Overbeek 1948; Overbeek 1977). The rate of adherence is maximal above a certain electrolyte concentration at which each gold particle collision yields an aggregate. This condition is termed "diffusion limited aggregation" or D.L.A because the rate of aggregation is limited mainly by gold particle diffusion. In D.L.A. the growth of the aggregates is initially very rapid but it soon slows down because single particles are depleted while slower diffusing, less numerous, aggregates appear progessively. In fact, the major part of the colour change takes only seconds, the size of aggregates increasing linearly with the logarithm of time elapsed (Lin et al. 1989).

---

[3] Unlike colloids with diameters 2 nm or more 1 nm particles are apparently unstable in the absence of electrolyte and can only be stabilized after complexing with proteins (Jan De Mey, pers. commun.).

The potency of cations in inducing D.L.A depends on their charge, so that di- and tri-valent cations are successively orders of magnitude more potent as inducers of particle aggregation. For example a given gold sol might show D.L.A. at 25 mM Na^+, 0.5 mM Ca^{2+} and 0.001 mM Al^{3+}. When the stabilization of gold particles by protein is assessed it is customary to use monovalent cations such as sodium in concentrations far in excess of that causing D.L.A. (e.g. 300 mM, see below).

Finally, there are other means of producing gold particle aggregation. For example, pyridine, a neutral molecule, probably causes aggregation by displacing adsorbed anions from the gold surface, thereby reducing the surface charges and increasing the likelihood that particle adhesion will take place. (Lin et al. 1989). Aggregation induced by ultracentrifugation is discussed below.

8.1.3 Preparation of Gold Colloids

Much emphasis has been placed on the need for clean glassware when preparing gold colloids (Faraday 1857; Roth 1983a). We usually select new flasks without marks or scratches on the glass that have not been used previously for other procedures. These are washed in double distilled water before use. With certain techniques, such as the citrate and tannic acid/citrate reductions described below, glass flasks and even plastic stirring bars may be used repeatedly to prepare the colloids. After each preparation the flasks are washed extensively with double glass-distilled water, dried and covered. When problems are encountered, siliconization of the glassware can be tried (e.g Faulk and Taylor 1971). This can be done using agents such as dimethyl-di-chlorosilane in 1 1 1,-trichlorethane (BDH,Poole, England) or Surfasil (Pierce, Rockford, Ill, USA).

Double glass-distilled water generally yields colloids with reproducible characteristics. All reducing solutions are made up immediately before use in distilled water using reagent grade chemicals and do not usually require filtration. The gold chloride solution is prepared directly from the whole amount supplied, because it is highly hygroscopic. A 1 or 5% solution (w/v) can be stored for years in stoppered glass flasks protected from light.

8.1.3.1 Methods

The two key processes that occur during gold colloid formation are particle nucleation and particle growth. The average particle size is largely determined by the number of "nuclei" at which particle growth starts and, the quantity of gold available for particle growth on these nuclei. The variation in particle size, on the other hand, is influenced by the relative timing of nuclei formation and particle growth. To obtain uniform-sized particles it is essential that rapid and uniform nucleation (which determines the total number of particles) is followed by uniform growth of all particles (which determines their size). Alternatively, if nucleation is not synchronous, the particles that nucleate first will grow to make relatively large particles while those that nucleate last will make relatively small particles. Although nucleation and growth are regulated by the temperature, pH and chemical nature of the reducing agent the precise mechanisms by which they exert their effects are unknown. Optimum conditions for preparing colloids used in immunocytochemistry can therefore only be determined experimentally (e.g.Slot and Geuze 1985).

Of the many published techniques for preparing colloidal gold only three are given in full here. Together, these give the researcher the possibility of preparing gold particles with a small particle size variation from about 2 to 150 nm diameter. These techniques are:

1. the citrate reduction method of Frens (1973),
2. the tannic acid/citrate reduction described by Slot and Geuze (1985) and,
3. thiocyanate reduction (Baschong et al. 1985)

Some other methods, such as the ascorbic acid reduction of Stathis and Fabrikanos (1958), are useful alternatives. Most tend to give a large variation in particle size but this can be reduced by purification of the complexes by gradient centrifugation (Slot and Geuze 1981; see below). The most important preparation methods are given in Table 1.

Citrate Gold (15–150 nm; Frens 1973). This technique produces gold particles with diameters of 12–15 nm and upwards. The particle size may be controlled by varying the final concentration of the citrate reducing agent (see Table 2). Increasing concentrations of citrate produce smaller and smaller particles until a size of approximately 12–15 nm is reached. The smaller particles have a smaller size variation so that on decreasing the particle diameter from 80 nm to 21 nm the CV drops from 15 to 11% (data from Goodman et al. 1981).

To prepare 12–15 nm particles 100 ml of 0.01% $HAuCl_4.2H_2O$ are heated in a glass flask until boiling, 5ml of 1% trisodium citrate $2H_2O$ are added rapidly with mixing. The solution initially turns blackish/blue due to nuclei formation

Table 1. Summary of methods for preparing gold colloids

Reducing agent	Particle sizes (nm)	Reference
Phosphorus	5–12[a]	Zsigmondy (1905)
Ascorbate	8–13[a]	Stathis and Fabrikanos (1958)
Citrate	15 to 150[b]	Frens (1973)
Borohydride	2–5[a]	Tschopp et al. (1982, Tschopp (1984)
Tannic acid/citrate	6	Muhlpfordt (1982)
Thiocyanate	1–3[a]	Baschong et al. (1985)
Tannic acid/citrate	3 to 15[b]	Slot and Geuze (1985)

[a] Paricles sizes are heterogeneous and vary over this range.
[b] Particle size can be regulated and homogeneous colloids of any size in this range can be prepared.

Table 2. Relationship between tri-sodium citrate concentration and particle size

Citrate (final conc. in %)	Particle size
0.015	25–30
0.010	40–45
0.006	70
0.0041	100
0.0032	150

Combined data from Frens (1973) and Goodman et al. (1981). These values should only be used as a guide since there can be significant variation in the final particle size from laboratory to laboratory.

Table 3. Relationship of amount of tannic acid to gold particle size in the Slot and Geuze (1985)

Particle size nm	Tannic acid Amount per 100 ml (ml of 1%)
3.5	5.00[a]
4.0	2.5[a]
5.5	1.00
6	0.5
7.5	0.25
10	0.08
11.5	0.05
14	0.025

[a] Use an equal volume of 25 mM K_2CO_3 in the reducing mixture.

but after about 10–15 min, completion of the colloid formation is indicated by a final change from violet/red to orange/red that is resistant to further boiling. When larger particles are prepared the colloids take up to 30 min to form.

Tannic Acid/Citrate Gold (3–15 nm; Slot and Geuze 1985). This is a variation on the citrate reduction technique and allows homogeneous sized gold particles of 3–15 nm to be prepared. Increasing amounts of tannic acid in the reducing mixture increases the number of nuclei that are formed before particle growth occurs, thereby reducing the final particle size.

The basic technique, to prepare 100 ml of colloid is as follows. Two solutions are made, a reducing solution and a gold chloride solution.

The reducing solution has (1) 4 mls of 1% (w/v) trisodium citrate. $2H_2O$ (2) variable volumes of 1%(w/v) tannic acid (3) distilled water added to make a final volume of 20 ml. The tannic acid is Aleppo tannin from nutgalls (Mallinkrodt, St. Louis, Mo., code 8835) (it is important to use this particular tannic acid). When 1 ml or more of tannic acid solution is used an equal amount of 25 mM K_2CO_3 is added to increase the pH. This allows formation of more nuclei, further reducing the final particle size (Table 3).

The gold chloride solution has 1 ml of 1% (w/v) $HAuCl_4.2H_2O$ made up to 80 ml with distilled water.

Both solutions are heated to 60°C and the reducing solution added to the vigorously stirred gold chloride solution. Stirring is very important, especially for particles in the smaller size range. Heating is continued until the resulting colloid boils gently. The colloids with the smallest particle sizes are formed rapidly but those with the largest particles may require up to 1 h of gentle boiling (with reflux) to complete their formation.

Further increases in gold particle size may be produced by adding more gold chloride after colloid formation. For example, an increase in the diameter of x% can be produced by adding $[(1.0 +x/100)^3 - 1^3]$ x (the original amount of gold choride added) to the colloid at 60°C (Slot, pers. commun.).

"Thiocyanate" Gold (2–3 nm; Baschong et al. 1985). The reduction procedure is a modification of the method of DeBroukere and Casimir (1948); 0.3 ml of 1 M NaSCN is added to 50 ml double distilled water containing 0.5 ml of 1% $HAuCl_4.2H_2O$ and 0.75 ml of 0.2 M K_2CO_3. This yellow colloid can be used immediately, yielding smaller particles of about 1.5 nm in diameter (Baschong and Wrigley 1989) or left overnight (12–15 h) in the dark, at room temperature, to produce 2.5 nm particles (CV approx. 15%). These colloids do not change colour after salt-induced aggregation but an increase in turbidity can be visualized by incident illumination on a dark background, or increased light scattering measured at 360 or 550 nm in a spectrophotometer. Sedimentation of these colloids can be achieved in most standard fixed angle centrifuge tubes at an average of 150000 Xg over 1 h.

8.1.3.2 Number and Concentration of Gold Particle/Complexes

It may be important to know the number or concentration of gold particles or gold complexes, for example, when protein-gold complexing or binding of protein-gold complexes to receptors are studied. Although various methods for counting gold particles have been proposed (e.g. X-ray microanalysis, Stols et al. 1980), most workers have employed a calculation method. This assumes, firstly, that all the gold salt has been reduced and, secondly, that the elemental gold is distributed throughout a population of spherical gold particles, the volume of which can be calculated (Frens 1973). Recently, an unbiased counting method that uses latex particles as a reference (Kehle and Herzog 1987) has shown however that this calculation method (see, for example, De Roe et al. 1987) underestimates the gold particle number. This new method has produced the following relationship between (1) gold particle concentration, (2) the size of the gold particle measured by electron microscopy and, (3) the absorbance of the colloid measured by spectrophotometry: A_{520} of 10^{12}/ml particles = 0.005 x r^2. This formula enables the concentration of gold particles and complexes to be calculated from the mean gold diameter, measured by electron microscopy, and the absorbance at 520 nm, measured in a spectrophotometer, without biased assumptions.

8.1.4 Factors that Influence Protein-Gold Complex Formation

Having prepared an appropriate gold colloid the next step is to prepare a complex between the gold particles and the affinity reagent to be used in the immunocytochemical staining sequence. The complexes are prepared by mixing the gold colloid with a solution of the affinity reagent, usually a protein or glycoprotein, which then adsorbs spontaneously to gold particles. It has become very clear recently that to prepare both stable, and active, complexes, the complexing conditions are critical. What follows is a discussion of how these conditions effect the protein-gold interaction. This should form a basis for selecting appropriate complexing conditions.

8.1.4.1 Protein Concentration

One of the principal factors affecting binding of protein to solid surfaces in aqueous solution is its concentration (Young et al. 1988). When the protein concentration is increased, over a series of individual binding experiments, the adsorption rate first increases steeply and then more slowly (see Fig. 1) (Ash 1978; Brash and Lyman 1969). Binding in the steep part of the curve is termed low concentration coverage. In this condition the affinity of adsorbed polymers is highest because multiple segments of the molecule are bound (Steinberg 1967; Soderquist and Walton 1980; Jonsson et al. 1987) and because distortion exposes the hydrophobic interior of the molecule to the surface (Norde and

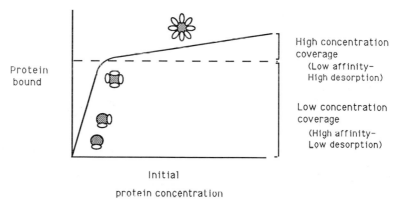

Fig. 1. Relationship between the amount of protein that is bound to a surface compared to the initial protein concentration during complexing. At low protein concentrations the amount of protein bound increases markedly as the protein concentration is increased (low concentration coverage). At some point the gradient of the curve becomes shallower but it does not flatten altogether (high concentration coverage). A hypothetical representation of complexes under different complexing conditions is superimposed. At low concentration proteins bind to the gold particles with a flat "side-on" orientation but at high concentrations this changes to an "end-on" interaction where affinity is reduced and desorption more marked

Lykema 1978; Jonsson et al. 1987; Makino et al. 1987; see discussion of protein-gold interaction below). As the protein concentration is further increased, increasing numbers of protein molecules bind to the surfae but at some point they start to crowd each other. Each molecule now binds to the surface through smaller segments and the affinity of the protein for the surface drops accordingly. The transition is represented by a "kink" in the binding curve (Soderquist and Walton 1980) which marks the beginning of a high concentration coverage region, a region where the binding curve becomes less steep (Fig. 1). A good example of the different affinities at high and low coverage comes from the protein A-gold system studied by Horisberger and Clerc (1985). In this case, the association constant of protein A for gold drops dramatically from 1.5 $\times 10^8 M^{-1}$ at low, compared to $8 \times 10^6 M^{-1}$ at high concentration coverage (see also Clerc et al. 1988; Young et al. 1988).

A favourable feature of low concentration complexing is that desorption of protein is much less than in the high concentration range (Soderquist and Walton 1980; Clerc et al 1988). For example, Horisberger and Clerc (1985) found that desorption of protein A from gold, measured over a 3-month period, increased from 1.6 to 47.6% as the number of molecules/gold particle was increased from 6.3 to 25 respectively (Horisberger and Clerc 1985)[4] Since the main effect of desorption is that free uncomplexed protein inhibits the binding of protein-gold complexes in the immunocytochemical labelling sequence, it therefore seems preferrable to prepare protein-gold complexes with low concentrations of protein.

Low concentration complexing may however have some potential disadvantages. For example inactivation of the protein could occur as a result of molecular distortion at the gold particle surface. Fortunately, for protein-gold complexes this effect seems to be rare (Horisberger and Rosset 1977; Horisberger and Vauthey 1984) despite the fact that some non-gold surfaces can produce striking inactivation of enzymes that is quite specific for low concentration coverage (Sandwick and Schray 1988).

One other potential problem with low concentration complexing is that a drop in activity can occur at very low protein concentrations, for example when there is less than one protein molecule per gold particle. This situation can easily be avoided, however, by simply raising the protein concentration which will increase the number of protein molecules bound to each gold particle.

Taking all these factors into account, the protein concentration during complexing should be kept low in order to reduce desorption but, to avoid complexes that contain too little protein, the concentration should probably be kept somewhere in the upper half of the low concentration coverage region. As yet, there is no simple test that can directly determine this concentration range, however, the minimal protein concentration which prevents electrolyte-induced aggregation turns out to be a rough guide to this condition (see below). This

---

[4]   In the case of IgG-gold, desorption remains minimal over a 6-week period (2.2 %, Geoghegan 1988).

concentration is probably close to the point where the gold first becomes fully coated with protein (see Fig. 1).

Many workers add a second, biologically inert, polymer after complexing (e.g. see Horisberger 1981). This appears to reduce aggregation of protein-gold complexes and may also improve their stability. Most often polyethylene glycol MW 20000 or Bovine Serum Albumin (BSA) has been used. The mode of action of these polymers is not entirely clear but they may prevent salt induced and/or ultracentrifugation induced aggregation (see below) by binding to bare areas of the gold surface.

8.1.4.2 Surfaces of Protein and Gold

The physico-chemical properties of the gold and protein surfaces will determine the non- covalent forces that hold them together: the gold surface is negatively charged but it is also hydrophobic, a property that probably results from adsorption of contaminants from the reducing mixture onto the native, hydrophilic gold (Smith 1980). The protein surface, in contrast, is hydrophilic and can carry uncharged, negative or positive regions, depending on the surface amino acid composition and the pH of the solvent. Together, these varied properties suggest a complex interaction involving electrostatic, hydrophobic and possibly other short acting intermolecular forces.

An attractive explanation for protein gold binding is that the positively charged regions on the protein bind to the negative gold surface (e.g. Faulk and Taylor 1971; Romano et al. 1974; Geoghegan and Ackerman 1977; Goodman et al. 1980, 1981). However, the observations that non-ionic polymers can still bind efficiently to gold particles (Horisberger 1983) and also, that the net charge of a protein does not seem to affect its affinity for the negatively charged gold (Horisberger and Clerc 1985), collectively suggest that electrostatic interactions cannot fully explain the protein gold interaction.

So what is the driving force for protein adsorption? Recent experiments on non-gold surfaces have provided some clues. A number of studies have shown that proteins bind equally, or even more extensively, to uncharged hydrophobic surfaces compared to charged hydrophilic ones (Soderquist and Walton 1980; Jonsson et al. 1987; Gölander and Kiss 1988) and adsorption is more extensive on a hydrophobic surface even when a hydrophilic surface has an opposite charge to that of the protein (Soderquist and Walton 1980). Similar results have also been obtained with fibrinogen, which orients not only its hydrophobic part towards negatively charged microcapsules but also its negatively charge-rich region (Makino et al. 1987). A hydrophobic driving force for protein binding to non-gold surfaces therefore seems most likely and appears to predominate over electrostatic interactions (Soderquist and Walton 1980). The latter seem to become important only when highly charged polymers such as poly-L-lysine bind to a surface possessing opposite charges (Gölander and Kiss 1988). The fact that low concentration adsorption on both hydrophobic gold particles and also non-gold surfaces show similar characteristics of high affinity and very low desorption strongly suggests that protein-gold binding is also mediated by hydrophobic interactions (Young et al. 1988).

One additional factor that may determine protein adsorption to surfaces is the length of the polymer chain. Its importance can be illustrated in binding studies using polymers of different lengths. For example, poly-L-alanine trimers adsorb poorly but longer polymers show dramatic increases in binding (Eirich 1977). Such observations might be explained by weak, short acting intermolecular forces (e.g. Van der Waals) which would become increasingly important as the polymer length increases. It is tempting to suggest that such forces drive the initial association of protein and gold, leading to the subsequent exposure and binding of hydrophobic parts of the protein.

Finally, a more specific interaction has been described between the gold surface and thiol groups of protein molecules (Bain and Whitesides 1988). This may account for the preferential association of small 1–2 nm gold particles with the hinge region of Fab fragments of immunoglobulins known to contain thiol groups (Baschong and Wrigley 1989).

In conclusion hydrophobic interactions are probably the principle type of interaction between protein and gold at low concentration binding which can be further modulated by electrostatic interactions and polymer length.

8.1.4.3 pH and Ionic Strength

In aqueous solution both the pH and ionic strength can influence the protein-gold interaction.

pH effects the distribution of charges on proteins but, as we have seen, charge interactions are probably not the major binding force between protein and gold. Therefore it is not surprising to find that pH has little effect on, either (1) the binding affinity of proteins for gold at both low and high concentration coverages (Horisberger and Clerc 1985; De Roe et al. 1987; Clerc et al. 1988) or, (2) on the stability of the protein-gold complex during storage (Clerc et al. 1988; Horisberger and Clerc 1985). In fact, pH has more marked effects on the amount of protein bound to a surface (Suzawa et al. 1982). As the pH approaches the isoelectric point of a protein, it becomes more compact and as a consequence, more molecules can "fit" onto a surface (Morrissey and Han 1978). Accordingly, more protein will complex to gold particles when the pH is close to the isoelectric point. This effect is illustrated very well with protein-A-gold complexing, where the number of molecules per 11 nm particle was shown to increase from 9 at pH 7.2 to 13 at pH 6.1 (Horisberger and Clerc 1985).

Ionic strength during protein-gold complexing is usually kept low (<10 mM monovalent cation) because of possible electrolyte-induced aggregation of the gold particles. Despite this potential problem there is clear evidence from studies of protein binding to non-gold surfaces that adsorption can occur in the presence of commonly used buffers such as phosphate buffer saline (see for example Gölander and Kiss 1988; Sandwick and Schray 1988). Most often the extent of binding is unaffected by high ionic strength (Jonsson et al. 1987;

Makino et al. 1987) but it can also be decreased (Soderquist and Walton 1980) or even increased (Jonsson et al. 1987). In the case of colloidal gold, proteins will also bind to the gold particles in the presence of buffer solutions such as phosphate buffered saline, and cacodylate buffer, provided the gold is added to a buffer solution that already contains the protein (Lucocq and Baschong 1986) – evidently, the formation of protein-gold complexes is much faster than electrolyte aggregation. Using this method, cytochemically active complexes can be prepared in commonly used buffer solutions (Lucocq and Baschong 1985). However, since the long term stability of these complexes has not been critically investigated and, in addition, increased desorption has been observed from non-gold surfaces after binding at high ionic strength (Soderquist and Walton 1980), protein-gold complexing in buffers should perhaps be reserved for special circumstances, e.g when proteins become unstable at low ionic strength.

Finally, it should be mentioned that gold colloids with smaller gold particles are less sensitive to electrolyte-induced aggregation (Frens 1972). This opens up the possibility of complexing after addition of buffers to the uncomplexed gold colloid. For example, 5 nm diameter gold particles are stable in 40 mM borate buffer pH 9 and can form complexes with immunoglobulins without excessive aggregation (Lucocq and Baschong 1986).

In conclusion, while pH has relatively minor effects on protein adsorption to surfaces including gold particles, it does influence the total amount that is bound with maximum amounts binding when the pH is close to the isolelectric point of the protein. In general, ionic strength should be kept low during complexing, although protein-gold complexes may be prepared in high ionic strength buffer solutions.

8.1.5 Preparation of Protein Gold Complexes

8.1.5.1 Selection and Adjustment of Complexing Conditions

The overall aims are to produce a complex with stable biological activity and single particle complexes. Taking into account the factors discussed above the following guidelines may be proposed.

1. Select the appropriate gold size for the experiment.
    Note that the particles should be easily visualised on the specimen and that the smaller gold complexes can produce higher amounts of specific labelling.
2. Adjust the pH and ionic strength of the protein and gold solutions.
    A useful starting point with a new protein is to select a pH close to the isoelectric point. Many workers use a pH just on the basic side of this value. When the isoelectric point is not known, a range of pH values can be tried (e.g 5–9 in 1-pH unit steps) but extremes of pH outside this range should be avoided because of colloid instability. Protein solutions can be adjusted to, or dialyzed against, 5–10 mM buffers such as sodium phosphate at the re-

quired pH for complexing. The gold pH may be adjusted by adding either 0.1 M HCl or 0.1 M NaOH (some workers use K_2CO_3 because it is less potent in causing electrolyte induced aggregation), or, small amounts of concentrated buffer. Buffers such as 0.2 M sodium phosphate may be added dropwise to vigourously stirred gold to a final concentration of 5–10 mM immediately before use to avoid slow electrolyte induced aggregation that can occur, especially with the large gold particles.

The pH of native gold is best measured with pH paper. pH electrodes can be used but the gold must be prestabilised before inserting the electrode because unstabilized gold colloid can impair the function of the electrode. One suggestion is to mix 10 µl of 5% BSA (Sigma fraction V), freshly prepared in distilled water, to a 5 ml aliquot of gold before inserting the electrode.

3. Determine the amount of protein required to stabilize the gold.

Electrolyte induced colour changes in red colloids are prevented by adsorption of protein to the gold particles (Zsigmondy 1901) an effect which forms the basis of a test for successful complexing.

In this test increasing amounts of a protein are mixed with colloidal gold in defined volumes at the appropriate pH and ionic strength. After a certain time, sodium chloride is added and the minimum amount of protein that prevents the colour change of electrolyte induced aggregation is noted. This is termed the minimal stabilizing amount, and for particle sizes of 3 nm and upwards it appears to be close to the low concentration coverage condition where the complexes are most stable (e.g. see Horisberger and Clerc 1985).

Visual Assay for Minimal Stabilizing Amount of Protein. This test is recommended for routine use and is performed as follows:

1. Serial 1/2 dilutions of protein are made in complexing buffer in volumes of 100 µl (approximately 20°C). Plastic or glass tubes, or alternatively, white plastic or porcelain blood typing wells (see Fig. 2) can be used.
2. 500 µl of gold sol (at appropriate pH) is added to each dilution and,
3. After 2 min 100 µl of 10% NaCl in distilled water is added.
4. 5–10 min later the colour of each dilution is compared with that of the colloid. With small sized colloids (3–5 nm) it may be necessary to wait up to 30 min in order to see significant colour changes. The tube with the lowest protein amount that prevents the electrolyte-induced colour change contains the minimal stabilizing amount (determined visually).

If more precision is necessary smaller steps in protein amount can be used. For example 1- or 2-µg steps below the minimal stabilizing amount determined in the above assay. Alternatively increasing amounts of protein can be added to a colloid until it becomes stable (Slot pers. commun.).

It should be noted that when particles less than 5 nm prepared by the tannic acid procedure are used, the colour changes are difficult to discern because the brown tannic acid masks the gold colour. This problem may be overcome by adding H_2O_2 to a final concentration of 0.2% (Slot and Geuze 1985). Alternatively the tubes can left overnight for the tannins to precipitate out.

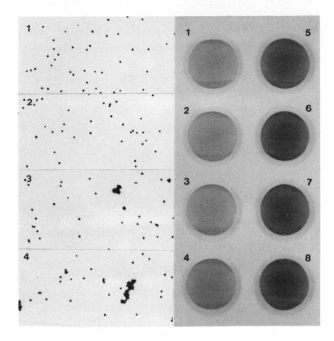

Fig. 2. Typical visual assay made with Bovine Serum Albumin (BSA) and 15 nm gold colloid made by citrate reduction. BSA was dissolved in 5 mM sodium phosphate (NaPi) buffer pH 5. To adjust the pH of the gold 1 ml of 0.2 M sodium phosphate buffer pH 5 was added dropwise with stirring to 39 ml of the gold colloid to give a final concentration of 5 mM. Serial dilutions of BSA were made in 100 µl aliquots of phosphate buffer and 500 µl gold added to each. After 2 min 100 µl 10% sodium chloride was added to each well and after 30 min the colour of each dilution was compared to the initial red. Well 1 to 8 contain 128, 64,32 16,8,4,2 and 1.0 µg BSA respectively. In this case, the fourth well would be taken as containing the minimal stabilizing amount, i.e 16 µg per 0.5 ml or 32 µg per ml of gold

Another way of assessing aggregation in these titrations is to apply the complexes to a plastic/carbon support film and observe them in a transmission electron microscope (Lucocq and Baschong 1986). The advantage of this method is that it can be performed on very small volumes and it may be more sensitive than the visual assay because it can detect small aggregates. However, it is not to be recommended for routine use because the amounts of protein that inhibit aggregation altogether may be much higher than those yielding stable complexes (at low concentration coverage). This test might, however, be useful if the precise aggregation state predicting stable complexes is known.

The procedure is as follows. Serial 1/2 dilutions of protein dissolved are made in buffer in a volume of 20 µl in plastic 1.5 ml Eppendorf centrifuge tubes and 20 µl of gold (appropriate pH) is added to each dilution. After 2 min 10 µl of 10% NaCl in distilled water are added and after a further 30 min an electron microscope grid coated with a plastic/carbon support film is placed on 10 µl removed from each tube (5 min). The grid is then washed on a 1 ml drop of distilled water and dried by draining, without blotting, on a filter paper.

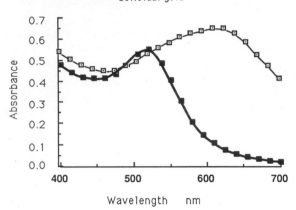

☐   Colloidal gold + NaCl

■   Colloidal gold

Fig. 3. Shift in absorbance spectrum of a gold colloid after electrolyte induced aggregation. Equal volumes of either distilled water or NaCl (final concentration 0.285 M) was added to a gold colloid consisting of 8 nm particles. After 10 min the absorbance spectrum was measured on a spectro-photometer

Two other methods of assessing aggregation deserve mention. These are: (1) spectrophotometry, which shows that the absorbance peak of red colloids is shifted from 515–540 nm to longer wavelengths when electrolyte induced aggregation occurs (see Fig.3) and, (2) laser spectroscopy (Chu et al. 1987; Lin et al. 1989), a specialized technique which gives information on the size of the aggregates and also the kinetics of aggregation.

8.1.5.2 Complex Preparation, Purification and Storage

The principle is to add native gold at the appropriate pH to a solution containing the affinity reagent (protein). The crude protein-gold complexes that are formed contain free uncomplexed protein and also aggregated complexes, both of which should be removed by a sutiable purification procedure[5], usually centrifugation.

The crude complex is prepared as follows:

1. The minimal stabilizing amount of protein from the visual assay is dissolved in 1 ml low molarity buffer, e.g. 10 mM phosphate buffer, at the appropriate pH for complex formation; some workers use distilled water without adjusting the pH. This solution can be placed in the bottom of the centrifuge tube in which the complex will be sedimented.
2. The appropriate amount of gold at the chosen pH is added rapidly[6].
3. After 2 min 10% BSA (w/v in distilled water) is added to the already formed complexes at a final concentration of 0.1%. This BSA solution can be

---

[5] The free protein competes with the gold complexes in the labelling sequence while the aggregates are responsible for poor resolution and diminished labelling efficiency.
[6] Alternatively, the protein can be added to an excess volume of stirred gold colloid.

prepared as a neutral solution, adjusted with NaOH. It can be centrifuged at 100000 g for 1 h to remove aggregates and stored at 4°C with 0.02% sodium azide (Slot pers. commun.). When working with lectins we use globulin free BSA (Sigma, Chemical Co.).

4. The crude complexes are centrifuged to concentrate them. When the gold particles are at least 15 nm in diameter they can be sedimented within 10 min on a bench Eppendorf centrifuge at approx. 14000 g (av). Smaller protein gold complexes do not sediment so rapidly and require ultracentrifugation which can be done in most standard sizes of ultracentrifuge tube in a fixed angle rotor at 100000 g(av) for 5 nm particle, 50000 g(av) for 10 nm and 20000 g(av) for 15 nm, each over 1 h.

5. After centrifugation, the largest aggregates are found in a tightly packed pellet on the side of the tube. Single particle complexes and small aggregates are in a loose pellet at the bottom. The supernatant is carefully removed, the loose pellet collected and adjusted to appropriate buffer conditions, e.g. PBS pH 7.2 in a final volume of about 1 ml, about 1/20th of the original volume.

In some cases, the complexes can be purified by repeated ultracentrifugation in a fixed angle rotor. However, the preferred method is to first concentrate the complexes by ultracentrifugation and then subject them to gradient centrifugation (Slot and Geuze 1981).

Gradient centrifugation (Slot and Geuze 1981) is performed in a 10 to 30% gradient of glycerol or sucrose prepared in a suitable buffer, such as phosphate buffered saline (PBS) but the buffer should be selected according to the requirements of the affinity reagent in the complex.

Bovine serum albumin (0.1%) is included.

1. The pellet from the first centrifugation is layered on top of the gradient and centrifuged in a suitable rotor e.g. SW40 (Beckman), 5 nm gold, 35 000 rpm and 10–15 nm, 20000 rpm, each for 45 min.

2. The single gold particle complexes are collected from the first 2–3 ml from the top of the red band in the gradient. The aggregates move faster through the gradient and are found further towards the bottom of the tube.

3. Gradient fractions can be stored at 4°C after addition of 0.02% (w/v) sodium azide. Protein A gold can remain active for more than one year when stored in this way. An alternative method is to freeze fractions from a glycerol gradient at −20°C or to adjust to 45% glycerol and store liquid at −20°C (Slot et al. 1988). Aliquots of the latter can also be stored frozen in liquid nitrogen indefinitely as can aliquots from sucrose gradients to which an equal volume of 2 M sucrose in PBS is added before freezing. Methods for lyophilization of protein gold complexes have been reported by Baschong and Roth (1985).

When ultracentrifuges are not available, osmosis can be used to concentrate the complexes, which can be further purified by gel filtration (Wang and Larsson 1985; Slot et al. 1988). The procedure for protein A-gold as suggested by Slot et al. (1988) is as follows. Take 20 ml of crude protein A-gold complex, containing 0.01% BSA, and put it in a dialysis bag. The bag is immersed in an excess volume of agitated 30% polyethylene glycol (20000 MW). The amount of water lost from the preparation can be followed by weighing the tube before and after. Normally, after approximately 2 h the 10–20 ×concentrated preparation is collected. A 14 ×0.7 cm column (ACA gel, LKB) is equilibrated with PBS pH 7.2 containing 0.1% BSA. Samples of about 25 µl of the concentrate are applied to the column. When run at a flow rate of 3 ml/h the red gold complex comes off after about 1 h. The separation of single particle complexes from aggregates may not be as good as with density gradient centrifugation.

Aggregation in the Complexes. Aggregation of protein-gold complexes can significantly reduce the final yield of complexes that contain single particles and when it is severe can prevent preparation of a complex altogether. One important cause is ultracentrifugation, after which aggregates appear as an dense spot at the bottom of fixed angled rotor tubes. The fact that this aggregation can occur under low ionic strength conditions suggests that the mechanism is distinct from electrolyte-induced aggregation. This type of aggregation is usually not a problem as long as gradient centrifugation can be used to remove the aggregates. Another problem is due to aggregation of the complexed protein molecules themselves. When viewed by transmission EM, the complexes characteristically show electron-lucent gaps between the parti-cles that represent the location of the protein molecules, in contrast to salt-induced aggregation where the gold particles are in direct contact with each other. This type of aggregation is especially marked with some lectins (*Ricinus Communis* I lectin) or with hydrophobic proteins such as the hyaluronic acid-binding link-protein (Eggli pers. commun.). One solution to this problem is to use a two step technique in which the biotinylated affinity reagent is then localised with streptavidin gold (Bonnard et al. 1984; see also Chap. 9).

Valency of Protein-Gold Complexes. The correct biological function of a protein may depend on its valency. For example, in endocytosis the correct routing of ligands bound to receptors often depends on monovalency of the ligand. Since most protein-gold complexes contain more than one protein molecule, they are multivalent and therefore may produce cross linking of receptors, triggering their internalization and artefactual re-distribution in the cell. Such effects have already been observed for insulin-gold and transferrin-gold complexes (e.g. Neutra et al. 1985; Griffiths et al. 1988).

One solution to this problem is to prepare complexes that behave as the monovalent ligand would do. The complexes are prepared from mixtures of the ligand and another protein that does not bind to the receptors, usually BSA. When the proportion of these two proteins is optimal, the concentration of the ligand on the gold surface is reduced sufficiently for the complexes to behave like the monovalent ligand. Such complexes have been made with insulin (Smith et al. 1988) and also Fab fragments of immunoglobulins (Baschong and Wrigley 1989).

8.1.5.3 Specific Protein-Gold Complexes

Most successful procedures for preparing gold protein complexes are similar to those outlined above: for the most commonly used gold reagents standard methods have now been established which are worth describing in detail, namely protein A–, immunoglobulin- and streptavidin-gold.

6Protein A-Gold. The protein A we use is from the Cowan strain of *Staphylococcus aureus* (Sigma, Poole Dorset, England) and is prepared at 1 mg in 1 ml of 10 mM NaPi buffer pH 6.0, the recommended pH for complexing (Horisberger and Clerc 1985). The stabilizing amount is 3–4 μg/ml of 5 nm gold (made using the tannic acid citrate procedure) decreasing to 2.0–2.5 μg for 9–10 nm particles (Slot pers. commun.). Enough protein A to stabilize 60 ml of colloidal gold (pH 6.0) is placed in 1 ml of sodium phosphate buffer (10 mM, pH 6.05) at the bottom of a polycarbonate centrifuge tube (45Ti rotor tube) and the colloidal gold rapidly added. Two min later 10% BSA, is added to a final concentration of 0.1%, the tubes topped up with distilled water and centrifuged to sediment the complexes. Centrifugation over 1 h in a 45Ti rotor tube requires mean values of 110000 *g* (5 nm), 60000 *g* (8 nm), and 20000 g (15 nm). After centrifugation, loose sediments are then loaded on to 10–30% gradients of glycerol or sucrose in PBS/0.1% BSA and centrifuged. We use polyallomer tubes in a SW40 rotor spun for one hour at 37000 rpm (5 nm), 25000 rpm (8 nm) or 15000 rpm (15 nm), each over 45 min. The top 3 ml of the coloured band on the gradient is collected and stored at 4 °C with 0.02% sodium azide and diluted immediately before use. As described above the glycerol concentration may be adjusted to approximately 50% or the sucrose to at least 1 M before freezing and storage in liquid nitrogen in 50 μl aliquots.

Fig. 4. Bovine serum albumin gold (8 nm) centrifuged into a 15 ml 10 30% glycerol gradient made in PBS. This gradient was centrifuged for 1 h at 50000 *g*(av). The single particle complexes were contained in the top 2–3 ml of the red band (arrowheads). The tube on the left is shown prior to the centrifugation. The tube volume is 13 ml and the amount of colloid added should give a band no wider than about 5 mm

Immunoglobulin Gold. Generally, pH 9 is recommended for complexing antibodies to gold because it is on the basic side of the range of isoelectric points of immunoglobulins (De Mey et al. 1981). However, findings from a number of laboratories suggest that useful complexes can also be prepared near to neutral pH (Romano et al. 1974; Baschong and Wrigley 1989). In the method of De Mey and co-workers (1981), affinity purified immunoglobulin solutions are adjusted to approximately 1 mg/ml and dialyzed against 2 mM borax-HCL buffer, pH 9. Protein aggregates can be removed by centrifugation at 100000 g, but this step may be unnecessary if gradient centrifugation is to be performed. The pH of the colloidal gold is adjusted to 9.0 with 0.2 M K_2CO_3 and the minimal stabilizing amount, determined visually, is added to the colloid. After 2 min BSA is added to a final concentration of 1% w/v and after centrifugation the sedimented complexes are resuspended in 20 mM Tris/HCl pH 7.5 containing 150 mM NaCl and 1% BSA. Centrifugation is either repeated or alternatively density gradient centrifugation can be performed.

Streptavidin Gold. The advantage of this labelling system is that avidin and streptavidin have a very high affinity for biotin (Green 1975). Streptavidin from *Streptomyces avidinii* (Chaiet and Wolf 1964) is preferred over avidin from eggs because the latter is "sticky", presumably due to its high isoelectric point (Finn et al. 1980; Guesdon et al. 1979) and difficult to complex to gold (Morris and Saelinger 1986; Roth 1983a; see Chap. 9).

The following procedure for preparing streptavidin-gold has been used successfully by Dr R. Parton (EMBL; see Brändli et al. 1990) 9 nm gold is prepared by the tannic acid/citrate method (Slot and Geuze 1985). 12 μg of streptavidin (Sigma Chemical Company) in 100 μl of distilled water is added to 1 ml of the gold solution adjusted to pH 6.6. The streptavidin-gold is then washed three times with PBS containing 0.2% BSA (w/v) (PBS/BSA) by centrifugation at 32000 g for 30 min to remove free streptavidin, and finally for 5 min at 13000 g to remove any aggregates.

Streptavidin-gold prepared in this way seems to be unstable. After more than a week of storage at 4°C (or longer when stored in 50% glycerol at −20°C) the labelling intensity is decreased, probably because free streptavidin, which has dissociated from the gold, competes for biotin molecules on the sections. The labelling intensity recovers when the gold complex is washed with 30 ml PBS/BSA and centrifuged for 30 min at 32 000 rpm (Beckman Ti70 rotor). Even after repeated centrifugation aggregation of the gold is not a problem.

Lectin, Glycoprotein and Enzyme Gold. The preparation of lectin and glycoprotein complexes (reviewed by Roth (1983a,b) follows closely the general principles outlined above. One important practical point is that when tannic acid/citrate gold is used for the production of lectin or glycoprotein gold complexes, non-specific labelling may be high, which can be avoided by treatment of the gold sol with hydrogen peroxide before complexing or by the use of ascorbic acid gold (Sata et al. 1989).

Concanvalin A appears to be inactivated by glycerol and sucrose used on gradients although in the case of sucrose the binding activity can be recovered by dialysis. *Ricinus communis* I lectin gold complexes may retain their binding activity in the presence of sucrose (Lucocq and Baschong 1986).

Bendayan has pioneered the use of enzyme gold complexes and procedures for their preparation and use are given in Bendayan (1985) and Chapter 9 respectively.

8.1.6 Ferritin

Ferritin is a large iron-storage protein (MW 750 kDa)) found in vertebrate spleen and liver, and to a lesser extent in bone marrow, kidney, testis and other tissues (Granick 1946; Harrison 1959). The spleen is the most abundant source, from which 0.1 g of crystalline ferritin can be obtained from each 100 g. The ferritin particle itself is 11 nm in diameter and consists of an outer protein shell, termed apoferritin (46 kDa) and an iron rich central core. The latter contains colloidal micelles of iron hydroxide and iron phosphate (Granick 1946) that are responsible for the electron density of the particle, measuring 5.5 nm in diameter.

Although ferritin can easily be visualized on the surface of cells, it is often difficult to distinguish from cell structures in ultra-thin sections contrasted with heavy metals. This is especially a problem in ultra-thin frozen sections and it is for this reason that colloidal gold, which is more electron-dense, has become more popular than ferritin. Gold has the additional advantage that, unlike ferritin, it is not naturally found in tissues.

Singer (1959) first proposed the idea of coupling ferritin covalently to antibodies in order to visualize them in the electron microscope (Singer and Schick 1961). In a two stage conjugation, these authors first coupled the cross linking reagent toluene 2,4-diisocyanate (TC) to ferritin at pH 7.5, removed the unreacted TC, and then added the ferritin-TC conjugate to IgG at pH 9, at which pH the second reactive group of TC is active. Finally, unreacted IgG was removed by gel filtration or centrifugation. Although this conjugation procedure is still used many workers now prefer to use glutaraldehyde instead of TC (Avrameas 1969).

The first application of this technique was to visualize antibodies against Tobacco Mosaic Virus (TMV) that were directly coupled to ferritin. By measuring the average distance from the surface of TMV to the ferritin particles, Singer and Schick concluded that the effective resolution was between 20 and 30 nm. Subsequently, this group was able to make an important contribution to membrane studies by directly coupling lectins to ferritin (Nicolson and Singer 1971). In recent years, most workers have found it more convenient to use ferritin indirectly, i.e. coupled to a secondary antibody (e.g. Geuze and Slot 1980).

8.1.6.1 Method for Conjugating Protein to Ferritin

(Kishida et al. 1975; modified by Dr Anne Dutton, University of California, San Diego, pers. commun.).

1. Between 20 and 30 mg ferritin in 2 ml of 0.1M NaPi pH 7.5 (phosphate buffer) are added to 1 ml 50% glutaraldehyde and the solution stirred for 30 min at 22°C.
2. This crude conjugate is then centrifuged at 12000 g (5°C) for 10 min to remove aggregated protein.
3. Activated ferritin is separated from glutaraldehyde on G50 column in phosphate buffer (G50 column 50×1.5cm) and the ferritin peak pooled using the absorbance at 440 nm.
4. Between 15 and 20 mg of this activated ferritin in phosphate buffer is added to 4–4.5 mg antibody and ^{125}I-labelled carrier protein. The total reaction volume should contain ferritin at 1.6 mg/ml, and antibody at 0.4 mg/ml. React for 72 h at 4°C.
5. The reaction is then quenched by making the reaction mixture 0.1 M in Tris HCl buffer pH 7.5 by adding sufficient 1.0 M Tris HCL buffer pH 7.5.and is then centrifuged at 10000 g for 10 min to remove aggregates.
6. The ferritin antibody conjugates are separated from the unconjugated protein on a Sepharose 6B column (2.5×90 cm) equilibrated and eluted in 0.1 M Tris-HCl buffer at 4°C at 8 ml/hr. The ferritin concentration can be determined by absorbance at 440 nm and the antibody protein concentration by radioactivity measurement. Pool the peak containing the conjugate and concentrate it by centrifugation at 180000 g for 1.5 h.

N.B. Both the final concentrations of ferritin and antibody in step 4 and the ratio of ferritin to antibody are important. Antibody preparations should be centrifuged before use at 100000 g for 1 h. The use of the ^{125}I-labelled carrier protein is not obligatory since the usable fractions can be selected in step 7 using the 440 nm elution pattern.

8.1.7 Imposil

Imposil is the commercial name for an iron-dextran preparation made and sold by Fisons Ltd. The Imposil particle consists of an electron-dense core (3 ×11 nm) of ferric hydroxide complexed to and surrounded by an electron-translucent shell of dextran. The overall size of the particle is 12×21 nm (Cox et al. 1972). As is evident from these dimensions, the Imposil particle is rod-shaped and hence easily distinguishable from ferritin, gold or any other spherical particle. It was precisely for this reason that Dutton et al. (1979) introduced Imposil as an electron-dense marker which could be used in double-labelling antibody studies. These workers conjugated Imposil covalently to antibodies. The basis for the procedure was to partially oxidize the dextran shell using periodate. The aldehyde groups thus formed were allowed to react with free amino groups in the antibody molecules to form Schiff bases. The latter were then reduced to stable secondary amine linkages at the same time that unreacted aldehyde groups were reduced to alcohols using borohydride.

The conjugates prepared consisted predominantly of one Imposil particle linked covalently to one antibody molecule together with some unconjugated Imposil that does not apparently interfere with labelling.

The combination of ferritin and Imposil was subsequently used by this group in a series of elegant double labelling experiments including one on the nature of plasma membrane-microfilament interactions using antibodies against actinin, tropomyosin and vinculin (Geiger et al. 1981; Tokuyasu et al. 1981, 1983).

Appendix. Silver Enhancement Procedure

This particular procedure uses silver acetate and may be performed in daylight or normal laboratory lighting conditions. Cleanliness is important and double-distilled water is used throughout the protocol until the fixation step (Skutelsky 1987; Hacker et al. 1988; Roth 1989).The procedure is performed as follows:

1. A 0.2% silver acetate solution is prepared (Solution A).
2. The specimens containing the gold particles (e.g. sections mounted on glass slides) are presoaked in an aliquot of 0.5% w/v hydroquinone in 0.5 M citrate buffer pH 3.8 (solution B). for 2–5 min.
3. The specimens are transferred into a clean vessel containing the developer –0.1% silver acetate, 0.25% w/v hydroquinone in 0.25 M citrate buffer pH 3.8 which is prepared immediately prior to use by mixing equal volume of solutions A and B. The developing time may be varied but a useful starting point for most preparations is 18 min. The preparations should be kept vertical or inverted, as precipitates are generated during the development.
4. Following a brief rinse in distilled water the specimens are immersed for 1–2 min either in a photographic fixative diluted according to the manufacturers instructions or 20% w/v aqueous sodium thiosulphate containing 15% w/v sodium sulphite (Springall et al. 1984)
5. After a final rinse in distilled water, the specimens may be contrasted with appropriate stains if necessary (e.g. for light microscopy nuclear fast red may be used).
   N.B. The developing process may be slowed by the addition of gum arabicum to the developing solution. A 50% w/v aqueous stock solution is prepared by dissolving raw acacia resin in distilled water over 5 days with intermittent agitation. Five parts hydroquinone/citrate buffer (solution B) is supplemented with between one to four parts gum arabicum solution before adding an equal volume of silver acetate solution (solution A) (Hacker et al. 1988). The gum arabicum may be removed form the preparations with water at 40°C after the photographic fixation step.

References

Ash SG (1978) Polymer adsorption at the solid/liquid interface. Colloid Sci 1:103–112

Avrameas S (1969). Coupling of enzymes to protein with glutaraldehyde. Use of the conjugates for the detection of antigens and antibodies. Immunochemistry 6:43–52

Bain CD, Whitesides GM (1988) Molecular-level control over surface order in self-assembled monolayer films of thiols on gold. Science 240:62–63

Baschong W, Roth J (1985) Lyophilization of protein-gold complexes. Histochemical J 17:1147–1153

Baschong W, Wrigley NG (1989) Small colloidal gold conjugated to Fab fragments or to immunoglobulin G as high resolution labels for electron microscopy. A technical overview. J Electron Microsc Tech 14:313–323

Baschong W, Lucocq JM, Roth J (1985) "Thiocyanate gold": small (2–3 nm) colloidal gold for affinity cytochemical labeling in electron microscopy. Histochemistry 83:409–411

Bastholm L, Scopsi L, Nielsen MH (1986) Silver-enhanced immunogold staining of semithin and ultra-thin cryosections. J Electron Microsc Tech 4:175–176

Beesley JE (1989) Colloidal gold: a new perspective for cytochemical marking. Royal Mircroscopical Society. Oxford University Press, Oxford

Bendayan M (1985) The enzyme-gold technique: a new cytochemical approach for the ultrastructural localization of macromolecules. In: Bullock GR, Petrusz P (eds) Techniques in Immunocytochemistry, vol 3. Academic Press, London, pp 180–201

Bienz K, Egger D, Pasamontes L (1986) Electron microscopic immunocytochemistry: silver enhancement of colloidal gold marker allows double labeling with the same primary antibody. J Histochem Cytochem 34:1337–1342

Bonnard C, Papermaster DS, Kraehenbuhl J-P (1984) The streptavidin-biotin bridge technique. Application in light and electron microscope immunocytochemistry. In: Polak JM, Varndell IM (eds) Immunolabelling for electron microscopy. Elsevier, Amsterdam, pp 95–111

Brändli AW, Parton RG, Simons K(1990) Transcytosis in MDCK cells: identification of glycoproteins transported bidirectionally between both plasma membrane domains. J Cell Biol 111:2909–2921

Brash JL, Lyman DJ (1969) Adsorption of plasma proteins in solution to uncharged, hydrophobic polymer surfaces. J Biomed Mater Res 3:175–189

Chaiet L, Wolf FJ (1964) The properties of steptavidin, a biotin binding protein produced by Streptomycetes. Arch Biochem Biophys 106:1–5

Chu B, Xu R, DiNapoli A (1987) Light scattering studies of a colloidal suspension of iron oxide particles. J Colloid Interface Sci 116:182–195

Clerc M-F, Granato DA, Horisberger M (1988) Labelling of colloidal gold with IgE. A quantitative study using monoclonal IgE anti-beta-lactoglobulin and evaluation of the biological activity of the gold complex with RBL-1 cells. Histochemistry 89:343–349

Cox JS, Kennedy GR, King J, Marshall PR, Rutherford D (1972) Structure of an iron-dextran coplex. J Pharm Pharmacol 24:513–517

Danscher G (1981) Localization of gold in biological tissue. A photochemical method for light and electron microscopy. Histochemistry 71:81–88

De Brabander M, Geuens G, Meydens R, Moeremans M, De Mey J (1985) Probing microtubule dependent intracellular motility with nanometre particle video ultramicroscopy (nanovid ultramicroscopy). Cytobios 43:273–283

DeBrabander M, Nuydens R, Gueuns G, Moeremans M, De Mey J (1986) The use of submicroscopic particles combined with video contrast enhancement as a simple molecular probe for the living cell. Cell Motil Cytoskel 6:105–113

De Brabander M, Nuydens R, Geerts H, Hopkins CR (1988) Dynamic behaviour of the transferrin receptor followed in living epidermoid carcinoma (A431) cells with nanovid microscopy. Cell Motil Cytoskel 9:30–47

de Brouckère L, Casimir J (1948) Influence des electrolytes sur la stabilité des hydrosols d'or. Bull Soc Chim Belg 57:547–554

De Mey J, Moeremans M, Geuens G, Nuydens R, De Brabander M (1981) High resolution light and electron microscopic localization of tubulin with the IGS (immunogold method). Cell Biol Int Rep 5:889–899

De Roe C, Courtoy P, Baudhuin P (1987) A model of protein-colloidal gold interactions. J Histochem Cytochem 35:1191–1198

Dutton A, Tokuyasu KT, Singer SJ (1979) Iron-dextran antibody conjugates. General method for simultaneous stained of two components high resolution immunoelectron microscopy. Proc Natl Acad Sci USA 76:3392–3396

Eirich FR (1977) The conformational states of macromolecules adsorbed at solid-liquid interfaces. J Colloid Interface Sci 58:423–435

Faraday M (1857) Experimental relations of gold (and other metals) to light. Philos Trans R Soc 147:145–181

Farrant JI (1954) An electron microscopic study of ferritin. Biochem Biophys Acta 12:564–570

Faulk WP, Taylor GM (1971) An immunocolloid method for the electron microscope. Immunochemistry 8:1081–1083

Finn FM, Titus G, Hoffman K (1980) Hormone receptor studies with avidin and biotinyl-insulin-avidin complexes. J Biol Chem 255:5742–5746

Frens G (1972) Particle size and sol stability in metal colloids. Kolloid-Z. u. Z. Polymere 250:736–741

Frens G (1973) Controlled nucleation for the regulation of the particle size in monodisperse gold suspensions. Nature 241:20–22

Geiger B, Dutton AH, Tokuyasu KT, Singer SJ (1981) Immunoelectronmicroscope studies of membrane microfialment interactions: distribution of apha actinin, tropomysoin and binculin in intestinal epithelial brush border and chicken gizzard smooth muscle cells. J Cell Biol 91:614–628

Geoghegan WD (1988) The effect of three variables on adsorption of rabbit IgG to colloidal gold. J Histochem Cytochem 36:401–407

Geoghegan W, Ackerman GA (1977) Adsorption of horseradish peroxidase, ovomucoid and anti-immunoglobulin to colloidal gold for the indirect detection of concanavalin A, wheat germ agglutinin and goat anti-human immunoglobulin G on cell surfaces at the electron microscopic level: a new method, theory and application. J Histochem Cytochem 25:1187–1200

Geuze III, Slot JW (1980) Disproportionate immunostaining patterns of two secretory protein in guinea pig and rat exocrine pancreas cells. An immunoferritin and fluorescence study. Eur J Cell Biol 21:91–100

Gölander C-G, Kiss E(1988) Protein adsorption on functionalized and ESCA-characterized polymer films studied by ellipsometry. J Colloid Interface Sci 121:240–253

Goodman SL, Hodges GM, Livingstone DC (1980) A review of the colloidal gold marker system. Scanning Electron Microsc 2:133–146

Goodman SL, Hodges GM, Trejdosiewicz LK, Livingston DC (1981) Colloidal gold markers and probes for routine application in microscopy. J Microsc 123:201–213

Granick S (1946) Ferritin: its properties and significance for iron metabolism. Chem Rev 38:379–403

Green NW (1975) Avidin. Adv Protein Chem 29:85–133

Griffiths G, Hoflack B, Simons K, Mellman I, Kornfeld S (1988) The mannose phosphate receptor and the biogenesis of lysosomes. Cell 52:329–341

Guesdon J-L, Ternynck T, Avrameas S (1979) The use of avidin-biotin interaction in immunoenzymatic techniques J Histochem Cytochem 27:1131–1139

Hacker GW, Grimelius L, Danscher G, Bernatzky G, Muss W, Adam H, Thurner J (1988) Silver acetate autometallography: an alternative enhancement technique for immunogold-silver staining (IGSS) and silver amplification of gold, silver, mercury and zinc in tissues. J Histotechnol 11:213–221

Harrison PM (1959) The structures of ferritin and apoferritin: some preliminary X-ray data. J Mol Biol 1:69–80

Hoefsmit ECM, Korn C, Blijleven N, Ploem JS (1986) Light microscopical detection of single 5 and 20 nm gold particles used for immunolabelling of plasma membrane antigens with silver enhancement and reflection contrast. J Microsc 143:161–169

Holgate CS, Jackson PS, Cowen PN, Bird CC (1983) Immunogold-silver staining: a new method of immunostaining with enhanced sensitivity. J Histochem Cytochem 31:939–944

Horisberger M (1981) Colloidal gold: a cytochemical marker for light and fluorescent microscopy and for transmission and scanning electron microscopy. Scanning Electron Microsc II:9–31

Horisberger M (1983) Colloidal gold as a tool in molecular biology. TIBS 8:395–397

Horisberger M (1984) Lectin cytochemistry. In: Polak JM, Varndell IM (eds) Immunolabeling for electron microscopy, chap 17. Elsevier, Amsterdam, pp 249–258

Horisberger M, Clerc M-F (1985) Labelling of colloidal gold with protein A. Histochemistry 82:219–223

Horisberger M, Rosset J (1977) Colloidal gold a useful marker for transmission and scanning electron microscopy. J Histochem Cytochem 25:295–305

Horisberger M, M Vauthey (1984) Labelling of colloidal gold with protein. Histochemistry 80:13–18

Inoué S, Bajer AS, Mole-Bajer J, Debrabander M, De Mey J, Nuydens R, Ellis GW, Horn E, Inoué TD (1985) Microtubules decorated with 5 nm gold visualised by video-enhanced light microscopy. J Cell Biol 101:146a

Jonsson U, Lundstrom I, Ronnberg I(1987) Immunoglobulin G and secretory fibronectin adsorption to silica. The influence of conformational chages on the surface. J Colloid Interface Sci 117:127–138

Kehle T, Herzog V(1987) Interaction between protein-gold complexes and cell surfaces: a method for precise quantitation. Eur J Cell Biol 45:80–87

Kishida Y, Olsen BR, Berg R, Prockop DJ (1975) Two improved methods for preparing ferritin-protein conjugates for electron microscopy. J Cell Biol 64:331–339

Lin MY, Lindsay HM, Weitz DA, Ball RC, Klein R, Meakin P(1989) Universality in colloid aggregation. Nature 338:360–362

Lucocq JM, Baschong W (1986) Preparation of protein colloidal gold complexes in the presence of commonly used buffers. Eur J Cell Biol 42:332–337

Lucocq JM, Roth J (1985) Colloidal gold and collidal silver. Metallic markers for light microscopic histochemistry. In: Bullock GR, Petrusz P (eds) Techniques in immunocyto-chemistry, vol 3. Academic Press, London, pp 204–234

Makino K, Ohshima H, Kondo T (1987) Interaction of poly(L-lactide) microcapsule surface with plasma proteins: reversal of zeta potential caused by fibrinogen. J Colloid Interface Sci 115:65–72

Morris RE, Saelinger CB (1986) Problems in the production and use of 5 nm avidin-gold colloids. J Microsc 143:171–176

Morrissey BW, Han CC (1978) The conformation of gamma-globulin adsorbed on polystyrene latices determined by quasielestic light scattering. J Colloid Interface Sci 65:423–431

Muhlpfordt H (1982) The preparation of colloidal gold particles using tannic acid as a additional reducing agent. Experientia 38:1127–1128

Namork E, Heier HE (1989) Silver enhancement of gold probes (5–40 nm): Single and double labeling of antigenic sites on cell surfaces imaged with backscattered electrons. J Electron Microsc Tech 11:102–108

Neutra MR, Ciechanover A, Owen LS, Lodish HF (1985) Intracellular transport of transferrin- and asioloorosomucoid-colloidal gold conjugates to lysosomes after receptor-mediated endocytosis. J Histochem Cytochem 33:1134–1144

Nicholson GL, Singer SJ (1971) Ferritin-conjugated plant agglutinins as specific saccharide stains for electron microscopy: application to saccharides bound to cell membranes. Proc Natl Acad Sci USA 68:942–945

Norde W, Lykema J (1978) The adsorption of human plasma albumin and bovine pancreas ribonuclease at negatively charged polystyrene surfaces. I Adsorption isotherms Effects of charge, ionic strength and temperature. J Colloid Interface Sci 66:257–265

Overbeek JTG (1977) Recent developments in the understanding of colloid stability. J Colloid Interface Sci 58:408–422

Pauli W (1949) Konstitution und Farbe des kolloiden Goldes. Helv Chim Acta 32:795–810

Romano EL, Stolinski C, Hughes-Jones NC (1974) An antiglobulin reagent labelled with colloidal gold for use in electron microscopy. Immunochemistry 11:521–522

Roth J (1983a) The colloidal gold marker system for light and electron microscopic cytochemistry. In: Bullock GR, Petrusz P (eds) Techniques in immunocytochemistry, vol 2. Academic Press, London, pp 217–284

Roth J (1983b) Application of lectin-gold complexes for electron microscopic localization of glycoconjugates on thin sections J Histochem Cytochem 31:987–999

Roth J (1989) Postembedding labeling on Lowicryl K4M sections: detection and modification of cellular components. Methods Cell Biol 31:513–551

Sandwick RK, Schray KJ (1987) The inactivation of enzymes upon interaction with a hydrophobic latex surface. J Colloid Interface Sci 115:130–137

Sandwick RK, Schray KJ (1988) Conformational states of enzymes bound to surfaces. J Colloid Interface Sci 121:1–12

Sata T, Lackie PM, Taatjes DJ, Peumans W, Roth J (1989) Detection of Neu 5 Ac (alpha 2,3) Gal (beta1,4) GlcNac sequence with Leukoagglutinin from Maackea amurensis: light and electron microscopic demonstration of differential tissue expression of terminal sialic acid in alpha2,3 and alpha2,6 linkage. J Histochem Cytochem 37:1577–1588

Scopsi L, Larsson L-I (1985) Increased sensitivity in immunocytochemistry: Effects of double application of antibodies and of silver intensification on immunogold and peroxidase-antiperoxidase staining techniques. Histochemistry 82:321–329

Shotton DM (1988) Review: video enhanced light microscopy and its applications in cell biology. J Cell Sci 89:129–150

Singer SJ, Schick AF (1961) The properties of specific stains for electron microscopy prepared by the conjugation of antibody molecules with ferritin. J Biophys Biochem Cytol 9:519–537

Skutelsky E (1987) The use of avidin-gold complex for light microscopic localisation of lectin receptors. Histochemistry 86:291–295

Slot JW, Geuze HJ (1981) Sizing of protein A-colloidal gold probes for immunoelectron microscopy. J Cell Biol 90:533–536

Slot JW, Geuze HJ (1983) The use of protein A-colloidal gold (pAg) complexes as immunolabels in ultra-thin sections. In:Cuello AC (ed) Immunohistochemistry, IBRO Handbook Series. Wiley, Chichester, England, pp 323–340

Slot JW, Geuze HJ (1985) A new method of preparing gold probes for multiple-labelling cytochemistry. Eur J Cell Biol 38:87–93

Slot JW, Geuze HJ, Weerkamp AJ (1988) Localization of macromolecular components by application of the immunogold technique on cryosectioned bacteria. Methods Microbiol 20:211–236

Smith RM, Goldberg RI, Jarett L (1988) Preparation and characterization of a colloidal gold-insulin complex with binding and biological activities identical to native insulin. J Histochem Cytochem 36:359–365

Smith T (1980) The hydrophilic nature of a clean gold surface. J Colloid Interface Sci 75:51–55

Soderquist ME, Walton AG (1980) Structural changes in proteins adsorbed on polymers surfaces. J Colloid and Interface Sci 75:386–397

Springall DR, Hacker GW, Grimelius L, Polak JM (1984) Investigation of immunogold-silver staining by electron microscopy. Histochemistry 83:545–550

Stathis EC, Fabrikanos A (1958) Preparation of colloidal gold. Chem Ind (Lond) 27:860–861

Steinberg G (1967) On the configuration of polymers at the solid-liquid interface. J Phys Chem 71: 292–330

Stierhof Y-D, Schwarz H (1991) Suitability of different silver enhancement methods applied to 1 nm colloidal gold particles: an immunoelectron microscopy study. J Electron Microsc Tech 17:336–343

Stols ALH, Schalken JJ, Stakhouders AM (1980) Quantitation of colloidal gold as a immunolabel with X-ray microanalysis. Electron Microsc 3:80–81

Suzawa TH, Shirahama R, Fujimoto T(1982) J Colloid Interface Sci 86:144

Tokuyasu KT (1980) Visualization of longitudinally orientated inetermediate filaments in frozen sections of chicken cardiac muscle by a new staining method. J Cell Biol 97:562–565

Tokuyasu KT, Dutton A, Geiger B, Singer SJ (1981) Ultrastructure of chicken cardiac muscle as studied by double immunolabeling in electron microscopy. Proc Natl Acad Sci USA 78:7619–7623

Tokuyasu KT, Dutton AH, Singer SJ (1983) Immunoelectron micrsocopy studies of desmin (skeletin) localization and intermediate filament organization in chicken cardiac muscle. J Cell Biol 97:1736–1742

Tschopp J (1984) Ultrastructure of the membrane attack complex of complement. J Biol Chem 259:7857–7863

Tschopp J, Podack ER, Müller-Ebernard HJ (1982) Ultrastructure of the membrane attack complex of complement: detection of the tetramolecular C9-polymerizing complex C5b-8. Proc Natl Acad Sci USA 79:7474–7478

Van Bergen en Henegouwen PMP, Leunissen JLM (1986) Controlled growth of gold particles and implications for immunolabelling efficiency. Histochemistry 85:81–87

Verwey EJW, Overbeek JTG (1948) Theory of the stability of lyophobic colloids. The interaction of sol particles having an electric double layer. Elsevier, New York

Wang B-L, Larsson L-I (1985) Simultaneous demonstration of multiple antigens by indirect immunofluorescence or immunogold staining. Novel light and electron microsocpical double and triple staining. Methods employing primary antibodies from the same species. Histochemistry 83:47–56

Yokota S (1988) Effect of particle size on labelling density for catalase in protein A-gold immunocytochemistry. J Histochem Cytochem 36:107–109

Young BR, Pitt WG, Cooper SL (1988) Protein adsorption on polymeric biomaterials 1.Adsorption isotherms. J Colloid Interface Sci 124:28–43

Zsigmondy R (1901) Die hochrote Goldlösung als Reagens auf Kolloid. Z Anal Chem 40:697–719

Zsigmondy R (1905) Zur Erkenntnis der Kolloide. G Fischer, Jena, Germany

Chapter 9

Non-Immunological High-Affinity Interactions
Used for Labelling

9.1 Lectins

The word lectin is used to describe a heterogeneous group of proteins from a
wide variety of biological sources that bind specific sugar molecules, especially
when the latter are covalently bound to protein or lipid. Most of the well known
lectins are of plant origin. Although the affinities of these proteins for their
ligands are relatively high they invariably show a range of specificities and
affinities for different sugar compounds (see Table 1).

While the affinity of lectins for monosaccharides is generally low, the
multivalency of most lectins results in significantly higher affinities, or, more
correctly, avidities, as the size of the oligosaccharide increases. Because of their
specificity, lectins have been used extensively as substitutes for antibodies to
determine the cellular location of oligosaccharides and polysaccharides. This
popularity can be explained by three reasons

- A wide variety of lectins, unlike most interesting antibodies, are commercially available in relatively pure form. They can also be purchased already
  coupled to fluorescein or particulate markers.
- Cells have large numbers of binding sites for many lectins, which results in
  impressive labelling patterns.
- Lectin activity can be specifically blocked by the appropriate monosaccharide or disaccharide. This serves as a convenient control for the specificity of
  the reaction.

Aside from these technical points, enormous interest in lectins has also been
generated by four striking biological observations related to these sugar-
binding proteins.

- Lectins agglutinate cells. For this reason they have also been referred to as
  haemagglutins (Barondes 1981; Sharon and Lis 1989).
- Some lectins are powerful mitogens, that is, they increase the rate of cell
  division in culture. The mechanism of mitogenesis is not understood but it
  appears likely that the lectins cross-link surface receptors resulting in
  transmembrane signalling that turns on secondary messenger systems
  (Möller 1980; Larsson et al. 1981).
- Malignant cells express different oligosaccharides on their cell surfaces when
  compared to normal cells which is reflected in striking differences in lectin

Table 1. Characterization of some commonly used lectins[a]

Lectin	Molecular weight	Isoelectric point	Specificity	Comments
Concanavalin A from the seeds of *Concanavalia ensiformis*	110 kDa (neutral pH; tetramer) 55 kDa dimer below pH 5.6 26 kDa monomer (pH 3)	5.5	αMan > αGlc > GlcNac[b]	Not glycosylated; Tetramer contains one Ca^{+2} and one Mn^{+2} per subunit. Reversably denatured by EDTA
Ricinus communis I (from castor beans) (toxin)	65 kDa (dimer) (A chain 32 kDa B chain 34 kDa	7.1	βGal and αGal > GalNAc	Highly toxic B chain-lectin A chain-toxin which blocks protein synthesis in a catalytic fashion
Ricinus communis II (from castor beans) (Agglutinin)	120 kDa (tetramer) (A chain 34 kDa and B chain 31 kDa	7.8	βGal > αGal α GalNAc > D Fuc	Slightly toxic
Wheat germ agglutinin (from wheat germ *Triticum vulgare*)	36 kDa (neutral pH; dimer) 18 kDa (pH 1.9; monomer)	8.5	GlcNAcβ1, 4GlcNAc[c] βGlcNAc > Neu5Ac	Not glycosylated
Lentil lectin from *Lens culinaris*	46 kDa (tetramer) (2 α subunits 5700 kDa) (2 β subunits 17500 kDa)	6.9	αMan > αGlc > GlucNac; also binds Methyl αMan Methyl αGlc	Contain Ca^{2+}, Mn^{2+} – irreversibly denatured with EDTA

Lectin	Molecular weight	pI	Specificity[b]	Remarks
Limulin, from horseshoe crab *Limulus polyphemus*	340 kDa (11 mer) 29 kDa (monomer)	5.0	Neu5Ac – α2,6 GalNAc > Neu5Ac > GlucNac	Requires 0.01 M Ca²⁺ (or Mn²⁺)
Helix pomatia agglutinin from snail	79 K (hexamer) 13 K (monomer)	6.5–10	GalNAc α1,3GalNAc > αGalNAc	There appear to be more than 12 different isoforms[d]
Ulex I from gorse seeds *Ulex europeus*	60–68 kDa (α 29 kDa) (β 31 kDa)	6.3	α-L-Fuc	Contains Ca²⁺ and Mn²⁺
Peanut agglutinin from *Arachis hypogea*	108 kDa (neutral pH; tetramer) 27 kDa (monomer)	5.9	Gal β1,3GlcNAc > α and β Gal	Not glycosylated
PHA (kidney bean) lectin from *Phaseolus vulgaris*	128 kDa (neutral pH; tetramer) 32 kDa (monomer)	5.5	Gal β1,4GlcNAc β1,2Man > Man	
Limax flavus[a+] lectin from slug	44 kDa dimer 22 kDa monomer (pH2)	9–9.5	Neu 5Ac; GlyNeu	

[a] Information from Muresan et al. 1982; Goldstein and Hayes 1978; Goldstein and Peretz 1986; Technical Bulletin of L'Industrie Biologique Francaise (IBF), 92115, Clichy, France[a+] Miller et al. J. Biol. Chem. 257, 7574–7580 (1982).

[b] Abbreviations Fuc – fucose; Glc – glucose; Gal – galactose; GalNAc – N-acetyl D-galactosamine; GlcNAc – N-acetyl D-glucosamine; Man – mannose; Neu5Ac – N-acetyl nearaminic (or sialic) acid. GlyNeu – n-glycolyr euraminic acid.

[c] GlcNac β1,4 GlcNac – is the disaccharide of Nacetyl glucosamine.

[d] Vretblad et al. 19- BPA 579m 52–56 (1979).

binding (e.g. Fischer et al. 1984). A difference is also often seen in the course of normal cell growth and development (Maylié-Pfenninger et al. 1980; Zieske and Bernstein 1982; Nemanic et al. 1983; Hazlett and Mathieu 1989).

- Many lectins are toxic to cells at high doses. A few, such as ricin, are extremely toxic at very low doses. The mechanism of ricin toxicity is now understood in some detail (see Neville and Hudson 1986; Olsnes 1987).

Although the potential use of lectins as affinity markers for labelling cells is widely appreciated in microscopic studies, it is perhaps more relevant here to raise a warning about the indiscriminate use of these reagents, as if their specificities were comparable to antibodies. In order to justify this concern, I have to provide some background information about sugars, oligosaccharides and glycoproteins.

9.1.1 Simple Sugar Chemistry

Most secreted proteins are glycoproteins, that is, they have variable amounts of sugar residues covalently linked to the polypeptide. There are at least 200 different basic types of sugar units, referred to as monosaccharides. Surprisingly, however, only 11 are found naturally in glycoproteins (Sharon and Lis 1981, 1989). Most of these are 6-carbon ringed sugars, referred to as hexoses, of which glucose is the best example. To simplify matters further, most of these sugars are in the D-isomeric configuration. There is only one hexose, fucose, as well as two pentoses, arabinose and xylose, which are found in the L configuration. In addition to these simple sugars there is a class of 9-carbon chain compounds, the neuraminic acids of which there are about 20 different derivatives. These are collectively referred to as sialic acids (Montreuil 1982).

This apparent simplicity in the construction of oligosaccharides in glycoproteins (as well as glycolipids) is, however, misleading. Carbohydrates are much more difficult to work with and to analyze than either proteins, nucleic acids or lipids. There are a number of reasons for this. First, the covalent bond between adjacent sugar molecules, referred to as the glycosidic bond, may be in one of two configurations, depending on whether the oxygen atom on carbon 1 is below (α) or above (β) the plane of the ring. Second, since more than two of the six carbons in hexoses are available for glycosidic linkages, branched forms are common. Finally, unlike proteins and nucleic acids which are made under direct control of the genetic code (each unit being identical), the oligosaccharides of glycoproteins and glycolipids are synthesized by a large family of glycosyl transferases. Even in one cell type a variety of different forms of one glycoprotein is frequently seen, a phenomenon referred to as microheterogeneity.

9.1.2 Glycoproteins

The side chains of oligosaccharide can only be covalently bound to protein by one of two possible linkages (see Montreuil 1982; Kornfeld and Kornfeld 1985 for reviews).

9.1.2.1 Linkages

N-Glycosidic Linkage. In this linkage, oligosaccharides are attached enzymatically via an N-acetyl glucosamine residue to asparagine residues in the polypeptide chain. Not all asparagine residues in the polypeptide are available for the above reaction. There are specific glycosylation signals found in the primary structure Asn-X-Thr/Ser, where X may be any amino acid except proline and aspartic acid. Furthermore, not all such sequences are necessarily glycosylated; there appear also to be strict requirements for a defined secondary structure.

O-Glycosidic Linkage. The second type of covalent bond is between N-acetyl galactosamine, galactose, xylose or N-acetylglucosamine and the hydroxyl groups of either serine, threonine, hydroxylysine or hydroxyproline.

In both types of linkages the structure of the polypeptide backbone determines primarily where and how many oligosaccharide side chains can be attached to the protein. As expected, there is a tremendous variability between different glycoproteins in the number and size of oligosaccharide side chains as well as in the degree of branching observed.

While the precise sites on the polypeptides where sugars are attached are determined by information in the polypeptides, the composition and structure of the oligosaccharide side chains are determined by a battery of enzymes in two major cell compartments, the endoplasmic reticulum and the Golgi apparatus. These enzymes are either glycosyl transferases that add on specific sugars, or hydrolases that take off specific sugars. The glycosyl transferases add monosaccharides, in the form of nucleotide bound precursors (e.g. UDP galactose), to the oligosaccharide that is bound to the polypeptide chain. Both these transferases and hydrolases act in a precise kinetic sequence to produce the final glycoprotein. The consequence of this is that the variability in oligosaccharide structure in glycoproteins, though potentially enormous in theory, is extremely restricted in practice.

9.1.2.2 Biosynthesis of Glycoproteins

In eukaryotic cells many integral membrane proteins and all secretory proteins are synthesized on rough endoplasmic reticulum (RER) bound ribosomes. Most of these proteins are glycoproteins. During synthesis the proteins are integrated into the RER membrane or translocated across it and glycosylation starts even

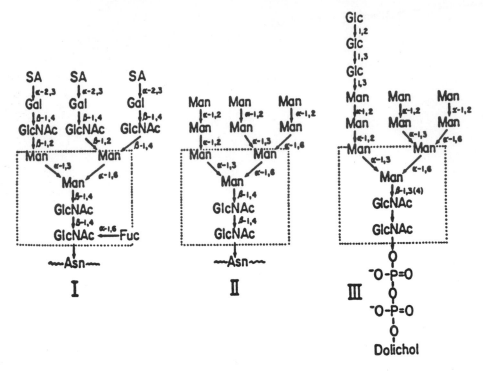

Fig. 1. Structure of a typical "complex" type (*I*), "high mannose" (*II*) and the dolichol intermediate (*III*) which is transferred onto asparagine residues on proteins during their transfer across the endoplasmic reticulum membrane. *Glc* glucsoe; *Man* mannose; *GlcNAc* N-acetyl D-glucosamine; *Fuc* fucose. (Hubbard and Ivatt 1981)

before completion of synthesis. For N-linked oligosaccharides some of the asparagine residues which are in the required Asp X-Thr/Ser sequence will acquire a pre-synthesized block of high-mannose oligosaccharides that is transferred from a lipid dolichol phosphate-oligosaccharide precursor (Fig. 1). A precise sequence of covalent modifications then occurs in a sequential fashion (Fig. 1). In the ER the core oligosaccharide is trimmed, that is, specific residues are removed, firstly two glucoses, then the third glucose followed by one (or two) mannoses. After transfer to the Golgi stack, where further trimming of mannose residues occurs (down to five mannoses), four terminal sugars, N-acetyl glucosamine, fucose, galactose and sialic acid, are added stepwise. Their precise positions in the oligosaccharide are determined by the enzymes involved (Kornfeld and Kornfeld 1985; see Fig. 2). The enzyme that adds the first of the terminal N-acetyl glucosamine residues has been localized to Golgi cisternae in the middle of the stack (Dunphy et al. 1985) while the galactosyl transferases that put on galactose and sialic acid (specifically that which puts it on in the α 2,6 linkage) and the *trans* Golgi network (Roth et al. 1985) have been shown to be concentrated in *trans* Golgi cisternae (Roth and Berger 1982).

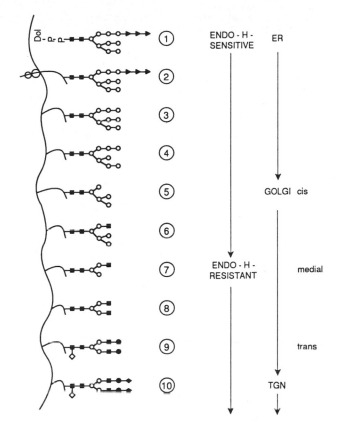

Fig. 2. Steps involved in N-linked glycosylation. The dolichol pyrophosphate precursor (*1*) is transferred as a complete unit onto some asparagine units of the newly synthesized protein during co-translational insertion. This scheme depicts a spanning membrane glycoprotein: the mechanism is identical for a soluble protein. The initial oligosaccharide contains two N-acetyl D-glucosamines (GlcNAc■), nine mannoses (O) and three glucose units (▲). In the ER the glucoses are removed by two glucosidases (I and II) and one to three mannose units are excessed by the ER mannosidase(s). The glycoprotein is then transported to the *cis* part of the Golgi complex where further mannoses are removed by the Golgi mannosidase I. The resultant $Man_5GlcNAc_2$ structure (*5*) is a substrate for GlcNAc transferase I in the medial Golgi. The product of this reaction (*6*) is a substrate for Golgi mannosidase II (medial Golgi) which removes two additional mannose residues. This trimming event makes the oligosaccharide (*7*) resistant to the enzyme endoglycosidase H. Thus, sensitivity or resistance to endo H is a widely used tool to track the movement of a glycoprotein through the Golgi complex. The cleavage by endo H can be detected as a shift in mobility of the glycoprotein on SDS gels. The precise location within the Golgi complex where the second GlcNAc and then two galactose (●) and the fucose (◊) are added has not been clarified. The available data make it seem likely that the site of galactosyl transferase (*trans* Golgi) is distinct from that of the sialyl transferase (TGN), which adds the terminal sialic acid (♦) (see Berger and Hesford 1985; Duncan and Kornfeld 1988). The sialyl transferase Golgi compartment (TGN) (Roth et al. 1985) is also the site to which recycling receptors are targeted to from the endocytic pathway (Duncan and Kornfeld 1988; Neefjes et al. 1988; Jin et al. 1989)

The pre-Golgi form of the oligosaccharide is often referred to as the "high-mannose" form while the late Golgi forms are often termed complex oligosaccharides.

The biosynthesis of the other large class of oligosaccharides, those linked to the hydroxyl groups of threonine or serine, (O-linkage) is much simpler. These involve direct and sequential addition of one monosaccharide at a time, bound to the appropriate nucleotide , to the growing polypeptide chain. There is no involvement of a lipid-linked intermediate and no trimming after synthesis. In contrast to the N-linked oligosaccharides, fewer generalizations can be made about the structure of the O-linked ones. They can vary from a single galactose residue in collagen to 205 disaccharide residues in bovine submaxillary mucin. The total number of sugar residues attached to a single polypeptide can vary from one in collagen to 85 in soluble blood group glycoproteins (Sharon and Lis 1981). Two rough generalizations can be made: they often contain a galactose-N-acetyl galactosamine disaccharide, and this disaccharide is commonly heavily sialylated. The terminal sugars are mostly sialic acid, N-acetyl galactosamine, galactose or fucose (for a recent summary of common structures see Paulson 1989).

Proteins may contain only N-linked oligosaccharides, only O-linked ones, or both together as, for example, in the glycoprotein of the red blood cell membrane glycophorin which has 15 O-linked and one N-linked oligosaccharides.

From what I have said so far, if lectins did have an antibody-like specificity for certain sugars in precise configurations in glycoproteins, as is often claimed, they could indeed be extremely powerful tools in cell and tissue biology. A number of considerations make it clear, however, that their potential must be far more limited. These are:

- With respect to N-linked glycoproteins, the situation is much more complex than the general scheme outlined above. Not all glycoproteins are processed in the same manner: some have a mixture of "complex" and "high mannose" chains in a single oligosaccharide structure, in other words, some high mannose chains are not processed during transport. For a given protein, processing of its oligosaccharides might occur differently in different tissues (Sheares and Robbins 1986). Others, such as rhodopsin, have all their mature oligosaccharides in an intermediate type of structure. Hence, it is difficult to correlate unequivocally a certain oligosaccharide structure with defined cellular compartments. Even more problematical is the fact that for any individual glycoprotein, a spectrum of different structures is usually obtained upon purification ("microheterogeneity", see Hubbard and Ivatt 1981). Further, a single glycoprotein may be glycosylated differently in different cell types, as has been convincingly shown for many viral glycoproteins. Further, the same glycosylation site may bear quite different oligosaccharide structures in different mutants of the same viral protein (see Hubbard and Ivatt 1981). Moreover, there are clear evolutionary differences between organisms with respect to oligosaccharide structures occurring on the same protein.

- In addition to oligosaccharides attached to proteins, there is a considerable amount attached to lipid molecules, which are also synthesized by the rough ER. In so far as they have been studied, the evidence suggests that the oligosaccharide structures on glycolipids are similar to those on glycoproteins (Rauvala and Finne 1979). Cytochemical methods are available, however, to specifically label glycolipids using gangliosides (Spiegel et al. 1982).

- A serious problem in working with oligosaccharide chains of glycoproteins is the fact that, with a few notable exceptions, it has proven difficult to ascribe a clearcut role to them, either with respect to the various modifications occurring during intracellular transport or to the final product. (For a review of recent progress see Paulson 1989.) This makes it difficult to use functional assays as polypeptides that function, for example, as enzymes.

- In contrast to what is often stated in introductory paragraphs to papers on lectin labelling, the specificity of most lectins is quite broad. There is not one which exclusively recognizes only one sugar in a precise configuration in an oligosaccharide. Hence, conA will bind mannose as well as glucose residues on glycoconjugates, whereas ricin binds terminal galactose as well as fucose, while wheat germ agglutin (WGA) binds both sialic acid and N-acetyl glucosamine (see Table 1). Further, while it used to be thought that lectin binding was specific for terminal non-reducing sugars in glycoconjugates, it is now clear that they may also bind mono- and oligosaccharides in internal positions (Montreuil 1982; Goldstein and Poretz 1986).

In recent years, it has become evident that glycosylated proteins are not only present in the biosynthetic and endocytic pathways but may also exist in proteins synthesized on free ribosomes. All of these proteins contain a simple, but unique modification, namely N-acetyl glucosamine O-linked to serine, and possibly threonine, residues (Holt et al. 1987, Hart et al. 1989). Two major classes of these proteins can now be put into this category, those that are present in the nuclear pores and those that are present in the nuclear matrix. Many of the latter which have been characterized have been now shown to be transcription factors (for a recent review see Hart et al. 1989). All of these O-linked GlcNac-proteins bind WGA. This fact should obviously be considered in any labelling studies using this lectin.

9.1.3 Lectin Labelling

At the histological level, lectins have been shown to be reproducible markers of certain cell types or regions of tissues, e.g. Murata et al. 1983; Alroy et al. 1984; Fischer et al. 1984. There has been enormous interest in the differences seen in lectin labelling patterns which appear during development in response to tumorgenesis and in a number of cases differences in labelling patterns become evident in certain diseases which are of importance for diagnostic purposes (see Lis and Sharon 1986a,b for references). There is no doubt that these differences are real and that lectins can play an important role in diagnostic medicine. The problem arises when one attempts to make molecular conclusions about the oligosaccharide structures responsible for the labelling. This is exemplified in the fruitless attempts at characterizing the structural basis for the striking difference in lectin labelling found between normal and malignant cells (Sharon and Lis 1989). In light of the considerations pointed out above, this is not surprising. In this sense, lectins may be more important in raising new questions than in answering them.

At the ultrastructural level, in addition to being used as probes for development, tumorigenesis and disease, lectins have been widely used as markers for intracellular organelles or for the plasma membrane. Hence, conA was recommended as a specific marker for the ER while ricin and WGA were indicated to be specific markers for the Golgi complex (Virtanen et al. 1980). This approach was extended at the ultrastructural level by Tartakoff and Vassalli (1983), who used a pre-embedding immunoperoxidase approach to show that conA selectively labelled RER and the *cis* side of the Golgi stack whereas WGA labelled only the trans elements of the Golgi. These authors claimed that these lectins could be used as specific markers for these compartments. Using the same lectins and thawed cryo-sections of Baby Hamster Kidney cells, as well as rat spermatids, however, we consistently found that conA labelled all elements of the secretory pathway as well as lysosome-like structures (Griffiths et al. 1982, and unpubl. data). This is, in fact, expected, since conA is capable of binding to both "high mannose" and "complex" oligosaccharides (Kornfeld and Ferris 1975). It is likely that addition of terminal sugars to N-linked oligosaccharide can mask the ability of conA (or other lectins) to bind to glycoproteins in some, but not all cases, resulting in a negative reaction. Similarly, WGA labelled plasma membrane, endosomes and lysosome-like structures in the cells we have tested. Again, this is not surprising, since lysosomal enzymes, as well as lysosomal membrane proteins , are known to contain complex oligosaccharides (von Figura et al. 1986; Kornfeld 1987). Note again that WGA will also bind to O-linked N-acetyl glucosamine residues present on cytosolic proteins that are especially enriched in the nuclear pore and the nucleus (Hart et al. 1989).

In an earlier study we showed that ricin (120) could be used to distinguish the *trans* part of the Golgi stack from the more proximal or *cis* part (Griffiths et al. 1982). A similar, though not identical, result was obtained by Sato and Spicer (1982) using the galactose specific peanut lectin. In our hands, WGA did not consistently do this. The *trans* labelling with ricin correlated with the localization of galactosyl transferase in the trans cisternae (Roth and Berger 1982, Strous et al. 1983). Nevertheless, in our experience ricin also labels the plasma membrane, endosomes as well as lysosome structures in a number of cell types. Hence it cannot be claimed to be a specific marker for the *trans* Golgi, but it may be useful as a crude marker to distinguish between different membrane compartments in some cases. Within the Golgi complex in particular, Roth and colleagues have used this approach to map different functions to different compartments (see Roth et al. 1983, 1984).

For some cell biological purposes, the use of lectins may be a very powerful tool for combined biochemical and morphological experiments. An excellent example relates to the use of lectin-resistant cell lines (see Lis and Sharon 1986b for a review). Due to their toxicity and specificity for sugars in oligosaccharides in the cell surface, many lectins have been used to select for mutants with altered cell surface glycoconjugates. The essence of this approach can be described for a lectin such as ricin which binds to terminal galactose residues by its "B" polypeptide chain. After binding, it is taken up by cells into the endocytic pathway from which a small proportion enters the *trans* Golgi network (van Deurs et al. 1988). Here a translocation event occurs whereby the toxic chain dissociates and passes through the membrane. Subsequently, the toxin can bind to ribosomes and arrest protein synthesis and kill the cells (see Olsnes 1987; Endo et al. 1987). Mutants which lack terminal galactose will not bind the lectin, will survive and can be selected. Such mutants have played an important role in the elucidation of the mechanisms of glycoprotein biosynthesis (see Kornfeld and Kornfeld 1985). One of the most widely used mutants include the clone 13 of CHO cells, selected using WGA, which has a defective UDP galactose transporter. Consequently, these cells are unable to transport this nucleotide sugar precursor into the lumen of the Golgi complex. Thus, despite having a normal galactosyl transferase, their glycoproteins are essentially devoid of galactose. If this cell is incubated with galactosyl transferase and the precursor sugar UDP galactose, galactose (which can be radioactively labelled) becomes incorporated into glycoconjugates on the cell surface. Using this approach, Duncan and Kornfeld (1988) could follow the pathway from the plasma membrane to the *trans* Golgi network where the conjugates now acquired sialic acid. In such a system a morphological approach using a sialic acid lectin, such as limulin, could, in principle, be used. In this case all the intracellular labelling would result from those molecules which have come from the cell surface and been exposed to the sialyl transferase. Similarly, the clone 15B of CHO cells, which has a defective N-acetylglucosamine (GlcNAc) transferase I is unable to add GlcNAc to the outer branches of its oligosaccharides. This, and similar mutants have been extensively used in a series of elegant in vitro experiments by Rothman and colleagues that have led to new molecular insights into the mechanism of transport through the Golgi complex (see Orci et al. 1989 for a list of references). The essence of these assays is to mix Golgi fractions from viral infected mutant and uninfected wild-type cells and look for complementation of function. The viral proteins will acquire the function present in the wild-type cell only if membranes of the two compartments have fused. Although evidently not used at the ultrastructural level, it is again foreseeable that lectins could have powerful uses in such a system, since the assays rely on mixing membrane fractions which lack terminal GlcNAc (and, consequently, subsequently added sugars Gal and sialic) with membrane fractions that have normal terminal sugars. In such cases, lectin binding is an "all or none" phenomenon. In the Rothman system, the mutant membranes could be distinguished immunocyto-chemically from the wild type membranes by virtue of their containing the viral proteins.

With specific exceptions such as those just discussed, the approach of making specific antibodies against purified preparations of organelles (Louvard et al. 1982; Reggio et al. 1984; Buckley and Kelly 1985), would appear to be a more powerful approach than the use of lectins for ultrastructural studies in cell biology. Similarly, antibodies, especially monoclonals, have been developed as specific cell surface markers for certain cell types in complex tissues such as brain (Raff et al. 1979 1984). Although the efforts needed to make such antibodies is often considerable, in contrast to the commercial availability of lectins, the ability to characterize the antigens, as well as the possibilities to clone and sequence them, makes this effort worthwhile.

9.1.3.1 Practical Details of Lectin Labelling

For electron microscopy two kinds of approaches have been used for lectin labelling. First, lectins have been directly conjugated to **ferritin**, (see de Petris 1978 for a review of the earlier literature). Most of the studies in the last 10 years, however, have used colloidal gold. While coupling lectins directly to gold is by far the simplest procedure in theory, it is not always straightforward in practice. The main problem appears to be to produce conjugates that are (1) free of aggregates and (2) stable upon storage. In many of the published micrographs using lectin-gold conjugates, large aggregates are often evident. Our earlier attempts to conjugate ricin and WGA to gold were never successful: either large aggregates were present or the conjugates would aggregate during their first overnight storage in the refrigerator. Nevertheless, in a number of studies reliable conjugates of many of the common lectins have been prepared, notably by the groups of Horisberger (see Horisberger and Rosset 1977; Horisberger 1979) and Roth (see Roth and Binder 1978; Roth 1982; Roth et al. 1983; Roth 1984; Deschuyteneer et al. 1988). For protocols to covalently couple lectins to ferritin see de Petris (1978); examples of labelling with such conjugates can be seen in the study by de Petris and Baumgartner (1982). In recent times, many of these reagents are available commercially. For example, recents report by Nakajima et al. (1988) and Hazlett and Mathieu (1989) show labelling obtained with lectin-gold conjugates from EY Laboratories (San Mateo, California, USA) and Polysciences (Warrington, PA, USA).

The second major approach for using lectins for labelling, for either light- or electron microscopic studies, has been to use antibodies against lectins followed by e.g. **protein A gold**. This was the approach that has worked best for our group (Griffiths et al. 1982, using antibodies prepared by Daniel Louvard, Institute Pasteur, Paris). It is not a trivial matter to produce antibodies against a class of molecules that are often highly toxic. The simplest solution would be to obtain them commercially, and antibodies to the common lectins are now available from a number of sources (e.g. Vector Laboratories, Burlingame, California, USA).

A third approach with lectins which has been commonly used for labelling studies is to use a two-step method in which the bound lectin is visualized by a glycoprotein that contains relatively high copies of the specific sugar on its oligosaccharides. The first example of this approach was by Bernhard and Avrameas (1971), who visualized conA using a mannose-rich glycoprotein, horseradish peroxidase, which could then be visualized by the DAB cytochemical reaction. A similar approach was also used by Roth et al.(1984),.who visualized the sialic acid-specific lectin from *Limax flavus* using the sialylated glycoprotein, fetuin, conjugated to colloidal gold.

A fourth possibility for labelling would be to biotinylate the lectin and visualize it with avidin-gold (Pino 1984). Biotinylated lectins are also commercially available from a number of sources (e.g. Vector, USA).

The standard control for all lectin labelling is to block the binding with the appropriate mono- or disaccharide. Because the affinity of these species by themselves is far lower than when they are part of an oligosaccharide structure it is necessary to use relatively high concentrations (0.1–0.2 M).

9.1.3.2 Use of Glycosyl Transferases and Glycosidases with Lectins

A novel approach for labelling intracellular compartments involving lectins was recently introduced by Lucocq et al. (1987). This group treated lowicryl K4M sections of liver tissue with a purified preparation of galactosyl transferase and its donor substrate UDP-galactose. The idea here is that any terminal N-acetyl glucosamine residue on glycoproteins can act as a substrate for galactosyl transferase. Thus, it would mimick a reaction that occurs normally in *trans* Golgi cisternae. The newly added sugar residues can subsequently be visualized using a suitable lectin, in the case of galactose, ricin, bound directly to colloidal gold. By comparison with sections untreated with galactosyl transferase, in which case the lectin will see the endogenous terminal gal residues, the additional labelling in enzyme/substrate-treated sections should reveal the sites of the terminal N-acetyl glucosamine. Thus, in the example given by Lucocq et al. (1987), in the absence of the enzyme/substrate treatment ricin labelled only *trans* Golgi cisternae whereas in its presence the ricin labelling extended throughout most of the Golgi cisternae. Furthermore, this group was able to identify the class of oligosaccharide to which the GlcNAc residues were attached. They preincubated the sections with endoglycosidases that specifically cleave N-linked oligosaccharides off the polypeptide chain. Treatment with these enzymes abolished the additional resin labelling, indicating that the GlcNAc residues were attached to N-linked oligosaccharide chains.

In principle this approach should also be possible with other glycosidase enzymes, some of which are commercially available. For example, treatment with neuraminidases can be used to remove terminal sialic acid residues, thus exposing sub-terminal galactose which could then bind ricin (Hazlett and Mathieu 1989).

9.2 Protein A

In their short review of protein A, Surolia et al. (1982) referred to this protein as "nature's universal anti-antibody". Indeed, this is a suitable description of a protein that has also become a universal tool of immunology and related disciplines, including immunocytochemistry. This protein, first described by Verwey (1940), is found naturally in the cell wall of most strains of *Staphylococcus aureus*, where it is covalently bound to the peptidoglycan. The highest producing strain of *S. aureus*, Cowan 1, contains 1.7% protein A by weight and this protein makes up almost 7% of the total mass of the cell wall (Surolia et al. 1982). The precise function of this protein for the bacteria, however, is still not clear.

This very stable and highly water-soluble protein is easily purified by digestion of bacterial lysates with a specific protease, usually lysostaphin, followed by ion-exchange gel filtration and affinity purification on IgG sepharose.

9.2.1 Properties of Protein A

From earlier work, protein A was considered to have a molecular weight of 42 kDa, to have three or four IgG binding domains and to be covalently bound to the bacterial cell wall via a small peptide (Sjödahl 1977a,b; Langone 1982a). However, from recent sequence analysis, it is clear that this description was imprecise: protein A has a signal sequence and all the characteristics of a bacterial membrane protein that spans the plasma membrane (Uhlén et al. 1984; Moks et al. 1986). On the ectoplasmic domain of the plasma membrane it has an 8-amino acid, proline-rich peptide that is repeated 12 times and which spans the cell wall layer. This is continuous with the amino-terminus of the molecule, which consists, in fact, of five homologous IgG binding domains of approximately 58 amino acids. The total molecular weight is about 58.7 kDa (Uhlén et al. 1984; Ljungquist et al. 1989). Each of these domains contains a single tyrosine residue that is exposed on the surface of the molecule in a structure that enables the molecule to be easily radio-iodinated (Goding 1978; Uhlén et al. 1984).

Protein A, which has no carbohydrate groups, is a cylindrical-shaped molecule whose three-dimensional structure has been worked out, as has the structure of its complex with the Fc domain of immunoglobulins (Deisenhofer 1981). Trypsin treatment has been used to separate the fragments and all its domains are apparently able to bind the Fc part of IgG with relatively high affinity. Nevertheless, when protein A binds IgG that is either free in solution or bound to a solid support, the data suggest that only two of the five potential binding domains are active. In other words, protein A is functionally bivalent. This is now supported by recent molecular biology approaches where the cDNAs coding for variable number of domains of protein A have been express in bacteria (Ljungquist et al. 1989). These results can be summarized by saying that one domain by itself binds one molecule of IgG whereas a polypeptide containing two of these domains can bind two molecules of IgG. However, when more than two (and up to five domains) are expressed as a single protein the complex can still only bind two IgG molecules. It appears that the presence of two large IgG molecules sterically hinder the accessibility of binding to further IgG molecules. Thus, in solution, a single protein A molecule is capable of cross-linking two IgG molecules. These complexes, for the most part, remain soluble (Langone 1982a). When the antibody is bound to antigen on a solid support, however, the situation may be different. Even in solution, the rate of binding of protein A to antibody that has bound antigen is significantly slower than to free antibody (Langone 1982a,b). When both were bound to a solid support, namely to antibody-coated red blood cells, the ratio of protein A to antibody appeared to vary between 0.5 and 1.3 (Gruhn and McDuffie 1979; Gray and Masouredis 1982). When protein A is bound to gold particles, it appears unlikely that it can bind more than a single IgG molecule; however, the stoichiometry of this binding has not been studied carefully. Conversely, a single IgG molecule seems unable to bind more than one protein A molecule when the latter is conjugated to gold. In support of this is the observation that, in indirect labelling such as rabbit antibody followed by protein gold, the gold particles are very rarely in pairs and almost never clustered.

Crystallographic studies have indicated that the protein binds specifically to the CH_2 and CH_3 domains of the Fc region; this binding has no effect on the conformation of the Fab domains nor, therefore, on the ability of the antibody molecule to bind antigen (Deisenhofer et al. 1978; Deisenhofer 1981). The affinity of binding of protein A to the Fc domains is 10^{-8} and 10^{-10} M and thus in the kDa range found for antibodies to antigens (Langone 1982a).

Protein A, which has an isoelectric point of 5.1, is an exceptionally stable molecule that is resistant to heating and whose activity is not affected by pH treatments ranging from 1–11.8. The effects of many harsh denaturing reagents such as 8 M urea and 6 M guanidine hydrochloride are reversible. Further, mild detergent treatment, such as 1% NP40, does not appear to affect its binding to IgG (Langone 1982a). Both the affinity and the stability of the binding of at least some IgGs to protein A may be affected by pH. For example, mouse IgGs tend to bind with higher affinities at pH 8–9 than at pH 7 (Ey et al. 1978). Conversely, rabbit IgG which binds protein A bound to a solid support at pH 7 with high affinity can be eluted at pH 5 (Langone 1982b).

9.2.2 Species Specificity

Protein A binds to antibodies of many, but not all species, although the affinity of binding to different antibody molecules varies widely. It binds primarily to the Fc domains of IgG molecules, especially those of human, rabbit, pig and guinea pig. Variable binding is also seen with some classes of IgGs from mouse, rat and other species but it shows no appreciable binding to chicken antibodies (Table 2; see Langone 1982a,b for a review). In addition to IgGs, individual

Table 2. IgG species that react with protein A for immunocytochemistry

Good binding	Poor binding	No binding
Rabbit	Goat	Avian
Pig	Most rat	
Guinea pig	Most mouse	
Human	Sheep	
Dog		
Mouse IgG 2A, 2B	Donkey/horse	
	Bovine	
Rat IgG 2C, IgG1		

antibodies from other classes of immunoglobulins may also bind protein A. For example, of 33 different human IgM antibodies tested, one-third bound protein A (Harboe and Folling 1974). Similarly, some human IgA and IgE, but apparently not IgD, may also bind protein A (Langone 1982a). From the available evidence that is of particular relevance to immunocytochemistry, the following observations can be listed:

- All rabbit, human, pig and guinea pig IgGs are excellent protein A binders.
- Chicken Igs does not bind protein A.
- Some mouse and rat IgG classes bind protein A with an affinity high enough for use in immunocytochemistry, but on average, very few of these antibodies are in the high-binding category.

It should be noted that the information shown in Table 2 is mostly derived from immunochemical assays. For immunocytochemistry, however, only high affinity interactions of protein A for antibodies are useful, a statement which is also true of antibodies themselves, with respect to antigen. Therefore, the information in Table 2 should be considered as a rough guide only. The easiest way to find out whether a particular antibody binds protein A or protein A gold for immunocytochemistry is to directly compare, using sections, a two-step labelling (antibody followed by protein A gold) with a three-step one (antibody followed by a second antibody that binds protein A, such as rabbit, then protein A gold). Note, however, that the three-step labelling will amplify the signal, giving rise to clusters of gold particles (see Chap. 7).

9.2.3 Binding to Fab Regions of Antibodies

In a few cases protein A has been shown to have some affinity for the Fab parts of antibodies (Young et al. 1984). When this affinity is high it can lead to significant problems in immunocytochemical studies. This has been our interpretation of spurious double-labelling results we have obtained with some antibodies (see p. 267). According to Langone (1982a), even low affinity interactions with Fab fragments, in addition to the Fc region, can complicate immunochemical data; they can also lead to precipitation of antibody-protein A complexes. According to M. Uhlén (Royal Institute of Technology, Stockholm), dog and bovine IgG are especially a problem in this respect. In the case of bovine IgG subclass II their evidence suggests that all of the binding with protein A is via the Fab domains.

9.2.4 Kinetics of Reaction and the Effect of pH

The binding of protein A to IgG either free in solution or bound to solid support appears to be very fast at temperatures between 4 and 37°C and equilibrium is usually reached in times ranging from a few minutes to less than 30 min. Longer times may be required, however, if less than excess amounts of protein A are used (Langone 1982a).

The dissociation of IgG (or IgG-antigen complexes) from protein A depends on time, temperature and pH. At neutral pH relatively little dissociation occurs between 4 and 37°C. However, in the presence of a competing excess of IgG or whole serum, 75–90% dissociation can occur after 30–60 min at 37°C, and even at 4°C, although significantly less (Kessler 1975; O'Keefe and Bennet 1980; Langone 1982a). Although this observation has apparently not been given any consideration, it is potentially very important for immunocytochemical experiments. Mildly acidic pH ≈ 5 can, as mentioned, also cause significant dissociation. A theoretical point to be considered in this respect is the possibility that acidic treatments after labelling, such as the use of pH 4 solutions of uranyl acetate, could remove some of the bound protein A. This may be another reason to advocate the use of a fixation step after labelling. According to some reports (Ey et al. 1978), the binding is higher at slightly elevated pH ≈ 8–9. For quantitative elution of protein A from IgG or antibody antigen complexes a number of different approaches have been used including low pH (2–3), guanidine hydrochloride, high salt ($MgCl_2$) or sodium thiocyanate. According to Langone (1982a), 0.1 M lithium diiodosalycilate appears to be the method of choice for preparative purposes.

9.3 Protein G

In recent years, this protein has attained swift recognition following its first isolation from cell walls of human group G strains of Streptococci (Björk and Kronvall 1984; Åkerström et al. 1985; Guss et al. 1986). It is a protein of about 63 kDa, which can be cleaved into two polypeptides of about 35 kDa. Although there are similarities between protein A and protein G, the latter appears to bind to a wider species range of antibodies than does protein A and, in most cases, its affinity for those antibodies is higher than that of protein A (Åkerström and Björk 1986). This is especially relevant for a number of classes of antibodies from rat, mouse and goat. From recent cDNA sequence data, it appears that protein G shares structural, but not sequence, homologies to protein A and differences are apparent between different bacterial strains: of the two strains sequenced, one had two IgG binding domains whereas the other strain had three such domains (Olsson et al. 1987). For IgGs from a number of species tested, the binding to protein G was highest at pH 4–5. In this respect, protein G is significantly different to protein A, which favours more neutral to alkaline pHs. This is evidence that the mechanism of binding of the two bacterial proteins appears to be different: amino acid analysis also shows significant differences between them. Further, the isoelectric point of protein G (< 3.5) is significantly more acidic than protein A (5.1) (Åkerström and Björk 1986).

A further difference between protein G and protein A is that protein G has a high affinity binding site for serum albumin. The latter can be proteolytically cleaved from the rest of the molecule with papain to yield a remaining 67 amino acid polypeptide that retains the normal IgG binding activity of the intact protein G (Nygren et al. 1988).

9.3.1 Use of Protein G for Immunocytochemistry

During the past 2 years, conflicting reports have appeared on the use of protein G complexes with colloidal gold for immunocytochemistry. Bendayan (1987) and Bendayan and Garzon (1988) reported that such complexes were quantitatively similar to, if not better than, protein A gold complexes for a number of different antibodies tested. In contrast, Taatjes et al. (1987) and J. Slot (University of Utrecht, pers. commun.) found protein G gold complex to be generally inferior to those formed with protein A. Further, chimaeras of protein A-protein G made by recombinant DNA technology (Eliasson et al. 1988, 1989) when complexed to gold were apparently not better than protein A gold complexes, J. Slot (pers. commun.). A recent report from Balslev and Hansen (1989) also compared a gold conjugate of a cloned protein G (see Fahnestock et al. 1986) with protein A gold. The optimal condition for binding in this study was found to be pH 4.5 with 6–8 µg/ml protein. In contrast to the earlier reports, these authors found that, overall, protein G was superior to protein A for two different rabbit antibodies tested on cryo-sections of pig intestine. This was also confirmed immunochemically using dot blots. Clearly, more studies are required in order to make more general conclusions about the potential use of protein G.

Recent studies from M. Uhlén's group (pers. commun.) indicate that protein G binds much more widely to the Fab region of different IgGs than does protein A. Whereas Fabs prepared from human IgGs bind insignificantly to protein A, they appear to bind strongly to protein G. This is also true for bovine and mouse. According to Uhlén, protein G has even been used to purify mouse monoclonal Fabs in some cases. No evidence is yet available for other species. These observations make it less likely that protein G will become as useful as protein A for double-labelling studies.

9.4 Avidin-Biotin Interactions

Avidin is a protein present in trace amounts (0.05%) in chicken egg white. The significance of this protein was first discovered over 60 years ago when it was found that rats fed with dried egg white developed a form of dermatitis (Boas 1927; Green 1975). As shown by György, however, the disease could be cured by dietary additions of biotin or vitamin H (see György 1954 for a review). The affinity of avidin for biotin (kDa 10^{-15}M) is now known to be one of the highest of all biochemical complexes.

Besides chicken egg-white, avidin is now known to be generally present in eggs and oviducts of birds, reptiles and amphibians. Its synthesis and secretion in chickens is regulated by sex-steroid hormones: only egg-laying hens produce significant amounts of avidin. More recent data indicate that higher animals may also secrete avidin, especially in response to inflammation (Korpela 1984a). It is now established that cultured fibroblasts and macrophages of chicken can secrete avidin (Korpela et al. 1983; Korpela 1984b). Besides eukaryotes, some species of bacteria are also able to synthesize a related molecule, streptavidin (see below).

The biological function most often attributed to avidin in eukaryotes is that it stops the growth of prokaryotes by removing an essential nutrient, biotin, from their surrounding medium (Green 1975). For an early comprehensive review of the use of the avidin-biotin complex in molecular biology see Bayer and Wilchek (1980).

9.4.1 Properties of Egg-White Avidin

Avidin is a tetrameric glycoprotein with a MW of 68 kDa. The four subunits are arranged with identical, and twofold symmetry. It is very soluble in water and salt solutions and is extremely stable over a wide range of pH and temperature (Green 1975). When complexed with biotin it is even more stable: whereas avidin itself is partially unfolded at 85°C, the avidin-biotin interaction is only destroyed by autoclaving at 132°C. Both in the free and biotin-bound form, avidin is also very resistant to denaturing agents and to proteolysis. In the presence of 3–6 M guanidine HCl, the tetramer dissociates into monomers. However, this treatment has no effect on the avidin-biotin interaction. Avidin is, nevertheless, unstable under oxidizing conditions, especially in strong light: this is thought to be due to the susceptibility of key tryptophane residues (Green 1975). These tryptophan residues are believed to be crucial for the binding to biotin. There is a single disulphide bond as well as a single tryptophane residue per avidin monomer. Hence, the avidin tetramer can bind a total of four biotin molecules. Despite the unusual high affinity, the interaction is nevertheless non-covalent (Green 1975). In addition to binding free biotin, avidin can also bind very tightly to the biotin prosthetic group of biotin-containing enzymes leading to loss of catalytic activity (Moss and Lane 1971). The fact that avidin has four binding sites for biotin means that it can be used to cross-link biotinylated molecules in labelling studies (see below).

Avidin is an exceptionally basic protein with an isolectric point around pH 10. This makes the protein a very "sticky" molecule, especially at neutral pHs, a potential problem in immunocytochemical studies. This can be especially serious for negatively charged molecules in tissues such as nucleic acids. According to Wood and Warnke (1981) and Hutchison et al. (1982) some commercial lots of avidin are especially sticky for chromatin. Because of its tryptophane residues avidin is fluorescent with a maximal absorption at 338 nm in its free state and 328 nm when bound to biotin. This can also be a problem for some immunofluorescence studies (Wood and Warnke 1981).

9.4.2 Streptavidin

In 1963 Chaiet et al. found that some species of the fungus *Streptomyces* produced an antibiotic that consisted of two components. First, a low molecular weight compound that inhibited biotin synthesis by bacteria and second, a higher molecular weight compound that became known as streptavidin.

Although streptavidin shows only 12% homology in amino acid sequence to egg-white avidin, the two proteins are clearly structurally related and have many similar properties (Green 1975). Hence, streptavidin is also a tetramer with each subunit binding one molecule of biotin. Its molecular weight (of approximately 60 kDa) is slightly smaller than avidin. However, there are some significant differences between the two molecules. First, streptavidin, unlike avidin, is not a glycoprotein and second, it has a slightly acidic isoelectric point (pH 6). Both these facts in theory should make the bacterial protein a less adhesive molecule than avidin. Further, streptavidin is not disulfide bonded.

For molecular details of recent data on the binding of avidin and streptavidin to biotin see Wilchek and Bayer (1989).

9.4.3 Biotin

Biotin is a small molecule of MW 244 that has also been referred to as vitamin H or coenzyme R. It is an essential vitamin which plays the role of carbon dioxide carrier in a number of key metabolic reactions that involve transfer of CO_2, in particular transcarboxylation, carboxylation and decarboxylation (see Moss and Lane 1971, for a review). It should be noted again that avidin-biotin binding results in a total loss of catalytic activity of all known biotin enzymes. In tissues biotin occurs mainly in a protein-bound form. Although it exists in all tissues, it is in relatively high concentrations in liver, kidney, pancreas and mammary gland. Its concentration is also elevated in tumours and in transformed cells. The molecule, which is very stable at room temperature, is soluble in water (0.22

g/l at 25 °C), more soluble in hot water and dilute alkali solution, and still more soluble in 95% ethanol (8 g/l) (Merck Index; ninth edition 1976).

9.4.4 Biotinylation of Proteins

The potential importance of the avidin-biotin reaction as a cytochemical tool arises directly from the fact that biotin can easily be covalently attached to essentially any protein or molecule that has a free amino group. It appears that the most important reactive groups for the biotinylation are lysine residues (Henderson et al. 1978). The use of biotinyl N-hydroxy succinimide (BNHS) to covalently tag biotin to proteins for electron microscopy was first carried out by Heitzmann and Richards (1974). Subsequently, this reaction was studied in some detail in an important paper by Guedson et al. (1981) who studied the effects of varying the degree of biotinylation on the biological activity of proteins, either enzyme activities or antibody-antigen reactions. For each protein they determined the number of free amino groups before and after the labelling reaction, thus estimating the number of molecules of biotin bound in each case. Their results showed clearly that a range of enzymes, antigens and antibodies could be efficiently biotinylated with 50–70% of the free amino groups substituted without appreciable loss of biological activity. In a few cases a slight loss of activity was measured at high levels of biotinylation. The method that they recommended is as follows.

A 0.1 M solution of BNHS is prepared by dissolving 34.1 mg BNHS in 1 ml of distilled dimethyl formamide. An appropriate volume of solution is then added to a solution of 10 mg/ml protein in 0.1 M NaHCO$_3$. As a guide as to the amount of the BNHS solution to add, the authors noted that in order to have a biotin to free amino group molar ratio of 1 at the initiation of the reaction they needed 57 µl for sheep antibody, 20 µl for glucose oxidase, 5 µl for horseradish peroxidase, 58 µl for alkaline phosphatase, 22 µl for β galactosidase and 88 µl for bovine serum albumin. The reaction is then allowed to proceed for 1 h at room temperature followed by dialysis at 4 °C against several changes of PBS. The authors then recommended adding an equal volume of double-distilled glycerol in order to have the convenience of storing the solution at −20 °C in ligand form (as for antibodies see Chap. 6). If it is necessary to remove the unreacted BNHS after the reaction this can be done by gel filtration on Sephadex G25 (Bonnard et al. 1984). The latter group also described the use of ^{14}C BNHS in order to be able to estimate the stoichiometry of binding.

Guedson et al. (1981) synthesized BNHS by a method described by Jasiewicz et al. (1976) using N-N-carbonyl diimidazole (Aldrich Co.). However, the commercial preparations of BNHS available from many companies can now also be used. For a slight modification of these methods, see also Hsu et al. (1981).

An example from our own work is the biotinylation of horseradish peroxidase (HRP) using the following protocol (Gruenberg et al. 1989). Twenty mg HRP in 9.5 ml of 0.1 M NaHCO$_3$/Na$_2$CO$_3$ (pH 9.0) are mixed with 11.4 mg of biotin-X-NHS (Calbiochem-Behring Corp. USA) in 0.5 ml dimethylformamide for 2 h at room temperature. This corresponds to a 50:1 molar excess of biotin over HRP. The unreacted active biotin groups are quenched with 1 ml of 0.2 M glycine at pH 8.0 and the solution is mixed for an additional 30 min. The mixture is dialyzed against several changes of PBS. The final biotHRP, 1.8 mg/ml, exhibits the same specific enzymatic activity as native HRP.

For an elegant example of the approach of biotinylating tubulin for kinetic studies on the nucleation of microtubules in vitro at the light microscopy level (using Texas Red Streptavidin), see Mitchison and Kirschner (1985).

9.4.5 The Use of a Linker in the Biotinylation Reaction

In some cases, the biotin on a molecule of interest may not be exposed enough to be able to bind the avidin complex. To overcome this problem, Kraehenbuhl and colleages have introduced the idea of using a neutral spacer molecule between the protein of interest and the biotin molecule (Bonnard et al. 1984). They used the reagent biotinyl-w-aminocapsoyl N-hydroxysuccinimide (BwACNHS) instead of BNHS in which an additional 7 atom linker is introduced between the biotin and the molecule of interest. The reaction is carried out by incubating the material to be biotinylated with 2 μmol of BwACNHS in 50 μl dimethylformamide for 2 h at room temperature. In the example given by Bonnard et al., cultured cells were biotinylated either with BNHS or with the BwACNHS and the binding of streptavidin was compared. Significantly better binding occurred when the reagent with the spacer was used. Whereas the optimal pH for binding in the latter case was 7.4, in the case of the BNHS biotinylation there was slightly higher binding at pH 4 than at pH 7.4.

9.4.6 Avidin-Biotin for Affinity Labelling Studies

9.4.6.1 General

The first use of the avidin-biotin reaction in affinity labelling was actually for ultrastructural rather than for light microscopical studies (Heitzmann and Richards 1974). In that paper, the authors non-specifically biotinylated bacterial membranes which were then labelled with avidin that had been conjugated to ferritin using glutaraldehyde. In subsequent years, while the avidin-biotin reaction has become a prominent tool in light microscopical studies, its use in EM studies has been rather limited. The possible reasons for this will be discussed below.

9.4.6.2 Labelling Protocols

For light microscopic studies Guedson et al. (1981) suggested two different kinds of labelling approaches using the avidin-biotin reaction. The first, which they referred to as the LAB technique, involved using a biotin-labelled affinity molecule such as an antibody or a lectin followed by avidin conjugated to an enzyme such as horseradish peroxidase (HRP). The second approach, which they called BRAB, involved three steps. In the first, biotin labelled affinity protein was applied followed by free avidin, which was then labelled in the third step with a second biotinylated marker protein. This approach is feasible because of the tetravalency of avidin which enables it to cross-link the two different biotinylated proteins. We can next discuss these approaches as they could be applied for EM studies.

9.4.6.3 The Two Step Approach

This is theoretically a very straightforward approach, where the primary affinity molecule is biotinylated and then labelled with an avidin-electron-dense marker conjugate. As mentioned, this was first accomplished by Heitzmann and Richards (1974), who described a protocol to covalently link avidin to ferritin using glutaraldehyde (see also Skutelsky and Bayer 1979). Subsequently, many workers have conjugated avidin to enzymes such as HRP: very few of these studies have been done at the EM level, however (for reviews see Korpela 1984a; van den Pol 1984).

Because of its relatively high electron density, it is clear that colloidal gold is theoretically the marker of choice for conjugation to avidin. However, the high isoelectric point of avidin makes it difficult to bind it to gold in a reproducible manner. The first description of a method to couple avidin to colloidal gold was by Tolson et al. (1981). Although there are many avidin-gold conjugates now commercially available in our experience very few of them are yet satisfactory for high resolution EM studies. Similarly, our own efforts to couple avidin to gold of various sizes have mostly produced preparations that have unacceptably high amounts of aggregates. The use of Streptavidin should solve this problem. A recent protocol which has been used successfully is described in the legend to Fig. 3. Bonnard et al. (1984) have also given a detailed description of the preparation of monodisperse Streptavidin-gold complexes.

9.4.6.3 Three-Step Protocols

As suggested by Guedson et al. (1981) for light microscopy, a three-step approach avoids the problem of conjugating avidin to markers. Although there is little literature available where such a protocol has been used, the approach can be discussed in a theoretical sense. The idea here is that the antibody used in the first step is biotinylated as is any marker protein which is conjugated to gold,

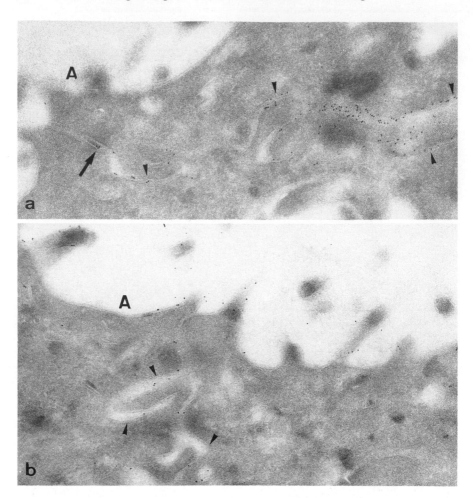

Fig. 3a,b. Examples of streptavidin-gold labelling of MDCK cells that had been surface biotinylated. A ricin-resistant cell line was biotinylated as described of the text and the cells, grown on filters were fixed with 8% wt/vol paraformaldehyde in 250 mM Hepes, pH 7.4, and cryo-sections were cut perpendicular to the plane of the filter as described by Parton et al. (1989). The thawed sections were transferred to grids and incubated on drops of PBS-BSA for 30 min, and then on 10 μl drops of streptavidin-gold (OD520 = 0.2) in the same buffer for 90 min at room temperature. After washing with PBS-BSA (six times, 10 min each), the sections were embedded in methyl cellulose-uranyl acetate. In (**a**) the cells were biotinylated basolaterally on ice, rinsed and then fixed immediately. In (**b**) the cells were treated as in **a** but then warmed up to 37°C for 3 h prior to fixation. Transcytosis of biotinylated membrane proteins from the basolateral surface (*arrowheads*) to the apical surface (*A*) is evident by the streptavidin-gold labelling (×44000). Control grids were treated as above, but were incubated with 0.5 mg/ml unconjugated streptavidin for 15 min prior to incubation on streptavidin-gold containing 0.5 mg/ml streptavidin. (From a study by Brändli et al. 1990. Courtesy of R. Parton, EMBL)

to be used in the third step. Avidin is used as a "sandwich" in the second step. The sequence of reaction would be (1) biotinylated antibody, (2) avidin, (3) biotinylated marker protein complexed to gold.

9.4.7 Biotinylation Studies with Living Cells

A recent innovation is the approach to tag biotin to the surface of living cells that may then be detected biochemically or immunocytochemically. At low temperatures (0–4 °C) this method can be used to specifically label the plasma membrane. Then, by warming up cells that were biotinylated at the cell surfaces to 20–37 °C, the endocytic compartments can be labelled. The approach can also be used to selectively tag the apical or basolateral domains of polarized cells grown on filters, for which the method was developed by Lisanti et al. (1988, 1989; see also Brändli et al. 1988; Busch et al. 1989). Lisanti et al. in fact described two different approaches for this purpose.

9.4.7.1 Sulfo-N-Hydroxy-Succinimide-Biotin

This compound (available from Pierce Chemical Co) is simply added to a suitable buffer such as PBS at a concentration of 0.5 mg/ml. The cells should be rinsed several times to get rid of serum proteins. For surface biotinylation the reaction is carried out on ice for 30 min and then the cells are rinsed with ice cold PBS.

9.4.7.2 Biotin Hydrazide

While the latter approach will biotinylate any free amino group on the cell surface, the biotin hydrazine method can be used to specifically biotinylate glycoprotein following periodite oxidation of sugars to form free aldehyde groups (Lisanti et al. 1989). Sodium periodate ($NaIO_4$ – 10 mM) is added to the cells in ice cold PBS in the dark. The cells are rinsed first three times with PBS and subsequently with 100 mM sodium acetate pH 5.5 containing 1 mM magnesium chloride and 0.1 mM calcium chloride. The cells can then be labelled with 2 mM biotin hydrazide (Pearce Chemical Co) in the same acetate buffer for 30 min on ice. The cells can then be incubated with PBS while agitated. In this reaction, the free aldehyde groups formed by the periodate oxidation react with the biotin hydrazide to form hydrazine groups (Heitzmann and Richards 1974). The sodium periodate and biotin hydrazide solutions should be freshly prepared. Note that this approach may also label glycerolipids (Henderson et al. 1978).

Following either of the above approaches the cells can be fixed and prepared for cryo-sections, or can be embedded in Lowicryl K_4M or LR white. The sections can then be treated with a suitable avidin or streptavidin-colloidal

Table 3. Streptavidin-gold labelling of Lowicryl (K4M and HM20), LR-gold and frozen sections of MDCK after apical surface biotinylation

Sections	Gold/µm of linear surface	Ref. for method
a) Cryo	18.6 ± 3.6	See Chapter 4
b) LR-gold	7.4 ± 0.7[a]	Berryman and Rodewald (1990)
c) K4M	6.2 ± 0.4[a]	W. Voorhout (University of Utrecht, pers. commun.)
d) HM20	1.3 ± 0.3[a]	W. Voorhout (University of Utrecht, pers. commun.)

[a] Before staining grids with uranyl acetate and lead citrate, the grids were incubated on 1% glutaraldehyde for 5 min and washed with distilled water to cross-link the streptavidin-gold to the sections. Otherwise a reduction in labelling of up to 30% was observed.

All the samples were from the same experiment and were fixed the same way using 1% glutaraldehyde in 200 mM cacodylate, pH 7.4; (b) was processed with a progressive lowering of temperature (PLT) protocol using acetone as the dehydrating agent. (c) and (d) were processed according to Voorhout (pers. commun.) using methanol as the substitution solvent.

N.B. The biotin group is expected to be fairly resistant to the denaturing effects of the fixatives, solvents, and resins. The differences in labelling are therefore likely to reflect the accessibility of the embedded biotinylated proteins to the streptavidin-gold, e.g. with HM20 the labelling may be restricted to the surface of the section, whereas with frozen sections labelling of antigens deep into the section can occur (R. Parton, EMBL, unpubl. data).

gold or other electron dense markers. For an example of such a labelling see Fig. 3.

In principle, any ligand which binds to, or is internalized by cells can be biotinylated before application.

9.4.8 Streptavidin-Gold

For the preparation of streptavidin-gold see Chapter 8.

An example of biotinylation of the surface of a ricin-resistant cell line of MDCK cells combined with streptavidin-gold is shown in Fig. 3. A quantative comparison of the labelling intensity seen with four different preparation methods is shown in Table 3.

9.5 Miscellaneous Labelling Approaches Relying on Non-Immunological High Affinity Interactions

9.5.1 The Enzyme Gold Method

An interesting labelling approach taking advantage of the affinity of enzymes for their substrates was introduced by Bendayan (1981). The principle of this method is to couple a purified preparation of a suitable enzyme to colloidal gold in order to bind the enzyme to its substrate present in sections. In a first paper, DNase and RNase were used to detect nucleic acids. A strength of this approach is that excellent controls are available, namely that the enzyme-gold complex can be blocked by co-incubation with the substrate, or with unlabelled enzyme. In a later publication, Bendayan (1984) studied the effect of different fixation and embedding protocols on the labelling with DNase and RNase gold complexes. Surprisingly, the author concluded that the best protocol for labelling with the RNase gold complex was epoxy resin, while glycolmethacrylate was found best suited for a DNase complex. In subsequent publications Lowicryl K4M sections were also used.

The enzyme-gold approach has recently been extended for a number of different enzymes, including glucosidase (Bendayan and Benhamou 1987), pectinases (Benhamou and Ouellette 1986), xylanase (Vian et al. 1983), exoglucanase (Benhamou et al. 1987), mannosidase (Londono and Bendayan 1987), as well as neuraminidase and hyaluronidase (Londoño and Bendayan 1988). For examples of labelling of Lowicryl K4M sections from this group see Fig. 4. For a detailed comprehensive review of this approach see Bendayan (1989).

9.5.2 Receptor/Ligand-Gold Complexes

Another high affinity interaction which can be taken advantage of for labelling studies is that between receptors and their ligands. This is especially true for cell surface receptors which can be detected simply by coupling their ligands to a suitable electron-dense marker, usually gold. In most of these studies the markers are first added to living cells at 4°C, a condition when no internalization occurs. The latter labelling can than be compared to labelling of different intracellular endocytic compartments after warming the cells to physiological temperatures (usually 37°C). Numerous examples can be found in the literature of the use of ligands such as transferrin (e.g. Neutra et al. 1985), asialoglycoprotein (Geuze et al. 1983), α2 macroglobulin (Yamashiro et al. 1984) and low density lipoproteins (Paavola et al. 1985)

There are also many successful studies where the ligand has been coupled to horseradish peroxidase (HRP), such as transferrin (Hopkins 1983; Willingham et al. 1984) or epidermal growth factor (Willingham and Pastan 1982).

In all such studies it is an established fact that the intracellular pathway of the internalized ligand when it is bound to an electron-dense marker may be different to that taken by the unbound ligand (see for example Neutra et al. 1985; Griffiths et al. 1988). This is most likely due to cross-linking of receptors and is especially true for relatively large (> 10 nm) gold particles that are likely to have a few ligands bound to each particle.

When sections of tissue are incubated with ligand-gold complexes the chances for success may be less than when the ligand is bound to living cells since, in the latter case, the process of receptor mediated endocytosis can serve

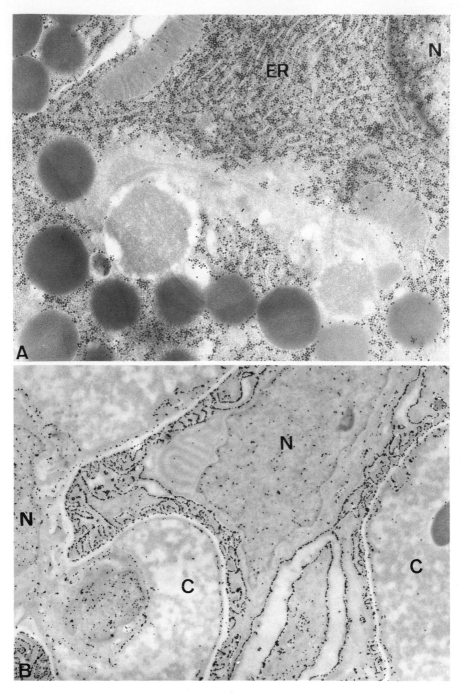

Fig. 4A, B

to concentrate the marker. Our own attempts at labelling cryo-sections of cultured cells with α2-macroglobulin- or transferrin-gold conjugates were not successful (unpubl. data).

Less common than the approach where the ligand is bound to the marker are those in which the receptor is bound. The obvious reason is that receptors are usually spanning membrane proteins that need detergents for solubilization and which are present in relatively low concentration in tissues. However, in some cases it may be possible to isolate soluble domains of these proteins. An example is given from our recent work on the cation-independent mannose 6-phosphate receptor (MPR), whose extracellular domain is naturally secreted into the medium/serum. These soluble fragments can then be purified on ligand columns, eluted and coupled to gold in order to detect the ligand, precursors of lysosomal enzymes containing the mannose 6-phosphate "address" signal (Ludwig et al. 1991).

9.5.3 Toxin-Gold Conjugates

Finally, mention should be made of a number of potent toxins which have been conjugated to colloidal gold for morphological studies. The lectin ricin has already been discussed. Other toxins that have been successfully conjugated to colloidal gold include cholera toxin (Montesano et al. 1982; Tran et al. 1987) and tetanus toxin (Montesano et al. 1982; Parton et al. 1988).

9.6 In Situ Hybridization

In recent years the approach of in situ hybridization has become an important tool in molecular biology for detecting defined sequences of RNA or DNA in cells and tissues. With a few exceptions, almost all the effort that has been put into this new technology has been concentrated at the light microscopy level. For this reason I will limit the discussion on this topic to a few pertinent remarks as well as give some key references.

Fig. 4A,B. Examples of the enzyme-gold technique on Lowicryl K4M sections. **A** shows labelling of a rat pancreas with α mannosidase (a commercial preparation) conjugated to 15 nm gold (15 μg/ml at pH 6). Most of the label is over the rough ER (*ER*) and nucleus (*N*) (×20000). For more details of this labelling see Londoño and Bendayan (1987). **B** shows labelling of rat kidney glomerulus with hyaluronidase gold (commercially prepared from bovine testis). The complexes were prepared at pH 7.5 and 1 μg/ml. The bulk of the labelling is over the plasma membranes as well as over nuclei (*N*), whereas no labelling is evident over the capillary lumen (*C*); ×10000. For more details see Londoño and Bendayan (1988). (Micrographs courtesy of M. Bendayan, University of Montreal, Canada)

The essence of this approach is to hybridize a piece of DNA or RNA onto the complementary stretch of base pairs in the relevant nucleic acid in situ. As in other aspects of nucleic acid molecular biology, the use of recombinant DNA technology now plays a key role in the in situ hybridization approach (for a discussion see Morrell 1989).

There is essentially no major conceptual difference between in situ hybridization and immunocytochemistry. In both cases the target molecule, e.g. a defined sequence of nucleic acid versus antigen, is specifically recognized by the probe, oligonucleotide vs. antibody. The chances of detecting the target molecule similarly depends in both cases on (1) the amount of the target molecule that is present and biologically active as well as accessible during the labelling reaction, and (2) the specificity and affinity of the probe for the target molecule under the conditions of reaction, e.g. whole mount or section.

It follows that for both light and electron microscopic studies, the same problems that have traditionally plagued immunocytochemistry and related high-affinity labelling techniques must also be tackled for the in situ approaches. Additionally, there are problems that are more specific for the latter. Thus, as for immunocytochemistry, the first problem is that of preserving the conformation of the nucleic acid in a form in which it is still accessible during the labelling reaction. With the theoretical exception of the freeze-substitution approach, it seems hard to imagine any protocol at the EM level for in situ hybridization, as for immunocytochemistry, that could avoid chemical fixation. In this respect the problems of cross-linking, or at least preserving, nucleic acids will probably necessitate new considerations (see Chap. 3). Similarly, as for immunocytochemistry one has the choice of using either a pre-embedding or a post-embedding/sectioning approach. The former has, until now, been more extensively applied for in situ hybridization studies at the EM level. As for immunocytochemistry, detergents have been used in order to increase accessibility of the reagents (Hutchison et al. 1982; Manuelidis et al. 1982; Webster et al. 1987; see Silva et al. 1989 for a review). The application of sections for carrying out in situ hybridization studies, until now, has been very limited indeed (Binder et al. 1986; Webster et al. 1987). Both the latter studies were done using Lowicryl K4M sections (see Binder 1987, for a review of the development of the technique). It seems surprising that thawed cryo-sections have not yet been used for EM in situ hybridization although cryo-sections are used routinely in many laboratories for light microscopic hybridizations (see Morrell 1989 for a recent review). For the same reasons that cryo-sections give the highest sensitivity for immunocytochemical labelling, it is to be expected, at least on theoretical grounds, that this procedure should also give the highest sensitivity for in situ studies at the EM level.

Whereas in immunocytochemistry one has the problem of visualizing the bound antibody, so in the in situ method must the bound probe be identified. In the earlier days of in situ hybridization radioactive probes and autoradiography were used extensively (see, for example, Hutchison et al. 1982). Recently, non-radioactive approaches have now become prominent, especially using biotin (Langer-Safer et al. 1982). Standard procedures are now available for tagging biotin to nucleotides/oligonucleotides (see Raap et al. 1989 for a review). The latter can then be visualized either with antibodies against biotin (followed by a suitable gold conjugate) or by avidin or streptavidin-gold conjugates (Hutchison et al. 1982; Binder et al. 1986; Silva et al. 1989). Avidin-peroxidase and avidin ferritin have also been used (Harris and Croy 1986), as have peroxidase-antibody conjugates (e.g. Manuelidis et al. 1982). As in immunocytochemistry, there is no obvious theoretical reason why multiple labelling steps cannot be applied with in situ hybridization, either using two different probes (and different visualizing systems) or by combining in situ hybridization with immunocytochemistry (for examples, see Silva et al. 1989; Singer et al. 1989).

Finally, most of the arguments/methods that are put forward with respect to quantitation of antigens (Chap. 11) will no doubt also apply to the in situ approach. Thus, as for immunocytochemistry, it seems likely that colloidal gold conjugates will become the particulate markers of choice. It is too early at this point to say what kinds of labelling efficiencies one can expect with this method but surely this information will have to come from the use of model systems similar to those that have been used for immunocytochemistry (see Chap. 11).

References

Åkerström B, Björck L (1986) A physiochemical study of protein G, a molecule with unique immunoglobulin G-binding properties. J Biol Chem 262:10240–10247

Åkerström B, Brodin T, Reis K, Björk T (1985) Protein G: a powerful tool for binding and detection of monoclonal and polyclonal antibodies. J Immunol 135:2589 2598

Åkerström B, Nielsen E, Björck L (1987) Definition of IgG- and albumin-binding regions of streptococcal protein G. J Biol Chem 262:13388–13391

Alroy J, Orgad U, Ucci AA, Pereira MEA (1984) Identification of glycoprotein storage diseases by lectins: a new diagnostic method. J Histochem Cytochem 32:1280–1284

Balslev Y, Hansen GII (1989) Preparation and use of recombinant protein G-gold complexes as markers in double labelling immunocytochemistry. Histochem J 21:449–454

Barondes SH (1981) Lectins: their multiple endogenous cellular functions. Annu Rev Biochem 50:207–231

Bayer EA, Wilchek M (1980) The use of the avidin-biotin complex as a tool in molecular biology. Methods Biochem Anal 26:1–9

Bendayan M (1981) Ultrastructural localization of nucleic acids by the use of enzyme-gold complexes. J Histochem Cytochem 29:531–541

Bendayan M (1984) Enzyme-gold electron microscopic cytochemistry: a new affinity approach for the ultrastructural localization of macromolecules. J Electron Microsc Tech 1:349–360

Bendayan M (1987) Introduction of the protein G-gold complex for high-resolution immunocytochemistry. J Electron Microsc Tech 6:7–13

Bendayan M (1989) The enzyme-gold cytochemical approach: a review. In: Hayat MA (ed) Colloidal gold: principles, methods, and applications, vol 2. Academic Press, New York, pp 117–147

Bendayan M, Garzon S, (1988) Protein G-gold complex: comparative evaluation with protein A-gold for high resolution immunocytochemistry. J Histochem Cytochem 36:597–607

Bendayan M, Benhamou N (1987) Ultrastructural localization of glucoside residues on tissue sections by applying the enzyme-gold approach. J Histochem Cytochem 35:1149–1158

Benhamou N, Ouellette GB (1986) Use of pectinases complexed to colloidal gold for the ultrastructural localization of polygalacturonic acids in the cell wall of the fungus Ascocalyx abientina. Histochem J 18:95

Benhamou N, Chamberland H, Ouelette GB, Pauze FJ (1987) Ultrastructural localization of beta-(1->4)-D-glucans in two pathogenic fungi and in their host tissues by means of an exoglucanase gold complex. Can J Microbiol 33:405–417

Berger EG, Hesford FJ (1985) Localization of galactosyl- and sialyltransferase by immunofluorescence: evidence for different sites. Proc Natl Acad Sci USA 82:4736–4739

Bernhard W, Avrameas S (1971) Ultrastructural visualization of cellular carbohydrate components by means of concanavalin A. Exp Cell Res 64:232–241

Berryman MA, Rodewald RD (1990) An enhanced method for post-embedding immunocytochemical staining which preserves cell membranes. J Histochem Cytochem 38:159–170

Binder M (1987) In situ hybridization at the electron microscope level. Scanning Microsc 1:331–338

Binder M, Tourmente S, Roth J, Renaud M, Gehring WJ (1986) In situ hybridization at the electron microscopy level: localization of transcripts on ultrathin sections of Lowicryl K4M-embedded tissue using biotinylated probes and protein A-gold complexes. J Cell Biol 102:1646–1653

Björk I (1972) Some physicochemical properties of protein A from *Staphylococcus aureus*. Eur J Biochem 29:579–584

Björk I, Kronvall G (1984) Purification and some properties of streptococcal protein G, a novel IgG-binding reagent. J Immunol 133:969

Boas MA (1927)The effect of desiccation upon the nutritive properties of egg white. Biochem J 21:712–725

Bonnard C, Papermaster DS, Kraehenbuhl J-P (1984) The Streptavidin-biotin bridge technique: application in light and electron microscopy immunocytochemistry. In: Polak JM, Varndell IM (eds) Immunolabelling for electron microscopy. Elsevier, Amsterdam, pp 95–112

Brändli AW, Hansson GC, Rodriguez-Boulan E, Simons K (1988) A polarized epithelial cell mutant deficient in translocation of UDP-galactose into the Golgi complex. J Biol Chem 263:16283–16290

Brändli AW, Parton RG, Simons K (1990) Transcytosis in MDCK cells: identification of glycoproteins transported bidirectionally between both plasma membrane domains. J Cell Biol 111:2909–2922

Buckley K, Kelly RB (1985) Identification of a transmembrane glycoprotein specific for secretory vesicles of neural and endocrine cells. J Cell Biol 100:1284–1294

Busch G, Hoder D, Reutter W, Tauber R (1989) Selective isolation of individual cell surface proteins from tissue culture cells by a cleavable biotin label. Eur J Cell Biol 50:257–262

Chaiet L, Miller TN, Tausig F, Wolf FJ (1963) Antibiotic MSD-235. Separation and purification of synergistic components. Antimicrob Agents Chemother 3:28–32

Deisenhofer J (1981) Crystallographic refinement and atomic models of a human Fc fragment and its complex with fragment B of protein A from *Staphylococcus aureus* at 29- and 28- resolution. Biochemistry 20:2361–2370

Deisenhofer J, Jones TA, Huber R (1978) Crystallization, crystal structure analysis and atomic model of the complex formed by a human Fc fragment and fragment B of protein A from *Staphylococcus aureus* Hoppe-Seyler's Z. Physiol Chem 359:975–985

De Petris S (1978) Immunoelectron microscopy and immunofluorescence in membrane biology. In: Korn ED (ed) Methods in membrane biology, vol 9. Plenum Press, New York, pp 1–20

De Petris S, Baumgartner U (1982) Spontaneous segregation of receptors for peanut and *Helix pomatia* agglutins to the uropod region of polarized lymphocytes. J Ultrastruc Res 80:323–338

Deschuyteneer M, Eckhardt AE, Roth J, Hill RL (1988) The subcellular localization of apomucin and nonreducing terminal N-acetylgalactosamine in porcine submaxillary glands. J Biol Chem 263:2452–2459

Duncan JR, Kornfeld S (1988) Intracellular movement of two mannose-6-phosphate receptors: return to the Golgi apparatus. J Cell Biol 106:617–628

Dunphy WG, Brands R, Rothman JE (1985) Attachment of terminal N-acetylglucosamine to asparagine-linked oligosaccharides occurs in central cisternae of the Golgi stack. Cell 40:463–472

Eliasson M, Olsson A, Palmcrantz E, Wiberg K, Inganäs M, Guss B, Lindberg M, Uhlén M (1988) Chimeric IgG-binding receptors engineered from staphylococcal protein A and streptococcal protein G. J Biol Chem 263:4324–4327

Eliasson M, Andersson R, Olsson A, Wigzell H, Uhlén M (1989) Differential IgG-binding characteristics of staphylococcal protein A, streptococcal protein G, and a chimeric protein AG. J Immunol 142:575–581

Elo HA, Korpela J (1984) The occurence and production of avidin: a new conception of the high-affinity biotin-binding protein. Comp Biochem Physiol 78B:15–20

Endo Y, Mitsui K, Motizuki M, Tsurugi K (1987) The mechansim of action of ricin and related toxic lectins on eukaryotic ribosomes. J Biol Chem 262:5908–5912

Ey PL, Prowse SJ, Jenkin CR (1978) Isolation of pure IgG1, IgG2A and IgG2B immunoglobulins from mouse serum using protein A-sepharose. Immunochem 15:429–436

Fahnestock SR, Alexander P, Nagle J, Filpula D (1986) Gene for an immunoglobulin-binding protein from a group G streptococcus. J Bacteriol 167:870–880

Fischer J, Klein PJ, Vierbuchen M, Skutta B, Uhlenbruck G, Fischer R (1984) Characterization of glycoconjugates of the gastrointestinal mucosa by lectins. J Histochem Cytochem 32:681–689

Forsgren A, Sjöguist J (1966a) "Protein A" from *Staphylococcus aureus* III. Reaction with rabbit γ-globulin. J Immunol 99:19–24

Forsgren A, Sjöquist J (1966b) "Protein A" from *S aureus* I. Pseudo-immune reaction with human γ-globulin. J Immunol 97:822–827

Geuze HJ, Slot JW, Strous GJSM, Lodish HF, Schwartz JL (1983) Intracellular site of asialoglycoprotein receptor-ligand uncoupling: double-label immunoelectron microscopy during receptor-mediated endocytosis. Cell 32:277–287

Goding JW (1978) Use of staphylococcal protein A as an immunological reagent. J Immunol Methods 20:241–253

Goldstein IJ, Hayes CE (1978) The lectins: carbohydrate-binding proteins of plants and animals. In: Tipson RS, Horton D (eds) Advances in carbohydrate chemistry and biochemistry, vol 35. Academic Press, New York, pp 128–340

Goldstein IJ, Poretz RD (1986) Isolation, physicochemical characterization, and carbohydrate-binding specificity of lectins. In: Liener IE, Sharon N, Goldstein IJ (eds) The lectins. Academic Press, Orlando, pp 35–250

Gray LS, Masouredis SP (1982) Interaction of ^{125}I-protein A with erythrocyte-bound IgG. J Lab Clin Med 99:399–409

Green NM (1975) Avidin. Adv Protein Chem 29:85–133

Griffiths G, Brands R, Burke B, Louvard D, Warren G (1982) Viral membrane proteins acquire galactose in trans Golgi cisternae during intracellular transport. J Cell Biol 95:781–792

Griffiths G, Hoflack B, Simons K, Mellman I, Kornfeld S (1988) The mannose-6-phosphate receptor and the biogenesis of lysosomes. Cell 52:329–341

Gruenberg J, Griffiths G, Howell KE (1989) Characterization of the early endosome and putative endocytic carrier vesicles in vivo and with an assay of vesicle fusion in vitro. J Cell Biol 108:1301–1316

Gruhn WB, McDuffie FC (1979) Measurement of immunoglobin binding to synovial fibroblast monolayers: comparison of Staphylococcal protein A binding to cytotoxic assay methods. J Immunol Methods 29:227–236

Guedson J-L, Ternynck T, Avrameas S (1981) The use of avidin-biotin interaction in immunoenzymatic techniques. J Histochem Cytochem 27:1131–1139

Guss B, Eliasson M, Olsson A, Uhlén M, Frej A-K, Jörnvall H, Flock J-I, Lindberg M (1986) Structure of the IgG-binding regions of streptococcal protein G. EMBO J 5:1567–1576

György P (1954) In: Sebrell WH Jr, Harris RS (eds) The vitamins, vol 1, Academic Press, New York, pp 527

Harboe M, Folling I (1974) Subclasses of IgM based on differences in reactivity with *staphylococcus aureus*. Scand J Immunol 3:878–879

Harris N, Croy RRD (1986) Localization of mRNA for pea legumin: in situ hybridization using a biotinylated cDNA probe. Protoplasma 130:57–67

Hart GW, Haltiwanger RS, Hold GD, Kelly WG (1989) Glycosylation in the nucleus and cytoplasm. Annu Rev Biochem 58:841–874

Hazlett LD, Mathieu P (1989) Glycoconjugates on corneal epithelial surface: effect of neuraminidase treatment. J Histochem Cytochem 37:1215–1224

Heggeness MH, Ash JF (1977) Use of the avidin-biotin complex for the localization of actin and myosin with fluorescence microscopy. J Cell Biol 73:783–788

Heitzmann H, Richards FM (1974) Use of the avidin-biotin complex for specific staining of biological membranes in electron microscopy. Proc Natl Acad Sci USA 71:3537–3541

Henderson R, Jubb JS, Whytock S (1978) Specific labelling of the protein and lipid on the extracellular surface of purple membrane. J Mol Biol 123:259–274

Holt GD, Snow CM, Senior A, Haltiwanger RS, Gerace L, Hart GW (1987) Nuclear pore complex glycoproteins contain cytoplasmically disposed O-linked N-acetylglucosamine. J Cell Biol 104:1157–1164

Hopkins CR (1983) Intracellular routing of transferrin and transferrin receptoptors in epidermoid carcinoma A431 cells. Cell 35:321–330

Horisberger M (1979) Evaluation of colloidal gold as a cytochemical marker for transmission and scanning electron microscopy. In: Monsigny M, Schrevel J (eds) Membrane glycoconjugates. Biol Cell 36:253–259

Horisberger M, Rosset J (1977) Colloidal gold, a useful marker for transmission and scanning electron microscopy. J Histochem Cytochem 25:295–305

Hsu S-M, Raine L (1981) Protein A, avidin and biotin in immunocytochemistry. J Histochem Cytochem 29:1349–1353

Hsu S-M, Raine L, Fanger H (1981) Use of avidin-biotin-peroxidase complex (ABC) in immunoperoxidase techniques: a comparison between ABC and unlabeled antibody (PAP) procedures. J Histochem Cytochem 29:577–580

Hubbard SC, Ivatt RJ (1981) Synthesis and processing of asparagine-linked oligosaccharides. Annu Rev Biochem 50:555–583

Hutchison NJ, Langer-Safer PR, Ward DC, Hamkalo BA (1982) In situ hybridization at the electron microscope level: hybrid detection by autoradiography and colloidal gold. J Cell Biol 95:609–618

Jasiewicz ML, Schoenberg DR, Mueller GC (1976) Selective retrieval of bition-labeled cells using immobilized avidin. Exp Cell Res 100:213–217

Jin M, Sahagian GG, Snider MD (1989) Transport of surface mannose 6-phosphate receptor to the Golgi complex in cultured human cells. J Biol Chem 264:7675–7680

Kessler SW (1975) Rapid isolation of antigens from cells with a staphylococcal protein A-antibody adsorbent: parameters of the interaction of antibody-antigen complexes with protein A. J Immunol 115:1617–1624

Kornfeld S (1987) Trafficking of lysosomal enzymes. FASEB J 1:462–468

Kornfeld R, Ferris C (1975) Interaction of immunoglobulin glycopeptides with concanvalin A. J Biol Chem 250:2614–2619

Kornfeld R, Kornfeld S (1985) Assembly of asparagine-linked oligosaccharides A. Rev Biochem 54:631–664

Korpela J (1984a) Avidin, a high affinity biotin-binding protein, as a tool and subject of biological research. Med Biol 62:5–26

Korpela J (1984b) Chicken macrophages synthesize and secrete avidin in culture. Eur J Cell Biol 33:105–111

Korpela J, Kulomaa M, Tuohimaa P, Vaheri A (1983) Avidin is induced in chicken embryo fibroblasts by viral transformation and cell damage. EMBO J 2:1715–1719

Kronvall G, Seal U, Finstad J, Williams RC Jr (1970) Phylogenetic insight into evolution of mammalian FcFragment of γ-globulin using staphylococcal protein A J Immunol 104:140–147

Langer PR, Waldrop AA, Ward DC (1981) Enzymatic synthesis of biotin-labeled polynucleotides: novel nucleic acid affinity probes. Proc Natl Acad Sci USA 78:6633–6637

Langer-Safer PR, Levine M, Ward DC (1982) Immunological method for mapping genes on *Drosophila* polytene chromosomes. Proc Natl Acad Sci USA 79:4381–4385

Langone JJ (1982a) Protein A of *Staphylococcus aureus* and related immunoglobulin receptors produced by streptococci and pneumonococci. Adv Immunol 32:157–252

Langone JJ (1982b) Applications of immunobilized protein A in immunochemical techniques. J Immunol Methods 55:277–296

Langone JJ, Boyle MD, Borsos T (1978) Studies on the interaction between protein A and immunoglobulin G. J Immunol 121:333–338

Larsson EL, Coutinho A, Martinez C (1980) A suggested mechanism for lymphocyte-T activation – implications on the acquisition of functional reactivities. Immunol Rev 51:61–91

Larsson E-L, Lindahl KF, Langhorne J, Coutinho A (1981) Quantitative studies on concanavalin A-induced, TCGF-reactive T cells. Am Assoc Immunol 127:1081–1090

Lis H, Sharon N (1986a) Lectins as molecules and as tools. Annu Rev Biochem 55:35–67

Lis H, Sharon N (1986b) Biological properties of lectins. In: Liener IE, Sharon N, Goldstein IJ (eds) The lectins. Academic Press, Orlando, pp 265–370

Lisanti MP, Sargiacomo M, Graeve L, Saltiel AR, Rodriguez-Boulan E (1988) Polarized apical distribution of glycosyl-phosphatidylinositol anchored proteins in a renal epithelial cell line: Proc Natl Acad Sci USA 85:9557–9561

Lisanti MP, Le Bivic A, Sargiacomo M, Rodriguez-Boulan E (1989) Steady-state distribution and biogenesis of endogenous madin-darby canine kidney glycoproteins: evidence for intracellular sorting and polarized cell surface delivery: J Cell Biol 109:2117–2127

Ljungquist C, Jansson B, Moks T, Uhlén M (1989) Thiol-directed immobilization of recombinant IgG-binding receptors. Eur J Biochem 186:557–561

Londoño I, Bendayan M (1987) Ultrastructural localization of mannoside residues on tissue sections: comparative evaluation of the enzyme-gold and the lectin-gold approaches. Eur J Cell Biol 45:88–96

Londoño I, Bendayan M (1988) High-resolution cytochemistry of neuraminic and hexuronic acid-containing macromolecules applying the enzyme-gold approach. J Histochem Cytochem 36:1005–1014

Louvard D, Reggio H, Warren G (1982) Antibodies against the endoplasmic reticulum and Golgi apparatus J Cell Biol 34:773–786

Lucocq JM, Berger EG, Roth J (1987) Detection of terminal N-linked N-acetylglucosamine residues in the Golgi apparatus using galactosyltransferase and endoglucosaminidase F/peptide N-glycosidase F: adaptation of a biochemical approach to electron microscopy. J Histochem Cytochem 35(1):67–74

Ludwig T, Griffiths G, Hoflack B. Distribution of newly synthesized lysosomal enzymes in the endocytic pathway of normal rat kidney cells. J Cell Biol 115:1561–1572

Manuelidis L, Langer-Safer PR, Ward DC (1982) High-resolution mapping of satellite DNA using biotin-labeled DNA probes. J Cell Biol 95:619–625

Maylié-Pfenninger M-F, Jamieson JD (1980) Development of cell surface saccharides on embryonic pancreatic cells. J Cell Biol 86:96–108

Mitchison TJ, Kirschner MW (1985) Properties of the kinetochore in vitro. Microtubule nucleation and tubulin binding. J Cell Biol 101:755–765

Moks T, Abrahmsén L, Nilsson B, Hellman Ul, Sjöquist J, Uhlén M (1986) Staphylococcal protein A consists of five IgG-binding domains. Eur J Biochem 156:637–643

Möller G (ed) (1980) T cell stimulating growth factors. Immunol Rev 51

Montesano R, Roth J, Robert A, Orci L (1982) Non-coated membrane invaginations are involved in binding and internalization of cholera and tetanus toxins. Nature 296:651–654

Montreuil J (1982) Glycoproteins. In: Neuberger A, van Deenen L (eds) Comprehensive biochemistry, vol 19B. part II. Elsevier, New York, p 1–188

Morrell JI (1989) Application of in situ hybridization with radioactive nucleotide probes to detection of mRNA in the central nervous system. In: Bullock GR, Petrusz P (eds) Techniques in immunocytochemistry. Academic Press, London, pp 127–146

Moss J, Lane MD (1971) The biotin-dependent enzymes. Adv Enzymol 35:321–443

Mota G, Ghetie V, Sjöquist J (1978) Characterization of the soluble complex formed by reacting rabbit IgG with protein A of *S aureus*. Immunochemistry 15:639–642

Murata F, Tsuyama S, Suzuki S, Hamada H, Ozawa M, Muramatsu T (1983) Distribution of glycoconjugates in the kidney studies by use of labeled lectins. J Histochem Cytochem 31:139–144

Muresan V, Iwanij V, Smith ZDJ, Jamieson JD (1982) Purification and use of limulin: a sialic acid-specific lectin. J Histochem Cytochem 30(9):938–946

Nakajima M, Ito N, Nishi K, Okamura Y, Hirota P (1988) Cytochemical localization of blood group substances in human salivary glands using lectin-gold complexes. J Histochem Cytochem 36:337–348

Neefjes J, Verkerk JMH, Broxterman HJG, van der Marel GA, van Boom JH, Ploegh HL (1988) Recycling glycoproteins do not return to the cis-Golgi. J Cell Biol 107:79–87

Nemanic MK, Whitehead JS, Elias PM (1983) Alterations in membrane sugars during epidermal differentiation: visualization with lectins and role of glycosidases. J Histochem Cytochem 31(7):887–897

Neutra MR, Ciechanover A, Owen LS, Lodish HF (1985) Intracellular transport of transferrin- and asialoorosomucoid-colloidal gold conjugates to lysosomes after receptor-mediated endocytosis. J Histochem and Cytochem 33(11):1134–1144

Neville DM, Hudson TH (1986) Transmembrane transport of diptheria toxins, related toxins ad colicins. Annu Rev Biochem 55:195–224

Nygren P-, Eliasson M, Abrahmsén L, Uhlén M, Palmcrantz E (1988) Analysis and use of the serum albumin binding domains of streptococcal protein G. J Mol Recognition 1:69–74

O'Keefe E, Bennett V (1980) Use of immunoglobulin-loaded protein A-bearing staphylococci as a primary solid phase immunoabsorbent in radioimmunoassay. J Biol Chem 255:561–568

Olsnes S (1987) Protein toxins – closing in on ricin action. Nature 328:474–475

Olsson A, Eliasson M, Guss B, Nilsson B, Hellman U, Lindberg M, Uhlén M (1987) Structure and evolution of the repetitive gene encoding streptococcal protein G. Eur J Biochem 168:319–324

Orci L, Malhorta V, Amherdt M, Serafini T, Rothman JE (1989) Dissection of a single round of vesicular transport: sequential intermediates for intercisternal movement in the Golgi stack. Cell 56:357–368

Paavola LG, Strauss JF III, Boyd CO, Nestler JE (1985) Uptake of gold- and [^{3}H]-cholesterol linoleate-labeled human low density lipoprotein by cultured rat granulosa cells: cellular mechanisms involved in lipoprotein metabolism and their importance to steroidogenesis. J Cell Biol 100:1235–1244

Parton RG, Ockleford CD, Critchley DR (1988) Tetanus toxin binding to mouse spinal cord cells: an evaluation of the role of gangliosides in toxin internalization. Brain Res 475:118–127

Parton RG, Prydz K, Bomsel M, Simon K, Griffiths G (1989) Meeting of the apical and basolateral endocytic pathways of the Madin-Darby canine kidney cell in late endosomes. J Cell Biol 109:3259–3272

Paulson JC (1989) Glycoproteins: what are the sugar chains for? Trends Biochem Sci 14:272–279

Pino RM (1984) Ultrastructural localization of lectin receptors on the surface of the rat retinal pigment epithelium: decreased sensitivity of the avidin-biotin method due to cell surface charge. J Histochem Cytochem 32:862–868

Raap AK, Hopman AHN, van der Ploeg M (1989) Hapten labeling of nucleic acid probes for DNA in situ hybridization. In: Bullock GR, Petrusz P (eds) Techniques in immunocytochemistry. Academic Press, London, pp 167–198

Raff MC, Fields KL, Hakomori S, Mirsky R, Pruss RM, Winter J (1979) Cell-type-specific markers for distinguishing and studying neurons and the major classes of glial cells in culture. Brain Res 174:283–308

Raff MC, Abney ER, Miller RH (1984) Two glial cell lineages diverge prenatally in rat optic nerve. Dev Biol 106:53–60

Rauvala H, Finne J (1979) Structural similarity of the terminal carbohydrate sequences of glycoproteins and glycolipids. FEBS Lett 97:1–8

Reggio H, Bainton D, Harms E, Coudrier E, Louvard D (1984) Antibodies against lysosomal membranes reveal a 100,000 mol-wt protein that cross-reacts with H^+K^+ ATPase from gastric mucosa. J Cell Biol 99:1511–1526

Roth J (1982) New approaches for in situ localization of antigens and glycoconjugates on thin sections: the protein A-gold (pAg) technique and the lectin-colloidal gold marker system. Proc 10th Int Congr Electron Microscopy, Hamburg, 3:245–252

Roth J (1983) Application of lectin-gold complexes for electron microscopic localizations of glycoconjugates on thin sections. J Histochem Cytochem 31:987–989

Roth J (1984) Cytochemical localization of terminal N-acetyl-D-glucosamine residues in cellular compartments of intestinal goblet cells: implications for the topology of O-glycosylation, J Cell Biol 98:399–406

Roth J, Berger EG (1982) Immunocytochemical localization of galactosyl transferase in HeLa cells: codistribution with thiamine pyrophosphatase in trans Golgi cisternae. J Cell Biol 92:223–229

Roth J, Binder M (1978) Colloidal gold, ferritin and peroxidase as markers for electron microscopic double labeling lectin technique. J Histochem Cytochem 26:163–169

Roth J, Brown D, Orci L (1983) Regional distribution of N-acetyl-D-galactosamine residues in the glycocalyx of glomerular podocytes. J Cell Biol 96:1189–1196

Roth J, Lucocq JM, Charest PM (1984) Light and electron microscopic demonstration of sialic acid residues with the lectin from Limax flavus: a cytochemical affinity technique with the use of fetuin-gold complexes. J Histochem Cytochem 32:1167–1176

Roth J, Taatjes DJ, Lucocq JM, Weinstein J, Paulson JC (1985) Demonstration of an extensive trans-tubular network continuous with the Golgi apparatus stack that may function in glycosylation. Cell 43:287–295

Ruel K, Joseleau JP (1984) Use of enzyme-gold complexes for the ultrastructural localization of hemicelluloses in the plant cell wall. Histochemistry 81:573–581

Sato A, Spicer SS (1982) Ultrastructural visualization of galactose in the glycoprotein of gastric surface cells with a peanut lectin conjugate. Histochem J 14:125–138

Sharon N, Lis H (1981) Glycoproteins: research booming on long-ignored, ubiquitous compounds. Chem Eng News, March 1981: 21–44

Sharon N, Lis H (1989) Lectins as cell recognition molecules. Science 246:227–246

Sheares BT, Robbins PW (1986) Glycosylation of ovalbumin in a heterologous cell – analysis of oligosaccharide chains of th cloned glycoprotein in mouse L-cells. Proc Natl Acad Sci USA 83:1993–1997

Silva FG, Lawrence JB, Singer, RII (1989) Progress toward ultrastructural identification of individual mRNAs in thin section: myosin heavy-chain mRNA in developing myotubes. In: Bullock GR, Petrusz P (eds) Techniques in immunocytochemistry. Academic Press, London, pp 147–166

Singer RH, Langevin GL, Bentley Lawrence J (1989) Ultrastructural visualization of cytoskeletal mRNAs and their associated proteins using double-label in situ hybridization. J Cell Biol 108:2343–2353

Sjödahl J (1977a) Repetitive sequences in protein A from Staphylococcus aureus. Eur J Biochem 73:343–351

Sjödahl J (1977b) Structural studies on the four repetitive Fc-binding regions in protein A from Staphylococcus aureus. Eur J Biochem 78:471–490

Sjöholm J, Bjerén A, Sjöquist J (1973) Protein A from Staphylococcus aureus XIV. The effects of nitration of protein A with tetranitromethane and subsequent reduction. J Immunol 110:1562–1569

Sjöquist J, Stålenheim G (1969) Protein A from Staphylococcus aureus IX. Complement-fixing activity of protein A-IgG complexes. J Immunol 103:467–473

Skutelsky E, Bayer EA (1979) The ultrastructural localization of cell surface glycoconjugates: affinity cytochemistry via the avidin-biotin complex. Biol Cellulaire 36:237–252

Skutelsky E, Goyal V, Alroy J (1987) The use of avidin-gold complex for light microscopic localization of lectin receptors. Histochemistry 86:291–295

Slot JW, Geuze HJ (1985) A new method of preparing gold probes for multiple-labeling cytochemistry. Eur J Cell Biol 38:87–93

Spiegel S, Skutelsky E, Bayer EA, Wilchek M (1982) A novel approach for the topographical localization of gylcolipids on the cell surface. Biochim Biophys Acta 687:27–34

Strous GJ, van Kerkhof P, Willemsen R, Geuze HJ, Berger EG (1983) Transport and topology of galactosyltransferase in endomembranes of HeLa cells. J Cell Biol 97:723–727

Surolia A, Pain D, Khan, MI (1982) Protein A: nature's universal anti-antibody. Trends Biochem Sci 7:74–76

Taatjes DJ, Chen T-H, Åckerström B, Björk L, Carlemalm B, Roth J (1987) Streptococcal protein G-gold complex: comparison with streptococcal protein A gold complex for spot blotting and immunolabelling. Eur J Cell Biol 45:151–159

Tartakoff AM, Vassalli P (1983) Lectin-binding sites as markers of Golgi subcompartments: proximal-to-distal maturation of oligosaccharides. J Cell Biol 97:1243–1248

Tolson ND, Boothroyd B, Hopkins, CR (1981) Cell surface labelling with gold colloid particulates: the use of avidin and staphylococcal protein A-coated gold in conjunction with biotin and fc-bearing ligands. J Microsc 123:215–226

Tran D, Carpentier J-L, Sawano F, Gorden P, Orci L (1987) Ligands internalized through coated or noncoated invaginations follow a common intracellular pathway. Proc Natl Acad Sci USA 7957–77961

Uhlén M, Guss B, Nilsson B, Gatenbeck S, Philipson L, Lindberg M (1984) Complete sequence of the straphylococcal gene encoding protein A. J Biol Chem 259:1695–1702

van den Pol AN (1984) Colloidal gold and biotin-avidin conjugates as ultrastructural markers for neural antigens. Q J Exp Physiol 69:1–33

van Deurs B, Sandvig K, Petersen OW, Olsnes S, Simons K, Griffiths G (1988) Estimation of the amount of internalized ricin that reaches the *trans*-Golgi network. J Cell Biol 106:253–267

Verwey WF (1940) A type-specific antigenic protein derived from staphylococcus. J Exp Med 71:635–644

Vian B, Brillouet JM, Satiat-Jeunemaitre B (1983) Ultrastructural visualization of xylans in cell walls of hardwood by means of xylanase-gold complex. Biol Cell 49:179–188

Virtanen I, Ekblom P, Laurila P (1980) Subcellular compartmentalization of saccharide moieties in cultured normal and malignant cells. J Cell Biol 85:429–434

von Figura K, Gieselmann V, Hasilik A (1986) Lysosomal enzymes and their receptors. A. Rev Biochem 55:167–193

Warnke R, Levy R (1980) Detection of T and B cell antigens with hybridoma monoclonal antibodies: a biotin-avidin-horseradish peroxidase method. J Histochem Cytochem 28:771–776

Webster HdeF, Lamperth L, Favilla JT, Lemke G, Tesin D, Manuelidis L (1987) Use of a biotinylated probe and in situ hybridization for light and electron microscopic localization of P_0 mRNA in myelin-forming Schwann cels. Histochemistry 86:441–444

Wilchek M, Bayer EA (1989) Avidin-biotin technology ten years on: has it lived up to its expectations? Trends Biochem Sci 14:408–412

Willingham MC, Pastan I (1982) Transit of epidermal growth factor through coated pits of the Golgi system. J Cell Biol 94:207–212

Willingham MC, Hanover JA, Dickson RB, Pastan I (1984) Morphologic characterization of the pathway of transferrin endocytosis and recycling in human KB cells. Proc Natl Acad Sci USA 81:175

Wood GS, Warnke R (1981) Suppression of endogenous avidin-binding activity in tissues and its relevance to biotin-avidin detection systems. J Histochem Cytochem 29:1196–1204

Yamashiro DJ, Tycko B, Fluss SR, Maxfield FR (1984) Segregation of transferrin to a mildly acidic (pH 6.5) para-Golgi compartment in the recycling pathway. Cell 37:789

Young WW, Tamura Y, Wolock DM, Fox JW (1984) Staphylococcal protein A binding to the Fab fragments of mouse monoclonal antibodies. J Immunol 133:3163–3166

Zieske JD, Bernstein IA (1982) Modification of cell surface glycoprotein: addition of fucosyl residues during epidermal differentiation. J Cell Biol 95:626–631

Chapter 10
Preembedding Immuno-Labelling

Two different kinds of preembedding strategies will be considered in this chapter. The first can be referred to as the "intact cell" approach and the second as the "cell ghost" approach. By the "intact cell" approach I refer to the classical immunoperoxidase approach where one aims to strike a balance between a minimal extraction of cellular components and a maximal penetration of antibody reagents. This is a problematic procedure for high resolution immunocytochemistry which should be considered as a last, rather than a first, method. That the use of this approach is especially suspect for membrane antigens seems almost too obvious to state; however, the vast number of publications that have used this procedure for precisely this purpose indicates that not everyone shares this concern[1]. An exception to this statement may be the use of immunoperoxidase for relatively low resolution EM studies. This is especially true for studies of cell connectivities in the nervous system where EM preembedding labelling has been used extensively, in combination with light microscopy and also serial section EM, to trace neuronal connectivities (see below). For a more general critique of this approach see Chapter 2.

There are very simple and reproducible methods for labelling the cell surface of tissue culture cells. These will not be considered here (for a discussion see de Petris 1978 and de Waele 1984). Rather, I shall concentrate on the technically more difficult goal of labelling antigens that are not easily accessible to incubation media. In practice this means both intracellular antigens and, in the case of many tissues, also extracellular antigens.

By the "cell ghost" approach I am referring to methods that are compatible with a significant extraction of lipids and soluble components of the cell. By definition, these techniques are aimed at cytoskeletal elements whose organization may withstand drastic detergent treatment. A definite advantage of this strategy, in contrast to the "intact cell" one, is that particulate markers such as gold may be used. Although complete accessibility of antibody for antigen cannot be guaranteed even by "total" detergent or solvent extractions, this approach appears more generally useful than the "intact cell" one.

---

[1] Admittedly this method is widely and successfully used for immunolabelling at the light microscope level. The resolution of the label can be extremely high. For example, individual microtubules can be resolved using immunofluorescence. However, interpretation of the labelling of membrane organelles must always be made with caution when one considers the harsh procedures (e.g. either 0.2% Triton X-100 following $\approx 3\%$ formaldehyde fixation for periods less than 30 min or methanol/acetone fixation) normally used to permeabilize cells.

Most reviews written on preembedding labelling have concentrated on providing a list of recipes that have empirically been found to give acceptable results rather than trying to understand the principles involved in the critical steps leading to and including the labelling reactions. The reason for this is that, compared with postembedding methods, preembedding methods are methodologically complex and difficult to ascribe "rules" to. Nevertheless, in reviewing the literature over the past decade, it is clear that many generalizations can now be made which should help a beginner to understand the main theoretical points in the preembedding approach. At a certain risk of oversimplification I shall attempt here to dissect these methods into their constituent parts and to analyze each part in some detail.

The basic protocol of any preembedding technique can be separated down into four parts:

1. Fixation
2. Permeabilization
3. Penetration and reaction
4. Post-fixation, embedding.

Without doubt, the first two steps are the most critical. The major difficulty is that the aims of fixation and permeabilization are diametrically opposed: the more the structure is destroyed the better the penetration and labelling, and vice versa. This is why these approaches are methodologically complex and always empirically determined. This may be less of a problem for the "cell ghost" approach, where complete delipidation of the cell with detergents or solvents may be acceptable.

Since the principles involved in fixation have been discussed in detail in Chapter 3, I will restrict further discussion here to a few relevant points and concentrate primarily on the **permeabilization** step that is essential to allow accessibility of antibodies and markers to all intracellular sites. Since the most commonly used tool to achieve this goal for EM studies is the use of detergents, it is necessary to discuss the essential chemical characteristics of detergents that are relevant to immunocytochemistry.

10.1 Permeabilization

10.1.1 Chemistry of Detergents

Amphipathic molecules are partly hydrophobic. Detergents are a special class of amphipathic molecules that are water-soluble. The effects of detergents on biological tissue can essentially be reduced to their effects on membranes which are themselves amphipathic since the lipid molecules in the bilayer have both hydrophilic parts exposed to an aqueous environment on two surfaces, and hydrophobic parts that make up the inner interior of a membrane. Membrane proteins inserted into this bilayer are classified as either "integral" or "spanning" when they have two hydrophilic domains on both sides of the bilayer connected by a stretch of hydrophobic amino acids within the membrane, or as "peripheral" when they have only one hydrophilic domain that usually faces the cytosol. In the latter case either a hydrophobic part(s) of the protein or a covalently attached lipid will serve to anchor the protein in the bilayer. Alternatively, the protein may be bound by non-covalent protein-protein interactions directly or indirectly to a spanning membrane protein. In recent years a new kind of peripheral membrane protein has emerged that is bound only to the outer leaflet of the plasma membrane bilayer via a glycophosphatidylinositol (GPI) lipid anchor (see Ferguson and Williams 1988 for review)

The effects of detergents on membranes, and in particular on membrane proteins, depend very much on the properties of the detergent, its concentration, temperature, ionic strength, and other factors (see Helenius and Simons 1975 and Helenius et al. 1979 for comprehensive reviews). However, some generalizations can be made with respect to increasing detergent "activity" which can be arbitrarily defined as the "strength" of the detergent. This "activity" will depend primarily on concentration, but also on other factors such as temperature and salt conditions. The first stage, at very low detergent activity, is that the amphipathic detergent molecules partition into the lipid bilayer. At higher detergent concentration lysis will occur, that is, the membrane will become permeable to ions and small molecules to which it is normally not permeable. At this stage some peripheral proteins may also be removed. When more detergent molecules are incorporated into the membrane, it undergoes a phase transition to a micellar state. This concentration of detergent is referred to as the critical micellar concentration (CMC). At this stage solubilization first occurs. At detergent concentrations just above the CMC mixed micelles will be formed, that is, micelles containing detergent, lipid and protein. At higher concentrations pure micelles, containing either proteins or lipid, will be formed along with pure detergent micelles. For these events, it should be noted that it is not only the concentration, as well as temperature, salt conditions, etc., of detergent that is critical, but also the ratio of detergent molecules to the amount of membranes or phospholipids within the system.

The best-characterized system for studying the effects of different detergents on membranes and on the underlying structures associated with those membranes is the Semliki forest virus (see Helenus and Simons 1975; Morein and Simons 1985). These studies give an excellent example of the different effects of different detergents on one simple biological system.

Semliki forest virus is a relatively simple enveloped virus consisting of a nucleocapsid (180 copies of a capsid protein and a single molecule of 42S RNA (the viral genome) which interacts with, and is enclosed by, a membrane bilayer consisting of typical plasma membrane lipids and 240 copies of the spike protein complex. The latter is a trimer consisting of two spanning membrane glycoproteins E1 and E2 and a third, non-membrane protein E3, which is a cleavage product of the precursor protein (p62) and which remains associated to the two membrane proteins by non-covalent protein-protein interactions.

The effects of the different detergents that were experimentally tested ranging from the harshest, SDS, to the gentlest, octyl glucoside, are summarized

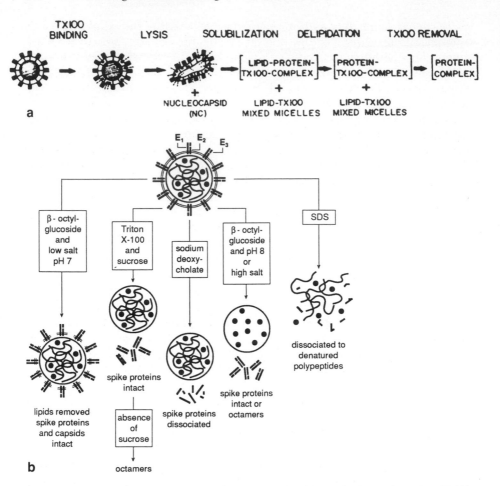

Fig. 1. A The stepwise dissociation of Semiliki Forest Virus (SFV) with increasing Triton X-100 (TX 100). The various stages in the dissociation are shown schematically. The bound Triton X-100 is indicated by (O), the *grey area* represents the internal nucleocapsid (*NC*) consisting of RNA and protein, and the *black spikes* represent the spike proteins which penetrate through the bilayer membrane. (Drawn from data described in Morein and Simons 1985). In **B** the effects of different detergents on SFV is shown. Note the wide variety in these effects, which can only be empirically determined. (Morein and Simons 1985)

in Fig. 1A. In Fig. 1B the effects of increasing concentration of one of these detergents, Triton X-100 is shown. A striking conclusion from these studies is, even for this simple system, that there is enormous variability in the effects observed for the different detergent conditions.

It is clear that when a membrane has been fixed by an aldehyde before the detergent treatment it will be more resistant to the effects of the detergent. Nevertheless, no systematic biochemical studies have been carried out on chemically fixed membranes. It should also be borne in mind that the commonly used aldehyde fixatives will effect very little cross-linking of lipids (see Chap. 3). What does appear likely, however, is that whether the membrane has been fixed or not, the concentrations of detergent required to allow antibodies and markers (usually 5–20 nm in diameter) to pass through any biological membrane is, by definition, a lytic one. At such concentrations it is possible that a significant amount of many peripheral membrane proteins will have been solubilized and, as far as any labelling is concerned, irretrievably lost. Even after fixation it is possible that some spanning membrane proteins may have been solubilized. In the absence of fixation this is obviously more likely to occur. For peripheral proteins in particular, the extent of solubilization will depend critically on the ionic strength: both low and high salt conditions are routinely used by biochemists to effect solubilization. The tendency for any membrane protein to become solubilized by detergent will be counteracted by the presence of protein-protein interactions either in the plane of the membrane itself or to cytoskeletal or extracellular matrix proteins. The best example of this is the erythrocyte membrane where the spanning membrane protein band 3, which functions as an anion channel, forms a complex with a peripheral cytoplasmic membrane protein, ankyrin, which is itself attached to another cytoskeletal protein, spectrin. The latter is stabilized further by interactions with actin (for a review see Bennett 1985). Clearly, such interactions will reduce the tendency of these proteins to become solubilized by mild detergent treatment. With harsher treatment, especially using ionic detergents, such as sodium dodecyl sulphate (SDS), there is an increased tendency for such strong protein-protein interactions to be disrupted. In practice, complete solubilization is only effected by boiling in SDS.

10.1.2 Non-Ionic Detergents – Triton X-100

For immunocytochemistry it is important to avoid ionic detergents since these are known to effect drastic conformational changes on proteins which are more likely to impair their antigenicity. Admittedly, some of these changes are reversible. Of the non-ionic detergents, Triton X-100 has been most widely used by biochemists to solubilize membrane protein since this tends to have little effect on their biological activity (exceptions are known, however; see Helenius and Simons 1975). Whereas the effects of Triton X-100 on membrane protein structure is considered to be "mild", its effects on membrane structure in general are very harsh indeed. The main reason is that even at low concentration, it will efficiently remove any lipids. Note again, however, that even when lipids are quantitatively removed a trilaminar unit-membrane may still be observed in conventional EM plastic sections (Korn 1968). Nevertheless, characteristic artefacts in such "membranes" after Triton treatment have been described, including artificial vesicle formation (Deamer and Crofts 1967, Fleischer et al. 1967). For this reason, Triton X-100 or similar detergents should be used in immunocytochemistry only for the "cell ghost" approach. For those studies where intact cell ultrastructure is desired, saponin has become the detergent of choice following the pioneering work of Seeman (1967), Seeman et al. (1973)

and Ohtsuki et al. (1978). For the theoretical reasons given below, the use of digitonin and some bacterial toxins should also be considered in this respect.

10.1.3 Saponin

According to the Merck Index (1976, ninth edition) the saponins are glycosides that are widely distributed in plants. They are made up of either a sterol or triterpene which is conjugated to a sugar moiety. Most commercially available saponins are a heterogeneous mixture of compounds. Unlike digitonin, no molecular weight is usually given.

As has been shown more extensively for the other plant glycoside, digitonin, the effects of saponin are mostly due to it making complexes with cholesterol in membranes which somehow leads to formation of pores. Since plasma membranes are relatively high in cholesterol it is widely thought that the primary target membrane for the saponins is the plasma membrane. Nevertheless, most, if not all, cellular membranes contain some cholesterol (van Meer 1989). Even membranes with very low cholesterol content such as the rough endoplasmic reticulum must be permeabilized by the typical concentrations of saponin used for preembedding studies (even after fixation) since in many studies at least some antibody molecules evidently reach their target antigens in this compartment.

A number of studies have investigated the effect of saponins on the release of soluble cytoplasmic proteins from unfixed cells. The aim of these type of studies has been to allow access to the cytosolic compartment without disturbing membrane-bound organelles and, in particular, secretion granules. In this way the effects of compounds such as calcium on the exocytic release of materials sequestered in secretory granules could be studied. This approach has been even more extensively used with digitonin and the bacterial toxins (see below). Noteworthy was the work by Brooks and Treml (1983), who studied the effects of saponin on the release of catecholamine upon stimulation in chromaffin cells. Initially, these workers looked at the time it took for trypan blue (MW 960) to enter cells that had been treated with 0.01% (wt/vol) saponin for 10 min. Significantly, while it took 130 s to see the dye entering the first cells the whole population became permeable within 150 s. Also significant was that this condition enabled the bulk of the large cytoplasmic protein lactate dehydrogenase (MW 134 kDa) to be released into the medium.

The ultrastructural effects of saponin have been extensively studied. Compared to a strong detergent such as Triton X-100, saponin is considered to be relatively "gentle" on membranes and on all ultrastructure in general. The studies of Seeman (1967) and Seeman et al. (1973) are especially relevant since they were the first to systematically study conditions that allowed the large tracer molecules, ferritin and colloidal gold, to enter into a model cell, the erythrocyte. Besides saponin, hypotonic lysis was also used to effect permeabilization. During hypotonic lysis, in the absence of fixation, Seeman (1967) showed that there was only a short transient period 15–25 s after the onset of lysis, during which the membrane was permeable to the markers. Thereafter, the membranes re-sealed and become impermeable again. If glutaraldehyde was present during the lysis, it was possible to fix the pores in the "open" position. The pores were, in fact, oval or slit-shaped structures approximately 10–20 nm long and up to 1 nm wide. In contrast, following saponin treatment no distinct pores were visible in thin sections of the membrane; nevertheless, the markers still entered the erythrocyte cytoplasm. What was observed with saponin were "invaginations" in some parts of the membrane. In a similar study, using negative staining methods, Ohtsuki et al. (1978) were able to see distinct ring-shaped structures about 8 nm in diameter in the membrane. Similar rings had been shown earlier in pure saponin-cholesterol micelles by negative staining (Lucy and Glauert 1964). Ohtsuki et al. (1978) concluded, however, that it was unlikely that these structures were the actual pore through which the tracer passed since they would need to be at least 12–15 nm for antibodies to be able to pass through them. It is possible, however, that the pores may have shrunk considerably during the negative staining procedure used to visualize them. It should be noted that even if it withstood the embedding procedure, a 15 nm pore would not be clearly visible in a routine thin plastic section ($\geq$ 30 nm thick).

An important finding in the study by Ohtsuki et al. (1978) was that, whereas saponin alone allowed ferritin to enter erythrocytes, for bone marrow and pancreatic exocrine cells, a combination of hypotonic fixation and subsequent saponin treatment was necessary for good penetration. The reason for this might have been that swelling of the cells after hypotonic treatment diluted the protein content of the cytoplasm. This would make the effective meshwork of the aldehyde-cross-linked cytoplasm larger, which would facilitate penetration. In other words, the effect they observed may have been due to an increased permeability of the cytoplasm rather than a direct effect on the membranes. Clearly, there are accessibility barriers other than membranes which must be overcome. In the latter study, for example, it was shown that secretory granules were impermeable to markers under all conditions tested, a consequence of the high concentration of protein within them. The idea of swelling the cytoplasm during the initial pre-fixation by hypotonicity followed by shrinking it again by a strong, hypertonic post fixation may have a more general applicability in preembedding labelling studies.

10.1.4 Digitonin

Besides saponin, another plant glycoside, digitonin, has been extensively used to permeabilize cells, especially in studies of exocytosis. Digitonin (MW 1229) reacts specifically and forms complexes with 3ß-hydroxysterols. In animal cells the most common 3ß-hydroxysterol is cholesterol. Since the plasma membrane is generally considered to be the organelle with the highest cholesterol content, digitonin is traditionally thought to bind primarily to this membrane. It should be noted, however, that the membranes of endosomes and lysosomes have similar concentrations of cholesterol to the plasma membrane (van Meer 1989). Although the effects of digitonin on these organelles has not been specifically studied, it seems likely that these organelles should also bind significant amounts of, and be permeabilized by, this compound.

The formation of complexes between digitonin and cholesterol somehow leads to the formation of large pores. In thin sections of cells treated with digitonin, the appearance of the plasma membrane is altered and characteristic disruptions are seen. This was the basis of an earlier cytochemical method to identify membranes rich in cholesterol (Ökrös 1968; Williamson 1969; Griffiths and Beck 1977). Digitonin (10–20 µM) can produce holes as large as 0.3–2 µm, according to one scanning EM study of chromaffin cells (Grant et al. 1987).

A number of biochemical studies have used digitonin usually in the concentration range 5–40 µM and for 15–30 min to preferentially release cytosolic proteins from cells without solubilizing proteins that are constituents of organelles such as secretory granules or mitochondria (e.g. Becker et al. 1980; Fiskum et al. 1980; Dunn and Holz 1982; Wilson and Kirschner 1983; Peppers and Holz 1986; Sarafian et al. 1987; Ahnert-Hilger et al. 1989a). Many of these studies have shown that a significant fraction of cytosolic proteins as large as lactate dehydrogenase (134 kDa) are lost after digitonin treatment. In a study by Kelner et al. (1986) the release of total cytosolic protein showed a plateau at 40 µM digitonin: higher concentrations had no further effect. This study also showed that between 10 and 40 µm the release of protein would be correlated with their molecular size – the larger the protein, the less likely they were to leak out. The situation is not necessarily this simple, however, since at lower concentrations (5 µm), some proteins were preferentially lost in a manner which was not dependent on their size. The complexity of using detergents such as digitonin to permeabilize cells is also evident in the ultrastructural study by Grant et al. (1987). A striking observation of these authors was the enormous variability between the effects of digitonin on adjacent chromaffin cells in a culture: while some cells were clearly heavily extracted their neighbours gave the appearance of being morphologically intact.

While it has not been as extensively used for immunocytochemistry as saponin, a number of workers have shown that digitonin can be used to enable antibodies to enter the cytosol of unfixed cells. Most of these studies have been restricted to the light microscopic level (Fiskum et al. 1980; Schäfer et al. 1987; Ahnert-Hilger et al. 1989a,b). A recent study by Perrin et al. 1987 extended this technology to the EM level showing a good immunoperoxidase localization of fodrin to areas beneath the plasma membrane of chromaffin cells. For this the authors treated their cells for 10 min with 10 µm digitonin followed by a 30-min fixation in 4% formaldehyde before incubation with antibody. I could find no reference to the use of digitonin on fixed cells in any of these studies.

10.1.5 Bacterial Toxins

In the past few years the use of two bacterial protein toxins for permeabilizing cells has become more prominent in biochemical studies of semi-intact cells. Two of these are especially interesting, namely the Streptolysin O (SLO) from β-haemolytic bacteria and the α-toxin from *Staphylococcus aureus* (Ahnert-Hilger et al. 1989b for a recent, comprehensive review). The attraction of such toxins is that when added to the cell culture medium they insert into the plasma membrane of cells and form stable, trans-membrane pores, functionally similar to the pore-forming protein complex of the complement system (C5b-C9). Whereas the α-toxin, which permeabilizes cells for low molecular weight compounds (1–2 nm pores), has little direct relevance for immunocytochemical studies SLO, which makes significantly larger pores, is potentially more important.

There is now convincing data that SLO can permeabilize cells to enable antibody molecules to enter the cytosol (Ahnert-Hilger et al. 1985, 1989b; Howell and Gomperts 1987) and conversely to allow large cytoplasmic markers such as lactate dehydrogenase (MW 134 kDa) to leak out. Examples of antibodies successfully tested in the latter references included anti-calmodulin and anti-synaptophysin. Calmodulin is a peripheral membrane protein while synaptophysin is a spanning-membrane protein in synaptic vesicles whose main epitopes are on the cytoplasmic side of synaptic vesicle membranes. While the studies cited above show clearly that the antibodies entered the cytosol, it must be emphasized that these experiments were not primarily designed for immunocytochemical studies per se. The major goal of these studies rather was to study cellular processes such as exocytosis. Thus, these toxins can enable the biological effects of normally impermeable molecules including antibodies on these processes to be studied. Further, under the conditions used while the plasma membrane is made leaky, many other intracellular membranes are evidently not: for example, contents of exocytic vesicles or intracellular calcium sources do not become soluble until key physiological processes are stimulated.

It follows that SLO is a promising reagent in immunocytochemical studies for permeabilizing cells and providing accessibility for those antibodies whose antigens are exposed in the cytosol. Whether any of these toxins can be used to permeabilize intracellular membrane organelles is less clear. Further studies are now required to see if their potential will be fulfilled in practice.

10.1.6 Permeabilization for EM Labelling in Practice

With the exception of a few studies that have used degitonin, most preemedding labelling studies in practice have used one of two different permeabilization approaches. First, either saponin, freeze-thawing or both, and second, Triton X-100, usually for the cell ghost approach.

10.1.7 The Saponin/Thick Section Approaches

The basic scheme as it has been used for cultured cells is as follows:

1. Fixation with low concentrations of aldehyde (usually below 0.2% glutaraldehyde). Often these low concentrations of glutaraldehyde are mixed with formaldehyde ($\approx$ 1–4%).
2. Permeabilization with saponin (up to 0.2% wt/vol)
3. Reaction with antibodies, usually in the presence of saponin.
4. Extensive post-fixation with glutaraldehyde followed by osmium tetroxide and routine embedding.

A variation of this approach for tissues is to cut thick (over 10 μm) sections using a "tissue chopper" or "vibratome" after fixation. Alternatively, cryostat sections have been used in combination with saponin. Note that, irrespective of how they have been prepared, the depth of penetration of reagents into these thick sections is limited even when detergents are used (Pickel et al. 1976; Chan-Palay and Palay 1977; Sternberger 1979). For example, a study using vibratome sections of brain by Piekut and Casey (1983) showed clearly that penetration of immunoreagents in their sections was no more than 8–9 μm on both surfaces of the section, despite long incubation times and the presence of Triton X-100. In this example, having sections thicker than 16–18 μm would serve no purpose with respect to the immunolabelling. Note that this assumes that one has accessibility to both sides of the sections: in cases where the sections are attached temporarily onto glass only one surface will be accessible. It is likely, however, that the reaction product of horseradish peroxidase may diffuse over significantly larger distances than these. Another example of penetration of no more than 5–10 μm with immunogold into tissue sections was reported by Triller et al. (1985).

Examples of papers where methods of this kind have been used successfully are Louvard et al. (1982), Guillouzo et al. (1982), Reggio et al. (1984), Tougard et al. (1982) and Gonatas et al. (1987). The same approach was also used successfully by Brown and Farquhar (1984), with the exception that the antigenicity of the membrane protein of interest, the mannose-6-phosphate receptor for lysosomal enzymes, was apparently very sensitive to glutaraldehyde (see Brown and Farquhar 1989, for a review of studies by this group). Fixation in this case was performed using the periodate-lysine-paraformaldehyde (PLP) mixture of McLean and Nakane (1974). As discussed in more detail in Chapter 3 (p.), the precise mechanism of this fixation mixture is not clear. Nevertheless, other workers have confirmed its usefulness in preembedding immunocytochemistry (for example, Lamberts and Goldsmith 1986).

10.1.8 The Freeze-Thaw Approach

The basic scheme here can be summarized as follows:

1. Fixation with high concentrations of glutaraldehyde (usually > 0.5%). The fixation may often be very short like minutes.
2. Treatment with borohydride (optional); see below.
3. Infusion with or without a low concentration of cryo-protectant, usually 10% dimethylsulfoxide (DMSO) or glycerol.
4. Freezing, in liquid nitrogen or isopentane.
5. Thawing, in the case of either cultured monolayers of cells or of thick (10–30 µm) frozen sections, in the case of tissues.
6. A "blocking" step followed by reactions with antibodies either with or without detergents (see Chap. 7).
7. Extensive aldehyde fixation, osmium and plastic embedding.

Amongst the first studies to use this strategy successfully were those of Leduc et al. (1969); Feldman et al. (1972) and Kuhlmann et al. (1974). For reviews of earlier studies, see Avrameas (1970) and Kuhlmann (1977). Later examples where this kind of approach has given good results can be found in the papers by Alexander et al. (1976), Kerjaschki and Farquhar (1983); Eldred et al. (1983); Courtoy et al. (1983); Sato and Spicer (1982); and Guillouzo et al. (1982).

In these groups of methods, freezing followed by thawing is the apparent physical perturbation that allows permeabilization. The freezing conditions used are such that vitrification is neither expected nor desired, and the glutaraldehyde cross-links presumably restrict the effects of ice crystal damage. Although no systematic study has been made, it seems likely that ice-crystal damage is the means by which both membranes and the cytoplasm are disrupted enough for reagents to penetrate. In the complete absence of cryo-protectants, cell and tissue damage is probably too great to facilitate EM localizations.

The idea of treating tissues with sodium borohydride after glutaraldehyde fixation stems from the observations of Weber et al. (1978) that the reaction of antibodies to tubulin was greatly enhanced when cells were treated with borohydride after the initial glutaraldehyde fixation. Sodium borohydride is a strong reducing agent of aldehydes, ketones and Schiff bases. It is an even more effective reducing agent in non-aqueous solvents. For this reason, Weber et al. (1978) dehydrated their cells after fixation and reacted with borohydride in 95% ethanol prior to rehydration and reaction with antibodies. An alternative method they used, where the borohydride is reacted in PBS, was, however, also successful, and this is the procedure which has become widely accepted. The precise role of the borohydride is far from clear. Eldred et al. (1983) have suggested that it reduces the Schiff bases formed by glutaraldehyde and thereby reduces the nitrogen-carbon double bonds to single ones. They speculated that the increased rotational movement of the latter would be less detrimental to the conformation of the antigen. This might explain the increased antigenicity in cases where glutaraldehyde directly affects the conformational structure of the antigen, but it does not readily explain how borohydride also increases the overall permeability of the cytosol (Willingham 1983).

For the use of sodium borohydride it should be noted that the stability of the solution depends critically on pH (T. Johnson, University of Fort Collins pers. commun.). At pH 7 the half-life of borohydride solution is about 10 s. Its stability increases logarithmically with increasing pH: hence at pH 8 it is ten times more stable than at pH7 and at pH 9 it is 100 times more so.

For most immunocytochemical purposes, to block auto fluorescence in light microscopy studies, a borohydride concentration of 0.01–0.1% at pH 8 should usually be sufficient. The solution is best stored at very high pH ($\approx$ 12) in the refrigerator. According to the Merck Index (1976, ninth edition) a saturated solution of this compound in 0.2% NaOH (= 44%) can be kept for several days. An alternative compound that has been used to block auto fluorescence is borane dimethylamine (2% wt/vol) (Collins and Goldsmith 1981), a compound which, like borohydride, reduces schiff bases.

10.2 Markers for Preembedding Labelling

10.2.1 Horseradish Peroxidase

Without doubt, the marker most widely used for pre-embedding labelling has been horseradish peroxidase (HRP), first used as a tracer by Straus (1957). Following the development of the diaminobenzidine cytochemical procedure for visualizing HRP by Graham and Karnovsky (1966), Nakane and Pierce (1967) and Avrameas and Bouteille (1968) introduced its use for immunocyto-chemistry. This was extended in particular by the contributions of Avrameas (1970) and Sternberger (see 1979 for a review). For a summary of approaches to couple HRP to antibodies, see Boorsma (1983).

HRP, a 40 kDa glycoprotein, is an enzyme that contains a heme group and, like other proteins containing this iron porphyrin prosthetic group, will catalyze the peroxidative reaction (Essner 1974). This reaction involves the reduction of hydrogen peroxide to water by a suitable electron donor: the one used routinely has been 3′,3′ diaminobenzidine (DAB). The oxidized form of DAB is a brown insoluble polymer which is visible in the light microscope. It is also electron-dense and becomes more so after reaction with osmium tetroxide. Other examples of heme-containing proteins include catalase and the non-enzymic proteins haemoglobin and cytochrome C. The latter has also been used as an immune-tracer (Kraehenbuhl and Jamieson 1974). The presence of any endogenous heme proteins in tissues is a problem which has to be contsidered when using any peroxidase technique (Essner 1974).

The most serious problem with the use of HRP and DAB for high precision labelling is the tendency for diffusion artefacts, as already mentioned. This phenomenon was systematically studied by Courtoy et al. (1983). These workers allowed cationic ferritin to bind electrostatically to anionic sites in the extracellular matrix of the kidney. Antibodies to ferritin were then added which were visualized by a peroxidase-conjugated second antibody. Since the antigen ferritin could be directly visualized, the resolution of labelling by the HRP technique could then be tested. Only when both the amount of immuno-adsorbed peroxidase and the time of exposure to DAB were limited, were discrete deposits of reaction product restricted to the periphery of ferritin particles. When labelling conditions were enhanced, the oxidized DAB was found to diffuse over large distances and to be re-adsorbed onto other, non-specific sites. It should be noted that the latter conditions are more likely to be used in practice, in order to have maximum amplification of the signal from the bound anibody. This paper clearly confirmed what was already suspected from earlier studies and is quite evident in a large number of published micrographs (see also De Camilli et al. 1983, who came to similar conclusions).

When the antigen of interest is enclosed in a membrane-limited organelle, the membrane tends to restrict the diffusion of the reaction product to that organelle. While this may facilitate identification of small labelled profiles, microdomains within that membrane compartment such as a budding vesicle, with a different composition, are unlikely to be recognized. Furthermore, since in practice the oxidized DAB tends to become adsorbed onto membranes, it is difficult, if not impossible, to distinguish between an antigen that is membrane bound in the lumen of a structure such as the endoplasmic reticulum from one that is free in the lumen. Finally, there are, to my knowledge, only a few convincing examples of labelling of antigens on the cytoplasmic side of membranes using HRP (e.g. Louvard et al. 1983). This again is probably due to diffusion artefact. Even if one was to accept the widely held opinion that the HRP methods, due to the amplification step(s), are more sensitive than particulate markers, the price to pay for this increase in sensitivity is a significant decrease in the precision of labelling. Methods which are designed to increase this amplification factor, such as using layers, or "sandwiches" of antibody (for example the peroxidase-anti-peroxidase technique of Sternberger 1979) only aggravate this problem. Under labelling conditions where the peroxidase labelling is relatively precise (e.g. Courtoy et al. 1983), the question of deciding what is reaction product and what is density due to osmium or other heavy metal staining "background" reaction becomes increasingly subjective. In contrast, a single gold particle on a membrane surface can be unequivocally identified and a statistically based approach can be used to determine if this is real signal or not (see Chap. 11). There are several arguments against the widely held opinion that the HRP-based preembedding methods are generally more sensitive than particulate markers.

- The real test of sensitivity is the number of antigens detected by the antibody. Both on sections and especially in the preembedding approach, this step is most likely to be the limiting factor. Structures in cells containing the antigens should, on average, have more chance to meet their antibody on the surface of the section than when that antibody has to transverse a dense meshwork of cytoplasmic and membrane counterparts. Whether or not one accepts this point, there seems no question that the secondary event, the application of the detecting complex, namely peroxidase conjugate versus gold conjugate, must have a much higher probability to detect the bound primary antibody on the surface of the section when compared to the preembedding approach. It follows that the labelling of sections must generally be more efficient than labelling of compartments in permeabilized cells and tissues. The fallacy in claiming that the enzymatic methods are more sensitive comes from the fact that the signal (but not the number of antigens detected) can be significantly amplified by enzymatic reaction. Note that there would be no problem in similarly amplifying the signal from a single bound antibody on a section by using multiple sandwich layers followed by a gold marker: the single gold particle (from protein A gold above) can be amplified almost indefinitely to fill the section with gold particles. The end result, of course, is that one gains no additional signal, or signal-to-noise ratio, but loses precision of labelling.

- There are many examples (mostly unpublished) of false-negative results with preembedding labelling using HRP-based methods. As discussed in Chapter 7, this can be true even for compartments from cells that are involved in secreting peroxidase molecules. In this case there was a failure to get sufficient penetration of DAB and H_2O_2. Even for these small molecules the meshwork of aldehyde cross-links must provide a significant penetration barrier; for antibody molecules this problem is obviously more serious.

10.2.2 Particulate Markers for Preembedding

10.2.2.1 "Intact" Cell Approach

Colloidal gold or ferritin have not, in general, been very successful for pre-embedding labelling of "intact cells" due to the inherent problems of penetration. An exception to this statement is the work of Willingham (1980, 1983), who has developed methods for permeabilization of tissue culture cells that allow immune-labelling with ferritin. The most widely used modification, referred to as the glutaraldehyde-borohydride-saponin (GBS) procedure (Willingham 1983) has combined good ultrastructural presentation with precise immunoferritin labelling of a number of different antigens, including labelling of the cytoplasmic surface of coated vesicles with anti-clathrin antibodies. In addition to the increase in precision of labelling over immunoperoxidase, the obvious advantage of this approach is that it enables the labelling to be quantified. Nevertheless, while this method may be suitable for some exposed cytoplasmic antigens, it cannot overcome the theoretical limitations that one can **never** be certain that equal accessibility to all compartments has been achieved: for certain organelles such as secretory granules or the nuclear matrix, it appears very unlikely that good accessibility could be the result of this treatment.

For additional studies where the group of Willingham successfully used ferritin in preembedding labelling studies of cultured cells, see Willingham et al. (1980), Willingham et al. (1981a,b) and Cabral et al. (1980).

10.2.2.2 "Cell Ghost" Approach

Since in this kind of procedure the structures of interest are, by definition, generally resistant to high concentrations of non-ionic detergents, they are theoretically more straightforward than any "intact cell" method. They are in practice, however, just as empirical.

The first approach, which can be taken as the extreme is for systems in which the cells can be extracted with Triton X-100 followed by labelling before fixation in glutaraldehyde and embedding. Intermediate filament bundles, for example, were labelled in this way with immunoferritin by Henderson and Weber (1981). For this procedure the time of exposure to the detergent appears to be absolutely critical with respect to what may be considered the classical compromise between adequately preserving morphology and getting sufficient accessibility for the immunoreagents. Thus, Langanger et al. (1984) empirically determined that for labelling stress fibres with antibodies against actin, α actinin and filamin the best time for epithelial cells was 40 s and 20 s for fibroblasts. This was again slightly varied in their subsequent publication (Langanger et al. 1986).

A more gentle approach is to mix the detergent with the fixative. This is obviously essential if the structure/antigen does not withstand direct detergent treatment in the absence of fixation. Thus Geuens et al. (1986, for tubulin), Langanger et al. (1984) (for α-actinin and filamin) and Mitchison et al. (1986) (for tubulin) used 1–2 min of 0.5–1% Triton X-100 mixed with 0.25–1% glutaraldehyde for 1–2 min. This was followed by a stronger (0.5%) glutaraldehyde fixation for 10–30 min followed by a harsher detergent treatment of 0.5% Triton X-100 for 30 min before immunolabelling. These examples simply point out that the procedures are indeed very empirical with respect to both the labelling conditions and to structural preservation. For an example of preembedding immunogold labelling of microtubules in a cultured cell followed by silver enhancement see Fig. 2.

Whereas the above studies were all concerned with cultured cells this approach can also be used for tissue slices, either made by vibratome/tissue chopper or with a cryostat. Thus, Kartenbeck et al. (1984), for example, used 20 μm cryostat sections of unfixed tissues that were air-dried prior to fixing with acetone at −20°C before immunolabelling (see Fig. 3). The structures of interest in this study, vimentin filaments and desmosomal plaques are evidently highly resistant to this relatively harsh treatment.

This cell ghost approach is also compatible with double-labelling as shown by the studies of Geuens et al. (1986), who used a double-indirect approach and Langanger et al. (1986), who used a double-direct method. Both studies used different sizes of colloidal gold. For a recent review of the cell ghost approach see Langanger and De Mey (1989). For an example of double-labelling see Fig. 4.

The study of Mitchison et al. (1986) is also worth mentioning with respect to another important point. In this study, the condition of the cells was carefully monitored during each step of the procedures by light microscopy. As for immunolabelling studies on sections this is recommended as it can greatly

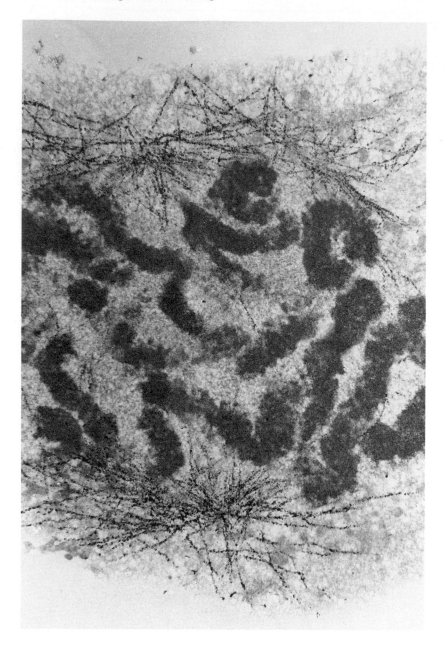

Fig. 2. Preembedding labelling of tubulin. PtK2 cell in early prometaphase. The cell was fixed with 0.5% glutaraldehyde in PBS, extracted with 0.2% Triton X-100 in PBS for 15 min after fixation, blocked with 0.8% BSA + 0.1% gelatin + 5% normal goat serum in PBS for 30 min. Microtubules were stained with a polyclonal rabbit anti-tubulin antibody; second antibody: Janssen AuroProbe One-goat anti rabbit, diluted 1:20 in PBS buffer, containing 0.8% BSA + 0.1% gelatin + 1% normal goat serum, incubation 2×45 min at 23°C. Postfixation 5 min with 2% glutaraldehyde. After washing in excess distilled water a silver amplification step has been

facilitate the determination of optimal conditions for both structural preservation and immunolabelling.

Finally, a potential advantage of the preembedding cell ghost approach over conventional sectioning techniques is that it is feasible to study thick sections and whole mounts of structures that are labelled throughout the depth of the preparations. In addition, stereo micrographs can be examined. For an excellent example see Schliwa and van Blerkom (1981).

10.3 Preembedding Studies of the Nucleus

It is important to stress that the cell nucleus may be an especially problematic organelle for immunoelectron microscopy studies. A number of general considerations make it seem likely that obtaining complete accessibility for antibodies to this organelle may be a more difficult task when compared to cytoplasmic organelles.

- Since the nucleus is almost always in the centre of the cell antibodies must traverse the whole cytoplasm before gaining access to nuclear antigens.
- The nuclear envelope and the nuclear lamins are formidable barriers even after extensive detergent treatment. The most likely site of entry for antibodies, the nuclear pores, may be less "open" after the usual pre-fixation step.
- Most parts of the nuclear matrix are much more dense than most parts of the cytoplasm. This can even pose problems for labelling with antibodies on sections (see Chap. 7).

An excellent study by Cooke et al. (1990) is worth discussing since it exemplified many of the problems facing studies on nuclear antigens. These workers were interested in an antigen (CENP-B) recognized by an antibody present in the sera of patients with autoimmune diseases. Earlier immunoperoxidase labelling studies had found that these antigens were found throughout the kinetochore region of the centromere, a result confirmed by these authors using a similar immunoperoxidase approach. Using an immunogold preembedding protocol, this study showed convincingly that the CENP-B antigen was, in fact, restricted to the heterochromatin region below the kinetochore and was absent from the kinetochore itself.

The authors' data argued convincingly that the artefactual immunoperoxidase result could be attributed to the well-known diffusion artefact with this

---

performed according to Danscher (1981). The amplification solution consisted of 3 ml gum arabic (stock: 25% in H_2O) 3 ml H_2O + 1.5 ml hydroquinone (stock: 0.85 g/15 ml) 1 ml citric acid (stock: 2.55 g of the monohydrate/10 ml) + 1.5 ml silver lactate (stock: 0.11 g/15 ml). The specimen has been incubated for 6 min at 23°C in a darkroom under red safelight. The cell has been flat embedded in Epon; semi-thin sections were made with a diamond knife and stained with lead citrate after Reynolds for 1 min (×71000). (Courtesy of Andreas Merdes and Jan de Mey, EMBL, Germany)

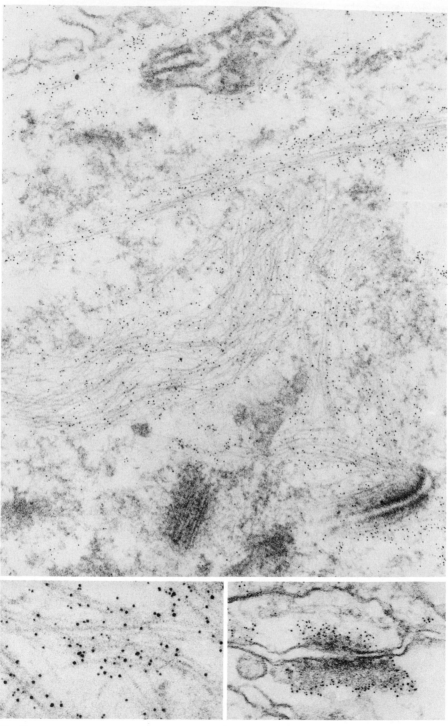

Fig. 3

technique. The success of the immunogold experiments apparently resulted in large part due to the use of the 1 nm gold probe conjugated to streptavidin (from Janssen Pharmaceuticals, Beerse, Belgium) to detect the biotinylated primary antibody. This was subsequently enhanced with silver before plastic embedding.

A similar false localization of a nucleolar antigen with the immunoperoxidase technique was shown in the study by Biggiogera et al. (1989).

10.4 Preembedding Labelling for Studies of the Nervous System

Although I have voiced my criticisms of the preembedding labelling approach for localizing intracellular antigens (the "intact cell" procedure), there is one area of biology where these approaches still appear to have general validity, namely for studying the connectivities of the nervous system (see Forssmann et al. 1981; Priestley and Cuello 1983; and Priestley 1984, for reviews). The goal of such studies is to fill a whole cell with an electron-dense particle or reaction product in order to be able to follow its often tortuous path along relatively long distances. The reaction product of horseradish peroxidase, usually the marker of choice in these studies, can be used to fill the entire cytoplasm of reactive cells. This makes it easier to identify very small cell profiles at low magnification than if particulate markers were to be used. In this case the diffusion artefact, a clear disadvantage for high precision labelling, is an advantage. The possibility of embedding tissues in conventional resin embedding media is also an advantage, especially when serial sections are needed. While the diffusion artefact usually makes unequivocal identification of labelled intracellular organelles very difficult, high resolution structural features like synapses of neighboring cells provide important additional information for tracing connections.

In studies of neuronal connectivities, one is often not concerned with fine-structural details within the nerve cell, but more with the ability to distinguish every part of the cell from the multitude of unlabelled cell profiles that surround it. For most of these studies an additional prerequisite is that the techniques used facilitate a localization at both the light- and electron-microscope level.

◀ ───

Fig. 3. Preembedding immunogold labelling of biopsy material of a human menigioma. The unfixed fresh tissue was frozen in isoheptane cooled by liquid nitrogen and ≈ 5 μm cryostat sections were placed on glass coverslips and fixed in acetone at −20°C for 2–5 min. The sections were then air-dried for 1–12 h in order to facilitate adhesion to the glass for the later steps. The sections were labelled with anti vimentin for 30 min, briefly rinsed in PBS, then labelled with a gold conjugate second antibody for 60 min. Unlike sections of fixed material, these sections need only be rinsed very briefly for 5–10 min after the gold. They were subsequently fixed with glutaraldehyde, osmium tetroxide and finally embedded in Epon. The glass slide was removed from the Beem capsule (that had been inverted over the section on the cover slip) using liquid nitrogen after polymerization of the plastic. Note the significant labelling of intermediate filaments. The *inset at bottom left* shows a high magnification part, while that on the *bottom right* shows labelling adjacent to a desmosome. For more details see Kartenbeck et al. (1984). (Courtesy of J. Kartenbeck, German Cancer Research Centre, 6900 Heidelberg, Germany)

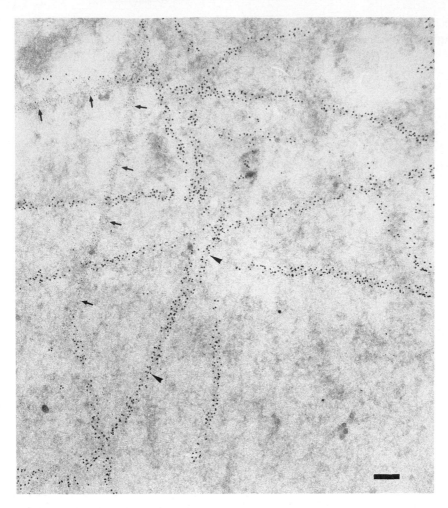

Fig. 4. Preembedding double-labelling of tubulin in MDCK cells grown on glass coverslips. Purified tubulin from *Paramecium* was microinjected into cells 2 min before a pre-fixation with a solution containing 0.3% glutaraldehyde, 0.5% Triton X-100, 80 mM Pipes, 5 mM EDTA, 2 mM $MgCl_2$ pH 6.8 for 2 min at RT. The cells were further fixed for 10 min with 0.3% glutaraldehyde in Pipes buffer and then extracted with 0.5% Triton X-100 for 30 min. The fixative was quenched with a freshly prepared solution of $NaBH_4$ (2 mg/ml) in PBS followed by a rinse in 0.1% Triton X-100 for 5 min. For the labelling the coverslips were covered with 1% fetal calf serum (FCS) for 20 min followed by the following steps: (1) Rabbit anti–*Paramecium* tubulin (1:50) in 1% FCS for 1 h; (2) mouse anti-α tubulin (1:200) for 1 h ; (3) goat anti-rabbit conjugated to 5 nm gold and (4) goat anti-mouse conjugated to 10 nm gold. After rinsing in buffer, the preparation was fixed again in 2% glutaraldehyde containing 0.2% tannic acid for 30 min, post-fixed in 2% OsO_4 for 30 min and "en bloc" stained in aqueous 0.5% uranyl acetate for 45 min. The cells were then embedded in Epon, sectioned (0.2–0.3 μm) and the microinjected cells were identified. Using this elegant, though elaborate, procedure, the newly synthesized microtubules containing the exogenous *Paramecium* tubulin can be labelled with the 5 nm gold (*arrows*) and can be distinguished from the endogenous parts of microtubules that contained MDCK tubulin (9 nm gold – *arrowheads*). ×36000, *bar* 0.2 μm. (Courtesy of C. Antony, Jaques Monod Institute, Paris, from a project done in collaboration with M.-H. Bré and E. Karsenti, EMBL)

According to Priestley (1984), the peroxidase-antiperoxidase immunocyto-chemical method (Pickel et al. 1975, 1976) is the most widely used preembed-ding method used for tracing neuronal connectivities. These methods mostly take advantage of the fact that different neuronal cell types express different neurotransmitters. For this, and for the other approaches mentioned above, it is necessary to make thick (mostly > 20 μm) sections that are incubated with the antibodies and markers. For this purpose, both vibratome-slices of unfixed tissue as well as cryostat sections of both fixed and unfixed tissues have been widely used (Priestley 1984). The problems of getting penetration of reagents into these slices are not different from any other preembedding approach and detergents such as Triton X-100 and less often, saponin are often used to facilitate penetration. For studies that compare different preparation protocols in some detail, see Eldred et al. (1983) and Lamberts and Goldsmith (1986). A potential problem with this approach is that, if the cell of interest is not sectioned by the vibratome or cryostat, it can only have access to the immunological reagents under harsh detergent conditions.

In contrast to fine structure localizations within cells, a number of feasible double-labelling preembedding protocols appear to exist for the study of neuronal cell connectivities. One of the first, introduced by van den Pol (1988; see also van den Pol and Decavel 1989), combined the use of colloidal gold with peroxidase. In order to facilitate access to the 30 μm vibratome sections, a relatively small gold particle (5 nm) coated with secondary antibody was used for the visualization of the first antibody. These gold particles were then enhanced with silver (see Chap. 8 and the legend to Fig. 2). Following a suitable blocking step to prevent cross-reaction of the two labelling systems, the sections were incubated with the second primary antibody that was then reacted with a biotinylated second antibody. The latter could then be visualized with an avidin-peroxidase complex. Besides being distinguishable at the EM level, these markers are also easily distinguishable at the light microscopic level.

Two recent publications have described the combination of diaminobenzi-dine, the conventional reagent used to visualize peroxidase activity, with a second benzidine marker. The first, by Levey et al. (1986) used benzidine dichloride while the second, by Norgren and Lehman (1989) used tetramethylbenzidine. In both cases, the reaction products were distinguishable in double-labelling studies at both the light- and electronmicroscopic levels. For other approaches see Lakos and Basbaum (1986), Silverman et al. (1983) and Pickel et al. (1986).

For additional extensive reviews of preembedding as well as other ap-proaches for labelling the nervous system see van den Pol and Decavel (1989), and van den Pol (1988).

10.5 The Use of Dinitrophenol IgG Conjugates

An interesting recent innovation for preembedding labelling was introduced by Pathak and Anderson (1989). Following a light pre-fixation, permeabilization and application of primary antibody to cultured cells, a secondary antibody that has been conjugated to multiple dinitrophenol (DNP) residues is applied. The specimen is then post-fixed, first with glutaraldehyde and then with osmium tetroxide and embedded in plastic (epon). The sections are then labelled with anti DNP and protein A gold. In the example studied by the authors, the low density lipoprotein (LDL) receptor system, this method was found to be more sensitive than either a peroxidase-preembedding protocol or the use of cryosections with indirect labelling.

10.6 Agarose Gel Method for Preembedding

An elegant system for labelling the cytoplasmic "outside" surface of vesicles in vitro was developed by de Camilli et al. (1983), who were specifically interested in showing that synapsin I was associated with the cytoplasmic surface of brain synaptosomes. The essential features of this method are that the vesicles are embedded in a thin slice of agarose allowed to gel between two cover slips. The agarose is porous and in this study allowed the penetration of antibodies and, subsequently, a ferritin conjugate. The labelled gel slices can then be embedded in an epoxy resin.

References

Ahnert-Hilger G, Bhakdi S, Gratzl M (1985) Minimal requirements for exocytosis: a study using PC12 cells permeabilized with staphylococcal alpha-toxin. J Biol Chem 260:12730–12734

Ahnert-Hilger G, Bader M-F, Bhakdi S, Gratzl M (1989a) Introduction of macromolecules into bovine adrenal medullary chromaffin cells and rat pheochromocytoma cells (PC12) by permeabilization with streptolysin O: inhibitory effect of tetanus toxin on catecholamine secretion. J Neurochem 52:1751–1758

Ahnert-Hilger G, Mach W, Föhr KJ, Gratzl M (1989b) Poration by α-toxin and streptolysin O: an approach to analyze intracellular processes. In: Tartakoff A (ed) Methods in cell biology, vol 31, Academic Press, New York, pp 63–90

Alexander CA, Hamilton RL, Havel, RJ (1976) Subcellular localization of B apoprotein of plasma lipoproteins in rat liver. J Cell Biol 69:241–263

Avrameas S (1970) Immunoenzyme techniques: enzymes as markers for the localization of antigens and antibodies. Int Rev Cytol 27:349–385

Avrameas S, Bouteille M (1968) Ultrastructural localization of antibody by antigen label with peroxidase. Exp Cell Res 53:166–176

Becker GL, Fiskum G, Lehninger AL (1980) Regulation of free Ca^{2+} by liver mitochondria and endoplasmic reticulum. J Biol Chem 10:9090–9012

Bennett V (1985) The membrane skeleton of human erythrocytes and its implications for more complex cells. Annu Rev Biochem 54:273–304

Biggiogera M, Fakan S, Kaufmann SH, Black A, Shaper JH, Busch H (1989) Simultaneous immunoelectron microscope visualization of protein B23 and C23 distribution in the HeLa cell nucleolus. J Histochem Cytochem 105:855–862

Boorsma DM (1983) Preparation of HRP-labelled antibodies. In: Cuello AC (ed) Immunohisto-chemistry. IBRO Handbook series Methods in the neurosciences, Wiley, Chichester, pp 87–109

Brooks JC, Treml ST (1983) Catecholamine release by chemically skinned cultured chromaffin cells. J Neurochem 40:468–473

Brown WJ, Farquhar MG (1984) The Mannose-6-phosphate Receptor for Lysosomal Enzymes is Concentrated in Cis Golgi Cisternae. Cell 36:295–307

Brown WJ, Farquhar MG (1989) Immunoperoxidase methods for the localization of antigens in cultured cells and tissue sections by electron microscopy. In: Tartakoff A (ed) Methods in cell biology. Academic Press, New York, pp 553–569

Cabral F, Willingham MC, Gottesman MM (1980) Ultrastructural localization to 10 nm filaments of an insoluble 58K protein in cultured fibroblasts. J Histochem Cytochem 28:653–662

Chan-Palay V, Palay S (1977) Ultrastructural identification of substance P cells and their processes in rat sensory ganglia and their terminals in the spinal cord by immunocytochemistry. Proc Natl Acad Sci USA 74:4050

Collins S, Goldsmith TH (1981) Spectral properties of fluorescence induced by glutaraldehyde fixation. J Histochem Cytochem 29:411–414

Cooke CA, Bernat RL, Earnshaw WC (1990) CENP-B: A major human centromere protein located beneath the kinetochore. J Cell Biol 110:1475–1488

Courtoy PJ, Hunt Picton D, Farquhar MG (1983) Resolution and limitations of the immunoperoxidase procedure in the localization of extracellular matrix antigens. Histochem Cytochem 31(7):945–951

Danscher G (1981) Locolization of gold in biological tissue. A photochemical method for light and electron microscopy. Histochemistry 71:81–88

Deamer DW, Crofts A (1967) Action of Triton X-100 on chloroplast membranes. J Cell Biol 33:395–409

De Camilli P, Harris SM Jr, Huttner WB, Greengard P (1983) Synapsin I (Protein I), a nerve terminal-specific phosphoprotein. II. Its specific association with synaptic vesicles demonstrated by immunocytochemistry in agarose-embedded synaptosomes. J Cell Biol 96:1355–1373

De Petris S (1978) Immunoelectron microscopy and immunofluorescence in membrane biology. In: Korn ED (ed) Methods in membrane biology. Plenum Press, New York, pp 1–201

De Waele M (1984) Haematological electron immunocytochemistry: detection of cell surface antigens with monoclonal antibodies. In: Polak JM, Varndell IM (eds) Immunolabelling for electron microscopy. Elsevier, Amsterdam, pp 267–303

Dunn LA, Holz RW (1982) Catecholamine secretion from digitonin-treated adrenal medullary chromaffin cells. J Biol Chem 258:4989–4993

Eldred WD, Zucker C, Karten HJ, Yazulla S (1983) Comparison of fixation and penetration enhancement techniques for use in ultrastructural immunocytochemistry. J Histochem Cytochem 31(2):285–292

Essner E (1974) Hemeproteins. In: Hayat MA (ed) Electron microscopy of enzymes, vol 2. van Nostrand-Rheinhold, New York, pp 1–33

Feldman G, Penaud-Laurencin J, Crassous J, Benhamou J-P (1972) Albumin synthesis by human liver cells: its morphological demonstration. Gastroenterology 63:1036–1048

Ferguson MAJ, Williams AF (1988) Cell surface anchoring of proteins via glycosylphosphatidylinositol structures. Annu Rev Biochem 57:285–320

Fiskum G, Craig SW, Decker GL, Lehninger AL (1980) The cytoskeleton of digitonin-treated rat hepatocytes. Proc Natl Acad Sci USA 77:3430–3434

Fleischer S, Fleischer B, Stoeckenius W (1967) Fine structure of lipid-depleted mitochondria. J Cell Biol 32:193–208

Föhr KJ, Scott J, Ahnert-Hilger G, Gratzl M (1989) Characterization of the inositol 1,4,5-trisphosphate-induced calcium release from permeabilized endocrine cells and its inhibition by decavanadate and p-hydroxymercuribenzoate. Biochem J 262:83–89

Forssmann WG, Pickel V, Reinecke M, Hock D, Metz J (1981) Immunohistochemistry and immunocytochemistry of nervous tissue. In: Heym C, Forssmann WG (eds) Techniques in neuroanatomical research. Springer, Berlin Heidelberg New York, pp 170–205

Geuens G, Gundersen GG, Nuydens R, Cornelissen F, Bulinski JC, DeBrabander M (1986) Ultrastructural colocalization of tyrosinated and detyrosinated α-tubulin in interphase and mitotic cells. J Cell Biol 103:1883–1893

Gonatas NK, Gonatas JO, Stieber A, Ternynck T, Avrameas S (1987) Detection of plasma cell immunoglobulins in tissue sections optimally fixed for ultrastructural immunocytochemistry. J Histochem Cytochem 35(2):189–196

Graham RC Jr, Karnovsky MJ (1966) The early stages of absorption of injected horseradish peroxidase in the proximal tubules of mouse kidney: ultrastructural cytochemistry by a new technique. J Histochem Cytochem 14(4):291–302

Grant NJ, Aunis D, Bader MF (1987) Morphology and secretory activity of digitonin- and α-toxin-premeabilized chromaffin cells. Neuroscience 23:1143–1155

Griffiths GW, Beck SD (1977) In vivo sterol biosynthesis by pea aphid symbiotes as determined by digitonin and electron microscopic autoradiography. Cell Tissue Res 176:179–190

Guillouzo A, Beaumont C, Le Rumeur E, Rissel M, Latinier M-F Guguen-Guillouzo C, Bourel M (1982) New findings on immunolocalization of albumin in rat hepatocytes. Biol Cell 43:163–172

Helenius A, Simons K (1975) Solubilization of Membranes by Detergents. Biochim Biophys Acta 415:29–79

Helenius A, McCaslin DR, Fries E, Tanford C (1979) Properties of detergents. Methods Enzym LVI:734–749

Henderson D, Weber K (1981) Immuno-electron microscopical identification of the two types of intermediate filaments in established epithelial cells. Exp Cell Res 132:297–311

Howell TW, Gomperts BD (1987) Rat mast cells permeabilized with streptolysin O secrete histamine in response to Ca^{2+} at concentration buffered in the micromolar range. Biochim Biophys Acta 927:177–183

Kartenbeck J, Schwechheimer K, Moll R, Franke WW (1984) Attachment of vimentin filaments to desmosomal plaques in human meningiomal cells and arachnoidal tissue. J Cell Biol 98:1072–1081

Kelner KL, Morita K, Rossen JS, Pollard HB (1986) Restricted diffusion of tyrosine hydroxylase and phenylethanolamine N-methyltransferase from digitonin-permeabilized adrenal chromaffin cells. Proc Natl Acad Sci USA 83:2998–3002

Kerjaschki D, Farquhar MG (1983) Immunocytochemical localization of the heymann nephritis antigen (gp330) in glomerular epithelial cells of normal Lewis rats. J Exp Med 157:667–686

Korn ED (1968) Structure and function of the plasma membrane: a biochemical perspective. J Gen Physiol 52:257–278

Kuhlmann WD (1977) Ultrastructural immunoperoxidase cytochemistry. Gustav Fischer, Stuttgart

Kuhlmann WD, Avrameas S, Ternynck T (1974) A comparative study for ultrastructural localization. J Immunol Methods 5:33–48

Kraehenbuhl JP, Jamieson JD (1974) Localization of intracellular antigens by immunoelectron microscopy. Int Rev Exp Pathol 13 1

Lakos S, Basbaum AI (1986) Benzidine dihydrochloride as a chromogen for single- and double-lable light and electron microscopic immunocytochemical studies. J Histochem Cytochem 34:1047–1056

Lamberts R, Goldsmith PC (1986) Fixation, fine structure, and immunostaining for neuropeptides: perfusion versus immersion of the neuroendocrine hypothalamus. J Histochem Cytochem 34:389–398

Langanger G, De Mey J (1989) Detection of cytoskeletal proteins in cultured cells at the ultrastructural level. In: Bullock GR, Petrusz P (eds) Techniques in immunocytochemisty, vol 4. Academic Press, Harcourt Brace Jovanovich Publ, New York, pp 47–66

Langanger G, De Mey J, Moeremans M, Daneels G, De Brabander M, Small JV (1984) Ultrastructural localization of a-actinin and filamin in cultured cells with the immunogold staining (IGS) method. J Cell Biol 99:1324–1334

Langanger G, Moeremans M, Daneels G, Sobieszek A, De Brabander M, De Mey J (1986) The molecular organization of myosin in stress fibers of cultured cells. J Cell Biol 102:200–209

Leduc EH, Scott GB, Stratis A (1969) Ultrastructural localization of intracellular immune globulins in plasma cells and lymphoblasts by enzyme-labeled antibodies. J Histochem Cytochem 17:211

Levey AI, Bolam JP, Rye DB, Hallanger AE, Demuth RM, Mesulam M-M, Wainer BH (1986) A light and electron microscopic procedure for sequential double antigen localization using diaminobenzidine and benzidine dihydrochloride. J Histochem Cytochem 34:1449–1457

Louvard D, Reggio H, Warren G (1982) Antibodies to the Golgi complex and the rough endoplasmic reticulum. J Cell Biol 92:92–107

Louvard D, Morris C, Warren G, Stanley K, Winkler F, Reggio H (1983) A monoclonal antibody to the heavy chain of clathrin. EMBO J 2:1655–1664

Lucy JA, Glauert AM (1964) Structure and assembly of macromolecular lipid complexes composed of globular micelles. J Mol Biol 8:727–748

McLean IW, Nakane PK (1974) Periodate-lysine-paraformaldehyde fixative. A new fixative for immunoelectron microscopy. J Histochem Cytochem 22:1077–1083

Mitchison T, Evans L, Schulze E, Kirschner M (1986) Sites of microtubule assembly and disassembly in the mitotic spindle. Cell 45:515–527

Morein B, Simons K (1985) Subunit vaccines against enveloped viruses: virosomes, micelles and other protein complexes. Vaccine 3:83–93

Nakane PK, Pierce GB jr (1967) Enzyme-labelled antibodies: preparation and application for the localization of antigens. J Histochem Cytochem 11:929–931

Norgren RB Jr, Lehman MN (1989) Double-label pre-embedding immunoperoxidase technique for electron microscopy using diaminobenzidine and tetramethylbenzidine as markers. J Histochem Cytochem 37:1283–1289

Ohtsuki I, Manzi RM, Palade GE, Jamieson JD (1978) Entry of macromolecular tracers into cells fixed with low concentrations of aldehydes. Biol Cell 31:119–126

Ökrös I (1968) Digitonin reaction in electron microscopy. Histochemie 13:91–100

Pathak RK, Anderson RGW (1989) The use of dinitrophenol-IgG conjugates to detect sparse antigens by immunogold labelling. J Histochem Cytochem 37:69–74

Peppers SC, Holz RW (1986) Catecholamine secretion from digitonin-treated PC12 cells. J Biol Chem 261:14665–14669

Perrin D, Langley OK, Aunis D (1987) Anti-α-fodrin inhibits secretion from permeabilized chromaffin cells. Nature 326:498–501

Pickel VM, Joh TH, Reis DJ (1975) Ultrastructural localization of tyrosine hydroxylase in noradrenergic neurons of brain. Proc Natl Acad Sci USA 72: 659–663

Pickel VM, Joh TH, Reis DJ (1976) Monoamine-synthesizing enzymes in central dopaminergic, noradrenergic and serotonergic neurons. Immunocytochemical localization by light and electron microscopy. J Histochem Cytochem 24:792–799

Pickel VM, Chan J, Milner TA (1986) Autoradiographic detection of [^{125}I]-secondary antiserum: a sensitive light and electron microscopic labeling method compatible with peroxidase immunocytochemistry for dual localization of neuronal antigens. J Histochem Cytochem 34:707–718

Piekut DT, Casey SM (1983) Penetration of immunoreagents in vibratome-sectioned brain: a light and electron microscope study. J Histochem Cytochem 31:669–674

Priestley JV (1984) Pre-embedding ultrastructural immunocytochemistry: immunoenzyme techniques. In:Polak JM, Varndell IM (eds) Immunolabelling for electron microscopy, chap 4. Elsevier, Amsterdam, pp 37–52

Priestley JV, Cuello AC (1983) Electron microscopic immunocytochemistry for CNS transmitters and transmitter markers. In: Cuello AC (ed) Immunohistochemistry, chap11. IBRO, Wiley, Chichester, pp 273–322

Reggio H, Webster P, Louvard D (1983) Use of immunocytochemical techniques in studying the biogenesis of cell surfaces in polarized epithelia. Methods Enzymol 98:379–395

Reggio H, Bainton D, Harms E, Coudrier E, Louvard D (1984) Antibodies against lysosomal membranes reveal a 100000-mol-wt protein that cross-reacts with purified H$^+$,K$^+$ ATPase from gastric mucosa. J Cell Biol 99:1511–1526

Sarafian T, Aunis D, Bader M-F (1987) Loss of proteins from digitonin-permeabilized adrenal chromaffin cells essential for exocytosis. J Biol Chem 262:16671–16676

Sato A, Spicer SS (1982) Ultrastructural visualization of galactose in the glycoprotein of gastric surface cells with a peanut lectin conjugate. Histochem J 14:125–138

Schäfer T, Karli U, Gratwohl E, Schweizer F, Burger M (1987) J Neurochem 49:1697–1797

Schliwa M, van Blerkom J (1981) Structural interaction of cytoskeletal components. J Cell Biol 90:222–235

Seeman P (1967) Transient holes in the erythrocyte membrane during hypotonic hemolysis and stable holes in the membrane after lysis by saponin and lysolecithin. J Cell Biol 32:55–70

Seeman P, Cheng D, Iles GH (1973) Structure of membrane holes in osmotic and saponin hemolysis. J Cell Biol 56:519–527

Silverman A-J, Hou-Yu A, Oldfield BJ (1983) Ultrastructural identification of noradrenergic nerve terminals and vasopressin-containing neurons of the paraventricular nucleus in the same thin section. J Histochem Cytochem 31:1151–1156

Sternberger LA (1979) Immunocytochemistry, 2nd edn. Wiley, New York

Straus W (1957) Segregation of an intravenously injected protein by "droplets" of the cells of rat kidney. J Biochem Biophys Cytol 3:1037–1045

Tougard C, Picart R, Tixier-Vidal A (1982) Immunocytochemical localization of prolactin in the endoplasmic reticulum of GH3 cells. Variations in response to thyroliberin. Biol Cell 43:89–102

Triller A, Cluzeaud F, Pfeiffer F, Betz H, Korn H (1985) Distribution of glycine receptors at central synapses: an immunoelectron microscopy study. J Cell Biol 101:683–688

van den Pol AN (1988) Silver intensification of colloidal gold or horseradish peroxidase for dual ultrastructural immunocytochemistry. In: Van Leeuwen, Buijs, Pool, Pach (eds) Molecular neuroanatomy. Elsevier

van den Pol AN, Decavel C (1989) Snyaptic interaction between chemically defined neurons: dual ultrastructural immunocytochemical approaches. In: Waterlood F, van den Pol AN, Bjorklund A, Hokfelt T (eds) Neuronal microcircuits -handbook of chemical neuroanatomy. Elsevier, Amsterdam

van Meer G (1989) Biosynthetic lipid traffic in animal eukaryotes. Annu Rev Cell Biol 5:247–275

Weber K, Rathke PC, Osborn M (1978): Cytoplasmic microtubular images in glutaraldehyde-fixed tissue culture cells by electron microscopy and by immunofluorescence microscopy. Proc Natl Acad Sci USA 75:1820

Williamson JR (1969) Ultrastructural localizations and distribution of free cholesterol (3-ß-hydroxysterols) in tissues. J Ultrastruct Res 27:18–133

Willingham MC (1980) Electron microscopic immunocytochemical localization of intracellular antigens in cultured cells: the egs and ferritin bridge procedures. Histochem J 12:419–434

Willingham MC (1983) An alternative fixation-processing method for preembedding ultrastructural immunocytochemistry of cytoplasmic antigens. J Histochem Cytochem 31:791–798

Willingham MC, Yamada SS, Pastan I (1980) Ultrastructural localization of tubulin in cultured fibroblasts. J Histochem Cytochem 28:453–461

Willingham MC, Rutherford AV, Gallo MG, Wehland J, Dickson RB, Schlegel R, Pastan IH (1981a) Receptor-mediated endocytosis in cultured fibroblasts: cryptic coated pits and the formation of receptosomes. J Histochem Cytochem 29:1003–1013

Willingham MC, Yamada SS, Bechtel PJ, Rutherford AV, Pastan IH (1981b) Ultrastructural Immunocytochemical Localization of Myosin in Cultured Fibroblastic Cells. J Histochem Cytochem 29:1289–1301

Wilson SP, Kirschner N (1983) Calcium-evoked secretion from digitonin-permeabilized adrenal medullary chromaffin cells. J Biol Chem 258:4994–5000

Chapter 11

Quantitative Aspects of Immunocytochemistry

11.1 General Comments

When one considers the wide range of EM studies that uses sections, both for structural and for immunocytochemical reasons, one is struck by the paucity of quantitative data. This is especially true in cell biology. It could even be surmised from this statement that quantitative methods are either not generally available or, perhaps, they have serious problems associated with them. Nothing could be further from the truth, however: reliable quantitative methods are available, are relatively easy to use, and are based on a sound mathematical foundation. In Chapter 1, I stated that when we observe a section in the EM we are primarily interested in size, position, shape, number and electron density of structures. For most biological purposes it is the size and the number of an object which are the most amenable and interesting features to quantify. The method that is used to estimate the number and other geometrical properties of three-dimensional structures from sections is **stereology**.

"Stereological methods are precise tools for obtaining quantitative information about three-dimensional microscopic structures based mainly on observations made on sections" (Gundersen et al. 1988a). In all cases, the methods are statistical and can be used to estimate the volume, surface, length (height) or number of objects. Imagine, for example, that we wanted to know the volume of mitochondria in a cell. Perhaps the most meaningful volume parameter to know would be the absolute volume of mitochondria per cell. With two exceptions[1], there is, however, no stereological method to estimate this parameter directly. What one would actually measure in practice would be the volume density or the volume fraction of mitochondria per cell. The reason for this is that most (but not all) stereological estimates are made as ratios by which one refers one structural parameter to a second one which is referred to as the reference space (see Table 1). In the example of the mitochondria, the volume density relates the volume of mitochondria to the volume of the reference space, in this case that of the cell. As explained in more detail below, the volume density in itself is a parameter that may not necessarily tell us very much about mitochondria. When, however, the cell volume (in μm^3, for example) is known we can estimate

---

[1] The exceptions are the Cavalieri principle (see p. 398) which would involve exhaustive serial sectioning and the recently introduced nucleator method (see p. 408).

Table 1. Summary of stereological notation for ratios

Dimension	Reference dimension			
	Component			
	Volume, V (cm^3)	Surface, S (cm^2)	Length, L (cm^1)	Number, N (cm^0)
Volume, V cm³	V_V cm⁰			
Surface, S cm²	S_V cm⁻¹	S_S cm⁰		
Length, L cm¹	L_V cm⁻²	L_S cm⁻¹	L_L cm⁰	
Number, N cm⁰	N_V cm⁻³	N_S cm⁻²	N_L cm⁻¹	N_N cm⁰

the absolute volume of mitochondria (in μm^3) from the volume density (the dimensionless ratio, e.g. $\mu m^3/\mu m^3$)

A similar argument holds for the second parameter with which we shall be discussing in some detail in this chapter, namely surface. Until recently there was no way to estimate directly the total surface area of a structure from sections.[2] What we can estimate using stereological procedures is surface density or the amount of surface measured relative to a second structural parameter. One commonly used surface density ratio is **surface to volume ratio** – that is the amount of surface of an object related to its own volume. Consider, for example, the surface density of mitochondria. As for volume density, the surface density is a parameter which, by itself, may not necessarily give us useful information. We cannot easily relate to a statement that the surface density (S_V) of mitochondria per volume of mitochondria is 20 $\mu m^2/um^3$. When, however, the reference volume is known in absolute terms (e.g. μm^3) then the S_V can be converted into absolute units of surface (μm^2). As we shall see in this chapter, this is information which has more general applicability.

In Table 1 (from Mayhew 1983, after Weibel 1979) a summary is provided of the interrelationships between a component of interest (volume, surface, length or number) and a suitable reference (which can also be volume, surface, length or number). For illustrative purposes in this table the ratios are expressed in centimetres. It is important for any stereological study that the ratio one selects to estimate is clearly defined at the outset:

A major aim of this chapter is to discuss ways by which we can quantitatively relate structural data to functional-biochemical data. In the example of the mitochondria, whereas volume density of mitochondria per cell or surface density of mitochondria (per mitochondrial volume) may in some cases be related to biochemical data, this is not generally the case. When, however, we know the total surface area and/or volume of mitochondria per cell we can

---

[2] The exception is again the nucleator method (see p. 408).

relate this information directly to both biochemical and immunocytochemical units. For biochemical studies the number of molecules or units of activity of a mitochondrial enzyme of interest that has been assayed, for example, can be related to the number of cells used for the assay. Knowing the absolute surface area, or volume, of mitochondria per cell enables us to correlate directly the number of biochemical units to the size of the structural units that contain them. In parallel, immunocytochemical experiments can be done to determine the number of immunogold particles per structural unit of mitochondria after labelling sections of these organelles with antibodies specific for the enzyme of interest. The attempt here is to relate the "biochemical unit" and the "immunocytochemical unit" to the same reference space of structure in order to be able to have a direct correlation. The ultimate goal of immunocytochemical studies is to estimate the concentration of the antigen directly from the density of the immunogold labelling.

For immunocytochemistry, the two structural parameters that are the most informative are those already covered in the example of the mitochondria, namely the volume and the surface of cells. The reasons for this will soon become apparent.

Before we delve into the "nuts and bolts" of simple stereological theory, it may be useful at this point to give a few hypothetical examples of the kinds of experimental situations we would ideally like to address using quantitative immunocytochemistry and the kinds of answers we are looking for.

Example A. Protein N, whose amount per cell is known, localizes by immunocytochemistry exclusively to the nucleoli. The following approach could be carried out:

1. The volume density of nucleoli per nucleus is estimated as well as the volume density of nucleus per cell. If we (independently) estimate the mean cell volume, we can then calculate the average absolute volume of one nucleolus.
2. Since the total amount of this protein per nucleolus is known, the concentration of the protein in the nucleolus can now be given (in number of molecules per unit volume) and this can then be correlated with the number of gold particles that label it. This enables the **labelling efficiency** to be calculated, that is the relationship of the number of gold particles obtained for a known number of antigens.
3. When a different cell type is labelled with anti-N, if one assumes the labelling efficiency to be the same as for the first[3], we could get an estimate of the number of N molecules in the nucleolus of this cell.

Example B. A membrane protein X is uniformly localised to the inner membrane of mitochondria using immunogold labelling.

---

[3] As will become clearer later in this chapter, in a strict sense this assumption may not be absolutely correct. It is my firm belief, however, that this approach can give valid and useful comparative data especially as a reference for complementary biochemical studies.

A rationale for absolute quantitation could be the following:

1. The density of gold label per surface area of inner membrane is estimated.
2. The total surface area of mitochondrial inner membrane per total cell volume is estimated (the surface density). From an independent estimate of the average cell volume, we can then calculate the total surface area of mitochondrial inner membrane per cell, in absolute terms. As referenced on p. 408, however, the recently introduced nucleator method enables this to be done directly.
3. By correlating the biochemical estimates of X per cell to the total surface area of inner membrane the average density of X in the inner membrane is estimated.
4. By correlating the latter to the number of gold particles the labelling efficiency is obtained, that is, the number of gold particles relative to the number of antigens.
5. In a different experimental situation where the biochemical data indicate a decrease in the concentration of X per cell, the immunogold labelling density could be used directly to obtain (at least) a rough estimate of the concentration of X in the inner membrane (by assuming that the labelling efficiency will be the same in both cases).

Example C. A soluble protein P which has been immune-localized to the matrix of peroxisomes is available in pure form. The following approach can give absolute estimations of antigen concentration.

1. The protein is embedded in a dense gel so that, when the complex is sectioned and labelled with antibodies and gold, the labelling will be restricted to the surface of the section. Under this condition the amount of label can be proportional to the concentration of antigen.
2. The labelling of the reference gel can then be correlated directly with the amount of gold particles over peroxisomes in a tissue. The tissue and reference gel, however, must have been processed in the same way as the tissue so that the conditions up to, and including, the labelling reactions are identical for both; under these conditions the labelling of both reference and tissue is restricted to the section surface. This enables the number of gold particles over the peroxisomes to be converted to absolute units of concentration simply by reading off from a standard reference curve.

Example D. Vimentin intermediate filaments in a cell have been labelled with anti vimentin antibodies and gold.

1. The density of gold labelling the filaments is estimated.
2. The total length of filaments per cell is estimated. From these two parameters the total number of gold particles per cell is estimated as well as the labelling efficiency.

In the remaining part of this chapter I shall try to convince the reader that these examples are not at all unrealistic. Techniques are available now that can give very useful quantitative information for such problems – as well as many others. The key is the combination of EM immunocytochemistry, biochemistry and stereology. It follows, therefore, that some knowledge of the principles of stereology is essential.

11.2 Basic Stereology

For most purposes, one need not become an "expert" in order to use stereology. As for any unfamiliar method, however, it is often advisable to check the validity of the approaches by seeking the advice of a specialist, especially when difficult or novel problems are tackled.[4]

For most, if not all, immunocytochemical projects the amount of immuno-gold labelling will be related to either the volume, surface area, or length of a structure as shown, in part, by the preceding examples. It goes without saying therefore that understanding the theoretical basis for these simple ratio parameters, as well as the practical methods used to estimate them, is critical to an application of **quantitative immunocytochemistry.**

11.2.1 General Introduction to Stereology

For this chapter, and at this stage in the development of quantitative methods for immunocytochemistry, we need to discuss three essential stereological parameters in some detail. These are volume density (symbolized as V_V in stereological terminology), surface density (S_V) and length density (L_V). The reason for this is that our goal here is to quantitate particulate markers with respect to volume, surface area or length of structures. (N.D. although the numerical density of organelles that exist in cells as discrete units, such as coated vesicles or secretion granules, is an interesting parameter to know, there appears to be no a priori need to be able to correlate the density of labelling directly to the numerical density of such organelles.)

Figure 1 introduces the general problem we wish to discuss. It shows a cube (the reference space) that contains an object X embedded within it. Imagine that this cube is sectioned by a perfect two-dimensional section or plane ("infinitely thin section") in a random position and direction such that any part

---

[4] For anyone who wishes to learn more about stereology it is highly recommended that this person joins the International Society for Stereology. For about US$ 15 per year a list is provided of names and addresses of current members as well as information about stereology courses. It is a policy of these societies that new members are encouraged to make contact with stereology experts in their region. For more information contact Dr. Haymo Kurz, Secretary/Treasurer ISS, Institute of Anatomy II, Albertstr. 17, D-7800 Freiburg, Germany.

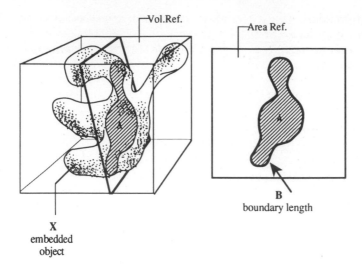

From a set of randomly located sections through the reference volume:

$$\text{Volume density of X } (Vv_{X, \text{ Ref.}}) = \frac{\Sigma \text{ Area A}}{\Sigma \text{ Area Ref.}}$$

If there is isotropic orientation of the sections as well:

$$\text{Surface density of X } (Sv_{X, \text{ Ref}}) = \frac{\Sigma \text{ Boundary length (B)}}{\Sigma \text{ Area Ref.}} \times \frac{4}{\Pi}$$

Fig. 1. A structure X is embedded in a cube, the reference space, and one (random) section is shown, as are the formulae for estimating volume density and surface density on random sections

of this cube has an equal probability of "meeting" a section. The essence of the stereological approach is to ask "what is the probability, in any random section through the cube, of slicing through the embedded object of interest". It becomes intuitively clear that:

- The larger the volume of X, the greater is the probability that a certain area of the profiles of X will be encountered in an average random section. Further, as the volume fraction of X increases, so, on average, does the total area (A) of X in any random section through the reference space. Note that volume fraction means the same as volume density (V_v).
- The larger the surface area of X the larger will be its total boundary length (B) over averaged, random sections. In other words, as its surface density in the reference space increases, in any random section that slices through the reference space the longer, on average, will be the length of the trace through its periphery. Note that the surface ratio of X to volume ratio of reference space is one example of a surface density parameter. (An additional caveat for determining surface (or length) densities is that in addition to randomiz-

ing section location within the structure the direction/orientation of sectioning is also critical. This will be discussed in more detail below).

- The larger the diameter of X the larger the probability that the section would contain at least one profile through X. (More strictly speaking it is the tangent or caliper diameter of the structure (the linear dimension perpendicular to the cutting plane) which determines the probability of the cutting plane passing through the object of interest (see Weibel 1979 pp. 13–19). For structures which are very long compared to their thickness, such as filaments, tubes or capillaries, their total length in the reference space can therefore be inferred from the number of profiles seen. (Note that the number of profiles is **not** related to the number of particles of X, but only to the total length of X particles; see Sect. 11.3.1.7, Numerical Density). The total length of filaments relative to the volume of reference space is the most common form of length density (L_v) measurement.

These statements have been proven mathematically many times over and it is not essential to fully appreciate the mathematics in order to be able to apply the formulae (for a mathematical background, see Weibel 1980). Somewhat unexpectedly, it turns out that the relevant formulae are very simple. They are also simple to apply provided one follows established guidelines or "rules" of stereology. The following description is meant to be a simple description of these rules. One of my goals here is to help the beginner overcome his fear of stereology, and in particular, of its application. It is essential to point out, however, that more advanced chapters and text books must be consulted before more sophisticated applications are carried out. Some key references will be given in the course of this chapter.

11.3 Estimation of Volume Density and Surface Density in Practice

11.3.1 Point and Intersection Counts

11.3.1.1 Volume Density

According to the principle of Delesse, a French geologist from the nineteenth century, the volume fraction of an object which is embedded in a larger structure (the reference space) is directly proportional to its areal fraction on random sections through the structure (Fig. 1). In other words, the ratio of the profile area of the object in the sections relative to the total area of the sections is identical in the mean to the ratio of the volume of the object, relative to the total reference space. Important to note here is that only by taking the average of the values from a sufficient number of random sections can one obtain the correct volume density : clearly some sections through the cube will miss the object entirely, while others will overestimate the ratio.

Estimation of volume density of nucleus (relative to total cell)

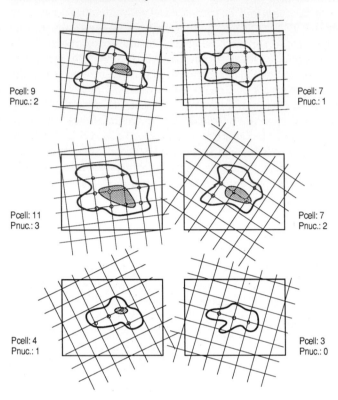

Pcell: 9
Pnuc.: 2

Pcell: 7
Pnuc.: 1

Pcell: 11
Pnuc.: 3

Pcell: 7
Pnuc.: 2

Pcell: 4
Pnuc.: 1

Pcell: 3
Pnuc.: 0

○ Total points on cells: ΣP cell = 41

· Total points on nucleus: ΣP nuc = 9
    Total points on cytoplasm: ΣP cyto = ΣP cell - ΣP nuc = 32

$$V_V \text{ (nuc, cell)} = \frac{9}{41} = 0.22$$

$$V_V \text{ (cyto, cell)} = \frac{32}{41} = 0.78$$

Fig. 2. The principle of point counting is illustrated. Six random sections are shown through a hypothetical cell pellet. The lattice grid has been laid randomly on the sections and the total number of points (= where the *test lines cross each other*) over the nucleus as well as the *total number of points* over the whole cell has been estimated (these points have been *circled*). Note that one section (*bottom right*) misses the nucleus completely. Such sections must be included because they correctly "weigh" the results in favour of the cytoplasm

A classical method of estimating the areal fraction of an object X embedded in a larger structure is to take random sections and to draw the profiles of each of the sections (both the total area of the section and the contours of the embedded object) on a sheet of cardboard. The profile of the object is then cut out with a pair of scissors in each case and all its cut profiles are weighed and compared to the weight of the total profile area of the sections (including the profiles of the object). The relative areas of the profiles of X to the total area obviously reflects the relative weights of the two sets of profiles. Clearly, however, this approach would be extremely tedious (but it is useful in helping to understand the principle involved). In practice, the easiest and most efficient method to estimate the ratio of the area covered by the object to the total area on each section is by point counting. That is a transparent lattice grid containing a system of test points (in most cases these consist of points of intersection of test lines) is placed over each of the micrographs. The most efficient kind of lattice grid is one having a **systematic** system of test points, that is where the test points are arranged in a regular array over the whole micrograph (Weibel 1980; see Mayhew 1983 for a lucid explanation). An important prerequisite is that the selection of micrographs must be free of bias (see "Sampling", Sect. 11.3.1.9). The number of points over the profiles of the structure is related to the total number of points over the section. The estimation of the areal density on the tissue section is thus performed with a statistical procedure of sampling point hits over profile and over total structure. Since, in most cases, a group of pictures are sampled in this way, the total number of points counted per individual picture need not be very large. This makes this approach quite efficient (see below). As an example, imagine we wish to know the volume density of the nucleus per cell (Fig. 2). The test grid is placed randomly over each of the random sections and the number of points over the nuclear profiles is counted, as are the total number of points over the entire cell profile (including the nucleus). The average of the ratio of the sum ($\sum$) of the numbers of points is then estimated.

$$\frac{\sum P_{(NUC)}}{\sum P_{(CELL)}} = \frac{\text{Area of nucleus}}{\text{Area of cell}} = \frac{\text{V nucleus}}{\text{V cell}} = V_V \text{ (nuc, cell)}.$$

V_V (nuc/cell) is, in other words, the volume density of the nucleus per cell. It is described by the symbol V_V, the superscript V indicating that we are dealing with a Volume parameter, while the sub-script V indicates that we are estimating a volume within a Volume. This ratio is expressed as a ratio of volumes, e.g.

$$\frac{\mu m^3}{\mu m^3} \; .$$

In general, volume density estimates are relatively simple to apply and have very few complications associated with them. It should be noted that the probability of a point (randomly placed in a cell) hitting the nucleus, depends solely on the volume ratio, $\dfrac{\text{Vol. nucleus}}{\text{Vol. cell}}$. Random sectioning is just a convenient way of putting points in cells at random positions. For V_V estimations object shape and orientation are irrelevant. The results are not, for example, affected by a particular orientation in space of the object relative to the reference space (anisotropy) – in contrast to surface and length density.

11.3.1.2 Surface Density

Surface density relates the amount of surface area of a structure such as membrane that is contained in a defined volume of reference space. Surface density estimation is slightly more complex than that of volume density and is concerned with estimating on random sections the relationship of the boundary length (or contour length) of those profiles of the objects (Figs. 1, 3) to the area of the profiles of the reference space. In three dimensions, this relates the surface to the volume. The simplest and the most reliable method to measure S_V is by making point counts (as for volume) as well as intersection counts. That is, a system with test points and **test lines** is placed over the images of the sections and the number of test points over the reference space is counted (as for V_v) as well as the number of intersections made between the lines of the test grid and the contour profile of the object. The parameter of interest here is the ratio of the number of intersections to the number of points, which relates the boundary length to the area in two dimensions, or surface area to volume in three dimensions.

A simple formula, based on ideas first put forward by Buffon in the late 18th century, enables S_V to be estimated from point and intersection counts:

$$S_V = \frac{2 \times I}{\text{total length of test line over the object}} \, ,$$

where I is the total number of intersections and the "total length of test line" is that indicated by the dense lines in Fig. 3. It is important for these measurements that all positions and orientations of the test lines in three-dimensional space are equally probable (i.e. the lines must be isotropic, see below). When a simple lattice grid is used to count the total number of intersections (I) through the boundary (which in cell biology usually means a membrane) of the object in both vertical and horizontal directions, as well as the number of test points (P) over the object (Fig. 3) the formula becomes:

$$S_V = \frac{2 \cdot I}{L} = \frac{2I}{P \cdot l(p)} \, ,$$

where L is the total length of test line in the reference space, $P \cdot l(p)$ is an estimate of L and $l(p)$ is the length of test line associated with each point in the test system. This formula holds for any test system. When a square grid is used this

Estimation of specific surface area of cells (i.e. relative to their volume)

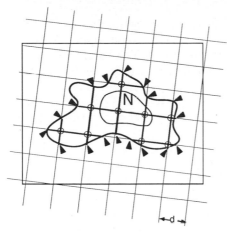

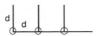

Note that for every point there are 2
lengths (d) of test line when
intersections are estimated in both
horizontal and vertical directions

○   Number of points on cell: P(cell) = 9

◄   Number of total intersections*: I(cell) = 18

let d = 1 μm (total test line length: ⎯⎯⎯ dense)

$$S_V(\text{cell:cell}) = \frac{2I}{P \times 2d} = \frac{I}{P \times d} = \frac{18}{9 \times 1\,\mu m} = 2\,\mu m^{-1}$$

The number of intersections must be counted in both vertical and horizontal directions for
this formula to be valid. When only one direction is used then

$$S_V = \frac{2I}{P \times d}$$

For reducing the number of intersections counted one may either use a larger mesh grid or
simply count in one direction only. Note that $P \times d$ is the total length of test line over the cell
profile. From the formula

$$S_V = \frac{4}{\Pi} B_A \qquad \text{(Weibel, 1979, p. 36)}$$

where
S_V = Surface density, (the amount of surface per reference space in this case volume (3D))
B_A = Boundary length density (the length of trace per area (2D))

$S_V = 2\,\mu m^{-1}$ means $2\,\mu m^2$ of surface per $1\,\mu m^3$ of volume

Notice however that from the above formula

$$B_A = 2 \times \frac{\Pi}{4} = \frac{\Pi}{2} = 1.57\,\mu m^{-1}$$

which means 1.57 μm linear trace per 1 μm² of area.

Fig. 3. The principle of intersection- and point counting is illustrated. The lattice grid is overlaid
on an a random section through a cell. The *arrowheads* indicate the intersections of the lines of
the lattice grid with the plasma membrane. The *circles* indicate those points that fall on the cell.
Note that the S_v of the nucleus (*N*) could also be determined using the same grid system. There are
six intersections and two points over the nucleus; thus S_v nuc = $3\,\mu m^{-1}$. For any test grid a critical
parameter is *d*, the distance between test points, which is expressed at the final magnification of
the micrograph (usually expressed in μm; remember the simple rule; at ×10000, 1 μm = 1 cm).
Note that when making intersection counts in only one direction (e.g. vertical) each point has *one*
linear test line equivalent of *d*; it is usually better to make intersection counts in both horizontal
and vertical directions (as in the example here) in which case each point corresponds to *two* linear
equivalents of *d*. The total test line length over the structure is indicated by a thickening of the test
lines

can.also be formulated as,

$$S_V = \frac{2I}{P \cdot 2d}, \quad \text{which simplifies to} \quad \frac{I}{P \cdot d},$$

where d is the distance between the points (Fig. 3). This formula holds true only when the lines and surfaces have a random orientation with respect to each other in three dimensions. Note that for every point over the structure there will be a corresponding length of test line d in both horizontal and vertical directions; hence the 2d in the formula (note, when only horizontal lines would be used, the formula would become,

$$S_V = \frac{2I}{P \cdot d}.$$

The units of d (which must be given on the scale of the specimen itself, i.e. must take into account the linear magnification) determine the units of S_v which are always given in length units $^{-1}$. Hence if d is given in μm, S_v is μm^{-1}. Note that this is equivalent to 1 μm^2 of surface area per 1 μm^3 of volume.

In practice, as for V_v, one counts points and intersections on a sufficient (see below) number of random sections. The estimation of S_v in practice is made from the sum $(\sum)$ of the number of intersections and points over the random sections.

Hence, for instance,

$$S_V = \frac{\sum 2I}{\sum P \cdot lp} \quad \text{or} = \frac{\sum I}{\sum P \cdot d}.$$

11.3.1.3 Length Density L_v

Imagine you want to estimate the total length of microtubules in a cell. As for volume and surface estimates, the first prerequisite would be to estimate the ratio of the length of microtubules with respect to cell volume. Thin sections are made which must be random in terms of location and orientation, as for surface, and the microtubules would essentially be observed in transverse section or "transect points" (this is strictly correct only for infinitely thin sections and infinitely thin or infinitely long microtubules, see below). The only practical procedure for ensuring this precondition is by using one of the unbiased methods of sectioning referred to above. The aim then is to relate the number of these profiles or transect points (Q in stereological notation) to the area of the section occupied by the cell A(cell)

$$Q_A = \frac{Q \text{ (profiles)}}{A \text{ (cell)}} ; \quad L_V = 2Q_A.$$

The length density L_v is thus twice the number of profiles per unit area. As for volume density, the area of the cell is simply estimated by point counting while the number of transverse sections through the filament is counted directly

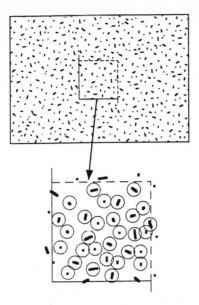

There are 30 profiles (circled) within the counting frame that do not touch the forbidden line (continuous)

If the counting frame is 1 μm^2:

$$Q_A = \frac{30}{1\ \mu m^2} = 30\ \mu m^{-2}$$

$$L_V = 2 \times Q_A = 2 \times 30\ \mu m^{-2} = 60\ \mu m^{-2}$$

Fig. 4. Example of a hypothetical section through a part of a cell showing profiles of filaments (or tubules). In a perfect (infinitely thin) section all the profiles would be seen as points. In a real section with finite thickness this would hold only for a part of the sectioned filaments. A counting frame obeying the "forbidden line" rule has been positioned randomly over the structures and the number of transects per frame (Q_A) are counted. This forbidden line counting frame of Gundersen (1977) avoids biased counts of particle number. *Above* A restricted field of vision of a section with some profiles. *Below* The same field at higher magnification to show the unbiased counting frame (Gundersen 1977). The *circled profiles* are estimated because they do not touch the forbidden line (drawn as a *continuous line*). In contrast, profiles which are within the frame as well as those that cross or touch the dotted line are counted

(Fig. 4). As for all stereological sampling it may be more efficient to count random sub-sets of profiles (see below).

11.3.1.4 The Problem of Anisotropy for Surface and Length Measurements

We have already alluded to the fact that for estimation of surface density it is a fundamental stereological requirement that the sections through the tissue, as well as the system of test lines that are used to probe the surface of interest, do not preferentially hit the surface from any one orientation (which would result in a bias). This is particularly important when the structure is anisotropic in its orientation, that is, it has, for instance, one axis which is preferentially orientated in a particular position(s) in space. A classical example would be muscle myofibrils in which the plasma membrane is preferentially orientated along the axis of the myofibrils. It is intuitively clear that the great majority of arbitrary sections through such a rod-shaped structure would give mostly transverse and oblique profiles.

A tissue can have more than one axis of anisotropy. In the case of muscle the Z bands have an axis of anisotropy which is perpendicular to the anisotropic axis of the myofibrils. With respect to rod-shaped structures such as myofibrils that are aligned in parallel, there is a far greater chance of sectioning them transversely and obliquely rather than longitudinally. If all the (arbitrary) sections were perpendicular to the axis of anisotropy it would give a condition that would fail to fulfil the fundamental requirement described above, the need for **isotropic** and **uniform** randoms sections. Besides surface density, the whole problem of anisotropy is equally problematic for length density measurements.

A few years ago the problem of anisotropy was perhaps the most intractable problem in stereology (e.g. see Weibel et al. 1979, 1980 for the state of the art 10 years ago). In recent years, however, a number of elegant new approaches have been introduced that have effectively solved this problem. The key to the solution has been to design strategies for sectioning the anisotropic tissue in an unbiased fashion (see below). A remarkable consequence of using these new methods is that it does not matter whether or not a structure of interest is anisotropic.

It should be noted that for the estimation of both volume density and numerical density the results are not affected by anisotropy.

11.3.1.5 Unbiased Methods of Sectioning for Determining S_v

As recently summarized by Gundersen et al. (1988a); Mayhew (1990) and Cruz-Orive and Weibel (1990) there are now four main practical approaches for producing sections of tissue that can provide unbiased estimated of surface (as well as length) density, irrespective of the degree of anisotropy in the structure. These are:

- Directionally independent isotropic uniform random (IUR) sampling; see also the orientator method of Mattfeldt et al. (1990).
- IUR-orientated sectioning by orthogonal triplet probes (ortrips) (see Mattfeldt et al. 1985), see also the orientator method of Mattfeldt et al. (1990).
- Vertical sectioning (see Baddeley et al. 1986; note that this procedure is not valid for (filament) length estimation) (see Figs. 11, 12).

For many biological problems associated with anisotropy, the introduction of the concept of vertical sectioning has become the method of choice for estimating S_v. The idea behind this method is the realization by Baddeley that since every straight line in three-dimensional space is contained within a "unique vertical plane", random vertical sections have the possibility to contain lines through an embedded anisotropic structure in all possible orientations. A vertical section is defined as any section plane that is perpendicular to some arbitrarily defined but identifiable reference plane – the horizontal plane (see also Cruz-Orive and Hunziker 1986; Gundersen et al. 1988a; Mayhew 1990). In the simplest instance, many tissues, such as epithelial cells, can be considered to have a "natural" vertical axis running from basal to apical and perpendicular to the basal domain. Similarly, for tissue culture cells an obvious vertical section is one perpendicular to the surface of the "horizontal" support (plastic) dish. If the tissue does not possess an obvious vertical axis the experimenter may (arbitrarily) define an axis referred to as "vertical". The latter fulfills the first absolute requirement for this method (Baddeley et al. 1986; Gundersen et al. 1988a). The second requirement is that on the horizontal plane the vertical sections must be random with respect to both position and rotation around the vertical axis. For tissue slices an useful method to effect this randomization is described in Baddeley et al. 1986; see also Gundersen et al. 1988. The final requirement is that a test system of cycloids must be used to determine S_V (see Cruz-Orive and Hunziker 1986; Fig. 5). Aside from the above references a very simple illustration showing the use of this approach to determine S_V of an everyday object (an apple) is given in Mayhew (1990).

- Isector. A simple and elegant method was recently described by Nyengaard and Gundersen (1991) in which the specimen is embedded in a home-made spherical mould made out of silicone rubber. The polymerized ball is rolled to randomize its position and re-embedded in a normal embedding mould. It is foreseen that this approach will become the method of choice for producing isotropic, uniform random (IUR) sections of small specimens.

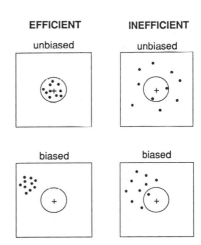

Fig. 5. A diagram using a "shooting range" to illustrate the difference between efficiency and bias. (Gundersen 1990)

Since all these methods of producing are relatively simple the stereologists now recommend using one of these approaches routinely for any estimation where anisotropy may be a problem. It should be emphasized that it is simpler to do this than it is to determine the degree of anisotropy.

11.3.1.6 The Need for Absolute Values

We have already pointed out that estimates of V_V and S_V in the absence of absolute values give only limited information. This can be more clearly demonstrated with an example. Imagine we estimate the V_V of nuclei in cells, and subsequently, the S_V of the nuclear (outer) membrane as shown in Figs. 2 and 3. If V_V is 0.25, it means that 25% of the cell volume is occupied by the nucleus. Note, that when the absolute volume of the cell is not known the information contained in this V_V value is limited. It can generally not be used in a comparative experiment. Imagine further that the cells are treated with a hypothetical drug which increases the volume of the nucleus but has no affect on total cell volume. By point counting, one determines that V_V increases from 0.25 to 0.5. If one is assured that the cell volume has not changed (and only then!) one can conclude that the absolute volume of the nucleus has doubled in this case. Note, however, that in the absence of independent data showing that the cell volume has not changed, doubling of V_V could be explained by:

- A doubling of nuclear volume while the cell volume stays constant (as described).
- A shrinkage of the cytoplasmic volume to half its normal value while the nuclear volume remains constant.
- Some combination of 1 and 2.

When, however, the cell volume is estimated in absolute terms for both conditions, it becomes simple to interpret the change in V_V.

The same problem holds true for S_V. Imagine that the S_V of the nucleus has doubled upon drug treatment. This could either mean that the absolute surface of the nuclear envelope has doubled with respect to the control, untreated nucleus or, conversely, that the volume of the nucleus has shrunk by half (without changing the surface). When, however, one knows the absolute volume of the cell in both cases, simply from the V_V estimates we can calculate the absolute volume of the nuclei in both cases. Then, simply by multiplying the absolute volume by S_V we get an estimate of the absolute surface of the nucleus in both cases, since:

$$\text{volume (V)} \times \frac{\text{surface}}{\text{volume}} (S_V) = \text{surface (S)}.$$

In this case the interpretation of the result becomes straightforward. Note, again, that only in the case of the absolute volume of the nucleus being unchanged in both conditions can we directly compare S_V estimates.

The same considerations also apply to combinations of length density, L_V with either V_V or S_V. For a more detailed discussion of this problem see Gundersen (1984).

11.3.1.7 Numerical Density

A significant part of the recent progress in stereological methods can be attributed to the introduction of fundamental new concepts which have led to efficient and unbiased practical methods for the estimation of the number of particles in a defined volume of tissue. Unlike earlier approaches, these methods require **no** assumptions about shape, size or distribution. The key breakthrough in this area was the introduction of the principle of the disector (Sterio 1984; see below) which requires two sections[5], a known distance apart, through the structure. A comprehensive discussion of these methods are outside the scope of this chapter and the reader is referred to two recent reviews by Gundersen et al. (1988b) and Cruz-Orive and Weibel (1990). Important here is to point out that this family of methods opens many fascinating new possibilities for biological research. They have, for example, enabled for the first time an unbiased estimate of the total number of neurons and glial cells in human brain (Pakkenberg and Gundersen 1988). Of cell biological interest is the potential of taking pairs of light microscopical (optical) sections of a tissue (Gundersen 1988b). Thus, from two micrographs taken at two different focus levels (a known distance apart), the number of cells or nuclei in a known volume can be estimated. The confocal microscope is especially suitable for some of these applications. In practice the only fundamental limitation of these new methods is an obvious one – all cells or particles of interest must be identifiable in all sections where they are counted. Typically, this means in sections where the nucleus is present, i.e. rather ideal conditions for identifying cells.

11.3.1.8 Statistics – Variations and Errors

Stereological measurements should aim to get as much quantitative information as is necessary from the minimum amount of work. Two key concepts describe the essence of good stereological sampling, namely "unbiasedness" and "efficiency". According to a recent review by Gundersen et al. (1988a) these concepts are used in a statistical sense and mean "without systematic deviation from the true value" and "with a low variability after spending a moderate amount of time", respectively (Fig. 5). It follows therefore that a consideration

---

[5] N.B. It is not always essential to have two sections, nor need their thickness be known since: (1) With the optical disector (see p. 402) you only need one (thick) section. (2) Disector thickness does not always have to be known (e.g. in those methods referred to as the fractionator, the selector and the nucleator; see Gundersen et al. 1988b and Cruz Orive and Weibel 1990 for an overview, as well as citations of the original references).

of statistics is critical to the design and interpretation of all stereological experiments. The following two problems must be considered:

- It is essential to design an experiment in such a way that the sections are randomly generated so that they represent uniformly all possible sections through our tissues/cells/structures, that is, free of bias (see below).
- A second consideration is to know the minimum cost and effort we need to invest in order to obtain statistically significant results.

The first thing to mention is that there always exists a certain amount of biological variation (variance) between structures (especially at the level of **organs** or **animals**) and that there is no way to avoid or influence this variation in a given experiment. The amount of variation differs from one system to the next, but it is important to note that this amount determines the smallest "window" of accuracy with which a certain parameter can be measured. In other words, if there is a 10% variation in the mean cell volume between animals in a population, the mean cell volume ($\bar{x}$) one estimates in a group of n animals (by whatever procedure) can **never** have a standard error (SEM) systematically less than $\dfrac{10}{\sqrt{n}}$% of $\bar{x}$. This is not a stereological problem, but merely a biological fact of life.

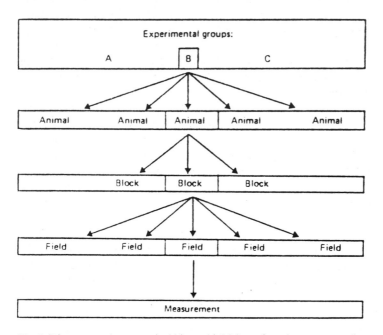

Fig. 6. Diagram to show a typical hierarchial (also referred to as a nested) sampling strategy for stereology. In this example five animals are selected (at random) per experimental group and three blocks are prepared of the tissue of interest from each animal. One section (only) is prepared from each block and five micrographs (fields) would be taken, giving a total of 15 micrographs per animal in this example. (Mayhew 1983)

As pointed out by Mayhew (1983), there are two fundamental reasons for failing to obtain the "correct" stereological result which he refers to as "accidental" (or random) error and "**systematic**" error (or bias). The former errors "arise because of variations in the biological material itself, in the way it is prepared for microscopical examination and in the way it is measured. Collectively these errors influence the reliability or precision (see Fig. 5) of the final estimate". This group of "errors" also influences the efficiency of estimation. Systematic errors influence the validity or **accuracy** of the final estimate, which is defined as the measure of the magnitude and directions of the departure from the "true" value (which is an unknown). Note that when there is no bias (and only then) precision and accuracy are synonymous.

Proper sampling is the only way of reducing the magnitude of potential "errors" in stereological studies as well as to minimize the time and effort (cost) required.

11.3.1.9 Sampling

A key element to the practical success of any stereological approach is that one does not measure everything one sees but selects or samples a (minimal) sub-population or fraction of structural elements on which the measurements are made. As already emphasized a uniform random sampling procedure must be used at all levels or else bias is introduced. For most stereological studies one must design a strategy that considers a number of different levels of sampling (hierarchy of sampling, see Fig. 7).

While it is difficult to give a valid, general rule of thumb about the minimum number of units required at different sampling levels, the following could be taken as a rough guide (Table 2).

At all levels sampling must be (uniform) random. With rare exceptions it is most efficient to perform the random sampling systematically, that is usually possible at all levels. Thus, micrographs can be taken of parts of sections that are immediately adjacent to a grid bar; for some parameters for structures that are not relatively frequent in the section it may be more efficient to move the section in one (arbitrary) direction until the next structure of interest is found and then photographed. This is true, for example, for estimating S_V of the Golgi complex (see p. 410).

In such a hierarchical system, sampling variances and measuring errors become superimposed on the biological variations between individuals. The simple take-home message is that the potential for variation/error decreases as we move down this hierarchy (see Gundersen 1980; Cruz-Orive and Weibel 1981; Gundersen and Østerby 1980, 1981; Mathieu et al. 1981; Gupta et al. 1982). Since the greatest contribution to the accumulative variation is most often that due to the biological variation between animals it is in most cases better to have fewer micrographs of a relatively large number of animals than to exhaustively sample a few animals with a relatively large number of micrographs. The figures in Table 2 give typical numbers for routine experiments (for more details see the above references). Many exceptions apply to this rule of

Formulae, definitions

For a simple variable, x with n independent observations:

Arithmetic mean $= \bar{x} = \dfrac{\sum x}{n}$

Variance $= s^2(x) = \dfrac{n}{n-1}(\overline{x^2} - \bar{x}^2)$

Standard deviation $= SD(x) = \sqrt{s^2(x)}$

Standard error (of the mean) $= SE(\bar{x}) = \dfrac{SD(x)}{\sqrt{n}}$

Coefficient of variation $= CV(x) = \dfrac{SD(x)}{\bar{x}}$

Coefficient of error of the mean $= CE(\bar{x}) = \dfrac{SE(\bar{x})}{\bar{x}}$

For an estimator of the ratio of the means of two random variables: $R = \dfrac{\text{mean } x}{\text{mean } y}$

Ratio of means $= \hat{R} = \dfrac{\bar{x}}{\bar{y}}$

$CE(\hat{R}) = \sqrt{CE^2(\bar{x}) + CE^2(\bar{y}) - 2r(x,y)\,CE(\bar{x})\,CE(\bar{y})}$

where $r(x, y) =$ coeff. of correlation between pairs

or:

$CE(\hat{R}) = \sqrt{\dfrac{n}{n-1}\left(\dfrac{\sum x^2}{(\sum x)^2} + \dfrac{\sum y^2}{(\sum y)^2} - 2\dfrac{\sum x \cdot y}{(\sum x)\cdot(\sum y)}\right)}$

$\sum x^2, \sum y^2 =$ sums of squares; $\sum x \cdot y =$ sum of multiplied pairs

Compound error: $CE(z) = \sqrt{CE_1^2 + CE_2^2}$

where z is the product of two independent estimators of two fixed parameters, with coefficients of error CE_1 and CE_2, respectively.

Fig. 7. Compilation of the most frequently required formulae for statistical evaluation of stereological data. (Taken in part from notes provided by Dr. H.J. Gundersen)

thumb; for example, if the population of animals/units under study is very similar then it may not be worthwhile to invest effort in studying a larger number of animals. Less than two at any level is never a good idea unless pronounced homogeneity has been shown to exist (Gundersen, pers. commun.).

Since stereology is inherently statistical in nature, the quality of the data is evaluated by standard statistical procedures. The experimenter sets the maximal degree of overall variation (precision) that is acceptable and the minimum amount of work necessary to get this precision is undertaken.

Table 2. Guide to minimum number of units required at different sampling levels

Level 1	Number of animals	≥ 5; note especially that this depends significantly on the biological variance and on the *aims* of the experiment; notably on the required sensitiveness or "discriminative power" of our experiment
Level 2	Number of organs per animals (Number of cell culture units)	1–2 or more in the case of some organs (at least 2)
Level 3	Number of blocks per organ	2–4; up to 10 if the organ is very inhomogeneous
Level 4	[a] Number of sections per block (Note that this level may necessitate its own hierarchy at different levels of magnification)	1 or several serial sections if number is to be estimated with disectors
Level 5	Number of micrographs per section	5–10; usually no more than 25
Level 6	Total number of points or intersections per animal	75–150 in each compartment of interest

[a] Note that this is not really a true sampling step.

As a simple example, imagine that a sphere is embedded in a cube and we are interested in the volume density of the sphere. The cube is sectioned randomly. Some sections will miss the sphere completely and will register a V_v of 0. Others will go through the centre of the sphere and will overestimate the "true" V_v. When the results are averaged over many sections, however, the estimate of V_v approaches the "true" V_v. Note again that the amount of biological variation within the system sets the limit for the confidence intervals with which we can determine mean values. When we are interested in the volume density of a single sphere in a cube the result can be very accurately determined; when, however, we are interested in a population of different spheres in cubes our estimate of each of them should be no more precise than the variation between the individuals. With respect to estimating the mean V_V in this example by **random** sampling our ability to reduce error due to limitations in our measurements will be reduced only by the square-root of the number (n) of sections measured $\left(\text{SEM} = \dfrac{\text{SD}}{\sqrt{n}} \right)$. Conversely, we can state that our ability to increase the **precision** with which we measure the parameter V_V will increase by the $\sqrt{n}$. In other words, to reduce the error by one-half (or conversely, to improve the accuracy of our estimates by a factor of 2), we need four times the number of sections. (As pointed out by Gundersen, pers. commun, when the sphere is sampled by a systematic series of sections this error is significantly reduced and SEM then approaches $\dfrac{\text{SD}}{n}$; see Gundersen and Jensen 1987). In practice, what is done is to accept, say, a 10% error and to count the minimum number of sections that gives a coefficient of variation (CV) that is less than this value.

There are statistical methods available for estimating the degree of variation (variance) contributed by the different levels of sampling (see above) as well as or for estimating the minimum number of blocks, sections, pictures, points and intersections that one needs for any given problem (see Gundersen and Østerby 1980, 1981; Gupta et al. 1982). The latter references deal exclusively with ratio estimators. For estimating the necessary number of sampling items when total quantities are the goal of the experiment see Pakkenberg and Gundersen (1988); West and Gundersen (1990). The formulae valid for systematic sampling are also presented in Gundersen and Jensen (1987). The most important statistical formulae that are required for evaluating routine stereological data are summarized in Fig. 7. Often, simple pilot experiments can be very useful for getting a rough idea of the sampling procedure that will be required (see the above references). The message which always comes across from the stereologists is that there is (surprisingly!) a clear tendency to count too much, that is to do more work than is necessary for a required confidence limit. The motto "do more less well" (Weibel; see Gundersen and Østerby 1981) is an apt one. Note again that it is better to count more pictures with less "points" and "intersections" per picture than to count a relatively few micrographs very precisely with a fine-mesh grid. It is even more efficient to use a lower number of micrographs from **more** animals (or tissue culture units).

11.3.1.10 Photographic Enlargement

At all levels of sampling at EM levels, the micrographs should be taken at the minimum magnification that allows clear visualization of a structure(s). This enables the largest possible area to be sampled. The picture should, ideally, be enlarged to the maximum size in order to facilitate the measurements. In practice, a compromise is made between making a significant enlargement and the limitation of what is practicable. The enlargement can either be made on prints or, more conveniently,the measurements can be made on projections from the negatives. This saves the cost and effort of printing. This is done using a projector that enlarges negatives in such a way that a transparent overlay can be superimposed on the images. Although machines are not, to our knowledge, commercially available, they can be built relatively easily by a good workshop. At least two models have been described – one for projecting 35 mm negatives (Weibel 1979) and one for projecting 6 ×10 cm EM plates (Griffiths and Hoppeler 1986) (Fig. 8). More details of these machines can be given upon request.

In some cases, it may be possible to make the estimations directly on a TV monitor attached to the light or electron microscope by using plastic grid overlays. This can save considerable cost and effort (Mattfeldt, pers. commun.; see Mattfeldt et al. 1990).

Fig. 8. Photograph of a machine designed by the EMBL workshop to enlarge 6 × 10 cm EM negatives by a factor ×4. The negatives are inserted into the slot (*arrow*) and the image is projected on the screen (*S*). Different transparent lattice grids can be placed on this screen

11.3.1.11 Use of Different Lattice Grids

On pages 355–379 of Weibel (1979), one can see a range of lattice grids which have been routinely used for different stereological estimations. For many purposes, a square lattice grid is convenient. For more sophisticated recent innovations see Gundersen and Jensen (1987). These grids can conveniently be xerographed or drawn out in large format either on drawing paper or by computer, and then photographically reduced or enlarged to the size required. The final transparent sheet may be printed on transparent potographic film or projected on an overhead projector photocopy film. For any measurement one aims to use the largest mesh grid (which means the least amount of counting) that will give statistically significant results, see Sect. 11.3.1.13, below. This size can be chosen by simple trial and error on a few pilot micrographs, but more sophisticated methods are available (Weibel 1979; Gundersen and Jensen 1987). Note that, contrary to expectation, it is **not** necessary for test systems to be drawn very accurately. An estimate of volume density using point counting is precise and unbiased as long as the points are arranged somewhat regularly and one knows the total number of points hitting the structure and the reference space (Gundersen, pers. commun.). For background into the theory of test systems see Cruz-Orive (1982) and Jensen and Gundersen (1982).

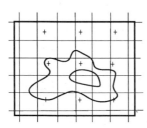

A

Single-lattice grid:
Appropriate:
For points or
intersections of cell
and/or nucleus.
Inappropriate:
For V_V (golgi, cell)

This example shows a relatively
high number of intersections
(arrowheads) for illustrative
purposes only. For the example
of the plasma membrane this
size of grid would lead to an
unnecessarily high number of
intersections being estimated.
Hence, for S_V estimates using
this grid one would either count
only intersections in one
direction (e.g. in this example
there would be 6 intersections of
the plasma membrane with the
horizontal lines); alternatively,
one would use a larger mesh
grid

B

Double-lattice grid:
Use coarse points $\oplus$
for cell and nucleus

Use fine points ↙
for small organelles
such as golgi,
mitochondria etc.
Note that each
coarse point $\oplus$ is
also a fine point.
$D = 2d$ or $D^2 = 4d^2$

(Nb. the point corresponds to
the precise point where the two
lines cross) the circle is used
merely to outline its position
more clearly.

C

Double-lattice grid:
An alternative
design. In this case
the + are the coarse
points and the
positions where the
lines cross give the
fine points. There
are thus 2 coarse
points and 9 fine
points on the total
cell profile.

Total coarse points of cell: $\Sigma P_D = 9$. This is equivalent to 9x4 or 36 fine points. (In this
example there are only 33 fine points over the cell; when averaged over many sections the
difference between the actual number of fine points and the number of coarse points x4
becomes insignificant.)
Total fine points on golgi: ΣP_d golgi = 2

$$V_V \text{ (golgi, cell)} = \frac{2}{36} = 5.5\%$$

Fig. 9A–C. Diagram to illustrate the use of single versus double-lattice grids (see also Weibel
1979). A single lattice grid might be appropriate for estimation of S_v and/or V_v of cell or nucleus
(as in Fig. 3). However, for the Golgi complex the lattice shown (in **A**) would not be efficient: in
order to count 100 points over the Golgi with such a lattice one would have to count over 1000
points over the total cell. A finer mesh is therefore required (**B**). This figure also shows the use of a
double-lattice grid to estimate V_v and S_v of cell or nucleus (course points – *circled*) and the fine
mesh lattice to estimate Golgi complex (*arrows*) in one step. An alternative type of design is
shown in **C**. For many purposes such a grid is visually easier to work with

11.3.1.12 Double-Lattice and Multi-Purpose Grids
for Sampling at Different Levels

In order to make the counting more efficient, a double (or multiple) lattice grid is often very convenient (Weibel 1979) (Fig. 9). This allows us to select, from the same transparency, a different grid mesh size for different structures on a micrograph. Hence, small structures can be sampled using a fine mesh grid (the "small" points), while the larger objects are counted using the coarse mesh ("large" points). This enables us to avoid over-counting the large structures while at the same time getting a reasonable sampling of the small structures. For the final calculation, the number of "large" points is multiplied by a factor (depending on how many "small" points are contained in the area associated with the "large" points) in order to be able to correlate the two estimates (Fig. 9).

An excellent multi-purpose grid designed by Gundersen and Jensen (1987) is shown in Fig. 10.

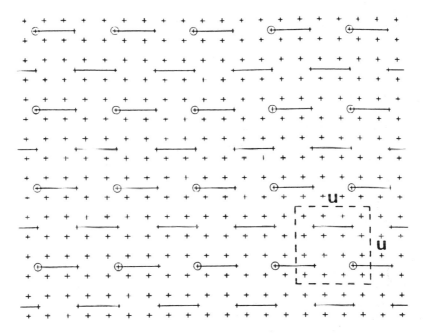

Fig. 10. The multi-purpose grid designed by Gundersen and Jensen (1987). It contains three different and very distinct point sets in the ratio 1:4:10 and two different sets of lines that are magnified or reduced for specific purposes. The authors' original description is as follows: "Tessellation of a quadratic unit with side length u making a test system with test lines and three sets of test points: encircled points (p1 = one per unit), all points at the ends of test lines (p2 = four points per unit) and isolated points (p3 = 16 per unit). The test system may be used for estimating B/A by counting all intersections, I between total profile boundary B and all test lines and counting all points, P of a suitable point set hitting all profiles: $[B/A] = (\Pi/2)\sqrt{p} \cdot I/\sqrt{P}$. p is either 1,4 or 16 depending on the choice of point set. The estimate of he total area of all profiles is $[A] = P.u^2/p$

11.3.1.13 Number of Points and Intersections

A consensus emerging from the stereologists (see Gundersen et al. 1988a) is that, for most experiments **at most** a total of 200 points or 200 intersections per animal or biological unit need to be counted. For example, if 20 micrographs are taken to estimate volume density of nuclei per cell, a grid that has a mesh size such that , on average, 10 points fall on the nuclei (fine points) and 10 (course) on the cells in a micrograph is sufficient.

It is a simple and useful exercise to compare the means and the coefficient of errors obtained using different grid sizes. For example, for nuclear V_v one could compare a grid size, giving, on average, about 10 points per picture with a very fine mesh one giving, say, 100 points per picture. Only then can the skeptical beginner be convinced of the 200 point/intersection "rule".

11.3.1.14 Section Thickness Effects: Measurement of Section Thickness

Stereological theory assumes that the estimations are made on infinitely thin sections (except for the disector method and volume-weighted mean volume estimations, see below). Any deviation from this statement may lead to significant errors under certain conditions. Obviously, although sections prepared for EM can be extremely thin (roughly 20–60 nm for plastic and 50–100 nm for routine cryo-sections) they still have a measurable thickness. The effect of this is that some structural parameters may be overestimated due to over-projections (for a detailed description of this so-called Holmes effect and the methods for correcting it – see Weibel 1979). In general, problems are to be expected only when the dimension of the structures to be measured is less than five to ten times the section thickness.

For many problems in quantitative electron microscopy, it is necessary to have an estimate of the section thickness. This is especially true for the "disector" method for estimating particle numbers or volumes (see below). A recent simple modification for estimating the thickness of plastic sections is described by Evans and Howard (1989; see also Bedi 1987). For thawed cryo-sections the "fold" method of Small (1968) can be used to obtain a reasonable indication when performed correctly (Griffiths et al. 1983). For more accurate estimations the mass-measurement method developed for hydrated cryosections (Eusemann et al. 1982) could be applied (Griffiths et al. 1984). For a critical summary of most available methods see de Groot (1988).

11.3.1.15 Section Compression

Significant compression (20–50%) always accompanies the sectioning procedure, probably irrespective of the embedding medium. However, in many, but clearly not all, cases the effect of this compression may be overcome when the sections are floated on an aqueous solution (Griffiths et. al. 1984; Jesior

1987). The important consideration here is whether a net compression has occurred when the section is viewed in the EM. The easiest ways to estimate this is to either measure the dimensions of the section and those of the block face or to measure the dimension of a structure in the section that was originally spherical in the tissue. In the rat liver, for example, the nuclei of the hepatocytes have been used for this purpose (Bolender 1974). If all structures are equally affected by compression the volume density estimations should not be affected. Estimations of S_v may, however, be overestimated. For more details see Weibel (1979).

11.3.1.16 Resolution Effects

Estimations of S_v, in particular, must be done at magnifications high enough to see all details of the boundaries otherwise the parameter can be severely underestimated (see Weibel 1979, p. 153; Weibel and Paumgartner 1978). [Note, however, that at increasing magnifications the detail (and boundary lengths) become greater and greater. Theoretically there is no way to know precisely what is the correct S_v. This is best exemplified by the famous "Coast of Britain effect" in that the length of the coast increases infinitely with increasing magnifications (see Weibel 1979, Fig 4.28)]. A striking example from real stereological studies comes from Weibel's group who have estimated the total alveolar surface area of human lung. From light microscopy the value obtained (≈ 80 m^2) significantly underestimated the more realistic value obtained with electron microscopy (145 m^2) (Gehr et al. 1978).

It is always important in stereological studies to obtain accurate estimations of magnification, both for light micrographs and electron micrographs; in other words one needs to calibrate the magnifications. For light microscopy this is achieved by a calibration slide which has ruler lines with known distances between them. This should obviously be done under identical conditions as for the micrographs, and, ideally, on the same film. Due to variations in factors such as electric current, the magnification of electron microscopes can also vary from day to day. A replica-grated EM grid whose lines are positioned a known distance apart should be photographed at the magnification used for the stereological studies.

11.3.1.17 Transparent Overlays Versus Machines

Until now we have only considered the use of transparent overlay lattice grids for point and intersection counting. Surely, the beginner may ask, much or all of these estimates can be made more quickly and accurately using a machine, either an image analyzer or a computer-assisted tracing device. With respect to the former, it seems that whereas image analyzers can be used for documenting relatively simple structures, at the present time they are of very limited use when it comes to analyzing routine electron micrographs. Similarly, for simple structures it may be useful in some cases to use a computer-aided tracing device.

Although it may surprise many, it is a general consensus amongst stereologists that simple transparent overlays are much more efficient (require less work) than the most sophisticated machines presently available. Take the example of estimating the boundary length (for S_V) of a complex organelle such as the Golgi complex. It is a relatively simple matter to count up to ten intersections (only!) with Golgi membranes on each of 20 micrographs. If, however, one has to follow each contour with a pen, it is significantly more work and more arbitrary decisions have to be made (which can lead to bias) in accurately following membrane profiles. This is especially true for regions where membranes are sectioned obliquely.

11.4 Estimation of Volume of Reference Space

11.4.1 Organ or Tissue

In order to obtain absolute values in stereology, it is often necessary to estimate the size of the reference space. For most cell biological purposes the mean cell volume is the most convenient reference, even when the cells are part of a complex tissue. For many studies, however, one may need to know the volume of an organ, or part of an organ. For large organs a simple approach is volume displacement after immersion in a liquid (Archimedes principle; see Weibel 1979). For many purposes, however, the principle first discovered by Cavalieri in the 17th century (and rediscovered recently by the stereologists; see Sterio 1984) has become the method of choice. In this procedure one makes thick slices through the whole organ of interest and it is an absolute prerequisite that the distance between sections (t) is known. The principle of the technique is to sum all the areas of all the section profiles (which can be determined, for example, by point counting) and multiply this by t. For practical details on how to prepare the organs for thick sectioning, as well as the theoretical background, see Gundersen and Jensen 1987 and Gundersen et al. (1988a). A good description is also given in Michel and Cruz-Orive (1988).

11.4.2 Estimating Mean Cell Volume

In most cell biological and immunocytochemical studies the most suitable reference space is the mean cell volume. There are a number of different stereological, as well as non-stereological approaches, which have been used to obtain this critical parameter.

11.4.2.1 Geometrical Considerations

By far the simplest (and probably least accurate) method is the one based on simple geometrical considerations. This is useful in those cases where a cell type has a regular shape, or at least approximates a regular shape. The simplest of these is a sphere. A number of cultured cells, which grow freely in suspension (or some cells which have been dissociated from tissues by enzyme treatment) are quite spherical. By measuring the mean of the cubed radii, it is straightforward to estimate the mean cell volume $\bar{v}$ using the formula:

$$\bar{v} = \frac{4}{3} \prod \bar{r^3},$$

where r is the real cell radius.

Note that when cells are highly variable in size for this formula to be valid, one must raise each measured radius to the third power and only then take the average $(\bar{v}^3)$.

In order to obtain an error-free estimate of the diameter, the situation is much more complex in practice, however. A relatively large number of cell profiles must be measured from sections and a particle size distribution histogram must be made from which the three-dimensional diameter is inferred by very complex mathematical and statistical procedures (see pp. 180–186, Weibel 1979). Occasionally, the nucleus may be spherical, as in the example of the (rat) liver hepatocyte. In this case, the nuclear volume has been used as a reference to estimate the mean cell volume (simply by reference to the volume density of cytoplasm per cell). Note, however, that for making rough calculations this is still an acceptable approach. For obtaining rapid, rough estimations of the volume of a spherical cell (or other particle), one simple method is to estimate the diameter of the largest profiles. By definition, these are profiles that must pass through the centre of the largest cell.

In a few cases, cells with more complex shapes, but which have a reasonable correlation to a well defined geometric shape have been estimated. This is the case for epithelial cells, both in vivo (Buschman and Manke 1981a,b) and in culture (von Bonsdorf et. al. 1985). In these cases the cells were approximated to a pyramid.

11.4.2.2 A Stereological Approach for Cultured Cells Growing on a Flat Support

For cultured cells that grow on plastic dishes a simple stereological approach has been designed to estimate cell volume that makes no assumptions about shape (Griffiths et. al. 1984). The principle is the following:

1. The number of cells per area of dish is counted. This gives the area of dish occupied by the average cell (A).
2. The cells are embedded in situ and sectioned perpendicular to the monolayer (vertical sections – which can also be used to estimate the S_v of the cell, and

A. Light Microscopy. Number of cells per area dish.

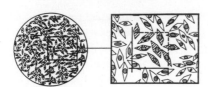

There are 9 cells (hatched) within the counting frame that do not touch the forbidden line. If the counting frame has an area of $10^4\ \mu m^2$ on average, one cell occupies an area of:

$$\frac{10^4\ \mu m^2}{9} = 1100\ \mu m^2/\text{cell}$$

B. Electron Microscopy. Area of cell: length of base (arrows). This gives the mean height of the monolayer.

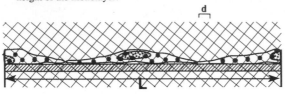

Vertical section (perpendicular to dish)

From point counting: total points cell ($\bullet$), $\Sigma P = 22$

if $d = 2\ \mu m$ and $L = 50\ \mu m$ the mean height H of the cells would be

$$H = \frac{P \times d^2}{L} = \frac{22 \times 4\ \mu m^2}{50\ \mu m} = 1.66\ \mu m$$

if one cell occupies an area of $1100\ \mu m^2$ and has an average height of $1.66\ \mu m$ the average volume of a cell must be: $1100\ \mu m^2 \times 1.6\ \mu m = 1760\ \mu m^2$

Fig. 11A,B. Diagram to illustrate a method for estimating the mean cell volume of cultured cells growing on monolayers. Note that mean height here refers to the average height across the monolayer, not the mean of the total heights. For more details see Griffiths et al. (1984, 1989)

therefore the absolute surface of the plasma membrane once the cell volume is known – see Baddeley et al. 1986; Griffiths et al. 1989). The area of the cell profiles (by point counting) is related to the length (L) of the monolayer base (by intersection counting or simply by using a ruler). This gives the mean height of the monolayer (H). The product of $H \times A$ gives the mean cell volume (for more details see Griffiths et al. 1984, 1989) (see Fig. 11).

11.4.2.3 Stereological Approach for Cells in Suspension

A centrifugation procedure introduced by Baudhuin (1974; see also Schwerzmann et al. 1986) for estimating the mean volume of mitochondria can be modified in order to estimate the mean cell volume of cells that are in suspension. As for the preceding method, this is especially useful because it makes no assumption about cell shape.

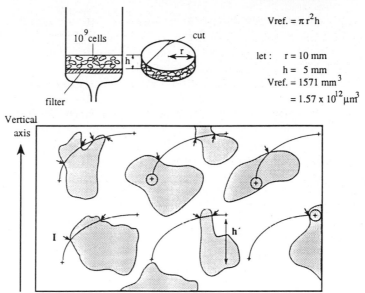

Reference volume:

$$V_{ref.} = \pi r^2 h$$

let : $r = 10$ mm

$h = 5$ mm

$V_{ref.} = 1571$ mm^3

$= 1.57 \times 10^{12}$ μm^3

Cycloid test system

Note that since this section is an excellent example of a 'vertical' section the cycloid test grid can be used. The large arrow indicates the vertical axis, h′ gives the height of the cycloid. In this example h′ = 1 μm.

Total points (+) on cell (circled): 3; Volume fraction of cells: $\frac{3}{12} = 0.25$

Total points (P) on ref. (including cells): 12

Total intersections (I) with cell surfaces (small arrows): 13

$$\text{Surface density} = \frac{2I}{P \cdot l(p)} = \frac{2 \cdot 13}{12 \cdot l(p)} = \frac{26}{12 \cdot 4 \ \mu m} = 0.54 \ \mu m^{-1}$$

length of cycloid per point, l(p) = 2 μm

n.b. the length of a cyloid test line is twice its height (h′)

Total cells: 10^9

$$\text{mean vol. cell, } \bar{v} \ (\text{cell}) = \frac{0.25 \cdot 1.57 \cdot 10^{12}}{10^9} = 392 \ \mu m^3$$

Total volume of reference space 1.57·10^{12} μm^3

Total surface of cells = 0.54·1.57·10^{12} μm^2 = 0.85·10^{12} μm^2

$$\text{Surface of one cell, } \bar{s} \ (\text{cell}) = \frac{0.85 \cdot 10^{12}}{10^9} = 850 \ \mu m^2$$

Fig. 12. Diagram to estimate mean cell (or any suitable particle) volume for cells that grow in suspension. A known number of cells is centrifuged in order to make a relatively even pellet of known volume that is embedded in plastic (e.g. Epon). Systematic sections are made in a vertical direction and, using the cycloid grid and counting points and intersections the volume and the surface of the cells can be estimated as described

The principle is as follows:

1. A known number of cells are centrifuged, or Millipore-filtered in order to make a cell pellet of defined shape so that its volume can be estimated.
2. Random, vertical sections of this pellet (defining the plane of the filter as the horizontal) are photographed (at the lowest magnification at which the cell profiles are visible) and, using the cycloid test system (Fig. 12), by simple point counting the volume density of the cells in the pellet is estimated, that is, the fraction of pellet that is occupied by cells. Since the total pellet volume is known, it now becomes possible to estimate the total volume of a known number of cells. By division this gives the mean cell volume. By estimating the number of intersections, the S_V and therefore total cell surface is obtained.

In practice, making a pellet of defined shape is not trivial. One procedure which did work in our laboratory (Dr. C Walter, unpubl. data) was to centrifuge aldehyde fixed cells through a millipore filter in a centrifuge tube in a freely suspended rotor (Fig. 12). This gives a disc-shaped pellet, whose volume is then $\Pi r^2 h$, where r is the radius and h is the average height (estimated from a number of random transverse sections).

An example of the use of a cycloid test system on a real example of a vertical section is illustrated in Fig. 13.

11.4.2.4. Disector Principle

An ingenious stereological method was introduced a few years ago (Sterio 1984) for estimating the number of particles in a defined volume of tissue. The method depends on making two parallel sections through the tissue, a known distance apart (or height, h, which must be smaller than the height of the smallest particles, see Fig. 14). The area of the sections (or a part of the sections) multiplied by h gives the "unit volume" of the tissue (or part of the tissue) in between the two sections. The approach then relies on the fact that looking at the "upper" section from above, each particle in the volume has only one discrete "top" or "cap". By counting the number of these caps in that volume, one obtains the desired parameter, namely particle number per volume of tissue. How does one obtain the number of particle caps? The simple and elegant answer is one simply counts those profiles which are present in the one section but not in the other one (Fig. 14). For this method it should be emphasized that, in general, an accurate estimation of section thickness is critical (see, however, footnote 5). Note again that at the light microscopy level a powerful alternative to making pairs of real, physical sections is to take **optical** sections by photographing a single thick section at two different optical planes, a known distance apart. The use of the confocal microscope will be a powerful tool in this respect (see Howard et al. 1985; Baddeley et al. 1987). A detailed discussion of the disector approach is outside

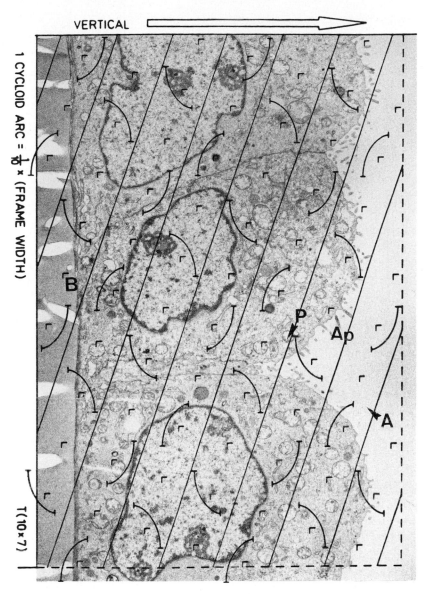

Fig. 13. Example of the use of vertical sections and the cycloid test system in practice. A plastic section of parts of three MDCK cells grown on a nucleopore filter is shown. *Ap* apical side; *B* basal (filter) side. The filter represents a natural horizontal axis; the perpendicularly aligned basal to apical axis then represents the vertical axis (*arrow*). The *arrow showing the vertical direction* should be carefully aligned with the vertical axis. The straight lines are *not* part of the test system and are simply there to facilitate the measurements. The volume of the reference space is counted by counting the test points (*P*) and the surface area of interest is estimated by counting the number of intersections with the cycloid arc (*A*). If, for example, one is interested in the S_V of the nucleus in the cell on the right there are five points and three intersections with the arc (the upper arc intersects twice). A simple formula given in Baddeley et al. (1986) enables an unbiased estimate of S_V to be calculated. Micrograph courtesy of Dr Robert Parton, EMBL. (Cycloid test system Fig. 8 of Baddeley et al. 1986)

Disector

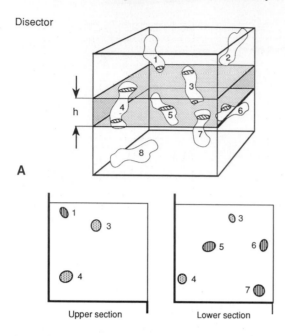

A

Upper section   Lower section

Fig. 14A

the scope of this chapter, and the reader is referred to the original article (Sterio 1984) and to recent reviews by Gundersen (1986) and Gundersen et al. (1988b) for more detail.

The point to emphasize here is that even for complex tissues with many different cell types, it is possible to estimate the number of any type of (identifiable) cell per unit volume. Knowing this parameter, one then simply needs to estimate, by point counting, the volume density of the cells of interest in the tissue. From the number per unit volume, and the fraction of the unit volume occupied by the cell of interest, one obtains the average number of cells in a defined volume of cells, which is by definition the mean cell volume (Fig. 16).

11.4.2.5 The Point-Sampled Linear Intercept Method

The area of a circle is Πr^2, where r is the radius. As described by Gundersen et al. (1988b), however, the area can also be estimated by measuring the distance l from any fixed point in the circle to the boundary, as long as the direction of measurement is isotropic. By taking the square of these and then averaging, $\sum \overline{l^2}$ (i.e. the mean of the square of these measurements) is an unbiased estimator of the area.

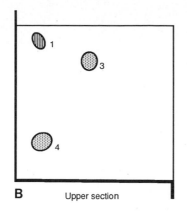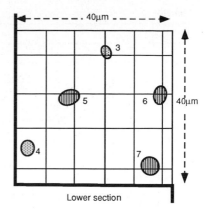

B   Upper section   Lower section

Disector volume = h·a = 10·40^2 = 16000 µm^3
Number of cells in two desectors, 2Q$^-$ = 4 (hatched)

$$N_V = \frac{2 \cdot Q^-}{2 \cdot h \cdot a} = \frac{4}{2 \cdot 16000} \, \mu m^{-3} = \frac{1}{8000}$$

Points on reference (a) = 20
Points on cells (n.b., all cells) = 3

Volume cells in ref., $V_V = \frac{3}{20} = 15\%$

Mean cell volume, $\bar{V}_{cell} = \frac{V_V}{N_V} = \frac{\frac{3}{20}}{\frac{1}{8000}} = 1200 \, \mu m^3$

Fig. 14A,B. The principle of the disector. (After Sterio 1984; see also the other references cited in the text). A There are eight particles in the cube, which is sectioned twice. In the lower section there are five sectioned profiles. In the upper section there are only three. These profiles (*5, 6* and *7*) are unique to the lower section. Looking from "above", therefore, these profiles would be counted since their "tops" or "caps" are unique to the volume of the disector (area of the section *shaded* in the picture multiplied by h). In this estimate, therefore, three unique particles are seen in the disector volume. The method can be made more efficient by looking in the "reverse" direction, that is, from "below". In this case, only one profile (no. *1*) is unique to the upper section that is not present in the lower one. Thus, in this estimate one obtains one particle per disector volume. The two estimates thus give an average of two particles per disector volume. Note that the forbidden line rule (Gundersen 1977) must be used for the estimations of particle number. In **B** the procedure is illustrated with *numbers*

The analogous principle can be used in order to measure the volume of any three-dimensional object from sections (see Cruz-Orive 1987; see Fig. 15), as long as these sections are oriented in all directions in three-dimensional space (isotropic). Alternatively the sections must be "vertical" (see p. 403). This powerful approach now represents a unique example of being able to get a three-dimensional parameter (the volume-weighted mean volume $\bar{v}_V$) directly from a two-dimensional section.

A

The particles hit by points are marked (A-G). The distance of the test line across the particle is estimated (arrows in A). Particle D is measured twice since two points lie on this particle. For particle F two measurements are taken corresponding to the two test points. Particles E and H are ignored since they are not hit by test points. Assuming the above figure has a final magnification -x 1500: then 1 mm corresponds to 0.67 μm.

	l (mm)	l (μm)	l^3 (μm^3)
A	10	6.7	300.8
B	11	7.3	394.4
C	6	4.0	64.0
D	13	8.7	651.0
	13	8.7	651.0
F	6	4.0	64.0
	6	4.0	64.0
G	4	2.7	19.7

$$\Sigma l_0^3 = 2208.9 \ \mu m^3$$

(The suffix zero indicates that the intercepts are point sampled, that is with a zero-dimensional probe)

$$n = 8$$

$$\text{mean} = \overline{l_0^3} = 276.1 \ \mu m^3$$

Volume - weighted mean cell volume

$$\overline{v}_V = \overline{l_0^3} \times \Pi/3$$

$$\overline{v}_V = 289.1 \ \mu m^3$$

Fig. 15A

The "volume-weighted" here refers to the fact that particles are sampled in proportion to their volumes. The larger particles are more likely to be sampled. The $\overline{v}_v$ will overestimate the "true" mean volume (in a number distribution $\overline{v}_N$) by a factor which depends on how much the size of the particles vary. When all particles are identical $\overline{v}_v = \overline{v}_N$ (for more details see Gundersen 1986; Cruz-Orive and Hunziker 1986). The essence of the point sampled linear intercept method is to place a system of test lines with test points over the section containing, for example, profiles of nuclei whose volumes we need to estimate. When a test point falls on a nucleus, the length of the test line through the point that intercepts that nucleus is then estimated. All directions of the test lines must be possible on random sections. The volume-weighted mean volume is:

$$\overline{v}_V = \frac{\Pi}{3} \cdot \overline{l_o^3}$$

B

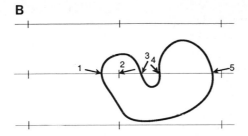

For non-convex particles the following procedure is followed.

1)   Measure the intercept length through the sampling point ($l_{0,0}$; 1 ---> 3)

2)   Measure the distances to the end points of all other intercepts from the sample point, in this case two $l_{0,1}$ (2 ---> 4) and $l_{0,2}$ (2-5); nb. it is always an even number. Note that the even numbered extra intercepts are added, the odd are subtracted in the formula below. Numerate them to both sides of the sampling point (numbered 0).

3) $$\overline{v}_V = \frac{\Pi}{3}\left(l_{0,0}^3 + 2\left(l_{0,2}^3 - l_{0,1}^3\right)\right)$$

in this case $l_{0,0} = 2$, $l_{0,1} = 2$ and $l_{0,2} = 3$

$$V = \frac{\Pi}{3}(2^3 + 2(3^3 - 2^3)\cdot) = \frac{\Pi}{3}(8 + 2(27 - 8))$$

$$\frac{\Pi}{3}(8 + 2\cdot19) = \frac{\Pi}{3}\cdot46$$

$$= 48.2 \text{ units}^3$$

Fig. 15A,B. A diagram to show how the point linear intercept method is used to measure the volume weighted mean volume $\overline{v}_V$ of particles from isotropic or vertical sections. The system of test lines is placed on the hypothetical micrograph as shown. In A the simple situation is shown where all particles are convex. Every time a point hits a cell the length of the intercept is estimated, that is, the length of test line that passes over the profile. This example is drawn to scale giving both the real distance on the figure (mm) and the real magnification, assuming that the image was enlarged ×1500 (e g low magnification EM). In B the more complex situation of a concave particle is illustrated

Note that for estimating the mean the symbol n refers to the number of points hitting the particles, not the number of particles. For sections through a pellet of cells that is isotropic the grid can be placed in the same orientation on each micrograph. If the cells are anisotropic, the orientation must be systematically varied from one picture to the next along the vertical axis (for details see the references cited in the text). Alternatively, vertical sections could be used

An example of the use of this method is given in Fig. 15 (the suffix zero indicates the dimension of the sampling probe which is a zero-dimensional system of points). In Fig. 15A an example is shown where all profiles are convex. When a profile shows a concave profile the more involved procedure shown in Fig. 15B should be followed.

Although the method is simple to apply in practice, there are strict rules which must be adhered to in order to obtain unbiased values. A comprehensive discussion of these is beyond the scope of this chapter. For more details of these "rules" as well as elegant descriptions of practical applications and limitations

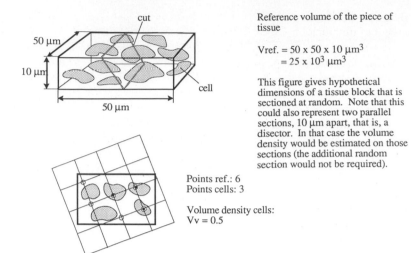

Reference volume of the piece of tissue

Vref. = 50 x 50 x 10 μm^3
      = 25 x 10^3 μm^3

This figure gives hypothetical dimensions of a tissue block that is sectioned at random. Note that this could also represent two parallel sections, 10 μm apart, that is, a disector. In that case the volume density would be estimated on those sections (the additional random section would not be required).

Points ref.: 6
Points cells: 3

Volume density cells:
Vv = 0.5

In this example there are 10 cells (or cell tops) in a disector 25 x 10^3 μm^3.
From point counting the volume density of the cells, V$_V$ = 0.5.
This means 50% of the tissue is occupied by cells.
Therefore, 10 cells occupy 0.5 x 25 x 10^3 μm^3 = 12.5 x 10^3 μm^3.

Volume of one cell: $\bar{v}$ cell = $\dfrac{12.5 \times 10^3\,\mu m^3}{10}$ = 1250 μm^2

Fig. 16. Diagram to illustrate the interrelationship between cell number, tissue volume and cell volume

of the method see Gundersen and Jensen 1985; Cruz-Orive and Hunziker 1986; Cruz-Orive 1987; Gundersen et al. 1988b and Mayhew 1989. (The principle of vertical sections is also described in these references as well as in the original reference by Baddeley et al. 1986). We have recently used this approach to estimate the volume-weighted mean volume of the nucleus in a baby hamster kidney cultured cell (Griffiths et al. 1989).

A recently introduced variation of this method can be used to estimate the absolute volume of any particle, be it the whole cell or a particle within the cell, provided the cell has one unique and identifiable particle which is used as a form of reference. Since the nucleus (or nucleolus) is one structure which is unique in most cells the method has been referred to as the nucleator (Gundersen 1988, Gundersen et al 1988b). The essence of this approach is to make a stack of (not necessarily serial) sections a known distance apart, i.e. a disector which is in effect a three-dimensional probe. Since the sections must sample the material in an isotropic fashion it is essential to use one of the isotropic sectioning modes referred to on p. 384, such as vertical sections. The original disector principle (above) is used as an unbiased method to select a particle (that is, those particles which show a profile in one section but not in an adjacent section). For the nucleator principle itself one needs a (set of) single section(s) that pass through the whole extent of the nucleus. On such profiles one puts a systematic system of test points such that a random point falling on the nucleus is selected. The distance from this point to the nuclear envelope (if nuclear volume is estimated) or to the plasma membrane (for cell volume) is then measured in a direction whose orientation is isotropic, uniform, random (IUR; this is ensured by a procedure whereby the orientation is determined by an angle whose magnitude is varied in a systematic manner – see Gundersen and Jensen 1985). This distance measures the intercept length and the mean (nuclear or) cell volume is

given by $\bar{v}_N = \dfrac{4}{\Pi} \cdot \overline{l^3 n}$.

The point to note here is that one uses the disector, three-dimensional probe, principle to **select** particles in an unbiased fashion (small particles have as much chance of being selected as do large particles). The point linear intercept method can then be used to estimate the mean cell volume, $\bar{v}_N$, that is an unbiased estimate of mean cell volume in the number distribution.

11.4.2.6 A Colloidal Gold Labelling Technique

Kehle and Herzog (1989) have recently introduced a novel approach using colloidal gold in order to estimate the surface area of cells in culture. Until now, we have discussed approaches where one first determines the cell volume and then estimate the surface area of cells simply by multiplying the volume by the surface to volume ratio (S_v) of the cell as estimated by point/ interception counts. In the Kehle and Herzog approach, one first estimates the absolute surface area of a cell. The principle of this method is to use a conjugate of radioactive ^{195}Au colloidal gold with any protein which will bind avidly to the cell surface, such as concanavalin A or other convenient lectin. In order to avoid any internalization of the label, the reaction must either be done at 4°C or after an initial fixation step. The specific radioactivity of the gold particles must first be estimated. For this, these authors have also developed a novel method based on mixing the gold with a suspension of latex beads of known size and concentration and placing these on an EM grid, thus gives the ratio of the number of latex beads to the number of gold particles (Kehle and Herzog 1987). Knowing the specific activity of the gold, the total number of gold particles bound per number of cells can be estimated simply by measuring the radioactivity in a gamma counter. This gives the average number of gold particles per cell. On random EM sections, the number of gold particles can then be related to the surface area of the plasma membrane. In the original publication, Kehle and Herzog used a semi-automatic image analyzing system to relate the surface area to the number of gold particles. It may, however, be both more simple and more efficient to relate the number of gold particles to intersection counts of random micrographs using the formula:

$$\text{Number of gold particles per } \mu m^2 \text{ surface} = \frac{N}{dtI},$$

where N = number of gold particles counted in the frame, d is the distance between points, I the number of intersections and t the section thickness (Griffiths and Hoppeler 1986). (This formula applies to the use of a square grid: appropriate modifications may be made when a different grid is preferred).

The total number of gold particles per cell (from the biochemical estimates) divided by the density of gold per surface area then gives the total surface area of the cell. From an independent estimation of S_v of the cell by point/intersection counts the cell volume can then be estimated.

11.5 Stereological Sampling in Practice

Once the mean cell volume is known a hierarchy of sampling procedures is usually used at different magnifications, or levels, to obtain the required estimates. The exact procedure obviously varies with the type of structures and the type of parameters required.

An example we have often dealt with is the absolute surface area of Golgi cisternal membranes in a baby hamster kidney cell (Griffiths et. al. 1983, 1989) whose mean cell volume was estimated by the procedure b in Sect. 11.4.2.7 to be approximately 1400 μm^3. A typical approach would be as follows:

Level 1. Count the number of cells per unit area of petri dish from light micrographs taken in a systematic manner.

Level 2. At relatively low magnification with the EM ($\times 2800$ primary magnification, magnified $\times 4$ on our enlarger) two ratios are estimated. First the average height of the monolayer (see above) and second, an estimation of cytoplasmic and nuclear volume. For the latter, the number of points over the nucleus was related to the number of points over the cytoplasm. The volume density of the cytoplasm of the BHK cell is approximately 0.75, which means 75% of the volume of the cell consists of cytoplasm, the rest nucleus. Hence the absolute volume of the cytoplasm is $0.75 \times 1400 \ \mu m^3 = 1050 \ \mu m^3$. (Conversely the nuclear volume is $0.25 \times 1400 = 350 \ \mu m^3$).

Level 3. At a primary magnification of x 8000 (final magnification $\times 32\,000$), the Golgi cisternae could be unambiguously identified. Random micrographs of the cytoplasm were taken (which means that every area in the cytoplasm should have an equal chance to be photographed). Using a double lattice grid the "coarse" points over the cytoplasm and the "fine" points over Golgi cisternae were counted (Fig. 10B). The volume density of Golgi complex per cytoplasm was thus estimated[6]. Since the Golgi complex occupies only a small fraction of the cytoplasm and is localized in a specific region next to the nucleus, its profiles are only seen, on average, in about one out of three sections. This leads to a larger error in the estimation than would be the case for a more abundant structure that is more homogeneously distributed. Hence a relatively large number of micrographs must be analyzed to obtain a coefficient of variation below 25%; in our study we accepted this high degree of variation (see Griffiths et. al. 1983). The results indicated that the Golgi cisternae represent roughly 1% of the cytoplasmic volume. In absolute terms this means $1050 \times 0.01 = 10.5 \ \mu m^3$ of volume.

Level 4. Immediately following the exposure of the level 3 micrographs, the section was moved, using the translation controls of the EM, in a systematic manner until the next Golgi stack was seen. What one does here, in effect, is to select the structure of interest in an unbiased fashion. This structure was then photographed at a relatively high magnification, ($\times 28\ 000$) which, after enlargement, enabled Golgi membrane profiles to be clearly identified.

The section thickness and the type of contrast available is critical here in order to visualise membranes clearly. On these enlarged micrographs a square lattice grid (A100 – Weibel 1979) was used to relate the intersections of the Golgi cisternal membranes to the points falling over the Golgi cisternae. (Ignoring those falling over the inter-cisternal space.) This enabled the surface density of Golgi cisternal membranes to be estimated.

The typical result for S_v is in the order 100 μm^{-1}. In other words, 1 μm^3 of volume of Golgi complex has 100 μm^2 of membrane surface if the sections were absolutely perpendicular to the plane of the membrane. However, since membranes are not always cut in this way a correction factor of $\dfrac{\Pi}{4}$ is required. Thus $100 \times \dfrac{\Pi}{4}$ or 79 μm^{-1} would be an estimate of the true S_V in practice (provided the sections are isotropic).

Since the average cell had 10.5 μm^3 of Golgi, the average cell has 10.5 (μm^3) $\times 79 \ \mu m^{-1} = 829 \ \mu m^2$ of Golgi cisternal membrane surface.

Note that S_v (surface density) is not identical to boundary length density (BA – the length of membrane boundary per area of micrograph), although both have the same units X^{-1}. They are related by the formula: $\dfrac{4}{\Pi} B_A$ (Weibel 1979, p. 36). The relationship between them is illustrated in Fig. 3.

---

[6] It is almost too obvious to state that the significance of any estimate of a complex structure such as the Golgi complex depends critically on one's ability to identify it. This, of course, is a biological rather than a stereological problem. While the use of defined markers, especially antibodies, can greatly facilitate this identification and should be used whenever possible, it is necessary to admit that there is always some need to make arbitrary decisions in identifying structures on sections. A useful tip when one is not sure about an "event" (e.g. is this a profile through the Golgi complex or not?) is to count only every **second** of these profiles and ignore the other.

11.5.1 Measurement of L_v in Practice

The examples above dealt with surface and volumes. If we were also interested in knowing the absolute length of microtubules (MT) in these cells the following approach could be taken:

1. A first prerequisite would be that the MT (which are rarely seen in our routine EM sections) would need to be clearly visualized. For this one could use a stain such as tannic acid or one could permeabilize the cells and immunolabel the microtubules with antibody and a gold conjugate.
2. After making isotropic, random sections micrographs could be taken at the minimum magnification at which the sectioned profiles of MT can be unambiguously seen. Presumably, this magnification would be in between levels 3 and 4 above (and would have to be empirically determined).
3. A lattice grid would be placed over the micrographs and points counted over the entire cytoplasm. In suitable "frames" (e.g. the area corresponding to one point, or a frame within this area in a double-lattice grid) the number of sectioned MT profiles would be counted (using an unbiased counting frame – see below) This would give the number (Q) of MT profiles (transects) per area of counting frame, (A). If A is given in μm^2, the number of profiles per area will be:

$$Q_A = \frac{Q}{A} \; \mu m^{-2} \quad \text{(Weibel 1979)}$$

and the length density $L_v = 2Q_A \; \mu m^{-2}$ (see Fig. 4).

This parameter gives the number of cross-sections or "transects" through microtubules per μm^2 of area of micrograph. Note however, that,

$$L_V = \frac{\text{number (dimensionless)}}{\mu m^2} = \frac{\text{length } (\mu m)}{\mu m^3} \; ; \quad \text{units}^{-2}.$$

In other words, $2Q_A(\mu m^{-2})$ also gives the length of MT per μm^3 of cytoplasm. Knowing the absolute volume of cytoplasm per cell, one can estimate the total length of MT per cell.

Two additional points should be noticed here. First, a general one about "counting frames". For reasons explained by Gundersen (1977), by counting all the particles (be they transects of filaments, gold particles or profiles of mitochondria) in the counting frame one will tend to overestimate the particles. A simple "forbidden line" rule (Figs. 4, 10) can be used to overcome this problem.

The second point concerns the specific problem of estimating tubule length density. The above protocol will strictly hold only if the tubules are isotropically orientated or if sections are IUR. When the latter is necessary, the orientator method of Mattfeldt et al. (1990) could be applied. The following strategy has been described by Dr. H. S. Gundersen (pers. commun.). It requires two assumptions about the tubules. First, the tubules must have diameters in the order of the section thickness (t) or more. (if the diameter is smaller, treat them as **filaments** of no real thickness) second, their profiles should be roughly circular.

1. In the case of tubules, estimate first the tubular S_V on vertical sections in the usual way.
2. Use small counting/sampling frames to sample systematically a number of profiles. Measure the smallest diameter, d, of all sampled profiles. Calculate the mean, d. Now $L_V \text{(tubules/ref)} \approx \dfrac{1}{\Pi \cdot \overline{d}} \cdot S_V \text{(tub/ref)}$.

Note that from the profile density in step 2, Q_A, one may calculate $L_V = 2 \cdot Q_A$ and thereby obtain an impression of the bias introduced by doing L_V estimation on vertical sections. (The bias is nil if the tubules are isotropic).

11.6 Quantitation in Immunocytochemistry

Just as quantitative morphometric estimates on sections of tissues extend the information gained from qualitative morphological data, so can a quantitative assessment of immunolabelling consolidate impressions gained from qualitative studies. In both cases, the quantitative approach has the potential to extract more biologically relevant information from the same preparations. Despite the fact that the emphasis in this chapter is on quantitation, it is nevertheless important to stress that a good qualitative result has more value than an ill-conceived quantitative one. As soon as one expresses a phenomenon in quantitative terms, the results are more likely to be taken as the "Gospel" than is the case for qualitative data, since it is usually simpler for most people to relate to numbers than it is to grasp the essential details of a qualitative morphological study. In contrast to the use of stereological methods for quantifying structure, quantitative aspects of EM immunocytochemistry are still in their infancy. The discussion in the rest of this chapter will reflect this fact. Nevertheless, significant progress has been made in the past few years, both in understanding some of the theoretical problems as well as in the development of practical procedures for quantifying immunocytochemical labelling.

For reasons explained already in Chapters 1 and 5, only the use of particulate markers and post-embedding procedures will be considered in this section.

Before we discuss the more difficult problems of labelling efficiency and estimation of antigen concentrations in sections, the first prerequisite is to describe the use of simple stereological procedures to quantitate the immuno-gold labelling.

11.6.1 Use of Point and Intersection Counting for Relating the Number of Gold Particles to the Volume, Surface or Length of Structures in Sections

It is essential to be able to relate the number of immunogold particles (N gold) to structural parameters in a quantitative fashion in order to estimate the label (density). The procedure used simply depends on the form of the labelled structure: N gold may be related to any of the following:

- The area of micrograph (number per μm^2 – which in three dimensions relates to the volume of structure, N gold/μm^3). An example would be an antigen that is found in the nucleus.
- The length of profile of a membrane (number/μm; which in three dimensions relates to surface, or N gold/μm^2, e.g. a plasma membrane antigen).
- The length of profile of a filamentous structure (e.g. microtubules) – (N gold/μm) (see Fig. 17).

In those cases where the labelling is found throughout the depth of the section, these two-dimensional parameters can be converted to real three-dimensional parameters by multiplying them by the section thickness factor (t). In the case of relating the number of gold particles to the surface this approach can only be considered as approximations (see legend to Fig. 18). Except for those cases where parts of very thin sections are selected where the membrane profiles are truly perpendicular to the plane of sectioning, it is necessary to multiply the number of gold/μm by the correction factor $\dfrac{4}{\Pi}$.

Area-Volume. The number of gold particles labelling an area of structure on a section (micrograph) that has been systematically sampled can be related to that area by point counting (Fig. 17A). If one needs to convert the area of section to the volume of organelle in the section, the section thickness factor t is required (Fig. 18). Note that this point counting approach is also suitable for quantitating gold on replicas, for example, of *planar* views of the plasma membrane (see Chap. 5)

Profile Length-Surface. The number of gold particles labelling a membrane profile that has been systematically sampled is related to the length of that profile by intersection counting as shown in Fig. 17. For converting profile length to surface area the section thickness factor t is again required (Fig. 18).

Profile Length-Filament. Whereas the methods for relating the number of gold particles to area or profile length are based on routine stereological principles, the method to relate the labelling to the length of filaments, such as microtubules, is not. As discussed above, according to stereological principles, infinitely thin sections through a population of filaments will appear as "points" on the sections. For quantitating immunocytochemical labelling of filaments (such as the examples from preembedding labelling shown in Figs. 3, 4, Chap. 10), it makes more sense in practice to sample systematically for areas of the

A. Number of gold particles per area of profile

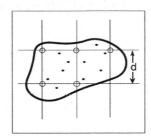

Number of gold: Q (gold) = 10
Number of points: P = 5
Distance between test lines: d = 1 μm

$$\text{Number of gold/area} = \frac{Q\,(\text{gold})}{P \times d^2}$$

$$= \frac{10}{5 \times 1\ \mu m^2}$$

$$= 2\ \text{gold}/\mu m^2$$

B. Number of gold particles per boundary length

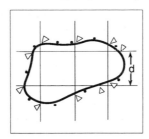

Number of gold: Q (gold) = 10
Number of intersection: I = 10
Distance between test lines: d = 1 μm

$$\text{Number of gold/boundary length} = \frac{Q\,(\text{gold})}{\frac{\Pi}{4} \times I \times d}$$

$$= \frac{10}{10 \times 1\ \mu m}$$

$$= 1\ \text{gold}/\mu m$$

C. Number of gold particles per length of filaments

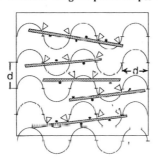

Number of gold: Q (gold) = 15
Number of intersections: I = 15
Distance between test lines: d = 1 μm

$$\text{Number of gold/length} = \frac{Q\,(\text{gold})}{I \times d}$$

$$= \frac{15}{15 \times 1\ \mu m}$$

$$= 1\ \text{gold}/\mu m$$

Fig. 17A–C. Diagram to show the principle of using point and intersection counts to estimate density of gold labelling relative to area (**A**), boundary length (**B**) or filament length (**C**). The formulae given in **A** and **B** are derived from Griffiths and Hoppeler (1986). The formula for **C** has been derived using similar principles, by Dr. Luis Cruz-Orive (pers. commun.)

section where the filaments are anisotropically orientated in parallel to the section plane (Fig. 17C). An implicit assumption here is that one is selecting filaments whose labelling is representative of the whole population. Systematic sampling can most easily be done by rotating the translational controls of the electron microscope in a systematic fashion, e.g. moving the section in one direction and photographing the next appropriate field that does not overlap with the last picture.

For relating gold numbers to volume or surface (as long as the pictures have been selected by unbiased procedures), the amount of gold labelling for a

A. Area - Volume

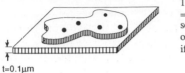

1 gold/μm^2 (area of section)
= 1 gold/0.1 μm^3 (volume or organelle in section)
or 10 gold/μm^3
if LE = 5% ---> 200 antigens/μm^3

t=0.1μm

B. Boundary length - Surface (membrane)

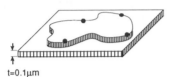

1 gold/μm (length of membrane boundary)
= 1 gold/0.1 μm^2 (membrane surface of organelle in section)
or 10 gold/μm^2
if LE = 5% ---> 200 antigens/μm^2

t=0.1μm

C. Filament length

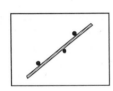

1 gold/μm (filament length)
if LE = 5% ---> 20 antigens/μm

Relating gold particles to volume, surface and length. In a strict sense this is valid only when, on sections, the antibody has complete access to the antigens throughout the depth of the section or when the antibody/marker has been applied before embedding (preembedding labelling) under conditions where all the antigen has access to the antibody marker.

Fig. 18A–C. The use of section thickness (t) to convert area to volume and boundary length to surface. The principle of labelling efficiency (*LE*) is also illustrated. Note that for surface estimations (**B**) it is either essential to sample only perpendicularly sectioned profiles of membranes (as shown) or the correction factor $\dfrac{\Pi}{4}$ must be multiplied with the value obtained. In both cases, these estimates can only be taken as rough approximations unless one is assured that the antibody/gold have complete access to the antigen throughout the depth of the sections

defined volume or surface area of structure in the section can simply be converted to a per cell basis. For an example of the former, imagine that there are, on average, five gold particles/μm^2 of nuclear profile on a section, 0.1 μm thick. This can be converted to five gold particles/0.1 μm^3 of nuclear volume or 50 gold/μm^3. If independent stereological estimates tell us that the absolute nuclear volume is 100 μm^3, the gold labelling is equivalent to 5000 gold particles per nucleus. For a membrane antigen imagine that, on average, five gold particles label 1 μm profile of plasma membrane. This converts to five per $\left(0.1 \times \dfrac{\Pi}{4}\right)$ μm^2 or five per 0.13 μm^2 membrane surface (assuming again, t = 0.1 μm) or 39 per μm^2. If the total plasma membrane surface per cell is 2000 μm^2, this amount of labelling is equivalent to 7800 gold particles per total plasma membrane per cell.

For relating the density of gold labelling anisotropically oriented filaments, it should be noted that the above approaches cannot be used, since the selection of the pictures with respect to the filament population is biased. Therefore, two independent estimates must be made. First, the density of labelling is estimated as described above (gold/μm filament). Second, the numerical density (I_v) of the filaments must be estimated on sections that comply with stereological rules of unbiasedness (Fig. 4). In this way the total length of filament per cell, for example, can be estimated. The latter, multiplied by the labelling density gives the total number of gold particles per total filament population per cell.

11.6.1.1 Exhaustive Counting Is Pointless

As described above for estimations of point and intersection counts, there is no need to count thousands of gold particles. It makes sense to take advantage of suitable lattice grids to sub-sample small enough regions. For example, imagine that the number of gold particles labelling a storage granule in a secretory cell must be estimated (Fig. 19). For the estimation of area of micrograph the "coarse" points can be used. For the estimation of gold particles, however, the frame area corresponding to a "fine" point can be estimated (Fig. 19).

11.6.2 Labelling Efficiency

The main reason why quantitative aspects of immunocytochemistry can still be considered to be in their infancy relates directly to the question of "*labelling efficiency*" (LE), that is, the proportion of antigen in a section that is recognised by antibody and marker or, more simply, the number of immunogold particles divided by a known number of antigens which they label. In the next section we shall look at the factors that affect this parameter.

11.6.2.1 Factors Affecting the Labelling Efficiency

Slot et al. (1989a, b) have recently listed three different groups of factors that can modify the LE in immunolabelling experiments.

- *Experimental variables*
  Also referred to as exogenous factors (Posthuma et al. 1984), these are the technical, procedure-related variables that can greatly affect the labelling density for one antigen on a defined structure. The main factors which have been shown to have an effect on LE include:

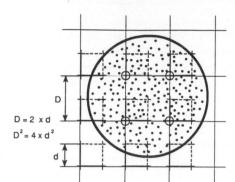

let $D = 1$ μm and $d = 0.5$ μm or $d^2 = 0.25$ μm^2
Total area of the structure = no. of points over the structure x $D^2 = 4$ x $D^2 = 4$ μm^2

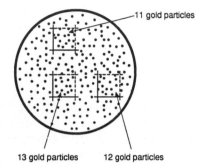

n.b. The particles touching the dense, forbidden line are ignored!

13 gold particles   12 gold particles

Σ of gold particles = 36 in an area of 3 x d^2 = 3 x 0.25 μm^2 = 0.75 μm^2 or 48 gold per μm^2 (or $1D^2$). Since the total area of the labelled structure is 4 μm^2 the total number of gold particles will be 4 x 48 = 192 (actual number of gold particles = 210). Note that when averaged over many figures this difference would become insignificant.

Fig. 19. Use of double lattice grids to sub-sample the number of gold particles falling on an organelle. The coarse points (*circles*) which are adequate to estimate the area of the profile in the section would lead to an unnecessarily high number of gold particles being counted in this example. For this purpose a smaller counting frame is used, as shown. (Note: the forbidden line rule should be applied; see Fig. 4)

1. The quality (titre) of the antibody and of the gold, or other particulate marker.
2. The conditions of chemical fixation.
3. The kind of embedding medium (Lowicryl, LR-white, Epon etc.) as well as other preembedding steps (including OsO_4 treatment and dehydration). In the cryo-sections the absence of an embedding medium ("normal" conditions) can be contrasted with sections of polyacrylamide embedded tissue (see below).
4. Overall labelling parameters: length of labelling; dilution of antibody; marker; temperature etc.

- Antigen-related variables

  Many antigens can be in different conformations depending, for example, on their location, position in the cell cycle and function. For example, secretory glycoproteins undergo an extensive series of post-translational modifications as they traverse the secretory pathway. Many cytosolic and nuclear proteins also undergo cyclic modifications, including phosphorylation and fatty acylation. Receptors and ligands may behave differently towards antibodies depending on whether they are free or bound. Cytoskeleton proteins such as tubulin or clathrin may be either free as sub-units or associated into larger filaments or lattices. Often these modifications can result in a total loss of immunoreactivity. Clearly, a detailed quantitative analysis of any antigen necessitates a consideration of the biology of the system. An additional factor which may have to be considered at high antigen concentration is the possibility of steric hindrance, that is, the presence of one bound antibody and/or gold particle may hinder the access of further antibody/gold particles to closely adjacent antigens (see below).

- Matrix density: surface relief

  For unembedded cryo-sections a serious factor which can cause differences in LE between different organelles is what Posthuma et al. (1985, 1986, 1987) have referred to as the matrix density. From looking at stereo images of immunogold labelled cryo-sections, this group has shown that the immuno-label can penetrate to variable amounts into the depth of an unembedded cryo-section. For an extreme example, imagine a section through a pellet of aldehyde-fixed but unembedded cells. It is a priori to be expected that most, if not all, of the outer surface of the plasma membrane in the section should be accessible to antibody (although some of it may be involved in attachment to the grid surface). In contrast, in a section through a dense secretion granule the label is exclusively on the upper (and occasionally the label may also have access to the lower) surface of the section (Posthuma et al. 1985). Recent studies by Stierhof et al. (1986) and Stierhof and Schwarz (1989) suggest that in well-preserved cryo-sections the majority of gold particles as small as 4 nm are restricted to one section surface (see, however, the discussion with reviewers at the back of the last reference for critical comments on these results).

  From theoretical considerations, it had earlier been anticipated that the presence of an embedding medium such as Lowicryl would result in no difference in LE between different structures. In support of this were observations showing that immunogold label is, indeed, restricted to the surface of Lowicryl sections, as shown by re-embedding labelled sections and sectioning them perpendicular to the original sectioning plane (Bendayan 1984b). Two independent sets of data have, however, failed to show the expected quantitative relationship over different structures using Lowicryl resins (Griffiths and Hoppeler 1986; Posthuma et. al. 1986, 1987). The simplest explanation would be to say that some penetration of the antibody, though not the gold conjugate, occurs into the Lowicryl section. An alternative theory was subsequently put forward, however, (Kellenberger et al. 1986, 1987, and pers. commun.). This suggests that the Lowicryl section surface, rather than being flat, offers considerable surface relief. Furthermore, this relief is different over different structures. The effect of this may be to effectively increase or decrease the surface area of structures in the section. As for the problem of estimating the degree of penetration into cryosections, it is difficult if not impossible to reliably quantify this phenomenon (see Kellenberger et al. 1986, 1987).

11.6.2.2 Can an Immunocytochemical Labelling Ever Approach 100% Efficiency?

Is there an inherent inefficiency in the labelling reaction? In other words, is there any theoretical reason why *all* the antigens in a structure should not be labelled with antibody under ideal conditions? A recent study shows, in fact, that this is possible. Prasad et al. (1990) mixed particles of the spherical, rota virus, with Fab fragments of antibodies directed against the VP4 spike membrane proteins of the virus. The mixture of antibody fragments and virus was suspended in a thin film between holes in an electron microscope grid, rapidly frozen (vitrified – see Chap. 5) and observed by cryo-electron microscopy. Using image reconstruction techniques the high resolution structure of the virus could be determined both in the presence and absence of the Fab fragments. The results showed that the 60 spikes of the virus each bound two Fab fragments (consistent with each spike being a dimer). This example must be taken as an extreme example since, in this approach, the virus is not fixed, the antibody fragments are not rinsed and non-bound antibodies are not visualized and can be ignored. Nevertheless, since the avidity of bivalent IgG molecules is much higher than Fab fragments, this would suggest that this situation should be achievable for other antigens under ideal conditions. It would be very interesting to see how the effects of chemical fixation would affect the accessibility of antibodies to antigens in such a model system.

11.6.2.3 Using Sections the Labelling Efficiency in Practice Is Always Significantly Below 100%.

As we shall see below, actual estimations of LE in model systems indicates that it is rarely above 10%, even in unembedded cryo-sections. This fact need not, however, prevent us from being able to use quantitative immunocytochemistry to estimate antigen concentration from labelled sections. The rationale can be illustrated by a hypothetical example. Imagine that a set of sections of a tissue have been labelled with an antibody against the antigen X followed by protein A gold. Imagine further that the LE is always 5%, that is, for every 100 copies of X in the section you will always get five gold particles. From a quantitative assessment of the labelling of the set of sections, and by relating this to stereological parameters of the surface, volume or length of the labelled structure within the section, it would be possible to determine the concentration of X directly from the section, that is, to give the number of molecules of X per surface, volume or length of the organelles of interest. In this way the concentration of the antigen in the same or different organelles could also be compared.

The above example remains hypothetical simply because in real life with routine sections for immunolabelling the LE varies from one antigen-antibody-marker combination to the next, and from one organelle to the next. However, two different kinds of strategies have emerged during the past few years which have the potential to give us valuable quantitative information about antigens of interest.

The first of these strategies makes no attempt to alter or control the labelling efficiency but simply tries to extract as much information as possible about the antigen. The second is a much more sophisticated approach, applicable for antigens that are available in pure form (or that are in known concentrations in "reference" compartments). The essence of this method is that it aims to make the LE the same over all organelles in the section (see below). Before we discuss these approaches it is relevant to discuss how LEs have been determined in practice.

11.6.3 Estimation of LE in Practice

In this section examples are given of actual estimations of LE for a number of different experimental situations ranging from unembedded cryo-sections to plastic (Lowicryl) sections. All of these examples were done using a two-step labelling procedure: the primary antibody was followed directly by protein A gold. Under this condition no amplification of the signal is expected. The question of amplification is discussed later in the chapter.

11.6.3.1 Semliki Forest Virus Membrane Proteins

We have used a viral model system in order to estimate the labelling efficiency using cryo- and Lowicryl sections (Griffiths and Hoppeler 1986 – see also Griffiths et al. 1989 for a correction). The essential features of this system were that the density of the spanning membrane spike protein complex of Semliki Forest virus was known in three different locations. These were the endoplasmic reticulum (ER), the Golgi complex and in the fully formed virus particles, which bud out through the plasma membrane. In order to determine these densities, two parameters had to be estimated, namely, first, the surface areas of the ER, Golgi complex (stereologically determined) and virus particles (whose size can be measured directly) and second, the biochemical concentration of viral membrane protein in those structures. Immunocytochemical labelling on cryo-sections had shown that the spike proteins were uniformly localized to the three types of structures (Green et al. 1981). We then asked, how many immunogold

Table 3. Companson of labelling efficiencies in different structures for the BHK-SFV system. (Griffiths and Hoppeler 1986; corrected in Griffiths et al. 1989)

	Corrected labelling efficiencies (in percent)	
	Cryo-sections	Lowicryl sections
ER	11.6	5.4
Golgi	5.1	2.6
Virus	12.0	1.2

particles could be related to these known densities of this antigen? This was done by calculating the number of antigens there would be in a defined volume of ER, for example, in a thawed cryo-section. The area of organelle in the section multiplied by the section thickness (t) gives the volume of organelle in the section (see Fig. 18)[7]. Knowing, from stereological estimates, the S_V of the membranes of interest enabled the amount of membrane surface to be estimated in the given area of section (micrograph). The strategy here was to assume a complete penetration of antibody and gold-protein A throughout the whole thickness of the section. While this assumption may be true for the extracellular side of the plasma membrane in sections of unembedded cells as discussed above it is unlikely to be valid for any internal membrane organelle. Any deviation from this assumption will simply have the effect of reducing the labelling efficiency.

In the case of the SFV spikes in the ER the results indicated a relatively high labelling efficiency of 12%; that is, on average, 12 gold particles labelled every 100 antigens throughout the depth of the section and no labelling of the remaining 88%. This is effectively the same as obtaining a 100% efficiency in 12% of the thickness of the section. Close inspection of the pattern of labelling clearly indicated that the great majority of the gold particles were single "hits", which agrees with the expectation of each bound antibody should only bind one protein A molecule.

Using the same approach for the Golgi complex and for virus particles on the cell surface the labelling efficiency was reduced to about 5%. The corresponding results for Lowicryl sections are given in Table 3.

As is evident in Table 3, the labelling efficiency can vary significantly from one compartment to the next. Presumably the main reason for this is the difference in penetration into different parts of the section, although other

---

[7] In this study only those parts of the sections were used where the ER or Golgi membranes were *perpendicular* to the plane of sectioning; this was ascertained by selecting the membrane profiles showing the trilaminer unit membrane appearance. The correction factor $\dfrac{4}{\Pi}$ was therefore not used.

factors such as steric hindrance (at high antigen concentrations) as well as differences in immunoreactivity in different compartments may also have played a role.

11.6.3.2 The Docking Protein (Hortsch et al. 1985)

The docking protein (DP) is an integral membrane protein of the rough ER which binds to the signal recognition particle (SRP). The latter is bound to the signal sequence of newly translated secretory and membrane proteins, in addition to the corresponding messenger RNA. This association of DP and SRP enables the translated protein to be vectorally transported across the membrane of the rough ER (see Hortsch and Meyer 1986).

We showed by immunocytochemistry that the DP was localised to the membranes of the rough ER, but not to those of the smooth ER in sections of rat liver (Hortsch et al. 1985). From the known surface density of the rough ER the amount of membrane in a defined volume of section (i.e. section area ×section thickness) was estimated (see footnote 7). This was then related to the number of immunogold particles that labelled the rough ER. The density of gold particles per surface area of ER membrane was estimated. By reference to our data on Semliki Forest virus we assumed a 10% labelling efficiency and calculated the number of DP molecules we should expect. The result, in fact, gave a reasonable correlation to an independent biochemical approach.

11.6.3.3 Ricin and Endocytosis
(van Deurs et al. 1988; see Griffiths et al. 1989, for a correction of this)

The lectin ricin binds to cell surface molecules, both glycoproteins and glycolipids, that contain terminal galactose residues. At 4°C, a condition which blocks endocytosis, it was estimated biochemically that there were 8×10^6 ricin binding sites on the surface of a BHK cell. These cells were fixed and cryo-sections were prepared and labelled with a rabbit anti-ricin antibody followed by protein A gold. On random cryo-sections the density of gold particles over the plasma membrane was estimated to be $86 + 11$ gold particles per μm^2. Since the absolute surface area of the BHK cell was already known to be 2200 μm^2 it could be estimated that a total of 1.9×10^5 gold particles labelled the biochemically-determined 8×10^6 ricin binding sites, giving a labelling efficiency of about 2% only. It is possible that steric hindrance may play a role at these high antigen concentrations (but see below). At 37°C, a part of the bound ricin was internalized with its bound glycoprotein (or glycolipid). As for the plasma membrane, the amount internalised (biochemically determined) could be related to the total number of gold particles in intracellular compartments. Again, the labelling efficiency was relatively low ($\approx 1\%$).

11.6.3.4 5′ Nucleotidase

Howell et al. (1987) estimated the efficiency of labelling for the plasma membrane enzyme 5′ nucleotidase by relating biochemical estimates of the number of molecules to the number of immunogold particles labelling cryo-sections. The estimated LE in this case was about 15%. Future studies are needed to see what kind of labelling efficiencies are obtained for soluble and for cytoskeletal antigens.

11.6.3.5 Soluble Proteins

The group of Slot and Geuze has established the LE for two model soluble proteins, amylase and superoxide dismutase (SOD) in extensive quantitative analyses over the past 5 years. The details of these pioneering studies will be described in more detail below. In Table 4 a summary is given of the LEs obtained for these proteins under different conditions using either cryo-sections or Lowicryl K4M sections. The cryo-sections of known concentrations of amylase were made either in a gelatin gel ("non-embedded") or in the same gel after embedding in polyacrylamide (PAA). The latter could be compared directly to gels embedded in Lowicryl K4M. The SOD labelling was done in liver tissue that had been extensively characterized biochemically and stereologically.

Table 4. LE values for suluble proteins

	Non-embedded (cryo)		Embedded	
	A	B	Polyacryl-amide/cryo	Lowicryl
Amylase (in gelatin)	$\leq 5\%$[1]	$\approx 1.0\%$[1]	$\approx 0.4\%$[2]	0.8%[1]
SOD (in liver hepatocytes)	20%[4]	2.5%[4]	0.8%[2,3]	

In A the values are given in the case of amlylase embedded in 5% gelatin, conditions in which the authors found penetration of label into the section interior. In B, for amylase this gives the value obtained when the gelatin concentration equals or exceeds 10%. Under this condition the matrix density is high enough so that the labelling is restricted to the section surface and linear with antigen concentrations. For SOD the high value of 20% (A) is given for areas of the hepatocytes where glycogen was originally present but removed during the specimen preparation resulting in extensive penetration of label into the section (this may be taken as an extreme condition giving, perhaps, the highest LEs possible). The value 2.5% in B for SOD gives the values in non-glycogen-containing areas (and can be taken as more representative of what may be expected in routine sections). Note the lowering of LE when both proteins are embedded, either in polyacrylamide or in Lowicryl.

References: 1. Posthuma et al. (1987); 2. Slot et al. (1989a); 3. Chang et al. (1988); 4. Chang and Slot, pers. commun. and Slot et al. (1989b)

11.7 Approaches for Quantitation Ignoring Labelling Efficiency

The rationale here is to accept the fact that the LE may vary between different antibodies – antigens – markers – organelles, but to assume that this variation is not significant with respect to the "window of accuracy" of the expected results. In the first instance these are simple situations where the LE may not be expected to vary significantly, namely in relative quantitation studies. In this approach no attempt is made to know the absolute concentration of an antigen and the LE is effectively ignored. For any sectioning procedure where a valid qualitative immunocytochemical signal has been obtained, there are many examples where counting gold particles can give meaningful results, even in the absence of supporting biochemical data. In these cases, the critical factor is that one compares two or more experimental situations where the conditions that determine the efficiency of labelling can be expected to be similar for both. The best example involves a comparison of the same structure under different experimental conditions: two hypothetical examples can be used to illustrate this point:

Example 1. Imagine that a peroxisomal matrix (soluble) protein X has been immuno-localized under "normal" conditions as well as after treatment with a drug which decreases its peroxisomal concentration. How would one carry out a quantitative assessment? Quite simply by relating the number of gold particles to the area of organelle in the section/micrograph using a simple stereological point counting procedure.

On micrographs of labelled sections, at a relatively high magnification where both peroxisomes and gold particles are clearly visible, a lattice grid is placed over the peroxisome. The number of points falling on the peroxisomes are counted as described earlier. Both treatments can be directly compared since the structures, as well as the conditions of labelling, are the same. (Admittedly, this is an assumption but in many examples a reasonable one).

Example 2. A spanning membrane protein which functions as a hormone receptor has been localized to the plasma membrane. After treatment with the hormone the number of molecules of this receptor on the plasma membrane decreases significantly. What is the magnitude of this decrease?

Random micrographs of the labelled plasma membrane are taken at the lowest magnification where the gold particles are visible. The density of gold labelling is estimated as shown in Fig. 18B.

The first example of this approach for relative quantitation in practice was to study the rate of intracellular transport of newly synthesized membrane glycoproteins of Semliki Forest virus after cycloheximide treatment. This treatment blocks protein synthesis but does not affect the outward movement of those proteins that have already been inserted into the membrane of the endoplasmic reticulum (ER). The number of immunoferritin particles over the

ER of cells not treated with cycloheximide was compared to the ER at different times after initiation of the treatment. Similarly, the Golgi stacks were compared to each other. No attempt was made to compare the ER to the Golgi stack directly. The kinetic analysis, later fully supported by biochemical data (and published in the same reference), indicated that the membrane proteins of this virus left the ER with a half-life of 7.5 min and the Golgi complex with a half-life of 22 min (Green et al. 1981).

The relative quantitation approach may also be used to compare the ratio of two antigens over different structures. An example from the literature comes from the work of Posthuma et al. (1984), who were interested in the ratio of the pancreatic enzymes, chymotrypsinogen to amylase concentration in secretion granules under different conditions. The ratios obtained by immunocytochemistry showed excellent agreement with those obtained by biochemical assays.

11.8 Absolute Quantitation Making Assumptions About Labelling Efficiency

In this approach one accepts that the LE may vary from one structure/antibody to the next and tries to obtain as much quantitive information as possible. As an extreme example imagine an unembedded cryo-section labelled with a rabbit antibody against a cytosolic soluble protein X and protein A gold that gives, on average, ten gold particles per μm^2 of area of section, 0.1 μm thick. In 1 μm area of section 0.1 μm thick there is 0.1 μm^3 of volume of cytoplasm. In 1 μm^3 of section this would correspond to 100 gold particles (Fig. 18). From this we can safely conclude that there must be at least 100 molecules/μm^3 of cytoplasm. If the cytoplasmic volume is known, this can be converted to a minimum estimate of total molecules per cell. If we can provide a rough estimate of LE, one can often obtain useful rough estimates of the number of molecules. The accuracy of such estimations are obviously directly related to the accuracy with which we can predict LE.

The attraction of this approach is that one acquires rough but useful estimates simply by using the same preparations one made for qualitative evaluation of immunolabelling. It has the advantage that one retains maximum sensitivity of the labelling technique (unlike the matrix gel method – see below). The method can be extremely rapid (it takes very little time to quantitate 20–40 micrographs). The more stereological information one has about the system the more information one obtains from the quantitative evaluation of the immunolabelling. However, even if no stereological data are available, it may be useful either to make assumptions about surface, volume or length parameters (for example by comparisons with published data in similar systems) or by making very rough stereological estimations. Cells or nuclei can always be approximated to spheres for an approximate estimation of their volume.

By assuming a range of LEs it is possible to estimate a range or "window" of antigen concentration. For antigens on the plasma membrane in unembedded cryo-sections, for example, since the actual LEs measured (as shown by the above examples) have ranged from 1–15%, these two values could be used to give an upper and lower value respectively for the density of antigen in the plasma membrane. Nevertheless, it is evident that the accuracy of this approach depends critically on the accuracy of guessing the LE. In the next section we shall look at a sophisticated new approach with the potential of estimating concentrations of antigens directly from counts of immunolabelling densities on sections.

11.9 Negation of the Labelling Efficiency Factor:
The Matrix Gel Method for Quantitating Soluble Antigens
Which Are Available in Pure Form

An interesting approach for quantitation has been introduced recently by the group of Slot and Geuze (see Slot et al. 1989a, b for reviews). The idea here is to section a double-block consisting of "reference gel" that has a known concentration of the antigen along with the cells/tissue in which one is interested in knowing both the distribution and the concentration of this antigen (Fig. 20). Two important prerequisites had to be fulfilled before such a system could be realized:

- It was essential to treat the reference gel and the tissue under identical conditions from fixation through all steps of immunolabelling.
- The access of the immunolabelling reagents to the antigens should be identical in both the reference and the tissue.

Whereas the first of these conditions is not difficult to fulfill, the second is far from trivial. As discussed above, the main problem appears to be that the matrix

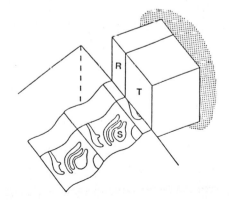

Fig. 20. Diagram to illustrate the matrix-gel method of Posthuma, Slot and colleagues. (Slot et al. 1989b). The composite block consists of a known concentration of antigen mixed with 10% gelatin (the reference R) sandwiched to tissue (T) embedded in polyacrylamide. The complex is fixed, sectioned and immunolabelled in identical fashion. Since the labelling of the antigen in the reference is linear with respect to concentration the concentration of antigen in the tissue can be directly estimated (for more details see the references cited in the text)

density (or surface relief) is different between different organelles. This leads to differences in LE between these compartments. In an elegant series of experiments, this group was able to establish conditions that fulfilled the above goal (Posthuma et al. 1987). In initial experiments they quantitated the labelling of cryo-sections of their model protein amylase embedded within different concentrations of gelatin, that served as a matrix. It was found that only when the gelatin concentration was high enough (10% wt/vol) was the labelling of the amylase restricted to the surface of the section. This was evident both by direct observation of stereo images and by the fact that above the critical 10% concentration of gelatin the amount of labelling was reproducible from one experiment to the next and did not differ between 10 and 20% gelatin. In contrast, at 5% gelatin there was variable penetration into the gel from one experiment to the next that gave labelling densities between two and six times higher than that seen with 10%. It should also be noted that in these studies it was essential to fix the gelatin blocks with a relatively high concentration of glutaraldehyde for at least 4 h in order to avoid loss of antigen from the sections.

In the next phase of these studies the authors introduced 30% polyacrylamide (PAA) as an embedding material prior to cryo-sectioning. This compound was chosen because, unlike gelatin, it can uniformly penetrate fixed cells. This had been shown earlier by labelling with an antibody against PAA (Slot and Geuze 1982). Amylase was then put in a gelatin matrix, as before, and embedded in PAA. Under this condition, the labelling was restricted to the section surface (irrespective of the concentration of gelatin used) and, more importantly, the density of labelling was both reproducible from one experiment to the next and proportional to the concentration of antigen. An interesting technical innovation was the introduction of multi-layered gelatin blocks whereby different concentrations of antigen could be present in each layer of gelatin. In this way one could compare the labelling of different concentrations of antigen in the same section (see Posthuma et al. 1987 for details). The data indicated that with this system the number of gold particles was proportional to the concentration of antigen over a wide range, from about 0.1 mg/ml to over 250 mg/ml. This argued strongly that steric hindrance was unlikely to be a significant problem with this system.

In the first real application of this approach this group studied the quantitative distribution of amylase in its native tissue, the pancreas (Posthuma et al. 1988). For this they used the two-layered blocks with a known concentration of amylase along with the rat tissue. Both layers were fixed in the same way and then embedded in 30% PAA and cryo-sectioned. The conditions for cryo-sectioning and labelling were then identical for both reference and tissue. The results showed, for example, that along the secretory pathway the most significant concentration increase for amylase occurred during its passage through early stages of the Golgi complex. To some extent the data supported an earlier study by Bendayan (1984a), who quantitated the same antigen in the same tissue using Epon sections after metaperiodate treatment. However, the latter study estimated only the amount of labelling signal without any attempt to relate this to real antigen concentrations. The significant point about the Posthuma et al. (1988) reference was that it enabled labelling densities (signal) to be related to the real antigen concentration in the section.

In a parallel study (Chang et al. 1988), this group also made an elegant quantitative analysis of the distribution of superoxidase dismutase (SOD), a soluble enzyme, in rat liver, which extended an earlier qualitative immunocytochemical study (Slot et al. 1986). The authors made parallel biochemical measurements of the activity of SOD per known weight of rat liver that contains a known number of parenchymal cells (estimated stereologically). The labelling of SOD was then quantitated with respect to the intracellular compartments of interest on cryo-sections of PAA embedded rat liver, and related both to the volumes of these compartments and to the amount of label obtained in the reference gel (containing a known concentration of SOD and embedded in PAA double-block as before). Thus, conditions of fixation, embedding, sectioning and labelling were again identical for both the tissues and the reference. From the micrographs, volume density estimates could be made of the organelles of interest. These could be related to the mean cell volume and, as for the biochemical data, to the initial starting weight of tissue. In this way the immunocytochemical data could be directly correlated to the biochemical data since they both referred to the same reference volume (and ultimately to the initial starting wet weight of liver). For the immunocytochemical labelling a "balance sheet" was made up where all the labelling over the organelles of interest was related to SOD concentration (by reference to the gel) and all added up. In this way they obtained a total liver estimate of SOD in mg/g liver from both immunocytochemistry and biochemistry. The results were virtually identical and a striking demonstration of the power of this new approach in correlating immunocytochemical, biochemical and stereological data.

For this approach it may actually not always be essential to have the antigen in pure form. The critical point is to know the concentration of antigen in a reference space. In collaboration with this group we are now looking at model systems using viral membrane proteins. The idea here is that the concentration of the membrane protein is known in one compartment and can therefore serve as a reference for determining the amount of antigen in other structures of interest.

11.9.1 Possible Limitations of this Approach

A limitation of this approach is that, by restricting the labelling to the surface of cryo-section the technique loses sensitivity. This has the effect of reducing the signal to noise ratio (see below). This may be a direct effect of polyacrylamide itself on antigenicity. These are serious problems for the many antigens whose concentration is low and whose detection, even under optimal conditions, is at the limit of (even) the cryo-section technique.

A theoretical problem to discuss is what is meant when one refers to the "section surface". The term section surface is defined here simply as that part of the section in which antigens are free to bind antibodies. It is a fundamental assumption of this approach that the surface of all parts of the section are equivalent in this respect. As we have discussed previously, however, for any embedding medium there is no physical description of the sectioning process which can describe exactly what happens when a knife cuts a 50–100 nm section from a block. We have already alluded to the possibility that, in Lowicryl sections, different structures appear to have different surface relief. We have no reason to expect that the surface of cryo-sections will be any different. Further, it remains to be seen whether the surface of the reference gel would always be directly equivalent to the surface of the embedded tissue. Nevertheless, the results of Posthuma et al. (1988) and Chang et al. (1988) strongly suggest that these considerations are unlikely to be a problem in practice. The key factor appears to be the need to have the same amount of surface antigen exposed to the antibody in each section, a condition that was clearly met in the above experiments.

A second (potential) problem concerns fixation. As a first approximation, the effects of the fixative on the antigen in both the reference gel and in the tissue can be expected to be similar in this method. Note, however, that while this may be true for extracellular antigens that are freely exposed to the buffer (as is the antigen in the gel), the situation intracellularly may be more complex. As we discussed in Chapter 3, the inability of commonly used buffers to cross membranes in significant concentrations, even after fixation, means that all intracellular components can be expected to be exposed, at least transiently, to a significant drop in pH. In other words, the fixation conditions may affect the antigen differently in compartments where the buffer concentration is low (intracellular) than in compartments where the buffer concentration is high (extracellular). At present, however, these arguments are strictly theoretical.

A practical limitation to this approach is the difficulty in getting uniform polymerization of the blocks after PAA infiltration. These blocks are also difficult to section in a reproducible manner. For this reason, Slot and colleagues are now looking into alternative possibilities for embedding tissue such that the main criterion is fulfilled, namely, that the LE should be identical in all parts of the section. As for Lowicryl K4M, their results with both LR white and LR Gold have not been consistent with this goal; specifically, they obtained significantly higher labelling for amylase when this was mixed into a 5% gelatin gel as opposed to a 10% gel before embedding in these resins. Of potential interest, however, has been the use of Lowicryl HM20. Initial studies where amylase/gelatin is fixed, cryo-protected with sucrose, rapidly frozen then dehydrated infiltrated at low temperatures in this resin have been promising (Slot, pers. commun.). This freeze substitution protocol was the only plastic embedding protocol (except Epon) that this group found that did not induce a considerable shrinkage of the tissue. The latter was a serious problem with conventional PLT embedding in Lowicryl K4M as well as LR gold.

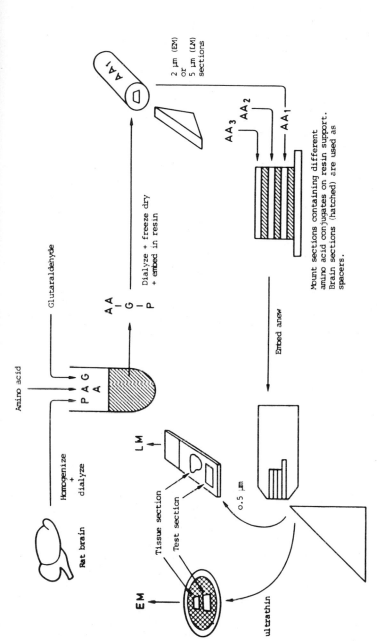

Fig. 21. Flow chart showing the main steps in the preparation of the test sections used to monitor the specificity of the immunocytochemical procedure at light- and electron microscopic levels in the system developed by Ottersen. Rat brains are homogenized and extensively dialyzed (to remove all free amino acids and other small molecules), and the macromolecular extract (*P*) is then reacted with an amino acid (*AA*) in the presence of glutaraldehyde (*G*). The complexes formed are embedded in resin. Sections are cut through the different blocks, each containing one particular amino acid conjugate, and stacked on top of each other, alternating with brain sections used as spacers. Cross-sections through this sandwich are then mounted on electron microscopic grids (for incubation together with ultra-thin sections) or on glass slides (for incubation together with semi-thin sections). Such test sections are incorporated in every immunocytochemical experiment, and permit a continuous survey of specificity. This is particularly important for the immunogold procedure, the outcome of which is very sensitive to inadvertent fluctuations in the incubation conditions. (Ottersen et al. 1991). For further details, see Ottersen (1987) and other references cited in the text. Courtesy of Dr. Ole Petter Ottersen, University of Oslo, Norway

11.10 Quantitative Model System Developed
for Small Molecular Weight Neurotransmitters

I have only recently become acquainted with the pioneering studies of Ottersen and collaborators on the immunolocalization of small molecular weight neurotransmitters, such as glutamate, GABA, glycine and taurine in the central nervous system (see Ottersen 1987 1989a,b). These studies are especially interesting because of the many technical difficulties faced when localizing small molecular weight, multi-functional compounds in complex tissues such as brain. In this work most of the hurdles faced were overcome in a very elegant fashion. Although many of the points of interest from this work are relevant for Chapter 3 (fixation), Chapter 4 (embedding) and Chapter 7 (specificity of labelling), I have chosen for the sake of completeness to describe most of their system in this chapter. The essential feature of this system is that, like the Posthuma and Slot model, it compares the labelling of a dilution series of the pure antigen in a reference gel to that in the tissues. Although not explicitly proven in this system, it is a reasonable assumption that the LEs will be very similar in both conditions.

Small molecular weight compounds along with lipids, are undoubtedly the most difficult to fix as well as to make antibodies against. While the details of the preparation of the antibodies can be read in the original papers by this group (see Storm-Mathisen et al. 1983) an important concept they introduced was the use of sections of a mixture of the test amino acid with a homogenate of brain tissue (see Figs. 21, 22). Importantly, the latter was first dialyzed in order to remove all free amino acids. The mixture of amino acid (in some cases mixed also with trace amounts of radioactive amino acids) and homogenate was then fixed in glutaraldehyde and embedded in an epoxy resin (Durcopan), in identical fashion to the brain tissue. In addition they prepared "sandwich"

---▶

Fig. 22. Test sandwich (prepared as described in Ottersen (1987) and shown schematically in Fig. 21) incubated in the same drops of immunoreagents as the tissue sections. The sandwich contains a series of different amino acid-glutaraldehyde/formaldehyde-brain protein conjugates (standard abbreviations for amino acids), separated by brain sections used merely as spacers. These reference sections were double labelled with antibodies against glutamine (Gln) with a 30 nm IgG conjugate and glutamate (Glu) and 15 nm IgG gold, in identical fashion to the brain sections, as described in the legend to Fig. 23. The conjugates appear as discrete electron dense bodies, some of which (*arrows*) are enlarged in the right panel. Note that the 30 nm gold particles (signalling glutamine immunoreactivity) are accumulated selectively over the glutamine conjugates whereas the 15 nm gold particles (signalling glutamate immunoreactivity) are localized over the glutamate conjugates. See Table 1 of Ottersen et al. (1991) for quantitative and statistical evaluation. *Bars* 2.4 µm (*left panel*); 0.5 µm (*right panel*). *Insert* (*top left*) photomicrographs of thin layer chromatograms (5 mm width) of soluble brain extracts (50 nl; *E* indicates site of application) that after separation were fixed with glutaraldehyde-polylysine and stained (peroxidase-antiperoxidase procedure) with an antiserum to glutamine (upper strip) or glutamate (lower strip). Note that the *labelled spots* show comigration with authentic glutamine and glutamate, respectively (*asterisks* indicate application sites for the amino acids)

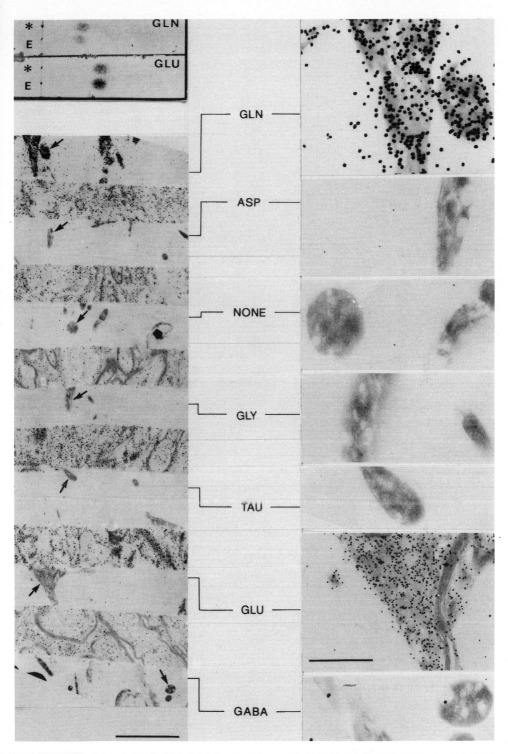

Fig. 22

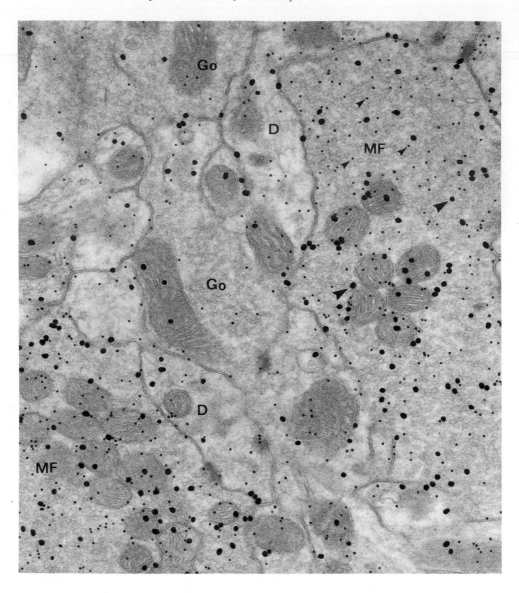

Fig. 23. An unpublished example of a section of rat cerebellar cortex granule cell layer (double-labelled for brain amino acids from the study by Ottersen et al. 1991). The brains were perfused with 2.5% glutaraldehyde and 1% formaldehyde in 0.1 M phosphate buffer. The tissue was treated with O_sO_4 (1%, 30 min), dehydrated in ethanol and propylene oxide and embedded in an epoxy resin (Duroupan, Fluka). The antisera against glutamine and glutamate were raised in rabbits following immunization with glutaraldehyde protein conjugates of the amino acids. For more details, including the evidence for specificity, see Ottersen and Storm-Mathisen (1984), as well as Fig. 22. The sections were double-labelled by the following steps, using the metaperiodate method of Bendayan and Zollinger (1983; see Chap. 4) for "unmasking" the effects of OsO_4 and the formaldehyde vapour method of Wang and Larsson (see Chap. 7) in order to block the reactivity of the first label before application of the second antibody. Large (30nm) particles

blocks containing multiple layers of test antigen. For example, up to eight different amino acids, each mixed with brain homogenate into a fixed tissue slice, could be embedded in multiple layers. In this way a section through this sandwich (reference block) would give, e.g. eight different zones, each containing a known amino acid (Fig. 22). Alternatively, different concentrations of one amino acid could be positioned into the sandwich. Using this model system the specificity of the antibodies could be assessed under conditions that closely parallel the situation on the real sections of brain (Fig. 23). They could ask, for example, whether their antibodies against glutamate show any significant cross-reaction against the closely related glutamine. In addition, the relationship between the amount of antigen and the degree of immunolabelling could be assessed. In a striking parallel to the work of Posthuma, Slot et al., this group could show a linear relationship between the amount of antigen and the degree of labelling; as for the amylase model system there appeared to be no effect of steric hindrance at very high antigen concentrations. For glutamate, the lowest concentration they could detect in this model system was 0.12 mmol/l, although the results were not always consistent between 0.12 and 0.42 mmol/l; for reference, the average glutamate concentration in the brain is 10 mmol/l. This system could then be used to estimate the concentration of antigen in defined regions of the brain. In one part of the cerebellum the results obtained with this quantitative immunocytochemical approach was in excellent agreement with known biochemical results. In other parts the estimates were considerably less than biochemical data. The authors discuss in detail the problem of fixation and point out data indicating significant losses of small molecular weight compounds after perfusion fixation. In such a case one might well be advised to question also the validity of the published biochemical data which are, after all, done after dissecting and homogenizing the tissue.

◀───

(*large arrowheads*) signal glutamine (the main precursor of transmitter glutamate); the small (15 nm) particles signal glutamate. Note that the small particles are concentrated in putative glutamatergic terminals (mossy fiber terminals; MF), whereas the large particles occur at highest concentrations over glial cells (*G*) which are believed to synthesize glutamine and to transfer it to the putative glutamatergic nerve endings. The Golgi cell terminals (*Go*), thought to use GABA as transmitter, are poor in glutamate as well as glutamine. *D* granule cell dentritic digits. Steps: (1) 1% HIO_4 in H_2O (7 min); (2) 9% $NaIO_4$ in H_2O (15 min); (3) 1% human serum albumin in H_2O (5 min); (4) first antiserum; (5) polyethyleneglycol (50 mg/100 ml) 0.05 M Tris buffer, pH 7.4, 5 min); (6) goat anti-rabbit IgG coupled to colloidal gold particles (Amersham), diluted 1:20 in the solution used in the previous step (2 h); (7) formaldehyde vapour (1 h, 80°C); (8) repetition of step 3; (9) second antiserum; (10) repetition of steps 5 and 6; (11) 2% uranyl acetate (20 min), followed by 1% lead citrate (2–5 min). In the experiments used for this figure and for Fig. 22, the first antiserum was against glutamine and the second was against glutamate (bound antibodies visualized by 30 nm and 15 nm particles, respectively). The glutamine antiserum was diluted 1:600 in Tris-phosphate-buffered saline (pH 7.4), and preadsorbed with 60 µM of aspartate-glutaraldehyde complexes. The glutamate antiserum was diluted 1:300 in the same buffer as above, and preadsorbed with 150 µM of glutamine-glutaraldehyde and 100 µM of aspartate-glutaraldehyde. Both antisera were applied to the sections for 2 h at room temperature. It is important to note that the reference system shown was treated in an identical fashion to the tissue. ×47000. (Courtesy of Dr. Ole-Petter Ottersen, University of Oslo)

Ultimately, the combination of sound biochemical studies done in parallel with an immunocytochemical approach such as the one developed by Ottersen will surely provide the answers.

In recent work this group has extended these studies by using double immunogold labelling of two closely related amino acid neurotransmitters, glutamate and glutamine (Ottersen et al. 1992). Besides the problem of preparing non-cross-reacting antibodies the desire to follow neurotransmitter amino acids is complicated by the fact that the same compounds exist as part of the normal metabolic pool of amino acids. For other relevant references from this group see Ottersen et al. 1990; Zhang et al. 1990; Blackstad et al. 1990; and Torp et al. 1991.

11.11 Sensitivity of Labelling: Limits of Detection

Closely related to the problem of labelling efficiency is the question of sensitivity. The interest here is to know what is the lowest concentration of antigen we can detect by an immunocytochemical technique. The main factor in practice that determines this limit is the amount of background labelling or the signal to noise ratio. If one has any signal that is above the noise it can, in theory, be detected. The question is – how much work is one willing to put into the analysis (for example, by counting thousands of micrographs even a minute difference between the signal and the noise could, in principle, be detectable!)? Data are available for both soluble proteins and for membrane proteins which can serve as a guide for discussing practical limits of detection.

11.11.1 Soluble Proteins

Using their matrix gel methods, Slot, Posthuma and colleagues have determined that the background labelling over sections of a gelatin gel without their antigen of interest, amylase, was 0.4 gold particles/μm^2. This value agrees well with our routine estimations of background labelling over the nucleus (for various membrane antigens) which is in the range 0.3–1 gold/μm^2. By referring to their standard curve which plots amylase concentrations against gold density this indicates that, for this antigen under the conditions used, the limit of detection is about 1000 molecules per μm^3 (Slot et al. 1989a). Since a 1 μM concentration of antigen contains 600 molecules/μm^3 (see Kellenberger et al. 1987), this suggests that the limit of detection in terms of concentrations is about 2 μM. Using similar calculations for labelling of Lowicryl K4M sections of bacteria, Stierhof et al. (1989) calculated that 10 μM could be the limit of detection. For a 100 kDa protein this would mean 10 μg/ml. It is remarkable that this is precisely the value predicted by Coons in 1956 (see Chap. 7).

11.11.2 Membrane Proteins

In a recent example from our work on the localization of the cation-dependent mannose 6-phosphate receptor (MPR) (Johnstone, Griffiths and Kornfeld, unpubl. data), we determined that the background labelling along profiles of the plasma membrane with two irrelevant antibodies was 0.02 ($\pm$ 0.008) gold particles/μm. Assuming a cryo-section thickness of 0.1 μm this value converts to $0.2 \times \dfrac{\Pi}{4}$ or 0.16 gold particles per μm^2 membrane (or 1 gold/6.25 μm^2).

For a fibroblast with a total surface area of 2000 μm^2, this would mean that we could expect 320 background gold particles per total plasma membrane. For the sake of comparison it should be noted that the maximal packing of membrane proteins in a membrane is in the order of 30000/μm^2 (see Griffiths et al. 1984).

11.12 Steric Hindrance

At very high antigen concentrations there must be a limit to the amount of labelling simply due to the limitation in packing the relatively bulky antibodies and gold particles into the restricted space on the section surface. The first serious discussion of this problem can be found in the excellent review by de Petris (1978) (which can be highly recommended as the most definitive discussion of quantitative aspects of immunocytochemistry of cell surfaces). This author points out that the maximum packing density of ferritin (13 nm centre-to-centre distance) on any surface should be about 6000/μm^2. When conjugated to antibodies or lectins the actual measurements on ferritin did not exceed 3000 particles per μm^2. A similar value is also obtained for protein A gold particles in two independent indirect labelling studies. First, the spike proteins of Semliki Forest virus are packed at the theoretical maximum packing density of 30000/μm^2. Using 5 nm gold we observed a density of labelling of 3500/μm^2 (Griffiths and Hoppeler 1986). Second, the labelling of cryo-sections of zymogen granules of the rat pancreas with anti amylase also gave an upper value just above 3000/μm^2 (Posthuma et al. 1986).

It should again be noted that in the gel method of Slot and Posthuma steric hindrance does not appear to be a significant factor in practice, since the density of label is proportional to the antigen concentrations up to 250 mg/ml (note that the total number of gold particles in this method is significantly reduced when compared to non-embedded cryo-sections). Similar results are obtained in the work by Ottersen (see above).

11.13 Amplification of Gold Labelling and Quantitation

All the considerations until now assume that, for proteins of average size, each immunogold particle will label one, and only one, antigen: in other words that no *amplification* of the labelling signal has occurred. This is the case, for example, when using rabbit antibodies followed by protein A gold. When, however, amplification steps are used, such as when a mouse monoclonal antibody is followed by rabbit anti-mouse IgG followed by protein A gold or an anti-mouse IgG gold conjugate, each antigen can be labelled by a group of gold particles (see Slot et al. 1988, 1989a, b for discussion, and Fig. 24). In this case

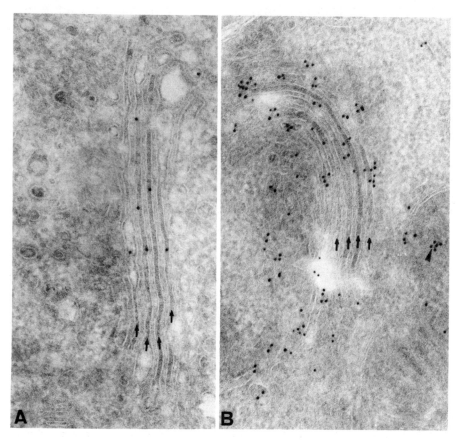

Fig. 24A,B. Cryo-sections of epithelial cells of rat duodenum labelled for the receptor to polymeric IgA. In **A** the antibody was labelled with protein A gold (two-step). Note the precise single-hit labelling within the electron-dense cisternae (*arrows*) of the Golgi complex. In **B** the antibody is labelled with an intermediate pig anti rabbit followed by protein A gold (three-step). The labelling density is higher, but the clustered pattern does not allow as precise a localization in the Golgi cisternae (*arrows*) as in **A**. Note the clusters are also evident over background structures such as mitochondria (*arrowhead*). **A** ×75500; **B** ×70000. Slot et al. 1989b)

each group or cluster should be considered as one "hit". The phenomenon is most easily seen over structures which should not contain the antigen, in other words, over "background" structures: note, however, that the size of the clusters over such background structures may not be a valid indication of the average cluster size relating to antigen since the "background" can be due to the first, second or third (gold) step. For antigens in relatively low concentrations it may be straightforward to count the number of clusters directly whereas it may be very difficult when the concentration is high. In the latter case it is simpler to count all (but see Fig. 19) the gold particles. It is essential to know the number of "hits", and the simplest approach is to estimate the average number of gold particles per cluster in areas where the labelling density is not prohibitively high. Once this is known, the total number of gold particles can be divided by the average cluster size to give the total number of primary antibody binding sites on the section.

In these considerations a distinction should be made between "signal" amplification, that is, simply getting more gold particles for the same number of antigens detected, and real amplification of labelling, that is, of seeing more antigen molecules. As shown recently by Slot et al. (1988, 1989b), it is possible for some three-step labelling protocols to actually detect more antigens when compared to a two-step protocol (the number of gold clusters in the former was significantly higher than the number of single "hits" in the latter). Noteworthy in this example, however, was that the background labelling was also increased by the three-step procedure while the resolution of the labelling decreased, as expected. Besides the spatial resolution the critical factor in these considerations is the signal-to-noise ratio (see Table 5 in Chap. 7 from Slot et al. 1989a and Fig. 21).

11.14 Signal-to-Noise Ratio and the Concentration of Antigen: Those Few Gold Particles!

Biochemists who have no difficulty in thinking in statistical terms when it comes to reading the amount of radioactivity in scintillation vials often seem to consider immunocytochemistry to be outside the rules of statistics and quantitative considerations! Consider, for example, a section labelled with an antibody against a surface receptor. The biochemist (or immunochemist), who has, with great effort, made an antibody, is often very disappointed when the immunocytochemical results indicate only a few gold particles on a structure of interest. Very often it is possible to estimate, at least approximately, how many antigens we can expect in a certain location. In all cases, for example, from a few micrographs, the average diameter of a cell can be estimated, approximated to a sphere (for example), and then a rough estimate of the surface area can be obtained. Let us say, for example, that a cell has 50000 molecules of receptors on the plasma membrane, whose size is 2000 μm^2. This immediately gives the information that the average density is relatively low, or 25 receptors per μm^2. As a rough guide, note again that an average cellular membrane will have 30000 spanning membrane proteins per μm^2 (of average molecular weight 50000 – Quinn et al. 1984). In other words, the receptor represents less than 0.001% of the total proteins in the plasma membrane.

Consider now, in an unembedded cryo-section of 0.1 μm thickness, a span of 10 linear microns along the plasma membrane. This length corresponds to 10 $\times 0.1$, or 1 μm^2 of membrane, which will contain on average, 25 copies of X. Even with 100% efficiency (which is unrealistic) you would obtain, on average, only 25 gold particles for this amount of plasma membrane. With the more realistic range of efficiencies expected, from 1–15%, one can expect only 0.25– 2.5 gold particles. Note that if these were ten fold less receptors (5000), even with an acceptable efficiency of 10%, you could only, on average, expect 0.25 gold particles per 10 linear microns, or perhaps only one gold particle over an entire cell profile! If the actual labelling observed is significantly higher than this, one must immediately suspect that something is seriously wrong either with the immunocytochemical labelling or with the original biochemical premise. By this approach, one avoids unexpected surprises and does not, for example, spend idle weeks trying different fixation protocols in a vain attempt to increase a signal which cannot be significantly increased!

This discussion leads to obvious implications. Just as the biochemist will consider the number of radioactive counts to be "real" when it is significant (on a statistical basis) relative to the background radioactivity (or noise) so then must statistical considerations determine the significance of relatively low labelling patterns. Hence, a relatively small number of gold particles that fall on a structure (the "signal") may be significant when compared to the number that falls on antigen-free structures (the "noise") in a sufficiently large number of micrographs (the precise number to be determined purely by statistical considerations). Obviously, this will also be related to control experiments where the signal to noise ratio is compared to that seen with control antibodies or, for example, to the amount of labelling over cells which do not contain the antigen (see Chap. 7). The relatively low levels of background that are obtained under routine conditions with both cryo-sections and the new generation resins means that these considerations are not just theoretical ones, but can be directly applied. This is even more important when one considers, paradoxically, that for antigens present in low concentrations the immunocytochemical techniques are probably best suited for quantitation. The first reason for this statement is our consistent (and not easily explained) finding that the lower the concentration of antigen the lower the ratio of background to signal. The second reason is that under conditions of low antigen concentration steric hindrance can be expected to be less problematical when compared to very high antigen densities. A 5–10% labelling efficiency on sections is not an unreasonable signal. It should always be remembered that the main problem is not that these EM techniques are so insensitive, but that the amount of antigen in a thin section is relatively small compared to biochemical samples. An ultrathin section is, after all, an ultra-thin section!

Finally, a direct consequence of these statements is that published micrographs of immunogold labelling data should not only be fairly representative with respect to the structures identified, but also with respect to the density of labelling. The temptation to show the rare micrograph with ten times more label than the average one should be avoided, because it is quantitatively misleading. In this respect, the scientific journals must in future be prepared to accept larger format micrographs that are labelled with relatively few gold particles, provided these micrographs are representative and, ideally, supported by quantitative estimates.

References

Anderson RGN, Brown MS, Goldstein JL (1977) Role of coated endocytic vesicle in the update of receptor bound low density lipoprotein in human fibroblast. Cell 10:351–364

Baddeley AJ, Gundersen HJG, Cruz-Orive LM (1986) Estimation of surface area from vertical sections. J Microsc 142:259–276

Baddeley AJ, Howard CV, Boyde A, Reid S (1987) Three-dimensional analysis of the spatial distribution of particles using the tandem scanning reflected light microscope. Acta Stereol 6(II):87–100

Baudhuin P (1974) Morphometry of subcellular fractions. Methods Enzymol 32:3–20

Bedi KS (1987) A simple method of measuring the thickness of semi-thin and ultra-thin sections. J Microsc 148:107–111

Bendayan M (1984a) Concentration of amylase along its secretory pathway in the pancreatic acinar cell as revealed by high resolution immunocytochemistry. Histochem J 16:85–108

Bendayan M (1984b) Protein A-gold electron microscopic immunocytochemistry methods, application and limitation. J Electron Microsc Tech 1:243–250

Blackstad TW, Karagülle T, Ottersen OP (1990) MORFOREL, a computer program for two-dimensional analysis of micrographs of biological specimens, with emphasis on immunogold preparations. Comput Biol Med 20:15–34

Bolender RP (1974) Stereological analysis of the guinea pig pancreas. 1. Analytical model and quantitative description of nonstimulated pancreatic exocrine cells. J Cell Biol 61:269–280

Bolender RP (1978) Correlation of morphometry and stereology with biochemical analysis of cell fractions. Int Rev Cytol 55:247–289

Buschmann RJ, Manke DJ (1981a) Morphometric analysis of the membranes and organelles of small intestinal enterocytes. J Ultrastruct Res 76:1–14

Buschmann RJ, Manke DJ (1981b) Morphometric analysis of the membranes and organelles of small intestine enterocytes. II. Lipid fed hamster. J Ultrastruct Res 76:15–26

Chang L-Y, Slot JW, Geuze HJ, Crapo JD (1988) Molecular immunocytochemistry of the CuZn superoxide dismutase in rat hepatocytes. J Cell Biol 107:2169–2179

Cruz-Orive LM (1982) The use of quadrats and test systems in stereology, including magnification corrections. J Microsc 125:89–102

Cruz-Orive LM (1987) Particle number can be estimated using a disector of unknown thickness: the selector. J Microsc 145:121–142

Cruz-Orive LM, Hunziker EG (1986) Stereology of anisotropic cells: application to growth cartilage. J Microsc 143:47–80

Cruz-Orive L-M, Weibel ER (1981) Sampling designs for stereology. J Microsc 122:235–242

Cruz-Orive L-M, Weibel ER (1990) Recent stereological methods for cell biology: a brief survey. Am J Physiol 258:148–156

De Groot DMG (1988) Comparison of methods for the estimation of the thickness of ultrathin tissue sections. J Microsc 151:23–42 de Petris S (1978) Immunoelectron microscopy and immunofluorescence in membrane biology. In: Korn ED (ed) Methods in membrane biology. Plenum Press, New York, pp 1–202

Eusemann R, Rose H, Dubochet J (1982) Electron scattering in ice and organic materials. J Microsc 128:239–249

Evans SM, Howard V (1989) A simplification of the "step" method for estimating mean section thickness. J Microsc 154:289–293

Gehr P, Bachofen M, Weibel ER (1978) The normal human lung. Ultrastructure and morphometric estimation of diffusion capacity. Respir Physiol 32:121–131

Green J, Griffiths G, Louvard D, Quinn P, Warren G (1981) Passage of viral membrane proteins through the Golgi complex. J Mol Biol 152:663–698

Griffiths G, Hoppeler H (1986) Quantitation in immunocytochemistry, correlation of immunogold labelling to absolute number of membrane antigens. J Histochem Cytochem 34:1389–1398

Griffiths G, Quinn P, Warren G (1983) Dissection of Golgi complex I. Monensin inhibits the transport of viral membrane proteins from medial to trans Golgi cisternae in baby hamster kidney cells infected with Semliki Forest virus. J Cell Biol 96:835–850

Griffiths G, McDowall A, Back R, Dubochet J (1984) On the preparation of cryosections for immunocytochemistry. J Ultrastruct Res 89:65–78

Griffiths G, Fuller SD, Back R, Hollinshead M, Pfeiffer S, Simons K (1989) The dynamic nature of the Golgi complex. J Cell Biol 108:277–297

Gundersen HJG (1977) Notes on the estimation of the numerical density of arbitrary profiles: the edge effect. J Microsc 111:219–223

Gundersen HJG (1980) Stereology – or how figures for spatial shape and content are obtained by observation of structures in sections. Microsc Acta 83:409–426

Gundersen HJG (1984) Stereology and sampling of biological surfaces. In: Echlin P (ed) Offprints from analysis or organic and biological surfaces. Wiley, New York, pp 478–506

Gundersen HJG (1986) Stereology of arbitrary particles. A review of unbiased number and size estimators and the presentation of some new ones, in memory of William R. Thompson. J Microsc 143:3–45

Gundersen HJG (1988) The nucleator. J Microsc 151:3–21

Gundersen HJG (1990) Stereology: the fast lane between neuroanatomy and brain function – or still only a tightrope? Acta Neurol. Scand (in press)

Gundersen HJG, Jensen TB (1985) Stereological estimation of the volume-weighted mean volume of arbitrary particles observed on random sections. J Microsc 138:127–142

Gundersen HJG, Jensen TB (1987) The efficiency osf systematic sampling in stereology and its prediction. J Microsc 147: 229–263

Gundersen HJG, Østerby R (1980) Sampling efficiency and biological variation in stereology. Mikroskopie 37:143–148

Gundersen HJG, Østerby R (1981) Optimizing sampling efficiency of stereological studies in biology: or "do more less well". J. Microsc 121:65–73

Gundersen HJG, Bendtsen TF, Korbo L, Marcussen N, Mller A, Nielsen K, Nyengaard JR, Pakkenberg B, Srensen FB, Vesterby A, West MJ (1988a) Some new, simple and efficient stereological methods and their use in pathological research and diagnosis. APMIS 96:379–394

Gundersen HJG, Bagger P, Bendtsen TF, Evans SM, Korbo L, Marcussen N, Mller A, Nielsen K, Nyengaard JR, Pakkenberg B, Srensen FB, Vesterby A, West MJ (1988b) The new stereological tools: disector, fractionator, nucleator and point sampled intercepts and their use in pathological research and diagnosis. APMIS 96:857–881

Gupta M, Mayhew TM, Sharma AK, White FH (1982) Inter-animal variation and its influence on the overall precision of morphometric estimates based on nested sampling designs. J Microsc 131:147–154

Holmes A (1927) Petrographic methods and calculations. Murphy, London

Hortsch M, Meyer DI (1986) Transfer of secretory proteins through the membrane of the endoplasmic reticulum. Int Rev Cytol 102:215–231-244

Hortsch M, Griffiths G, Meyer DI (1985) Restriction of docking protein to the rough endoplasmic reticulum: immunocytochemical localization in rat liver. Eur J Cell Biol 38:271–279

Howard V, Reid S, Baddeley AJ, Boyde A (1985) Unbiased estimation of particle density in the tandem scanning reflected light microscope. J Microsc 138:203–212

Howell KE, Reuter-Carlson U, Devaney E, Luzio JP, Fuller SD (1987) One antigen, one gold? A quantitative analysis of immunogold labeling of plasma membrane 5'-nucleotidase in frozen thin sections. Eur J Cell Biol 44:318–327

Jensen EB, Gundersen HJG (1982) Stereological ratio estimation based on counts from integral test systems. J Microsc 125:51–66

Jesior J-Cl (1986) How to avoid compression. II. The influence of sectioning conditions. J Ultrastruct Mol Res 95:210–217

Kehle T, Herzog V (1987, 1989) A colloidal gold labeling technique for the determination of the surface area of eukaryotic. Eur J Cell Biol 48:19–26

Kellenberger E, Villiger W, Carlemalm E (1986) The influence of the surface relief of thin sections of embedded, unstained biological material on image quality. Micron Microsc Acta 17:331–348

Kellenberger E, Dürrenberger M, Villiger W, Carlemalm E, Wurtz M (1987) The efficiency of immunolabel on Lowicryl sections compared to theoretical predictions. J Histochem Cytochem 35:959–969

Madsen S, Ottersen OP, Storm-Mathisen J, Sturman JA (1990) Immunocytochemical localization of taurine: methodological apsects. In: Pasantes-Morales H, Martin DL, Shain W, Martin del Rio R (eds) Taurine: Functional neurochemistry, physiology and cardiology. Wiley, New York, pp 37–44

Mathieu O, Cruz-Orive LM, Hoppeler H, Weibel ER (1981) Measuring error and sampling variation in stereology: comparison of the efficiency of various methods for planar image analysis. J Microsc 121:75–88

Mattfeldt T, Möbius H-J, Mall G (1985) Orthogonal triplet probes: an efficient method for unbiased estimation of length and surface of objects with unknown orientation in space. J Microsc 139:279–289

Mattfeldt T, Mall G, Gharehbaghi H, Möller P (1990) Estimation of surface area and length with the orientator. J Microsc 159:301–318

Mayhew TM (1983) Stereology: progress in quantitative microscopical anatomy. In: Navaratnam V, Harrison RJ (eds) Progress in anatomy, vol 3. Cambridge Univ Press, pp 81–112

Mayhew TM (1989) Stereological studies on rat spinal neurons during postnatal development: estimates of mean perikaryal and nuclear volumes free from assumptions about shape. J Anat 162:97–109

Mayhew TM (1990) The surface area of an object revisited – but from random directions. J Theor Biol 144:259–265

Mayhew TM, Cruz-Orive L-M (1973) Stereological correction procedures for estimating true volume proportions from biased samples. J Microsc 99:287–299

Michel RP, Cruz-Orive L-M (1988) Application of the Cavalieri principle and vertical sections method to lung: estimation of volume and pleural surface area. J Microsc 150:117–136

Müller AE, Cruz-Orive LM, Gehr P, Weibel ER (1981) Comparison of two sub-sampling methods for electron microscopic morphometry. J Microsc 123:35–42

Nyengaard JR, Gundersen HJG (1991) The isector. A simple method for making isotropic uniform random sections from specimens. J Microscopy (submitted)

Ottersen OP (1987) Postembedding light- and electron microscopic immunocytochemistry of amino acids: description of a new model system allowing identical conditions for specificity testing and tissue processing. Exp Brain Res 69:167–174

Ottersen OP (1989a) Quantitative electron microscopic immunocytochemistry of neuroactive amino acids. Anat Embryol 180:1–15

Ottersen OP (1989b) Postembedding immunogold labelling of fixed glutamate: an electron microscopic analysis of the relationship between gold particle density and antigen concentration. J Chem Neuroanat 2:57–66

Ottersen OP, Storm-Mathisen J (1984) Glutamate- and GABA-containing neurons in the mouse and rat brain, as demonstrated with a new immunocytochemical technique. J Comp Neurol 229:374–392

Ottersen OP, Storm-Mathisen J, Madsen S, Skumlien S, Strmhaug J (1986) Evaluation of the immunocytochemical method for amino acids. Med Biol 64:147–158

Ottersen OP, Storm-Mathisen J, Bramham C, Torp R, Laake J, Gundersen V (1990) A quantitative electron microscopic immunocytochemical study of the distribution and synaptic of glutamate in rat hippocampus. Prog Brain Res 83:99–114

Ottersen OP, Zhang N, Walberg F (1992) Metabolic compartmentation of glutamate and glutamine: morphological evidence obtained by quantitative immunocytochemistry in rat cerebellum. Neuroscience 46:519–534

Pakkenberg B, Gundersen HJG (1988) Total number of neurons and glia cells in human brain nuclei estimated by the disector and the fractionator. J Microsc 150:1–20

Posthuma G, Slot JW, Geuze HJ (1984) Immunocytochemical assays of amylase and chymotrypsinogen in rat secretory granules. Efficacy of using immunogold-labeled ultrathin cryosections to estimate relative protein concentrations. J Histochem Cytochem 32:1028–1034

Posthuma G, Slot JW, Geuze HJ (1985) The validity of quantitative immunoelectron microscopy on ultrathin sections as judged by a model study. Proc R Microsc Soc 20-IMS

Posthuma G, Slot JW, Geuze HJ (1986) A quantitative immuno-electron microscopical study of amylase and chymotrypsinogen in peri- and tele-insular cells of the rat exocrine pancreas. J Histochem Cytochem 34:203–207

Posthuma G, Slot JW, Geuze HJ (1987) The usefulness of the immunogold technique in quantitation of a soluble protein in ultrathin sections. J Histochem Cytochem 35:405–410

Posthuma G, Slot JW, Veenendaal T, Geuze HJ (1988) Immuno-gold determination of amylase concentrations in pancreatic subcellular compartments. Eur J Cell Biol 46:327–335

Prasad BV, Burns JW, Marietta E, Estes MK, Chiu W (1990) Localization of VP4 neutralization sites in rotavirus by three-dimensional cryo-electron microscopy. Nature 343:476–479

Quinn P, Griffiths G, Warren G (1984) Density of newly synthesised membrane proteins in intracellular membranes. II Biochemical studies. J Cell Biol 98:2142–2147

Schwerzmann K, Cruz-Orive LM, Eggman R, Sanger A, Weibel ER (1986) Molecular architecture of the inner membrane of mitochondria from rat liver: A combined biochemical and stereological study. J Cell Biol 102:97–103

Slot JW, Geuze HJ (1982) Ultracryotomy of polyacrylamide embedded tissue for immunoelectron microscopy. Biol Cell 44:325–331

Slot JW, Geuze HJ (1984) Gold markers for single and double immunolabelling of ultrathin cryosections. In: Polak JM, Varndell IM (eds) Immunolabelling for electron microscopy. Elsevier, Amsterdam, pp 129–142

Slot JW, Geuze HJ, Freeman BA, Crapo JD (1986) Intracellular localization of the copper-zinc and manganese superoxidase dismutase in rat liver parenchymal cells. Lab Invest 55:363–371

Slot JW, Geuze HJ, Weerkamp AJ (1988) Localization of macromolecular components by application of the immunogold technique on cryosectioned bacteria. Methods Microbiol 20:211–236

Slot JW, Posthuma G, Chang LY, Crapo JD, Geuze HJ (1989a) Quantitative assessment of immuno-gold labeling in cryosections. In: Verkleij AJ, Leunissen JLM (eds) Immuno-gold labelling in cell biology. CRC Press Inc, Florida, pp 135–156

Slot JW, Posthuma G, Chang L-Y, Crapo JD, Geuze HJ (1989b) Quantitative aspects of immunogold labeling in embedded and in nonembedded sections. Am J Anat 185:195–207

Small JV (1968) Measurement of section thickness. In: Bocciarelli DS (ed) Abstracts Fourth European Regional Conference on Electron Microscopy, vol 1. Tipografia Poliglotta Vaticana, Rome, pp 609–610

Sterio DC (1984) The unbiased estimation of number and sizes of arbitrary particles using the disector. J Microsc 134:127–136

Stierhof Y-D, Schwarz H (1989) Labeling properties of sucrose-infiltrated cryosections. Scanning Micros Suppl 3:35–46

Stierhof Y-D, Schwarz H, Frank H (1986) Transverse sectioning of plastic-embedded immuno-labeled cryosections: morphology and permeability to protein A-colloidal gold complexes. J Ultrastruct Mol Struct Res 97:187–196

Storm-Mathisen J, Leknes AK, Bore AT, Vaaland JL, Edminson P, Haug FMS, Ottersen OP (1983) First visualization of glutamate and GABA in neurones by immunocytochemistry. Nature 301:517–520

Torp R, Andiné P, Hagberg H, Karagülle T, Blackstad TW, Ottersen OP (1991) Cellular and subcellular redistribution of gkutamate- and taurine-like immunoreactivities during fore-brain ischemia: a semiquantitative electron microscopic study in rat hippokampus. Neuroscience 41:433–447

van Deurs B, Sandvig K, Peterson OW, Olsner S, Simons K, Griffiths G (1988) Estimation of the amount of internalized ricin that reaches the trans Golgi network. J Cell Biol 106:253–267

von Bonsdorf CH, Fuller SD, Simons K (1985) Apical and basolateral endocytosis in Madin-Darby canine kidney (MDCK) cells grown on nitrocellulose filters. EMBO J 4:2781–2792

Wang B-L, Larsson L-T (1985) Simultaneous demonstration of multiple antigens by indirect immunofluorescence or immunogold staining. Novel light and electron microscopical double and triple staining method employing primary antibodies from the same species. Histochemistry 83:47–56

Weibel ER (1979) Stereological methods, vol I. Practical methods for biological morphometry. Academic Press, New York

Weibel ER (1980) Stereological methods, vol 2. Theoretical foundations. Academic Press, New York

Weibel ER, Paumgartner D (1978) Integrated stereological and biochemical studies on hepatocyte membranes. II. Correction of section thickness effect on volume and surface density estimates. J Cell Biol 77:584–597

Weibel ER, Stäubli W, Gnägi HR, Hess FA (1969) Correlated morphometric and biochemical studies on the liver cell. 1. Morphometric model, stereological methods, and normal morphometric data for rat liver. J Cell Biol 42:68–91

West MJ, Gundersen HJG (1990) Unbiased stereological estimation of the number of neurons in the human hippocampus. J Comp Neurol 296:1–22

Zhang N, Walberg F, Laake JII, Meldrum BS, Ottersen OP (1990) Aspartate-like and glutamate-like immunoreactivities in the inferior olive and climbing fibre system: a light microscopic and semiquantitative electron microscopic study in rat and baboon (*Papio anubis*). Neuroscience 38:61–80

Chapter 12

An Overview of Techniques for Labelling at the EM Level

The aim of the final chapter is to summarize, and to put into a more general perspective, some major points we have stressed throughout this book and to finish with an attempt to compare, in two tables, the strengths and weaknesses of the three different approaches we have discussed, namely, cryo-sectioning, plastic sectioning and preembedding labelling.

The first "take-home message" I would like to leave the reader with is that before any EM techniques are seriously undertaken two independent sets of experiments must be considered. The first of these is absolutely obligatory, namely an immunochemical characterization of the antibody. Without an independent demonstration that the antibody in question recognizes the antigen(s) of interest it is pointless to show any immunocytochemical data. The second, the use of light microscopy (LM), may not be essential for EM localization studies, but it is often very beneficial. Note again that as long as the preparation technique for LM does not extract the antigen, in those cases where one fails to get a clear and specific labelling by LM there is usually no point in continuing to the EM level. The additional advantages are:

- Rapid screening of antibodies.
- Assessing conditions of preparation (fixation, use of detergents etc.).
- Sample size: the availability of a significantly larger section surface means one is more likely to section through rare cell types or structures.
- Identification: it enables the large battery of light microscopical stains which aid in the identification of cellular structures to be used in conjunction with immunocytochemical labelling.

For many purposes, however, especially when dealing with tissue culture cells, it is often more efficient to go straight to the EM level using an immunogold approach. Besides the obvious increase in resolution of structures, the use of electron microscopy has two additional powerful advantages over the light microscope. The first, a consequence of the significant increase in resolution, is that it is far easier to assess the specificity of the antibody. The second is that the gold detection methods, it could be argued, are methodologically simpler than any of those (including gold) that are used for light microscopy. The main point here is that there is almost no electron dense particle in a cell that can even remotely be mistaken for a gold particle. In contrast, with any light microscopic detection system, whether it be fluorescence, coloured reaction product or phase-dense silver deposits, there is always the danger that the signal comes from an endogenous reaction (autofluorescence, endogenous peroxidase, etc.).

Table 1. Use of sections for immunogold labelling at EM level[a]

	Cryo 1 day	Resin[b] > 2–3 days
Time required		
Ease of sectioning	Simple, but a few extra days are required to learn the basic technique	Simple – methods available in any EM lab
Access of antibody to antigens in section	Mostly restricted to section surface but can penetrate some structures	Surface only
Quantitiation	Different labelling efficiencies (LE) between different organelles in the section unless polyacrylamide embedding is used	Lowicryl K4M, LR white and LR gold have shown different LE between compartments. HM20 may be superior in this respect (after freeze-substitution)
Sensitivity:		
Large antigens that are easy to cross-link by fixtive	++	+/++
Small molecules (not easily cross-linked	+/–	++ (resin helps to fix these antigens9
Lipid antigens	Danger of lateral diffusion along membranes after fixing	Freeze substitution – technique of choice
Accessibility to contrasting agents/enzymes	Since there is no resin all stabilization is due to fixative cross-links. This should be an advantage for experiments where enzymes (such as proteases or nucleases) are used to digest components on the section. The accessibility for contrasting agents is also far greater in the absence of resin.	Resin can severely retard the movement of heavy metals (and enzymes etc.) into sections. This is especially a problem for the more hydrophobic resins
Serial sections	Few sections only	Many serial sections routinely possible
Toxicity	None (except fixatives and uranyl acetate)	Resins are toxic. The most serious effects are skin allergies and related problems.
Storage and shipment of tissue blocks	Indefinite in liquid nitrogen. Difficult to transport	Indefinite at room temperature. Easy to send from lab to lab
Storage of sections	At most, overnight (on liquid)	Indefinite at room temperature

[a] Both approaches require additional equipment when compared to normal epon embedding. The costs of a cryo-chamber for cryo-sectioning or of, the commercial UV polymerization machine for Lowicryl resins as well as the freeze-substitution units are roughly equivalent. Both approaches are excellent methods for labelling sections for light microscopy as well as for directly comparing the same section at the LM and EM levels.

[b] For reasons explained in Chapter 4 this table mostly excludes the use of Epon resins.

Controlling for these phenomena is thus an additional variable in light microscopic studies.

The move to the higher resolution EM approach is, nevertheless, done at a cost. The main disadvantages of EM over LM are:

- There is (usually) less antigen available for detection per unit area of "field of view". One consequence of this is that the conditions that give optimal labelling are likely to be more stringent for EM than for LM.[1]
- For EM labelling it also becomes more important to keep the antigen precisely in its original structure.[2] Whether an antigen, for example, is in the lumen of the ER or between closely opposed cisternae may be very important for an EM localization but it would hardly be noticed at the LM level. This puts more stringent demands on the cross-linking (fixation) process for EM methods. Any increases in cross-linking, however, are likely to reduce the accessibility of antibody to antigen.

In Tables 1 and 2, I have attempted to compare the advantages and disadvantages of the three main approaches for doing immunocytochemistry at the EM level. Such a comparison should be taken as a very rough guide only since it is, by necessity, an over-simplification. The main problem with such a table is that it severely underplays the complex interaction, introduced in Chapters 1 and 2 between the "labelling" goals, that is, to detect the antigen and the "fine structural" goals, namely the desire to position the antigen with respect to well-preserved, and clearly identifiable, cell structures.

For many problems it may obviously be important to use more than one approach in order to strengthen an immunocytochemical result. This may be either because it is simply more convincing to obtain the same result with two different approaches, for example, it reduces the possibility for artefact or because two different approaches can complement each other. Thus, one may want to combine the generally more sensitive labelling approach of cryo-sections with the ability to reconstruct the three-dimensional appearance of a structure from serial sections after preembedding or post-embedding of a plastic resin.

When used in combination with other methods, even the immunoperoxidase approach can be a useful method at the EM level. I have emphasized repeatedly

---

[1] Imagine that one localizes a nuclear antigen using EM sections. If by immunogold labelling one obtains five gold/μm^2 of nuclear profile (against a background labelling of 0.5 gold/μm^2) this is not an unreasonable signal; subtle changes in the conditions of labelling, be it a new batch of antibody or differences in fixation, can, however, quickly bring the specific signal down towards background levels. For light microscopy, especially when one is dealing with whole tissue culture cells, one has the possibility to recognize the whole population of antigens in that nucleus. Even conditions which reduce the number of antigens by many orders of magnitude may not necessarily show striking effect on the signal. This is especially true if methods for amplifying the signal, i.e. from the bound primary antibody are used.

[2] Note that this statement holds for sections of in vitro preparations of organelles as well as for sections of whole cells or tissues. The problem of maintaining the interaction of organelles with their neighbours must be considered separately.

Table 2. Advantages and disadvantages of preembedding methods requiring a permeabilization step

---

Advantages

1. When compared to the section methods (Table 1), this approach is more likely to label some antigens throughout the whole section thickness. This makes it a better method to look at thick sections or whole mounts (and stereo images).
2. Removal of cytoplasmic components by the permeabilization step facilitates the identification of some organelles, especially the cytoskelon.
3. Can be carried out using technology available in any standard EM lab.

Disadvantages

1. It is often very difficult to achieve the required compromise between maintaining cell structure and getting accessibility of antibodies to cell structures. False negative results due to lack of accessibility are always a potential problem with the approach. Extraction of many components inevitable.
2. When HRP-related techniques are used, diffusion of reaction product is a well known artifact. Endogenous peroxidase activity must be controlled.
3. For immunogold approaches, this approach is only quantifiable if accessibility to all cell compartments is assured. For HRP the situation is worse and it can often be quantitatively misleading.

---

throughout this book that with respect to the labelling goals I cannot see a single advantage of immunoperoxidase over immunogold and plenty of disadvantages. However, with respect to the fine structural aspects this approach may nevertheless be useful especially when one can fill a compartment, or even a whole cell in the case of neurons, with the HRP-reaction product.

Subject Index